AF182251

Materials Engineering: Properties, Design and Applications

Materials Engineering: Properties, Design and Applications

Edited by
Sally Renwick

Willford Press
www.willfordpress.com

Published by Willford Press,
118-35 Queens Blvd., Suite 400,
Forest Hills, NY 11375, USA

Copyright © 2018 Willford Press

This book contains information obtained from authentic and highly regarded sources. Copyright for all individual chapters remain with the respective authors as indicated. All chapters are published with permission under the Creative Commons Attribution License or equivalent. A wide variety of references are listed. Permission and sources are indicated; for detailed attributions, please refer to the permissions page and list of contributors. Reasonable efforts have been made to publish reliable data and information, but the authors, editors and publisher cannot assume any responsibility for the validity of all materials or the consequences of their use.

Trademark Notice: Registered trademark of products or corporate names are used only for explanation and identification without intent to infringe.

ISBN: 978-1-68285-425-9

Cataloging-in-Publication Data

Materials engineering : properties, design and applications / edited by Sally Renwick.
p. cm.
Includes bibliographical references and index.
ISBN 978-1-68285-425-9
1. Materials. 2. Engineering. I. Renwick, Sally.
TA403 .M38 2018
620.11--dc23

For information on all Willford Press publications
visit our website at www.willfordpress.com

Willford Press

Printed in China.

Contents

Preface

Materials engineering is the branch of engineering which studies the identification and design of materials. Some of the examples of materials that are studied in this field are metals, ceramics, nanomaterials, polymers, etc. It is an interdisciplinary field with applications in diverse industries like ceramics, aviation, medicine, electronic engineering, etc. This book elucidates new techniques and their applications in a multidisciplinary approach. It consists of contributions made by international experts. Reaserchers and students actively engaged in field of materials science and engineering will find this book of crucial and unexplored concepts.

The information shared in this book is based on empirical researches made by veterans in this field of study. The elaborative information provided in this book will help the readers further their scope of knowledge leading to advancements in this field.

Finally, I would like to thank my fellow researchers who gave constructive feedback and my family members who supported me at every step of my research.

Editor

1

A Methodology for Quantitatively Characterizing the Dispersion of Nanostructures in Polymers and Composites

S. Pfeifer and P.R. Bandaru*

Department of Mechanical and Aerospace Engineering, Materials Science Program, University of California, Room 258, Engineering 2, San Diego, La Jolla, CA 92093-0411, USA

We seek to provide a metric for quantifying the degree of dispersion of nanostructures, such as carbon nanotubes in polymeric and other matrices, motivated by the technological importance of nanostructure-based composites. Our proposed measure of dispersion uses a quadrat-based sampling algorithm and a metric $d(P||Q)$, to correlate randomness to the dispersion. This allows a quantitative comparison of the given distribution (say, P) to a preferred distribution or pattern (say, Q). We provide examples from our own studies and those from the literature on the application of the metric.

Keywords: Nanostructures, Polymer Matrix Composites, Statistical Properties/Methods, Surface Analysis

1. Introduction It is often necessary to quantitatively measure the degree of dispersion of microscopic and nanoscopic entities in a macroscopic polymer matrix. As an example for illustrating the applicability of such a notion, composites constituted of carbon nanotubes (CNTs) placed in a polymer [1,2] have been widely proposed for electromagnetic interference shielding,[3–5] in high sensitivity infrared sensors,[6] structural applications,[7,8] etc. It has been widely accepted that the composite properties would be optimal when the CNTs are uniformly dispersed within the polymer matrix.[9] More generally, agglomeration of nanostructures/CNTs—in the bulk, surfaces/films, or on fibers/fiber surfaces—is undesirable as it promotes non-uniformity in the measured properties, whereby the characteristics would be a function of which part of the sample was measured. The uniform dispersal and bonding of CNTs in a polymer may confer unique properties to the composite, e.g. through the postulated formation of an interphase region,[10] enhanced charge carrier scattering,[11] etc. Similar considerations also apply to the dispersion of other structures such as nanoparticles,[12] e.g. used in polymer composite foams [13] where aggregation and bundling can lead to poor interfacial bonding of the structures with the polymer matrix. Bundling is not unexpected due to the strong van der Waals bonding prevalent in such structures. This in turn can cause variable and diminished properties in the composite. While single-walled CNT (SWCNT) and multi-walled CNT (MWCNT)-based composites have been reported [14–16] to have enhanced elastic modulus and ultimate tensile strength, it has been frequently seen that beyond a certain loading, fillers can be deleterious (e.g. at ~0.6 vol% in phenol/SWCNT composites [17] or polypropylene/SWCNT composites [16]), presumably due to bundling of the CNTs. In our own studies, we have seen a decrease in the work of fracture of a SWCNT-RET (reactive ethylene terpolymer) composite at ~0.1 vol% loading fraction of the nanotubes.[8]

It was also proposed that the nanostructure surfaces and interfaces could be functionalized through the use of suitable coupling agents [18,19] and made to interact more homogeneously with the polymer matrix. However, it is practically difficult to uniformly disperse nanostructures as the very same characteristics that confer their unique properties (e.g. high specific surface area) also encourage mutual attraction. It has also been shown [20] that commonly used homogenization techniques such as ultra-sonication/blending could destructively reduce CNT length to diameter aspect ratio. While maintaining uniformity in dispersion is difficult and is presently an active research topic, it would nevertheless be pertinent

*Corresponding author. Email: pbandaru@ucsd.edu

to understand *quantitatively* or define more definitively the degree of dispersion of nanostructures, such as CNTs, within a polymer. This paper sets out to provide such a metric suitable for quantifying the dispersion. The merit arises from the strong theoretical foundations used, where well-established information theory principles, invoking the entropy, can be adapted to understand and correlate randomness in the dispersion thereby allowing a comparison of the given distribution to a preferred distribution or pattern. We also suggest that our technique can be used for describing the dispersal of any minority phase within a majority phase.

At the very outset, the visual examination of a micrograph is inadequate to gauge the uniformity of a particular dispersion of the nano-/micro-structures within a matrix, e.g. the presence/absence of agglomerates can both indicate similar dispersion uniformity. A variety of qualitative, semi-quantitative and quantitative methodologies have then been used for the quantification of the dispersion. We first discuss ASTM (American Society for Testing and Materials) standard (D2663),[21] along with the three test methods invoked for the dispersion of carbon black type fillers in polymeric matrices. One version (Test Method A) incorporates visual inspection or linear intercept methods,[22] where the dispersion level is compared with a series of five photographic standards/comparison charts [23] and then rated numerically from 1 (poor dispersion) to 5 (very good dispersion). Alternately, in a more quantitative methodology (Test method B), the total cross-sectional area of the additive agglomerates with a spatial extent of greater than 5 μm is counted and subtracted from the known added content of the additive. The percentage of additives below the 5 μm limit is given as a measure of the dispersion, as follows: >97%—'high' dispersion, 92–97%—'intermediate' dispersion and <92%—'low' dispersion. In another semi-quantitative method (Test Method C), the cut surface of a certain sample is traced with a stylus and the amount of roughness caused by the agglomerates is translated to measure the dispersion.

Alternately, fundamental statistical measures could also be used for a dispersion index. For example, the average number (/mean) number of particles and the deviation from the mean (i.e. through the variance as in the χ^2 distribution-based tests) in a given unit could be considered. However, such measures may not be valid for low filler content and often assume an underlying Gaussian distribution. While a better measure could be obtained through considering a higher order moment about the mean, i.e. through measuring the skewness, which may be in the range of zero (for well-dispersed and uniformly distributed particles) to infinity (for large aggregates) there is an implicit reference to an expected mean number of particles per unit/quadrat which is a priori unknown. An alternate measure (valid for 0–20% CNT loading in polymer matrices) considering both the *spread* of the additive (considering a *dIndex*, varying between 0 and 0.5) as well as the *size* of the additive agglomerate (through a *sIndex*, varying between 0 and 0.5) based on a quadrat [23] methodology was recently proposed.[24] The overall dispersion index was considered to be the arithmetic average of the two indices and varied between zero (maximally ill dispersed) and one (maximally well dispersed). However, the assignment of a range between zero and one, as well as considering a 'maximum particle agglomerate size' is still arbitrary and not well founded. Moreover, the notions of a range and minimal or maximal dispersion are not well defined and clearly quantifiable.

In much of the literature, the measurement of dispersion in a particular micrograph area is quantified by dividing the net area into cells (/quadrats) of equal area. While there is a basic degree of uncertainty in choosing the optimal size of the quadrat (as this choice determines the dispersion index number), it still seems to be the most meaningful basis. Alternate measures, e.g. the free-space length, L_f,[25] which correlates to the 'characteristic size of the unreinforced polymer domains' and defined as the 'width of the largest randomly placed square for which the most probable number of intersecting particles is zero' are still equivalent to measuring the mean length of the largest quadrat square likely to contain zero particles, and have a potential drawback in that a large L_f could be indicated for low volume fraction nanostructure fillers. The Morisita index,[26] which was also proposed as an alternate measure relatively independent of the quadrat size, compares images with similar objects and was shown to be inadequate for nanocomposites with similar degrees of filler mixing.[22]

The methods discussed above are based on assumptions of underlying statistical distributions (typically assumed to be normal/Gaussian), and inherently lack an objective theoretical foundation, e.g. the relevant equations are not of the logarithmic form. Such a requirement is mandated from information theory,[27] which is proposed in this paper as an attribute to be used to relate the degree of uniformity in dispersion to the probability of locating CNTs.[28] The specific advantages of our approach are evident in that we were able to (a) successfully categorize images and image distributions that could not be characterized by other indices, such as the Morisita methodology,[22] (b) produce a single metric, for quantifying the dispersion, and (c) could compare a given image to another image or a given/preferred distribution. The drawbacks may be related, as discussed earlier, to the issue of obtaining an optimal quadrat size, choosing between accuracy and computational time. An optimal quadrat square should indeed be chosen to adequately characterize the nanostructure distribution, i.e. larger quadrats can make a clustered distribution look uniform since such quadrats tend to have nearly the same number of nanostructures/CNTs, while smaller quadrats may not contain any CNTs—also see discussion in Section 3.

2. Principles of the Approach A suitable dispersion metric should enable a comparison with either an idealized dispersion pattern or alternatively with a desired probability distribution (e.g. uniform, Poisson distribution, etc.). From a more quantitative perspective, one example of a well-defined metric involves the use of the *quadrat* approach [22,23] which can be applied at any nanostructure fraction. In such a method, for example, the CNTs are located from a visual micrograph obtained, e.g. through transmission/scanning electron microscopy (TEM/SEM)-based images of a composite cross-section. The image is then subdivided into several squares/*quadrats* and the number of pixels corresponding to the CNTs within each quadrat counted. A perfectly uniform arrangement of the dispersants within the matrix would then exhibit an equal number of centers in each quadrat, so that deviations may be gauged, e.g. by calculating the relative entropy or the distance between the distributions.

A measure of the average randomness (/entropy) of a given composite constituted from the nanostructure (filler)—polymer (matrix) distribution is given by the average of the possible arrangements of filler (f) and matrix (m) entities (with $f + m = 100\%$ or unity) and is proportional to the total number of their arrangements, i.e. $f!m!/(f + m)!$, which can be simplified through the Stirling formula to $f\ \ln(f) + m\ln(m)$, where ln is the natural logarithm. The entropy of the filler-matrix ensemble for a given distribution, say P, could then be defined through [27]

$$H(P) = f_P \ln(f_P) + m_P \ln(m_P). \tag{1}$$

Comparing such a distribution 'P' to another given nanostructure-polymer dispersion, say 'Q', whose entropy is given by $H(Q) = f_Q \ln(f_Q) + m_Q \ln(m_Q)$, we can derive a comparison function to denote the relative entropy between these two distributions, $D(P||Q)$, through [27,29]

$$D(P||Q) = H(P) - H(Q). \tag{2}$$

This interpretation of the *relative entropy* is then illustrated with respect to how a particular nanostructure distribution, say corresponding to 'P', differs from a preferred/standard distribution corresponding to 'Q', e.g. a Poisson or uniform (square, hexagonal lattice, etc.) distribution.[24,25] In other words, $D(P||Q)$ measures the 'inefficiency of assuming that the distribution is Q, when the true distribution is P'.[27,29] A corollary of the previous statement is that when a particular distribution 'P' approaches the preferred distribution 'Q', the relative entropy should approach zero (which is the logarithm of 1). However, $D(P||Q)$ is not a complete metric since the equality $D(P||Q) = D(Q||P)$ usually does not hold.[23] We postulate an average, i.e. $d(P||Q)$ as follows [26]:

$$d(P||Q) \equiv \frac{1}{2}[D(P||Q) + D(Q||P)] \geq 0. \tag{3}$$

It is apparent that $d(P||Q) = d(Q||P)$ and when the distributions are equivalent, i.e. $d(Q||Q) = 0$.

3. Implementation of the $d(P||Q)$ Metric to Gauge the Uniformity of a Given CNT Distribution Within a Polymer Matrix We will next outline the methodology for utilizing the above principles in quantifying the deviation of a given CNT distribution in a polymer from a standard distribution. Initially, an algorithm (implemented in MATLAB™) was used to generate 10,000 randomly positioned quadrats, imposed on a micrograph. Although the algorithm could have simply placed a quadrat centered at each pixel within the image, the use of such a random sample of quadrats saves computational time when evaluating larger micrographs, while providing a satisfactory representation of each image. Also, this allows the direct comparison between differently sized micrographs since the conventional approach of using a fixed quadrat grid would require a different number of quadrats for differently sized micrographs. Additionally, a fixed quadrat grid can result in significantly different dispersion metrics depending on where the grid is superimposed over the micrograph.[23,30] We note that a sufficiently large quadrat square should be chosen to adequately characterize the actual CNT distribution, e.g. if no CNT lies within a quadrat, the quadrat size should be increased until the expanded quadrat contains at least one pixel from a CNT. This is essential for Equations (2) and (3) to be well defined.

It is reiterated that the choice of quadrat size is very important in gauging the dispersion and producing a reliable d-metric. An optimal quadrat square should indeed be chosen to adequately characterize the nanostructure distribution, i.e. larger quadrats can make a clustered distribution look uniform since such quadrats tend to have nearly the same number of nanostructures/CNTs, while smaller quadrats may not contain any CNTs.

Consequently, while there is no straightforward way to determine the optimal quadrat size and shape, several empirical rules have been previously suggested, e.g. that approximately twice the mean area of a particle is suitable as the square quadrat size [31] should be used. For our study, an initial quadrat area of 45 pixels × 45 pixels, corresponding to ~3.5 times the average particle diameter, related to the midpoint of a range of initial quadrat sizes, that gave acceptable and reproducible dispersion measurement trends, was chosen. We could modify this initial quadrat size slightly without changing the relative ranking of dispersions in our study. The general procedure, which could be employed for gauging the d-metric, was as follows:

(A) Given a particular image, the numerical value of the degree of darkness of a pixel (which is the smallest element of an image) is identified with 8-bit resolution—range from 0 (black) to 255 (white). While our algorithm can also be adapted for colorized images, by assigning bin numbers to different colors, we focus on gray scale images to reduce the complexity.

(B) The pixel darkness values are obtained for the nanostructure (e.g. CNT) and the polymer through the SEM/TEM micrographs using the ImageJ® application.

(C) The pixel value threshold for distinguishing the nanostructure from the contrasting polymer background is determined from an examination of the micrographs. In most cases, a 150-pixel darkness value could distinguish CNTs from the polymer background. *It is to be expected that other images corresponding to those produced by different equipment/imaging conditions may require a different threshold.*

(D) The initial length of each quadrat square (in terms of the number of pixels) should then be set to correspond to *at least* (twice) the minimum nanostructure size, for a given magnification. As previously discussed, a small quadrat size (corresponding to a large number of quadrats on the micrograph as well as many empty quadrats) result in excessive computation time while a large quadrat size may yield inaccurate results, e.g. they may make a clustered distribution look uniform.

(E) On another note, the quadrat size should be sufficiently large to capture the nanostructures/CNTs that vary in diameter either within an image or when the magnification changes without being so large as to mask clusters of inclusions. For instance, a 30 × 30 pixel quadrat has a greater than 99.9% chance of containing at least one CNT pixel with perfectly distributed CNTs in a 1 vol% CNT-polymer constituted composite. Consequently, such an initial quadrat size should work well with imperfectly distributed CNTs across a range of loadings. Periodic boundary conditions should be applied to avoid preferential sampling near the image boundaries.

As in the proper reporting of any scientific measure, the dispersion uniformity must be gauged over a number of micrographs (of samples prepared under nominally identical conditions) under consistent conditions (e.g. at a given value of magnification, reported threshold value, etc.). The utilized MATLAB® m-files comparing an (i) image to a distribution and (ii) two mutual images have been appended to the *Supplementary Information.*

Our program was then configured to yield the area fraction of the CNTs within each of the 10,000 quadrats. The CNT distribution probabilities, p_i—related to probability of finding CNTs within the ith quadrat, were found through using the area a_i (in units of pixels) of the CNTs within the ith quadrat of area A_i, through

$$p_i = \frac{a_i/A_i}{\sum_{i=1}^{n} a_i/A_i}. \tag{4}$$

The denominator normalizes the probability over the total number (n) of quadrats, so that $\sum_{i=1}^{n} p_i = 1$. It is to be noted that the p_i are related to the filler (i.e. $p_i = f_{i,P}$) or the matrix (i.e. $p_i = m_{i,P}$) in a given distribution, P. Equation (4) can also be used as an estimator of a probability function for deterministic patterns, e.g. (i) $p_i = 1/n$, for a uniform distribution. However, of the five types of two-dimensional Bravais lattices, the hexagonal lattice could be considered the most well-dispersed since it has the largest number of equidistant nearest neighbors.[32] Consequently, for the purpose of a preferred/standard distribution a hexagonal lattice (HEX) was chosen. It is to be emphasized that our proposed approach and algorithm can be applied to any distribution, and we chose the HEX distribution only for illustrative purpose. The Uniform and the HEX distributions are similar but not identical ways to characterize dispersion. The Uniform distribution, for example, expects the same number of particle pixels circumscribed within each quadrat and that it would always be unnecessary to expand a quadrat to encompass at least one pixel. On the other hand, quadrats superimposed in a random fashion over HEX will not always contain the same number of particle pixels and a quadrat placed over an empty zone in HEX will require enlargement to encapsulate at least one particle pixel. Generally, a user is free to select an *ideal* distribution according to their needs and/or applications. For example, one user may have the flexibility to call Figure 2(a) 'ideal' for their application. Another user may prefer a HEX pattern. A third may want a UNIFORM distribution or perhaps a more stratified pattern (say, for a waveguide). Our algorithm lets the user specify their version of 'ideal' for their particular application. Although there can be different versions of what constitutes ideal, we view this flexibility as a desirable attribute of the algorithm.

It may also be desirable to change the density of the chosen/preferred pattern to represent a higher or lower volume fraction of the nanostructure dispersion. We selected an area fraction and particle diameter in HEX to illustrate a reasonable example of an 'ideal' image suitable for comparison in a study. However, the user is free to account for particle loading, magnification and particle diameter in the selection of the ideal image or distribution, e.g. by using larger or smaller diameter particles (and/or with more or less area fraction) to represent the volume (area) fraction of inclusions of a particular diameter in a nanocomposite. Our algorithm, also allows the user to specify an equal number of particles in each quadrat as an ideal—instead of using an idealized pattern. This may be desired since deterministic patterns such as HEX may exhibit a *picture perfect* level of dispersion that may not needed for some applications. In such ways, images of nanostructures dispersed in polymers can be compared with a preferred point pattern/lattice, using Equation (4) as an estimator of the probability function. The a_i/A_i may also be considered a reasonable estimator of the CNT volume fraction.

For calculating the a_i, we had to consider carefully from digital images/micrographs the pixels that do represent CNTs. Generally, digital images are created by a quantization and sampling process, which mathematically represents an image through a matrix of real numbers. Each matrix element corresponds to a pixel which is the smallest element of the image and is identified by its position within the image and a numerical value representing the degree of darkness of that pixel.[33] In our processed images, the numerical value (with 8-bit resolution) could range from 0 (black) to 255 (white). We measured CNT and polymer pixel values within our micrographs using the ImageJ application (http://rsbweb.nih.gov/ij/) and determined, for our case (Figure 1) that a 150 pixel value threshold could distinguish CNTs from the contrasting polymer background. Both MATLAB™ and ImageJ recognize pixels by their numerical value, and it is to be expected that other images corresponding to those produced by different equipment/imaging conditions may require a different threshold. The initial length of each quadrat square was then set to be 45 pixels which is approximately 3.5 times the average CNT diameter at 1250× magnification (the exact conversion between pixels and CNT diameter would vary with the magnification [22,34]). Ten measurements of $d(P||Q)$, with respect to the HEX pattern, were averaged for any particular image/pattern under test to estimate the population average and lower the standard error estimate (as indicated in Table 1).

We then applied the above methodology to test the proposed dispersion metric, to both our own pattern images of CNTs dispersed in an epoxy polymer (RET) as well as those published in the literature.[22,25] The RET (Elvaloy 4170) was constituted of (1) polyethylene, (2) a polar methyl-methacrylate group and (3) epoxide functional groups. While (1) and (2) contribute to mechanical elastomeric characteristics and corrosion

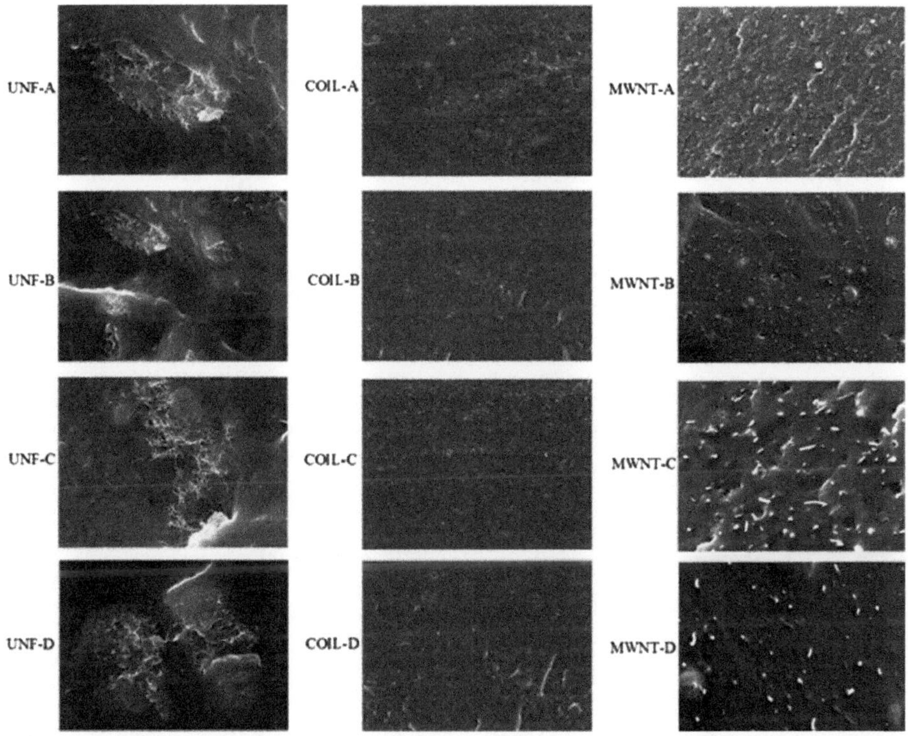

Figure 1. The uniformity of dispersion of carbon nanotubes (CNTs) dispersed in a RET polymer matrix, decreases from the top to the bottom in these SEM micrographs and a quantitative measure can be obtained through a d-metric analysis—Table 1. *Left column*: Increasingly poor dispersion of unfunctionalized CNTs (increasing from the top, UNF-A to the bottom, i.e. UNF-D). *Center column*: Increasingly poor dispersion of coiled CNTs (increasing from the top, COIL-A to the bottom, i.e. COIL-D). *Right column*: Increasingly poor dispersion of MWCNTs (increasing from the top, MWNT-A to the bottom, i.e. MWNT-D).

Table 1. The d-metric for the images in Figure 1 indicates a quantification of the degree/uniformity of dispersion.

$\langle d(\text{HEX}\|\|\text{Pattern})\rangle$	$\langle d(\text{HEX}\|\|\text{Pattern})\rangle$	$\langle d(\text{HEX}\|\|\text{Pattern})\rangle$
$\langle d(\text{HEX}\|\|\text{UNF-A})\rangle = \mathbf{5.805}(0.007)$	$\langle d(\text{HEX}\|\|\text{COIL-A})\rangle = \mathbf{0.858}(0.004)$	$\langle d(\text{HEX}\|\|\text{MWNT-A})\rangle = \mathbf{0.275}(0.001)$
$\langle d(\text{HEX}\|\|\text{UNF-B})\rangle = \mathbf{5.815}(0.009)$	$\langle d(\text{HEX}\|\|\text{COIL-B})\rangle = \mathbf{1.259}(0.011)$	$\langle d(\text{HEX}\|\|\text{MWNT-B})\rangle = \mathbf{0.623}(0.004)$
$\langle d(\text{HEX}\|\|\text{UNF-C})\rangle = \mathbf{6.092}(0.011)$	$\langle d(\text{HEX}\|\|\text{COIL-C})\rangle = \mathbf{1.721}(0.015)$	$\langle d(\text{HEX}\|\|\text{MWNT-C})\rangle = \mathbf{3.084}(0.012)$
$\langle d(\text{HEX}\|\|\text{UNF-D})\rangle = \mathbf{6.577}(0.009)$	$\langle d(\text{HEX}\|\|\text{COIL-D})\rangle = \mathbf{2.831}(0.007)$	$\langle d(\text{HEX}\|\|\text{MWNT-D})\rangle = \mathbf{4.690}(0.007)$

Note: The numbers in bold indicate the d-metric values, the standard deviation from 10 measurements is indicated in the parenthesis.

resistance and are critical to the utility of RET as a hot-melt adhesive and coating, the epoxide group has high reactivity [19] and is amenable for effective anchoring of the constituent ring bonds with functional groups (e.g. $-OH$, $-COOH$, $-NH_2$, etc.) on the CNTs.[35] As the functional groups are associated with defects on the CNTs and are randomly dispersed, isotropic bonding of the nanotubes with the polymer matrix was implied and expected to yield relatively uniform CNT dispersion. Both pristine and –COOH functionalized SWCNTs (average diameter of 1.5 nm, length range ~5–20 μm), MWCNTs (average diameter of 140 nm, length range ~5-9 μm), as well as coiled CNTs [36,37] were used. Further details of the dispersal procedure and structural, electrical, and electromagnetic characterization have been reported previously.[3]

Generally, considerable clumping reflective of CNT agglomeration was observed when unfunctionalized CNTs (to the left of Figure 1) were mixed into the polymer. We then observed that the general strategy of employing mutual chemical reaction between functional groups on the CNT and the polymer through covalent functionalization of the nanotube surface [38] resulted in a relatively more uniform dispersion of SWCNTs in the polymer over a wide range of nanotube volume fractions, i.e. from 0.2 to 4.5 vol% (for functionalized coiled nanotubes and multi-walled nanotubes, in the center and to the right of Figure 1, respectively). It was noted that TEM images have poor contrast between SWCNTs and the surrounding matrix and the SEM images have adequate subject contrast between the inclusions and the surroundings to demonstrate the efficacy of our dispersion algorithm.

We now characterize the extent of uniformity in the CNT dispersed RET polymer through the $d(P||Q)$ metric, as applied to SEM images of the distribution of unfunctionalized and functionalized CNTs in the polymer (Figure 1). The images in Figure 1 represent the SEM micrographs of fracture surfaces of the samples. Contrast was used to distinguish the CNTs (which are predominantly metallic) from the background roughness. More specifically, we measured CNT and polymer pixel values within our micrographs using the ImageJ® application and distinguish CNTs from the contrasting polymer background, through appropriate digital filtering. The use of

Table 2. The d-metric for the images in Figure 2 indicates a quantification of the degree/uniformity of dispersion.

$\langle d(\text{Uniform}\|\|\text{Pattern})\rangle$
$\langle d(\text{Uniform}\|\|\text{a})\rangle = \mathbf{0.021}(0.001)$
$\langle d(\text{Uniform}\|\|\text{b})\rangle = \mathbf{0.066}(0.001)$
$\langle d(\text{Uniform}\|\|\text{c})\rangle = \mathbf{0.153}(0.003)$
$\langle d(\text{Uniform}\|\|\text{d})\rangle = \mathbf{1.354}(0.007)$

Note: The numbers in bold indicate the d-metric values, the standard deviation from 10 measurements is indicated in the parenthesis.

more sophisticated edge-detection subroutines may also be used to distinguish roughness from the nanostructures. We used such schemes and found that our algorithm was insensitive to the roughness using the 150 pixel value threshold mentioned earlier.

Table 1 shows the $d(\text{Image}||\text{HEX})$ metric comparing each image in Figure 1 to a standard hexagonal pattern. Table 1 then shows the *d-metric* could be a quantitative measure of the extent of dispersion, yielding progressively larger values for images that exhibit greater clustering/poor dispersion and deviating more from the chosen hexagonal lattice standard. The numerical values are indicative of the number of bits representing the difference/distance between the given and the standard distribution.[27]

We also compared the utility of the $d(P||Q)$ metric with other results from the literature (Figure 2—taken from [25] and Figure 3—taken from [22]). In the former paper, the importance of reducing particle size to increase the degree of matrix/polymer reinforcement was discussed.[30] We analyzed the micrographs indicating the nanoparticle dispersions in Figure 2, taken from the paper by Khare and Burris,[25] using our d-metric approach. While the general procedure was discussed earlier in the section, through (A)–(E), we outline the specific methodology. For example, in Figure 2(a), subsequent to the assignment of pixel values, i.e. 0 (black) and 255 (white), multiple quadrats each of size 50 nm square (corresponding to twice the length scale of the dots) was superimposed on and covered the image. The choice of 50 nm was dictated by reproducible dispersion metric results in addition to a reasonable computation time,

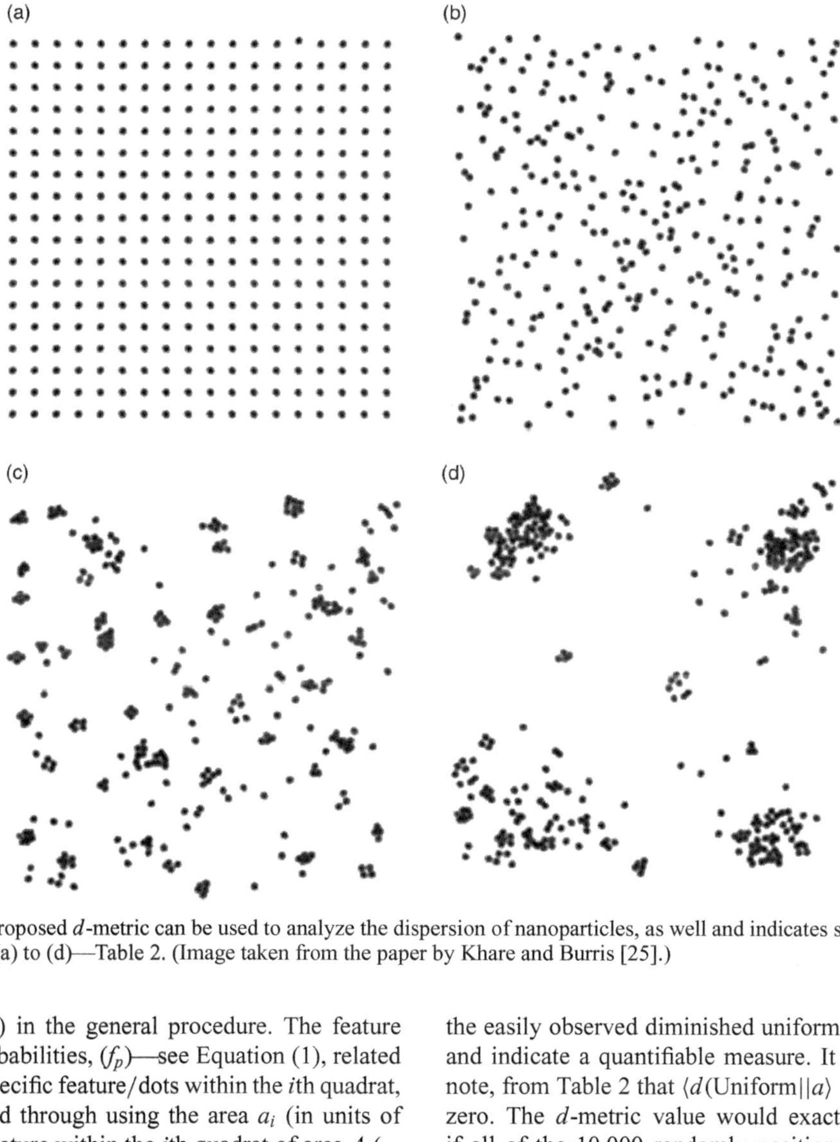

Figure 2. Our proposed d-metric can be used to analyze the dispersion of nanoparticles, as well and indicates steadily decreasing uniformity from (a) to (d)—Table 2. (Image taken from the paper by Khare and Burris [25].)

per (D) and (E) in the general procedure. The feature distribution probabilities, (f_p)—see Equation (1), related to finding the specific feature/dots within the ith quadrat, were determined through using the area a_i (in units of pixels) of the feature within the ith quadrat of area $A_i(= 50\,\text{nm} \times 50\,\text{nm} = 2500\,\text{nm}^2)$ through Equation (4). The H (P) was obtained from Equation (1). Concomitantly, H (Q), with Q as a hexagonal distribution was computed (using a 50 nm quadrat size). The d-metric was estimated from Equations (2) and (3). Ten measurements of $d(P||Q)$, with respect to the HEX pattern, were averaged for any particular image/pattern under test to estimate the population average and lower the standard error estimate. In the other micrographs—Figure 2(b)–(d), an identical method to that discussed above was again used to obtain the respective d-metric values. The obtained results are indicated in Table 2. The comparison is now to a uniform distribution, with an implicit assumption that this is the desired distribution. The steadily increasing d-metric values from the top to the bottom, in the order (a) $<$ (b) $<$ (c) $<$ (d), are in accordance with the easily observed diminished uniformity of dispersion and indicate a quantifiable measure. It is interesting to note, from Table 2 that $\langle d(\text{Uniform}||a\rangle = 0.021$ and not zero. The d-metric value would exactly be zero only if all of the 10,000 randomly positioned quadrats (see the beginning of Section 3, in our paper) happen to encompass/circumscribe an equal number of particles (/dark pixels representing the particles). In more detail, in Figure 2(a), some randomly placed quadrats may superimpose over the gaps between the particles while other randomly positioned quadrats superimpose over the particle pixels. Consequently, the number of particle pixels circumscribed by each quadrat will not always be the same and accounts for $d(\text{Uniform}||a) \neq 0$.

In yet another study taken from the literature, the dispersion of alumina nanoparticles in a polyethylene terephthalate (PET) polymer matrix was investigated [22] through sample cross-sections taken from the composite (as given in Figure 3). The authors' analysis of the degree of dispersion was considered inadequate, as there was no obvious discrimination between

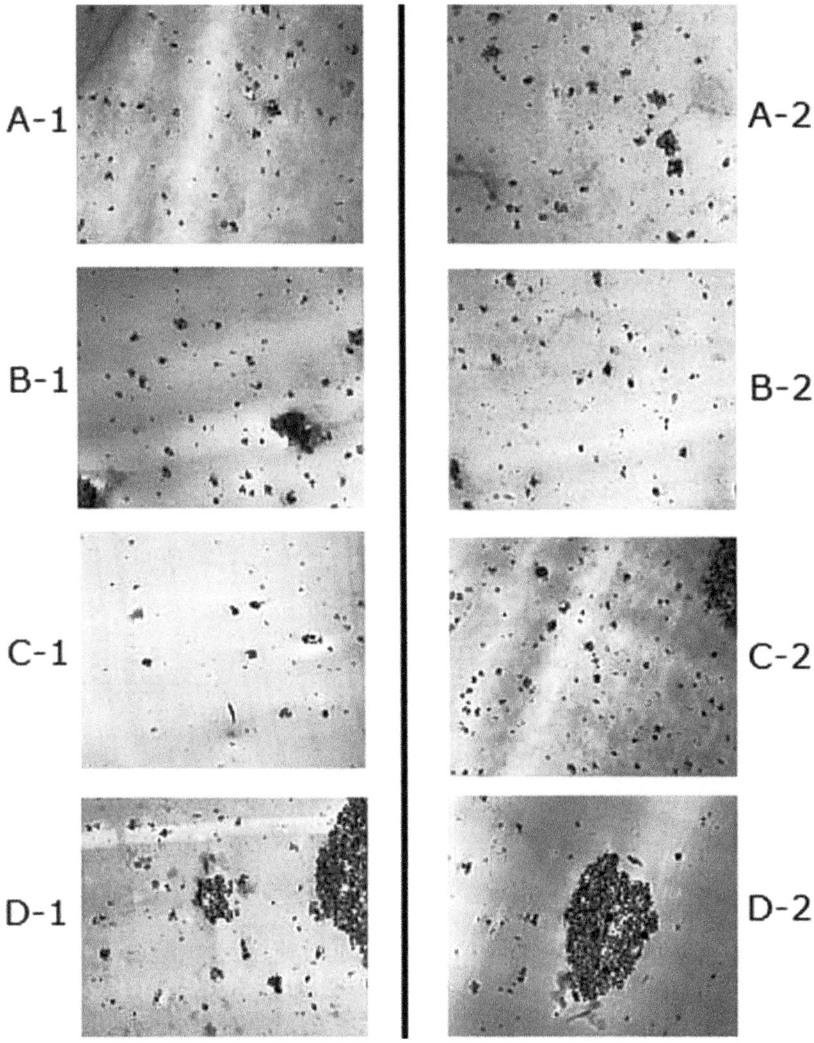

Figure 3. Another example of the application of the d-metric to the dispersion of alumina nanoparticles in PET polymer through the analysis of a TEM image from the literature (Image taken from the paper by Kim et al. [22]). The non-uniformity of dispersion increases from A1 to D1 and A2 to D2. The dispersion metric results are presented in Table 3.

single nanoparticles vs. an agglomeration of contiguous nanoparticles and in addition, no comparison was made to a standard/preferred pattern. We then applied our d-metric-based approach to Figure 3, with the results indicated in Table 3. The comparison is again to a hexagonal pattern distribution. The steadily increasing d-metric values from the top to the bottom, for both columns, now indicate definitive and well-founded values for the

Table 3. The d-metric for the images in Figure 3 indicates a quantification of the degree/uniformity of dispersion.

$\langle d(\text{HEX}\|\|\text{Pattern})\rangle$	$\langle d(\text{HEX}\|\|\text{Pattern})\rangle$
$\langle d(\text{HEX}\|\|\text{A-1})\rangle = \mathbf{0.293}(0.002)$	$\langle d(\text{HEX}\|\|\text{A-2})\rangle = \mathbf{0.306}(0.001)$
$\langle d(\text{HEX}\|\|\text{B-1})\rangle = \mathbf{0.468}(0.002)$	$\langle d(\text{HEX}\|\|\text{B-2})\rangle = \mathbf{0.322}(0.001)$
$\langle d(\text{HEX}\|\|\text{C-1})\rangle = \mathbf{0.710}(0.005)$	$\langle d(\text{HEX}\|\|\text{C-2})\rangle = \mathbf{0.630}(0.002)$
$\langle d(\text{HEX}\|\|\text{D-1})\rangle = \mathbf{0.717}(0.002)$	$\langle d(\text{HEX}\|\|\text{D-2})\rangle = \mathbf{1.282}(0.004)$

Note: The numbers in bold indicate the d-metric values, the standard deviation from 10 measurements is indicated in the parenthesis.

degree of dispersion. While it may initially be surprising to note that the similar values of the d-metric values for Figure C-1 and Figure D-1, one way of rationalizing the values is that the large gaps between particles in the former are as far (in terms of the average relative entropy or distance) from pattern HEX as the large agglomerated clusters in the latter figure.

Our methodology may also be adapted for three-dimensional images. Cross-sections of two-dimensional section scans at varying depths can be combined into a collage and our algorithm applied without modification. Alternatively, each cross-section can be evaluated individually and the d-metric dispersions evaluated at increasing cross-sectional depths. Such an evaluation can be easily accomplished, for example, by plotting the d-metric on statistical control charts.[39] The algorithm also can be adapted to three-dimensional imaging techniques through replacing quadrats with cuboids and area fractions with volume fractions.

4. Conclusions We have shown conclusively that the d-metric, based on Equation (3), can be used to satisfactorily describe nanostructure dispersion in polymer composites. The metric was applied to micrographs of CNT-polymer composites, taken from our own studies as well as from previous literature, and yields a measure of the degree of uniformity relative to a preferred/standard distribution. The proposed measure incorporates a firm mathematical basis and has the advantage that the deviation of a given distribution from a standard distribution can be quantitatively gauged.

Acknowledgements The encouragement and financial support from the National Science Foundation (Grant CMMI 1246800) is acknowledged. Discussions and interactions with S. Park are deeply appreciated.

References

[1] Moniruzzaman M, Winey KI. Polymer nanocomposites containing carbon nanotubes. Macromolecules. 2006;39:5194–5205.
[2] Ajayan PM, Schadler LS, Giannaris C, Rubio A. Single-walled carbon nanotube-polymer composites: strength and weakness. Adv Mater. 2000;12:750–753.
[3] Park S-H, Theilmann P, Asbeck P, Bandaru PR. Enhanced electromagnetic interfrence shielding through the use of functionalized carbon nanotube-reactive polymer composites. IEEE Trans Nanotechnol. 2010;9:464–469.
[4] Park S-H, Thielemann P, Asbeck P, Bandaru PR. Enhanced dielectric constants and shielding effectiveness of, uniformly dispersed, functionalized carbon nanotube composites. Appl Phys Lett. 2009;94:243111.
[5] Chung DDL. Electromagnetic interference shielding effectiveness of carbon materials. Carbon N Y. 2001;39: 279–285.
[6] Pradhan B, Setyowati K, Liu H, Waldeck DH, Chen J. Carbon nanotube-polymer nanocomposite infrared sensor. Nanoletters. 2008;8:1142–1146.
[7] Bekyarova E, Thostenson ET, Yu A, Itkis ME, Fakhrutdinov D, Chou T-W, Haddon RC. Functionalized single-walled carbon nanotubes for carbon fiber-epoxy composites. J Phys Chem C. 2007;111:17865–17871.
[8] Park SH, Bandaru PR. Improved mechanical properties of carbon nanotube/polymer composites through the use of carboxyl-epoxide functional group linkages. Polymer (Guildf). 2010;51:5071–5077.
[9] Breuer O, Sundararaj U. Big returns from small fibers: a review of polymer/carbon nanotube composites. Polym Compos. 2004;25:630–645.
[10] Todd MG, Shi FG. Complex permittivity of composite systems: a comprehensive interphase approach. IEEE Trans Dielectr Electr Insul. 2005;12:601–611.
[11] Smith RC, Liang C, Landry M, Nelson JK, Schadler LS. The mechanisms leading to the useful electrical properties of polymer nanodielectrics. IEEE Trans Dielectr Electr Insul. 2008;15:187–196.
[12] Fiedler B, Gojny FH, Wichmann MHG, Nolte MCM, Schulte K. Fundamental aspects of nano-reinforced composites. Compos Sci Technol. 2006;66:3115–3125.
[13] Lee LJ, Zeng C, Cao X, Han X, Shen J, Xu G. Polymer nanocomposite foams. Compos Sci Technol. 2005;65:2344–2363.
[14] Kim KH, Jo WH. A strategy for enhancement of mechanical and electrical properties of polycarbonate/multi-walled carbon nanotube composites. Carbon N Y. 2009;47:1126–1134.
[15] Yeh MK, Tai NH, Lin YJ. Mechanical properties of phenolic-based nanocomposites reinforced by multi-walled carbon nanotubes and carbon fibers. Compos Part A. 2008;39:677–684.
[16] Lo'pez Manchado MA, Valentini L, Biagiotti J, Kenny JM. Thermal and mechanical properties of single-walled carbon nanotubes–polypropylene composites prepared by melt processing. Carbon N Y. 2005;43: 1499–1505.
[17] Tai NH, Yeh MK, Peng TH. Experimental study and theoretical analysis on the mechanical properties of SWNTs/phenolic composites. Compos B. 2008;39:926–932.
[18] Ni C, Chattopadhyay J, Billups WE, Bandaru PR. Modification of the electrical characteristics of single wall carbon nanotubes through selective functionalization. Appl Phys Lett. 2008;93:243113.
[19] Tasis D, Tagmatarchis N, Bianco A, Prato M. Chemistry of Carbon Nanotubes. Chem Rev. 2006;106: 1105–1136.
[20] Pfeifer S, Park SH, Bandaru PR. Analysis of electrical percolation thresholds in carbon nanotube networks using the Weibull probability distribution. J Appl Phys. 2010;108:24305.
[21] Subcommittee: D24.71. ASTM D2663-08 (Standard test methods for carbon-black dispersion in Rubber) [Internet]. Am Soc Test Mater Digit Libr. 2014. Available from: enterprise.astm.org.
[22] Kim D, Lee JS, Barry CMF, Mead JL. Microscopic measurement of the degree of mixing for nanoparticles in polymer nanocomposites by TEM images. Microsc Res Technol. 2007;70:539–546.

[23] Rogers A. Statistical analysis of spatial dispersion: the quadrat method. London, UK: Pion Ltd.; 1974.
[24] Haslam MD, Raeymaekers B. A composite index to quantify dispersion of carbon nanotubes in polymer-based composite materials. Compos Part B. 2013;55: 16–21.
[25] Khare HS, Burris DL. A quantitative method for measuring nanocomposite dispersion. Polymer (Guildf). 2010;51:719–729.
[26] Morisita M. Iq index, a measurement of dispersion of individuals. Res Popul Ecol (Kyoto). 1962;4:1–7.
[27] Cover TM, Thomas JA. Elements of information theory. Hoboken, NJ: Wiley-Interscience; 2006.
[28] Johnson D, Sinanovic S. Symmetrizing the Kullback-Leibler distance. Houston, TX: Rice University; 2001.
[29] Leon-Garcia A. Probability and random processes for electrical engineering. 2nd ed. Reading, MA: Addison-Wesley; 1994.
[30] Cencov NN. Statistical decision rules and optimal inference. Providence, RI: American Mathematical Society; 1982.
[31] Curtis JT, McIntosh RP. The interrelations of certain analytic and synthetic phytosociological characters. Ecology. 1950;31:434–455.
[32] Feng D, Jin G. Introduction to condensed matter physics. Singapore: World Scientific Publishing Co. Plc. Ltd; 2005.
[33] Gonzalez RC, Woods RE. Digital image processing. 2nd ed. Upper Saddle River, NJ: Prentice-Hall, Inc.; 2002.
[34] Dabak AG. A geometry for detection theory. Department of Electrical and Computer Engineering Houston, TX: Rice University; 1992.
[35] Love CT, Gapin A, Karbhari VM. Interfacial adhesion in multi-walled carbon nanotube/reactive ethylene ter-polymer composites. Baltimore, MD: Society for the Advancement of Material and Process Engineering; 2007.
[36] Bandaru PR, Daraio C, Yang K, Rao AM. A plausible mechanism for the evolution of helical forms in nanostructure growth. J Appl Phys. 2007;101:94307.
[37] Wang W, Yang K, Gaillard J, Bandaru PR, Rao AM. Rational synthesis of helically coiled carbon nanowires and nanotubes through the use of Tin and indium catalysts. Adv Mater. 2008;20:179–182.
[38] Sun L, Warren GL, O'reilly JY, Everett WN, Lee SM, Davis D, Lagoudas D, Sue HJ. Mechanical properties of surface-functionalized SWCNT/epoxy composites. Carbon N Y. 2008;46:320–328.
[39] Chambers DS, Wheeler DJ. Understanding statistical process control. 2nd ed. Knoxville, TN: SPC Press; 1992.

2

Melting of Al Induced by Laser Excitation of 2p Holes

Yudi Rosandi[a,b], Fairoja Cheenicode Kabeer[c,d], Yaroslav Cherednikov[b], Eeuwe S. Zijlstra[c], Martin E. Garcia[c], Nail A. Inogamov[e] and Herbert M. Urbassek[b]*

[a]*Department of Physics, Universitas Padjadjaran, Jatinangor, Sumedang 45363, Indonesia;* [b]*Fachbereich Physik und Forschungszentrum OPTIMAS, Universität Kaiserslautern, Erwin-Schrödinger-Straße, D-67663 Kaiserslautern, Germany;* [c]*Theoretische Physik, Universität Kassel and Center for Interdisciplinary Nanostructure Science and Technology (CINSaT), Heinrich-Plett-Straße 40, 34132 Kassel, Germany;* [d]*Fritz-Haber-Institut der Max-Planck Gesellschaft, Faradayweg 4–6, D-14195 Berlin, Germany;* [e]*Landau Institute for Theoretical Physics, Russian Academy of Science, Prospekt Akademika Semenova 1-A, 142432 Chernogolovka, Moscow, Russia*

Novel photon sources—such as XUV- or X-ray lasers—allow to selectively excite core excitations in materials. We study the response of a simple metal, Al, to the excitation of 2p holes using molecular dynamics simulations. During the lifetime of the holes, the interatomic interactions in the slab are changed; we calculate these using WIEN2k. We find that the melting dynamics after core hole excitation is dominated by classical electron–phonon dynamics. The effects of the changed potential surface for excited Al atoms occur on the time scale of 100 fs, corresponding to the Debye time of the lattice.

Keywords: Laser Materials Processing, XUV Irradiation, Molecular Dynamics, Al

Ultrafast laser-induced melting of metals has been studied theoretically in the past by computer simulation and many aspects are by now well understood.[1–6] Such studies consider the action of a visible-light (VIS) or ultraviolet (UV) laser, which conveys the photon energy to the conduction electrons of the metal. Electron–phonon coupling then equilibrates the energy with the lattice, which is heated and melts. Aspects such as the superheating of the lattice, the nucleation of the liquid phase, or the influence of pre-existing lattice defects, for example, grain boundaries, have been studied.

Apart from this 'conventional' energy transfer via the conduction electron gas, another route of heating the lattice was detected in several materials; the accompanying melting process was termed *non-thermal melting*, as lattice disordering and melting occur so quickly—on the timescale of several 100 fs—that the lattice is not in (local) thermodynamic equilibrium. The process operates via the changed interactions that lattice atoms experience as a consequence of the laser-excited electrons. It has first been found in covalently bonded materials crystallizing in the diamond structure such as C or Si [7–10] and, more recently, also in a wider class of semiconducting and semimetallic materials, for example, in InP [11] and Bi.[12] On the microscopic scale, nonthermal melting can be distinguished from conventional melting by fractionally diffusive atomic pathways.[13]

In the present paper, we inquire how laser-induced changes of the interatomic interaction potential of a simple metal, Al, influence the melting process. VIS or UV lasers will not be able to induce such changes.[14] However, extreme ultraviolet (XUV) pulses with an energy of 75 eV are able to selectively excite 2p core holes in Al atoms. We study their effect on the subsequent lattice dynamics.

We assume the laser pulse to have a Gaussian time distribution with maximum at time $t_0 = 0$ and width σ,

$$S(t) = \frac{E_0}{\sqrt{2\pi}\sigma} \exp\left[-\frac{1}{2}\left(\frac{t-t_0}{\sigma}\right)^2\right]. \qquad (1)$$

Here E_0 is the total energy absorbed by the electrons per unit volume. Due to the large penetration depth of 75 eV

*Corresponding author. Email: urbassek@rhrk.uni-kl.de

photons of around 50 nm,[15] we assume the excitation to occur homogeneously in our Al slab. We choose here $E_0 = 0.0214$, 0.0428, and 0.0856 eV/Å^3; since the atomic volume of Al is $\Omega = 16.39$ Å^3, the lowest energy is equivalent to an energy density of $\epsilon = 0.35$ eV/atom. This is around the melting threshold of Al.[16,17] XUV laser pulse widths of $\sigma = 4$, 40, and 400 fs will be considered.

The laser is assumed to excite Al atoms from their ground state to an excited state Al^{2p}, where a hole in the 2p level has been created. Since the excitation energy is 73.1 eV,[18] the concentration of excited atoms corresponding to the energy density E_0 amounts to $p = 0.45\%$.

The lifetime of the 2p hole equals roughly 40 fs.[19,20] De-excitation occurs mainly via Auger decays, since radiative processes are too slow.[19] We assume that the excitation energy of the 2p hole, 73.1 eV, is put into the conduction band after de-excitation.

As indicated earlier, in our simulations, excited Al atoms, Al^{2p}, are diluted in the specimen. It is therefore reasonable to assume that they only interact with ground-state Al atoms, Al^0. The interaction is calculated using the all-electron full-potential linearized augmented plane-wave code WIEN2k [21][1]: inside a supercell with side lengths $2a \times 2a \times 2a$, where a is the lattice parameter of Al, we create a core hole by removing a $2p_{3/2}$ electron from one Al atom. This electron is added to the conduction band in order to maintain charge neutrality.[22,23] In Figure 1, which shows part of our supercell, the thus excited atom has been colored red. One of the nearest neighbors is then displaced along the line connecting it with the excited åtom, as indicated by a purple arrow. The total energy is compared to the total energy of the same supercell with no core hole as a function of the interatomic distance between the two aforementioned atoms. Following our discussion about the nature of many-body interactions in Al below, we assume that the dilute excitation of core holes in Al mostly affects the pair interaction between the excited atom and its nearest neighbors. Neglecting other effects, our computed energy equals the change in the pair potential $\Delta V(r)$ between an excited and a ground-state atom. Our result is plotted in Figure 1. We see that it gives an attractive contribution to the interatomic forces; that is, excited atoms attempt to create lattices with smaller lattice constant. Beyond 3 Å it decreases slightly by 0.02 eV. We believe that this small decrease is due to a slightly incomplete screening within our 32-atom supercell. As shown in Figure 1, we have not included this change in ΔV for $r > 3.0$ Å in our fit.

Figure 2 demonstrates schematically how the excitation of a core hole leads to lattice heating. The potential curves are provided for the so-called effective potentials (see Equation (5)). An initially cold atom moves from the ground-state potential to the excited-state potential

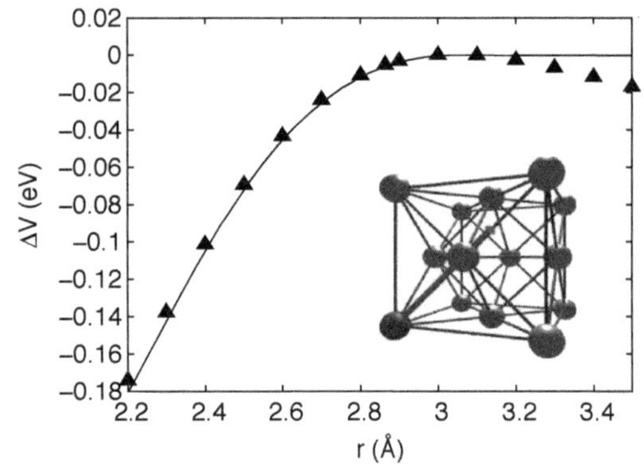

Figure 1. Difference $\Delta V(r)$ between the interaction of an excited atom and a ground-state atom to that between two ground-state atoms. Data points show our results obtained with WIEN2k and the solid line shows the fit to the WIEN2k data that has been used in our molecular dynamics simulations. The inset shows part of the 32-atom supercell used to determine the interaction between a core-hole-excited and a ground-state Al atom: one $2p_{3/2}$ electron has been excited from the red atom to the conduction band. In addition, one of its nearest neighbors is displaced along the line connecting it with the excited atom, as indicated by the purple arrow.

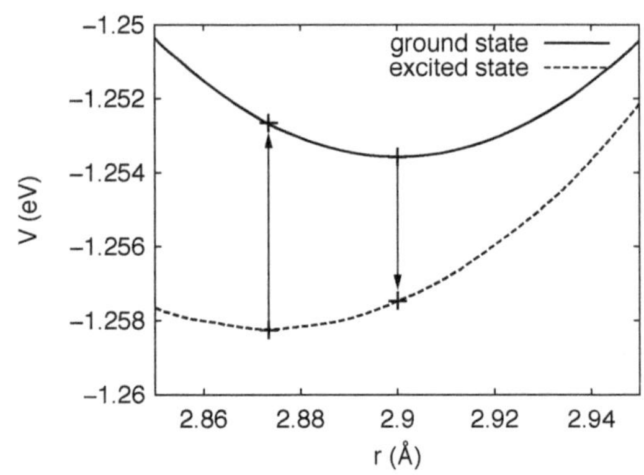

Figure 2. Sketch of the excitation and de-excitation process. The potentials plotted are the effective potentials, Equation (5), for Al^0-Al^0 (ground state) and Al^0-Al^{2p} (excited state). Both transitions are idealized to occur at the respective potential minima; in our simulations, the de-excitation process will occur from a position that may vary for each individual event, due both to thermal motion and to the individual excitation lifetime.

and now feels an attractive force. While moving toward its new equilibrium position, it acquires kinetic energy which will ultimately lead to thermal heating of the lattice. After the lifetime of the excitation, the potential changes back to the ground-state curve; for simplicity, in the figure we assume this to happen at the minimum of the excited potential surface. Thereafter, the atom will move back toward its equilibrium position, again acquiring kinetic energy which eventually heats the lattice.

We employ a molecular dynamics (MD) scheme coupled to the two-temperature model (TTM) to describe the time evolution of the Al thin film; the coupled code is available.[24] We choose an Al crystal with a (100) surface as the target material. The simulation crystallite consists of 131,072 atoms, arranged in 64 monolayers ($d = 13$ nm). The lateral size amounts to 164 nm^2, arranged in 64×64 monolayers. We employ lateral periodic boundary conditions on the sides of the crystal, while the top and bottom surfaces are left free. A many-body interaction potential of the embedded-atom-model (EAM) type is employed for Al [25] with cutoff radius $R_c = 5.558$ Å. This state-of-the-art potential reproduces the melting properties of Al, such as the melting temperature and the latent heat of melting within 3%.[25]

In EAM potentials, the energy of an atom i is composed of the pair interaction $V(r_{ij})$ depending on the distance r_{ij} of atom i and its neighbors j and a many-body term $F(\rho_i)$,[26,27]

$$E_i = \frac{1}{2}\sum_j V(r_{ij}) + F(\rho_i). \qquad (2)$$

Here the electron density at site i, ρ_i, enters as the sum of contributions of all neighboring atoms, $\rho_i = \sum_j g(r_{ij})$. In our system, this expression describes the interaction between ground-state atoms Al^0; the functions V, F, and g are provided in [25]. In our approach, the mixture of Al^{2p} and Al^0 atoms is described by keeping the many-body terms F and g unchanged; however, the pair potential is modified to $V(r_{ij}) + \Delta V(r_{ij})$ in the interaction of an Al^{2p} atom i and an Al^0 atom j (or vice versa).

For small changes in the system, the many-body EAM potential may be approximated by an *effective* pair potential, $V_{\text{eff}}(r)$.[26,28] This can be shown as follows. The energy E_i, Equation (2), does not change under the transformation

$$\tilde{F}(\rho) = F(\rho) - c\rho, \quad \tilde{V}(r) = V(r) + 2cg(r), \qquad (3)$$

with constant c. Now let ρ_0 denote the value of the electron density at a ground-state Al^0 atom in the equilibrium lattice and choose $c = F'(\rho_0)$. Then, a Taylor expansion of $F(\rho)$ around ρ_0 shows that

$$\tilde{F}(\rho) = \tilde{F}(\rho_0) + \tfrac{1}{2}(\rho - \rho_0)^2 F''(\rho_0) + \cdots \qquad (4)$$

up to terms of third or higher order in $\rho - \rho_0$. Thus, small changes in the electron density ρ around the equilibrium value will not influence E_i nor the atom dynamics in this representation. The pertinent pair potential $\tilde{V}(r)$ gives then the *effective* pair potential

$$\tilde{V}(r) = V_{\text{eff}}(r) = V(r) + 2F'(\rho_0)g(r). \qquad (5)$$

This argument shows that for small changes in the EAM system, the dynamics can be described by an effective pair potential; this fact justifies our assumption that dilute excitations in Al can be described by a change in the pair potential. The corresponding effective potential for a dilute Al^{2p} surrounded by Al^0 is then $V_{\text{eff}}(r) + \Delta V(r)$; both effective potentials are plotted in Figure 2.

The excitation of Al atoms is treated as a random process: we excite Al atoms in the slab with a time-dependent probability distribution proportional to the laser profile (Equation (1)). While excited, the interaction of Al^{2p} atoms to ground state Al^0 atoms is governed by a different interaction (Figure 1). De-excitation occurs randomly with an exponentially distributed individual lifetime around a mean lifetime of 40 fs. Then, the interaction switches back to the ordinary Al^0-Al^0 interaction, while the energy of 73.1 eV is given to the electronic system.

The TTM, which has been introduced more than 30 years ago for describing electron emission from metals,[29] is commonly used to characterize the energy transfer between the electronic system and the lattice. In this model, the electron and atomic temperatures are connected by a coupling factor that controls the electron–phonon relaxation time. In our case, where an ultrathin Al film is considered, the electron temperature swiftly homogenizes due to the fast electron diffusion. The electron temperature T_e is calculated from

$$C_e(T_e)\frac{dT_e}{dt} = -G(T_e)[T_e - T_a] + \tilde{S}(t), \qquad (6)$$

where C_e denotes the electronic heat capacity and G the coupling factor. $\tilde{S}(t)$ is the source term. In contrast to VIS or UV laser excitations, where the source simply equals the laser profile, Equation (1), in our case, $\tilde{S}(t)$ equals a sum of Dirac delta functions, $(73.1\,\text{eV}) \cdot \delta(t - t')$, with t' the times when Al^{2p} core holes decay. The electronic heat capacity and the coupling depend on the electron temperature as described in [30]. T_a denotes the average atom temperature in the sample.

Atom dynamics is calculated by MD. We characterize the state of the system using the local temperature $T_{a,i}$ of atom i. It is determined as an average over all N_i atoms in a sphere (volume V_c) of radius R_c, equal to the cutoff radius of the potential, around atom i [5,31]:

$$\frac{3}{2}N_i k T_{a,i} = \sum_{j=1}^{N_i} \frac{m}{2}(\mathbf{v}_j - \mathbf{V}_i)^2, \qquad (7)$$

where m is the atom mass, k is the Boltzmann constant, j enumerates the atoms in V_c, $\mathbf{v}_j$ is their velocity, and $\mathbf{V}_i$ is the average velocity of all atoms within this sphere.

Electronic energy is given to the atoms by scaling the thermal velocities of the atoms at every MD time step. That is, energy is given only as *thermal energy* to the atoms. The energy input to atom i, ΔE_i, is calculated from the temperature difference of the atomic and

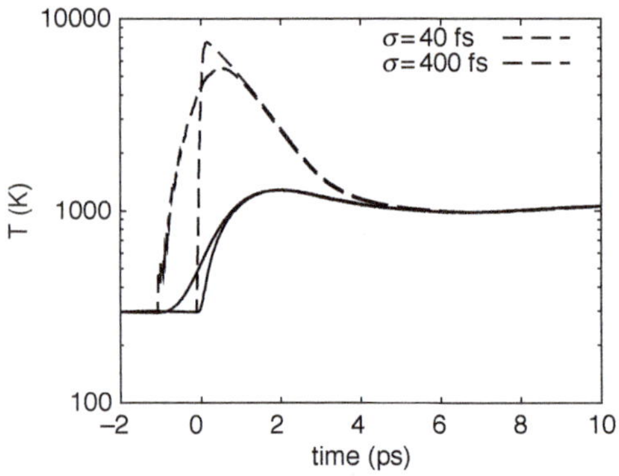

Figure 3. Temporal evolution of the electron (dashed line) and atom (solid line) temperatures in the XUV laser-excited slab ($p = 0.45\%$) for two values of the pulse width σ.

electronic system at the corresponding time step Δt,

$$\Delta E_i = G(T_e)[T_e - T_{a,i}]\Delta t. \tag{8}$$

The simulations are started with a system equilibrated at 300 K at a time of $t = -5$ (-7) ps for $\sigma = 40$ (400) fs. At this time, the TTM solver is inserted after every MD computation step and solves Equation (6), providing the change of electron temperatures for the next MD time step. Details are provided in [17]. The atom trajectories are followed up to time $t = 20$ ps.

In Figure 3, we compare how the electronic and atomic temperatures evolve as a function of time after irradiation. The high initial electron temperature of 7744 K decreases within around 5 ps to its equilibrium value of 1042 K. Simultaneously the atom temperature T_a increases and reaches a maximum after around 2 ps. Thereafter, temperature slightly decreases due to the global adiabatic film expansion and the latent heat of the melting process.

Since the long-time thermodynamics in the films is identical for the two pulses studied (Figure 3), we show as an example atomistic snapshots of the dynamics in the melting film for the $\sigma = 400$ fs laser pulse (Figure 4).

The atoms have been colored according to the local-order parameter, which describes whether the structure is crystalline or molten. It is calculated according to [1,2,32]. In short, we define the local-order parameter η_i of atom i as

$$\eta_i = \left| \frac{1}{Z} \sum_{j=1}^{Z} \frac{1}{6} \sum_{k=i}^{6} \exp(i\boldsymbol{q}_k \cdot \boldsymbol{r}_{ij}) \right|^2, \tag{9}$$

where Z is the number of first- and second-nearest neighbors. The six wave vectors $\boldsymbol{q}_k$ are characteristic of the fcc structure and are chosen such that $\exp(i\boldsymbol{q}_k \cdot \boldsymbol{r}_{ij}) = 1$ for all vectors $\boldsymbol{r}_{ij}$ connecting an atom i to its first- or second-nearest neighbors j in the perfect fcc lattice. The (global)-order parameter η is then the average over the local-order parameters of all atoms in the system.

Values larger than a threshold value—here taken as 0.05—indicate a crystal, and smaller values a liquid. It is seen that at 2 ps the aluminum film still has maintained a high degree of crystallinity. Melting then starts from the surface and moves inward. Since the excitation was chosen such that the film is just about to melt, we see that even at 5 ps a small part of the film still has retained some order in the center of the film. We note that the fact that melting starts from the surface (as inhomogeneous melting) is in agreement with previous studies of ultrafast laser-induced melting.[3,17]

We explore the effect of a higher excitation density in Figure 5. In addition to the case of a concentration of core holes of $p = 0.45\%$ considered up to now, we now also use values of $p = 0.90$ and 1.80%. Correspondingly also the energy density increases from $\epsilon = 0.35$ to 0.70 and 1.40 eV/atom, moving far above the melting threshold. A laser pulse width of only 4 fs was used in order to separate the excitation and de-excitation of the core holes clearly in time from the lattice dynamics. We see that due to higher excitation density melting occurs faster—the melting time decreases from 4 over 2 to 1 ps. This effect has been observed previously in simulations of ultrashort-pulse melting of Cu thin films.[5]

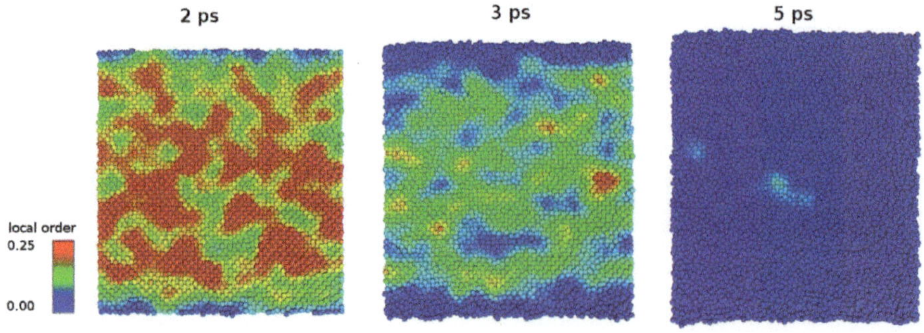

Figure 4. Snapshots of the Al film at times of 2, 3 and 5 ps after XUV laser irradiation with a pulse length of 400 fs. Color denotes the local-order parameter.

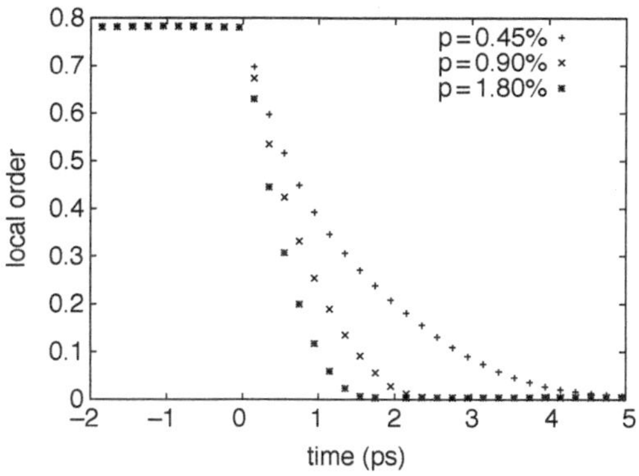

Figure 5. Time evolution of the local-order parameter in the XUV-laser-irradiated slab for various concentrations p of core holes and a laser pulse width $\sigma = 4$ fs.

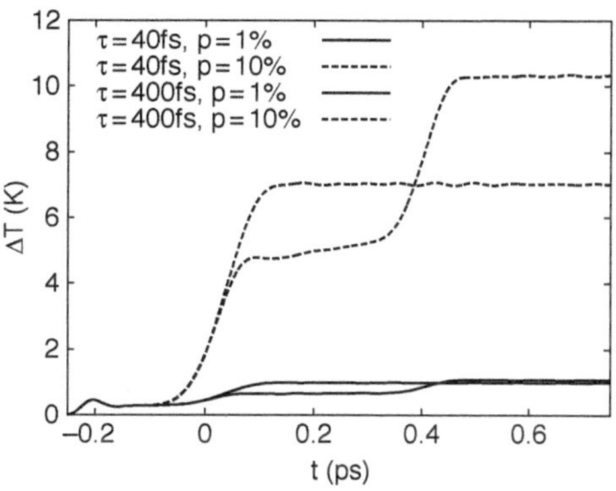

Figure 6. Change in lattice temperature caused by core-hole excitation and de-excitation of Al atoms. Results are shown for the actual lifetime of the core hole of $\tau = 40$ fs and also for an artificially increased lifetime of $\tau = 400$ fs. Results are provided for two values of the concentration p of core holes in the lattice for an XUV laser pulse of 40 fs width.

In order to understand in more detail the action of the lattice heating by core holes, we first discuss with the help of Figure 2 the size of the energy transfer to the lattice that is to be expected due to the sudden potential changes; note that the numbers provided are valid for the effective potentials described earlier (Equation (5)). Initially atoms are on the ground-state potential and have equilibrium distances of 2.90 Å. After the sudden excitation they move toward the new equilibrium value of 2.87 Å, delivering a maximum of 0.7 meV to thermal energy. Let us assume that the atoms are de-excited when they are in their new equilibrium location; after returning to the ground-state potential surface they deliver 0.92 meV. The total energy transfer to the lattice amounts to 1.62 meV per bond; this adds up to 19.4 meV when taking into account that each Al^{2p} atom has 12 Al^0 nearest neighbors. The actual situation is more complex, since the exact position from which atoms are de-excited is not fixed, due both to thermal motion and to the individual excitation lifetime; so the above numbers are upper estimates.

In order to quantitatively assess the lattice heating by core holes, we performed a model simulation in which we suppress artificially the heating of the electron gas by setting the electron–phonon coupling $G = 0$ in Equation (6). Thus, the only action of the excitation of the core electrons is via the changed interatomic potential ΔV (Figure 1). Figure 6 shows the temperature gain in the lattice due to the excitation and de-excitation processes. In order to emphasize the effect, we increased the concentration of excited atoms in the target up to 10%; in addition, we used deterministic lifetimes in order to clearly separate the effects of excitation and de-excitation. The small temperature increase before the start of the laser pulse is due to incomplete target relaxation.

Results are displayed for the actual lifetime of 2p holes of $\tau = 40$ fs. For demonstration purposes, the simulation was also performed for an artificially increased lifetime of $\tau = 400$ fs—larger than the Debye time of the solid. This separates the excitation process (during the laser pulse, at 0 ps) and the de-excitation process (at 0.4 ps). The figure shows that both processes add to target heating due to the potential switching. For the natural lifetime, the effects of excitation and de-excitation overlap; the combined effect is smaller than for the 400 fs demonstration case, since atoms are already de-excited while they are still in the vicinity of their original position and have not gained much kinetic energy. The time scale of the effect is 100 fs—around the Debye time. We conclude that for a concentration of $p = 10\%$ of excited atoms, the energy transfer to the lattice corresponds to 10 K, or equivalently to roughly 1 meV per atom in the lattice. This means each Al^{2p} contributed about 10 meV during excitation and de-excitation to lattice heating, in good agreement with the discussion above.

We study the melting of Al induced by a short-pulse excitation of 2p core-hole states. Our model includes both the heating of the lattice by conduction electrons (electron–phonon coupling) and the heating induced by the changed potential energy surfaces of ground-state and excited Al atoms.

We find that for the specific case of Al the melting dynamics is dominated by classical electron–phonon dynamics. The effects of the changed potential surface for XUV-excited Al atoms are minor. We note that this study is to the best of our knowledge the first of its kind and that the influence of dilute excitation of core holes in other materials is unknown.

The lattice heating induced by the changed potential surface for excited Al atoms occurs on the time scale of 100 fs, corresponding to the Debye time of the lattice. We expect that the effect can be amplified by using temporally structured laser pulses, such as double-peak laser pulses, which repeatedly induce potential surface changes of the Al atoms.

Acknowledgments We acknowledge discussions with H.C. Schneider and S. Kaltenborn. YR is grateful for funding from the Indonesian Directorate General of Higher Education (DIKTI), contract no. 393/UN6.R/PL/2015, and appreciates the computational resources provided by the computer cluster 'Elwetritsch' of the University of Kaiserslautern. ESZ and MEG thank the Deutsche Forschungsgemeinschaft for financial support through the project GA 465/16-1/ZI 1307/1-1 and the ITS in Kassel for computing power. NI thanks support from the Russian Science Foundation grant 14-19-01599.

Disclosure statement No potential conflict of interest was reported by the authors.

Note

1. In our calculations, the muffin-tin spheres centered around the Al atoms have a radius of $r_{\mathrm{MT}} = 1.09$ Å. Linearized augmented plane waves with $r_{\mathrm{MT}}K < 6.0$, where $K = |\mathbf{K}|$ is the absolute value of the $\mathbf{K}$-vector of the plane wave, have been included in the basis set. The Brillouin zone of the 32-atom supercell has been sampled by a $10 \times 10 \times 10$ $\mathbf{k}$-grid. We have used the local density approximation.

References

[1] Ivanov DS, Zhigilei LV. Combined atomistic-continuum modeling of short-pulse laser melting and disintegration of metal films. Phys Rev B. 2003;68:064114.
[2] Upadhyay AK, Urbassek HM. Melting and fragmentation of ultra-thin metal films due to ultrafast laser irradiation: a molecular-dynamics study. J Phys D: Appl Phys. 2005;38:2933–2941.
[3] Lin Z, Zhigilei LV. Time-resolved diffraction profiles and atomic dynamics in short-pulse laser-induced structural transformations: molecular dynamics study. Phys Rev B. 2006;73:184113.
[4] Ivanov DS, Zhigilei LV. Kinetic limit of heterogeneous melting in metals. Phys Rev Lett. 2007;98:195701.
[5] Upadhyay AK, Urbassek HM. Effect of laser pulse width on material phenomena in ultrathin metal films irradiated by an ultrafast laser: molecular-dynamics study. J Phys D: Appl Phys. 2007;40:3518–3526.
[6] Lin Z, Leveugle E, Bringa EM, Zhigilei LV. Molecular dynamics simulation of laser melting of nanocrystalline Au. J Phys Chem C. 2010;114:5686–5699.
[7] Shank CV, Yen R, Hirlimann C. Time-resolved reflectivity measurements of femtosecond-optical-pulse-induced phase transitions in silicon. Phys Rev Lett. 1983;50: 454–457.
[8] Stampfli P, Bennemann KH. Theory for the instability of the diamond structure of Si, Ge, and C induced by a dense electron-hole plasma. Phys Rev B. 1990;42: 7163–7173.
[9] Stampfli P, Bennemann KH. Time dependence of the laser-induced femtosecond lattice instability of Si and GaAs: role of longitudinal optical distortions. Phys Rev B. 1994;49:7299–7305.
[10] Zier T, Zijlstra ES, Garcia ME. Silicon before the bonds break. Appl Phys A. 2014;117:1–5.
[11] Bonse J, Wiggins SM, Solis J. Dynamics of femtosecond laser-induced melting and amorphization of indium phosphide. J Appl Phys. 2004;96:2352–2358.
[12] Sciaini G, Harb M, Kruglik SG, Payer T, Hebeisen CT, Meyer zu Heringdorf F-J, Yamaguchi M, Horn-von Hoegen M, Ernstorfer R, Miller RJD. Electronic acceleration of atomic motions and disordering in bismuth. Nature. 2009;458:56–59.
[13] Zijlstra ES, Kalitsov A, Zier T, Garcia ME. Fractional diffusion in silicon. Adv Mater. 2013;25:5605–5608.
[14] Recoules V, Clérouin J, G Zérah , Anglade PM, Mazevet S. Effect of intense laser irradiation on the lattice stability of semiconductors and metals. Phys Rev Lett. 2006;96:055503.
[15] Carron NJ. An introduction to the passage of energetic particles through matter. New York: Taylor and Francis Group; 2007.
[16] Rosandi Y, Urbassek HM. Ultrashort-pulse laser irradiation of metal films: the effect of a double-peak laser pulse. Appl Phys A. 2010;101:509–515.
[17] Rosandi Y, Urbassek HM. Melting of Al by ultrafast laser pulses: dynamics at the melting threshold. Appl Phys A. 2013;110:649–654.
[18] Bearden JA, Burr AF. Reevaluation of X-ray atomic energy levels. Rev Mod Phys. 1967;39:125–142.
[19] Keski-Rahkonen O, Krause MO. Total and partial atomic-level widths. At Data Nucl Data Tables. 1974;14:139–146.
[20] Almbladh C-O, Morales AL, Grossmann G. Theory of Auger core-valence-valence processes in simple metals. I. Total yields and core-level lifetime widths. Phys Rev B. 1989;39:3489–3502.
[21] Blaha P, Schwarz K, Madsen GKH, Kvasnicka D, Luitz J. WIEN2k, an augmented plane wave plus local orbitals program for calculating crystal properties. Austria: Vienna University of Technology; 2001.
[22] Zijlstra ES, Cheenicode Kabeer F, Bauerhenne B, Zier T, Grigoryan N, Garcia ME. Modeling of material properties after ultrashort laser and XUV excitation. Appl Phys A. 2013;110:519–528.
[23] Kabeer FC, Zijlstra ES, Garcia ME. Road of warm dense noble metals to the plasma state: *Ab initio* theory of the ultrafast structural dynamics in warm dense matter. Phys Rev B. 2014;89:100301(R).
[24] The MD code. Available from: http://merapiphysikuni-klde/objectmd/
[25] Ercolessi F, Adams JB. Interatomic potentials from first-principles calculations: the force-matching method. Europhys Lett. 1994;26:583–588.
[26] Carlsson AE. Beyond pair potentials in elemental transition metals and semiconductors. In: Ehrenreich H, Turnbull T, editors. Solid state physics. Vol. 43. Boston (MA): Academic Press; 1990. p. 1.
[27] Daw MS, Foiles SM, Baskes MI. The embedded-atom method: a review of theory and applications. Mater Sci Rep. 1993;9:251–310.
[28] Foiles SM. Application of the embedded-atom method to liquid transition metals. Phys Rev B. 1985;32: 3409–3415.
[29] Anisimov SI, Kapeliovich BL, Perel'man TL. Electron emission from metal surfaces exposed to ultrashort laser pulses. Sov Phys JETP. 1974;39:375–377.
[30] Lin Z, Zhigilei LV, Celli V. Electron-phonon coupling and electron heat capacity of metals under conditions of strong electron-phonon nonequilibrium. Phys Rev B. 2008;77:075133. Available from: http://wwwfaculty virginiaedu/CompMat/electron-phonon-coupling/

[31] Colla ThJ, Urbassek HM. Visualization of keV-ion-induced spikes in metals. Radiat Eff Defects Solids. 1997;142:439–447.

[32] Morris JR, Song X. The melting lines of model systems calculated from coexistence simulations. J Chem Phys. 2002;116:9352–9358.

3

Core Structure and Motion of Metadislocations in the Orthorhombic Structurally Complex Alloy $Al_{13}Co_4$

Marc Heggen* and Michael Feuerbacher

Peter Grünberg Institut, Forschungszentrum Jülich GmbH, D-52425 Jülich, Germany

The atomic structure of metadislocation cores in the phase o-$Al_{13}Co_4$ is investigated using aberration corrected scanning transmission electron microscopy. Metadislocations with Burgers vectors of $-b/\tau^4$ [0 1 0] and b/τ^3 [0 1 0] having six and four stacking faults, respectively, are found. They are associated with separate phason defects, which escort the movement of the metadislocations through the material. A first partial atomic model for metadislocation glide motion, taking Cobalt atom jumps into account, is developed. Atom jumps take place in pairs along various crystallographic directions. Typical jumps occur between atomic columns separated by 1.5 Å.

Keywords: Complex Metallic Alloy, Metadislocation, Dislocation Core, Scanning Transmission Electron Microscopy

Metadislocations are large and structurally highly complex defects mediating plasticity in various complex metallic alloys (CMAs).[1,2] Likewise, their atomic mechanisms of motion are highly complex as they involve a large number of atomic jumps for an elementary step. As yet, there is no comprehensive model for metadislocation motion on the atomic scale for any CMA. Recently, a first atomic model for motion was developed considering merely the Pd atoms in ε_6-Al–Pd–Mn, a CMA phase with 318 atoms per unit cell. Repetitive jump patterns along various crystallographic directions were found, which involve about 100 atomic positions per individual step of metadislocation motion.[3] The case of metadislocation motion in ε_6-Al–Pd–Mn, however, is particularly complex due to an extraordinary high number of atoms involved and due to the fact that metadislocation motion in this material takes place by pure climb. For this reason, a detailed microstructural investigation of less complex metadislocations involved in a simpler glide mode and the development of an atomic model for these dislocations may support the comprehension of structurally related but more complex metadislocations.

A CMA phase, promising in these terms is o-$Al_{13}Co_4$, which possesses an orthorhombic structure with space group Pmn2_1, lattice parameters $a = 8.2$ Å, $b = 12.3$ Å and $c = 14.5$ Å, and 102 atoms per unit cell.[4,5] The main structural features of o-$Al_{13}Co_4$ are pair-connected pentagonal prismatic channels extending along the [1 0 0] direction.[5] Figure 1(a and b) show a structure model of o-$Al_{13}Co_4$. Yellow and green spheres represent cobalt and aluminium atomic positions, respectively. Figure 1(b) shows a projection along the [1 0 0] direction. The orthorhombic unit cell is shown in blue. The projected structure can be described by a tiling consisting of regular pentagons and rhombs.[6] Figure 1(b) shows an experimental micrograph taken by high-angle annular dark field scanning transmission electron microscopy (HAADF-STEM) of o-$Al_{13}Co_4$ along [1 0 0]. Due to the 'Z-contrast' conditions, columns containing heavy atoms are emphasized. White dots represent columns containing heavy cobalt atoms. Contrast of the aluminium positions is below the noise level of the micrograph. The pentagon rhomb tiling is superimposed. Rhombs are arranged in characteristic rows along the [0 1 0] direction with alternate orientation (yellow and blue in Figure 1(c)). The structure of monoclinic $Al_{13}Co_4$ (denoted m-$Al_{13}Co_4$) is homeotypic to that of monoclinic $Al_{13}Fe_4$, and structurally closely related to o-$Al_{13}Co_4$.[5–8] The tiling description of m-$Al_{13}Co_4$ is characterized by a parallel arrangement of the pentagon and rhomb elements using just a single type of rhombs.

The existence of metadislocations in the phase o-$Al_{13}Co_4$ was first postulated due to geometric considerations [9] and later verified experimentally.[10,11] First macroscopic uniaxial plastic deformation experiments

*Corresponding author. Email: m.heggen@fz-juelich.de

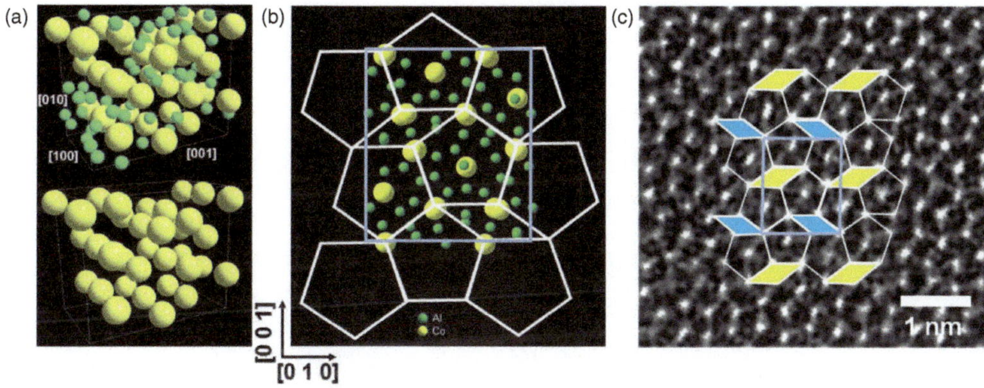

Figure 1. (a) Structure model of o-$Al_{13}Co_4$ showing the Cobalt (yellow) and Aluminium (green) atomic positions (upper image) and just the Cobalt position (lower image) within one unit cell (white lines). (b) Structure model of o-$Al_{13}Co_4$ along [1 0 0] showing Cobalt (yellow) and Aluminium (green) atomic positions within the unit cell (blue rectangle). The structure is described by a pentagon/rhomb tiling. (c) HAADF-STEM micrograph of o-$Al_{13}Co_4$ with pentagon/rhomb tiling showing rhombs with alternate orientation (yellow and blue).

and microstructural investigations using transmission electron microscopy (TEM) were performed on single crystalline o-$Al_{13}Co_4$. Deformed samples exhibited localized zones of inhomogeneous deformation on their surfaces which correspond to slip traces along (0 0 1) planes. TEM investigations revealed a high density of [1 0 0] dislocations attached to (0 0 1) planar defects.[10]

A subsequent microstructural investigation [11] identified the dislocations as metadislocations. In contrast to metadislocations in ε-phases, however, metadislocation motion in o-$Al_{13}Co_4$ takes place by [0 1 0] glide on (0 0 1) planes. This finding was confirmed by recent microstructural investigation on o-$Al_{13}Co_4$ deformed below the ductile–brittle transition temperature using microcompression.[12] Closer inspection of the glide planes by aberration-corrected electron microscopy revealed (0 0 1) slabs of a monoclinic phase in the orthorhombic $Al_{13}Co_4$ matrix [11] partially compensating the strain of the metadislocation core. The Burgers vector of the metadislocation investigated in [11] was determined as b/τ^3 [0 1 0], where $\tau \approx 1.618$ is the irrational number of the golden mean.

Although different tiling models for metadislocations in orthorhombic $Al_{13}Co_4$ were proposed, e.g. with and without the presence of escort defects,[1,11,13] the atomic structure of a metadislocation core and its surrounding are unknown as yet. However, detailed knowledge on the atomic core structure is a prerequisite for the development of an atomic scale model for metadislocation motion. Such a model would be of large interest, as it would represent the first atomic model for metadislocation glide in complex materials. In the present work, the atomic core structure of a metadislocation in the o-$Al_{13}Co_4$ phase is analysed using probe-corrected STEM. Using this information, a first atomic model for metadislocation motion is developed.

We investigated plastically deformed single crystalline o-$Al_{13}Co_4$ samples, which have been uniaxially deformed to about 6% strain. Detailed information on the plastic deformation tests and sample preparation can be found in [10,11]. Atomic-resolution STEM was performed in a probe-corrected FEI TITAN 80-300 employed with a HAADF detector. ‘Z-Contrast’ conditions were achieved for sufficiently thin specimen areas using a probe semi-angle of 25 mrad and a detector inner collection angle of 70 mrad. Under these conditions, the image intensity associated with an atom column is approximately proportional to the square of the mean atomic number of the atoms in the column for sufficiently thin samples.[14]

Figure 2(a) shows a high-resolution HAADF-STEM micrograph of a metadislocation in the phase o-$Al_{13}Co_4$. A tiling of the surrounding area is superimposed and a full tiling of the electron microscope image is given in Figure 2(b). The core is outlined as green polygon. At the right-hand side of the core, a slab of the monoclinic $Al_{13}Co_4$ phase is located, which is characterized by a parallel arrangement of rhombs (yellow) within the pentagon/rhomb tiling. This observation is in accordance with earlier observations, reporting (0 0 1) planar defects in as-cast and deformed $Al_{13}Co_4$, which were identified by means of X-ray diffraction and high-resolution TEM as monoclinic slabs in an orthorhombic matrix.[6,11] Since the ideal orthorhombic $Al_{13}Co_4$ phase contains alternating rows of rhombs along the [0 0 1] direction (Figure 1(c)), two subsequent rows of rhombs oriented in parallel constitute a (0 0 1) stacking fault. In this sense, the metadislocation in Figure 2 is associated with six stacking faults (dashed red lines). The stacking faults extend to the right-hand side, which indicates the direction of motion: the metadislocation shown moves from right to left, and the stacking faults are located in its wake.

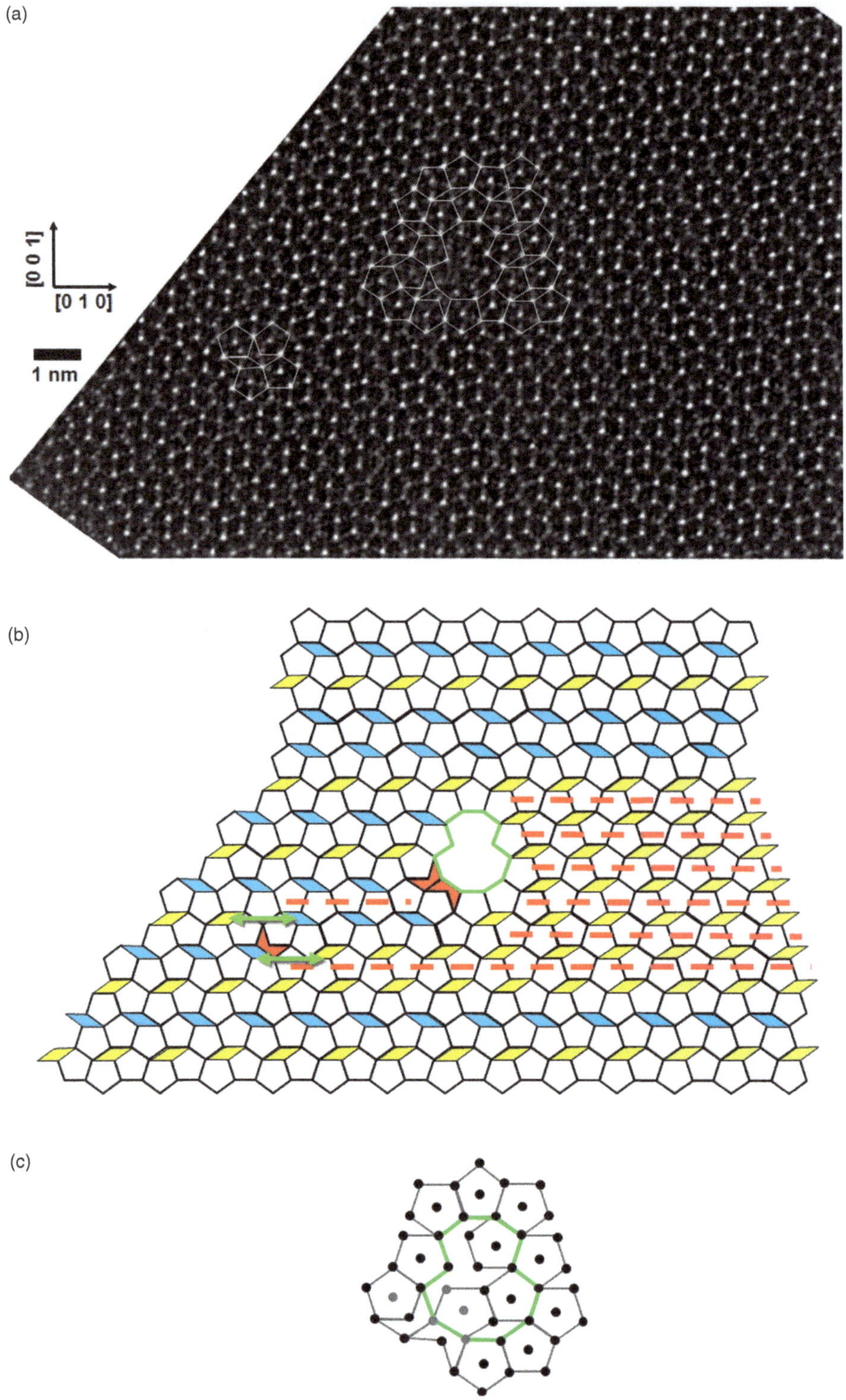

Figure 2. (a) HAADF-STEM micrograph of a metadislocation core and tiling description (white). The image shows slight contrast variation, possibly due to thickness variation, local straining or covering of the specimen surface. The image was subjected to rotation, which exhibits the initial scanning frame as inclined edges at the left, and to slight fast Fourier transform (FFT) filtering to reduce high-frequency noise. (b) Full tiling of the micrograph in (a). Metadislocation core (green polygon), phason defects (red polygons) and stacking faults (red dashed lines) are indicated. (c) Observed and derived Cobalt positions (black and grey circles) within and around the metadislocation core (green) extracted from the micrograph in (a).

A Burgers circuit around the core yields a Burgers vector of $\vec{b} = -b/\tau^4$ [0 1 0] which is an irrational fraction of the lattice constant. The observed Burgers vector is smaller by a factor of τ than the metadislocations observed in the o-$Al_{13}Co_4$ phase before.[11] Figure 2(b) furthermore shows three red tiles, two of which are located directly at the left-hand side of the core and a third one at a distance of about 5 nm left from the core. The red tiles are identified as phason defects [9] (also referred to as phasons).

Recently, a new type of metadislocation was observed in the complex T-Al–Mn–Pd phase.[15,16] This metadislocation consists of a complex core and three separated phason defects, which were referred to as escort defects. Upon deformation, the escort defects move ahead and locally transform the T-phase structure for accommodation of the dislocation core. Metadislocation core and phason defects leave a slab of R-phase in their wake, which can be interpreted as six stacking faults. It is obvious that the $Al_{13}Co_4$ observations in the present work are very similar to the T-Al–Mn–Pd case. In both cases, metadislocations are escorted by phason defects and cause a structural transformation from an orthorhombic to a monoclinic phase in a slab of several nanometre thickness in their wake. Furthermore, in both cases, the transformed slab can be interpreted as the introduction of six stacking faults.

In contrast to the case of the T-phase, two phason defects in o-$Al_{13}Co_4$ (Figure 2(a) and 2(b)) are directly connected to the core. Accordingly, it can be argued that they are not separated defects but part of the core, which is then larger and non-symmetric. In this sense, the shape of the polygon describing the metadislocation core is not strictly unique. We, however, decided for the symmetric representation since there is no strict requirement for the associated phasons to be connected to the core. From a geometrical point of view, the phasons can be separated from the core and are still able to fulfil their function in the arrangement.

The function of the phason defects can best be seen at the example of the left phason, which is separated from the metadislocation core. It inverts the stacking sequence of two neighboured (0 0 1) planes of rhomb tiles such that the orientation of the rhombs changes from the left to the right-hand side of the phason (arrows in Figure 2(b)). Interpreted in terms of stacking faults (dashed lines), the function of the phason is to provide a redirection: stacking faults, which are extending ahead of the core, are redirected by phason defects, so that all structural changes induced by the movement of the dislocation are in its wake. The same mechanism of phason defects redirecting stacking faults and escorting a metadislocation core upon movement was observed in the T-Al–Mn–Pd phase.[15,16]

Figure 2(c) represents the distribution of cobalt columns at the metadislocation core experimentally observed as bright dots in Figure 2(a). At the right-hand side inside the core, the cobalt positions can be clearly identified and matched by two pentagons. At the lower left-hand side, however, the distribution of bright dots is somewhat ambiguous, which may be due to the presence of the strain field or kinks along the dislocation line. Here, the location of cobalt atoms was interpreted in terms of a suitable location within the pentagon/rhomb tiling. They are indicated in grey in Figure 2(c). A clear difference between the metadislocations in o-$Al_{13}Co_4$ and T-Al–Mn–Pd is thus present in the atomic distributions within the cores. While the atomic structure within a metadislocation core in T-Al–Mn–Pd shows a distinct mirror symmetry with respect to the (0 0 1) plane, our observations reveal a non-symmetric intensity distribution within the metadislocation core in o-$Al_{13}Co_4$.

Figure 3(a) shows a HAADF-STEM micrograph of another type of metadislocation with superimposed tiling. Again, the core is associated with phason defects (red tiles) and stacking faults (red dashed lines). In this example, four stacking faults are observed on the left-hand side of the core and the dislocation moves to the right into the ideal o-$Al_{13}Co_4$ phase region. Further bypassing stacking faults (pink dashed lines) which are not associated with the metadislocation are observed at the bottom of the image. A Burgers circuit around the core yields a Burgers vector $\vec{b} = b/\tau^3$ [0 1 0], which coincides with that in [11]. On the left-hand side of the metadislocation core, the stacking faults are, however, not densely stacked as in Figure 2(b); a line of yellow rhombs intermits the parallel stack of blue rhombs. In this sense, this example does not correspond to the most compact and simple elementary arrangement, which may be due to the presence of external disturbances such as the bypassing stacking faults. Nevertheless, the characteristic features of the metadislocation, the number of associated stacking faults and the corresponding Burgers vector length are invariable.

Figure 3(b)–(d) illustrates tiling descriptions of simple, elementary metadislocations in o-$Al_{13}Co_4$ using a pentagon/rhomb tiling. Metadislocations with 4, 6 and 10 stacking faults are shown. For simplicity, just the metadislocation core and the stacking faults are shown (dashed red lines), phason elements redirecting stacking faults are not considered. They have Burgers vectors of b/τ^3 [0 1 0], $-b/\tau^4$ [0 1 0] and b/τ^5 [0 1 0]. Simultaneous existence of a sequence of Burgers vectors, scaled by factors of τ and the double-Fibonacci number of stacking faults or phason planes is characteristic of metadislocations in various materials.[2,15–19]

The investigation of the atomic metadislocation core structure in the present investigation offers a basis towards the development of an atomic model for metadislocation glide motion in the o-$Al_{13}Co_4$ phase. We use the tiling description (Figure 2(b)) and its decoration with cobalt atoms derived from the experimental images (Figure 2(c))

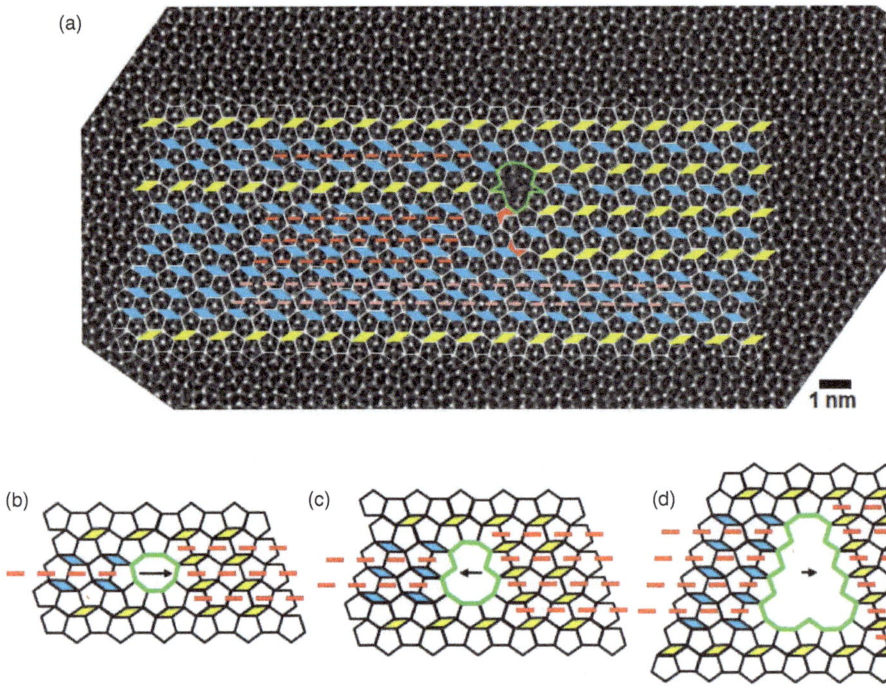

Figure 3. (a) HAADF-STEM micrograph of a metadislocation with superimposed tiling showing a complex core structure (green) phason defects (red) and stacking faults (red dashed lines) and stacking faults not associated to the metadislocation (pink dashed lines). (b–d) Tiling descriptions of elementary metadislocations with four, six and ten stacking faults. Burgers vectors are indicated as black arrows (not to scale).

and shift the location of the metadislocation by one unit along the direction of movement.

Figure 4 shows the initial (grey) and final (red) tiling and the respective cobalt positions. Initial and final positions of the cobalt columns correspond in most areas around the metadislocation core (small deviations are present due to the elastic strain field around the core). Within the core and at the location of the phason defects, atomic jumps between initial and final positions occur (black arrows). The atomic jumps take place along various crystallographic directions and are generally short—typical jump distances are 1.5 Å. Four jumps take place between cobalt columns at the phason element on the left and 14 jumps occur in and around the metadislocation core. It can be noticed that jumps often occur paired and in a symmetric way. At the location of the phason element, for instance, two jumps at the left occur in a $[0 \quad \tau - 3 \quad -\sqrt{2+\tau}]$ direction, whereas on the right two jumps occur inversely in a $[0 \quad \tau - 3 \quad \sqrt{2+\tau}]$ direction. It should be kept in mind that the presented model only takes into account the cobalt positions, since the involvement of the aluminium atoms is experimentally not accessible by our HAADF-STEM approach. Density functional theory or molecular dynamics simulations will be required for the further development towards a full model including the aluminium atoms as well.

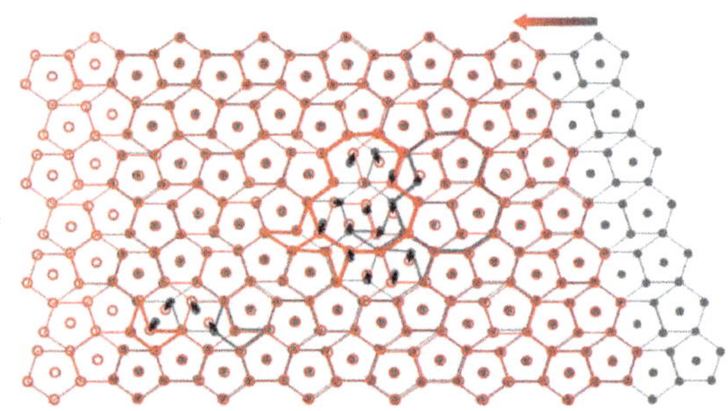

Figure 4. Atomic model for metadislocation motion considering Cobalt atomic positions (circles). Final (red) and initial (grey) positions are displaced by one unit cell along the direction of movement (red and grey arrow). Atomic jumps between initial and final positions are indicated by black arrows.

In summary, we have investigated the atomic structure of metadislocation cores in the phase o-$Al_{13}Co_4$. Two metadislocations with Burgers vectors of $-b/\tau^4$ [0 1 0] and b/τ^3 [0 1 0] are found. They are associated with six and four stacking faults, respectively. Like for metadislocations in other complex alloy phases, the sequence can be geometrically extended,[2,15–19] but only two instances of the series are experimentally observed here. The metadislocations are associated with separate phason defects, which escort the metadislocation cores though the material as they move. This mechanism of metadislocation glide using escort defects is very closely related to the plastic deformation mechanism in the complex phase T-Al–Mn–Pd.[15,16] In both cases, metadislocations and escort defects create a structural transformation from an orthorhombic to a monoclinic phase in a slab of several nanometre thickness behind the metadislocation.

The atomic core structure extracted in the present investigation was furthermore used to create a first atomic model for metadislocation motion considering the cobalt atomic positions. Atomic jumps take place in pairs along various crystallographic directions and typical jump distances of 1.5 Å are found.

Acknowledgements We thank C. Thomas and M. Schmidt for producing the sample material. This work was supported by Deutsche Forschungsgemeinschaft (PAK 36).

References

[1] Klein H, Feuerbacher M. Structure of dislocations and stacking faults in the complex intermetallic ξ′-(Al–Pd–Mn) phase. Phil Mag. 2003;83:4103–4122.

[2] Feuerbacher M, Heggen M. Metadislocations. In: Hirth JP, Kubin L, editors. Dislocations in solids. Vol. 16. Amsterdam: Elsevier; 2010. p. 109–170.

[3] Heggen M, Feuerbacher M. Metadislocation core structure and atomic model for metadislocation motion. Acta Mater. 2013;61:3851–3857.

[4] Gödecke T, Ellner M. Phase equilibria in the aluminium-rich portion of the binary system Co–Al and in the cobalt/aluminium-rich portion of the ternary system Co–Ni–Al. Z Metallkd. 1996;87:854–864.

[5] Grin J, Burkhardt U, Ellner M, Peters K. Crystal structure of orthorhombic Co_4Al_{13}. J Alloys Comp. 1994;206:243–247.

[6] Saito K, Sugiyama K, Hiraga K. Al13M4-type structures and atomic models of their twins. Mat Sci Eng A. 2000;294–296:279–282.

[7] Hudd RC, Tailor WH. The structure of Co_4Al_{13}. Acta Crystallogr. 1962;15:441–442.

[8] Grushko B, Wittenberg R, Bickmann K, Freiburg C. The constitution of aluminum-cobalt alloys between Al_5Co_2 and Al_9Co_2. J Alloys Comp. 1996;233:279–287.

[9] Feuerbacher M, Heggen M. On the concept of metadislocations in complex metallic alloys. Phil Mag. 2006;86:985–990.

[10] Heggen M, Deng D, Feuerbacher M. Plastic deformation properties of the complex metallic alloy phase $Al_{13}Co_4$. Intermetallics. 2007;15:1425–1431.

[11] Heggen M, Houben L, Feuerbacher M. Metadislocations in the orthorhombic structurally complex alloy $Al_{13}Co_4$. Phil Mag. 2008;88:2333–2338.

[12] Walter C, Wheeler JM, Barnard JS, Raghavan R, Korte-Kerzel S, Gille P, Michler J, Clegg WJ. Anomalous yielding in the complex metallic alloy $Al_{13}Co_4$. Acta Mater. 2013;61:7189–7196.

[13] Heggen M, Feuerbacher M. Metadislocations: the case of pure glide. MRS Proc. 2013:1517:1-12. doi:10.1557/opl.2012.1754

[14] Nellist PD, Pennycook SJ. The principles and interpretation of annular dark-field Z-contrast imaging. Adv Imag Electron Phys. 2000;113:147–203.

[15] Heggen M, Houben L, Feuerbacher M. Plastic deformation mechanism in complex solids. Nat Mater. 2010;9:332–336.

[16] Heggen M, Houben L, Feuerbacher M. Metadislocations in the complex metallic alloys T-Al–Mn–(Pd, Fe). Acta Mater. 2011;59:4458–4466.

[17] Feuerbacher M, Heggen M. Elastic energy of metadislocations in complex metallic alloys. Acta Mater. 2012;60:1703–1711.

[18] Engel M, Trebin HR. A unified projection formalism for the Al–Pd–Mn quasi-crystal Xi-approximants and their metadislocations. Philos Mag Lett. 2005;85:2227–2247.

[19] Feuerbacher M, Balanetskyy S, Heggen M. Metadislocations in xi-AlMnPd. Acta Mater. 2008;56:1849–1856.

Flow Stress in Submicron BCC Iron Single Crystals: Sample-size-dependent Strain-rate Sensitivity and Rate-dependent Size Strengthening

Rui Huang[a], Qing-Jie Li[b], Zhang-Jie Wang[a], Ling Huang[a], Ju Li[a,c,d], Evan Ma[a,b] and Zhi-Wei Shan[a]*

[a]Center for Advancing Materials Performance from the Nanoscale (CAMP-Nano) & Hysitron Applied Research Center in China (HARCC), State Key Laboratory for Mechanical Behavior of Materials, Xi'an Jiaotong University, Xi'an 710049, People's Republic of China; [b]Department of Materials Science and Engineering, Johns Hopkins University, Baltimore, MD 21218, USA; [c]Department of Nuclear Science and Engineering, Massachusetts Institute of Technology, 77 Massachusetts Avenue, Cambridge, MA 02139, USA; [d]Department of Materials Science and Engineering, Massachusetts Institute of Technology, 77 Massachusetts Avenue, Cambridge, MA 02139, USA

Through *in situ* scanning electron microscope microcompression tests, we demonstrated that the strain-rate sensitivity of body-centered cubic single crystal iron pillars will be reduced by one order when the pillar size was reduced from 1000 to about 200 nm. In addition, size-strengthening exponent exhibited obvious strain-rate dependence. We propose that the observed behavior is a result of the high stresses required to induce curvature bowout of dislocation arms at small sample or grain sizes, which overwhelms the lattice friction stress contribution and diminishes the role played by the mobility difference between edge and screw dislocations.

Keywords: single-crystal α-Fe, size effect, strain-rate sensitivity, Orowan equation, source truncation hardening

The flow stress of BCC metals is well known to be dependent on the strain rate imposed. The activated process underlying this strain-rate sensitivity (m) can be understood as follows [1–10]: unlike FCC metals in which both screw and edge dislocations have a planar core structure, the screw dislocations in BCC metals have a multi-plane core structure. The resulting difficulty in their slip makes the movement of screw dislocations much slower than edge dislocations and hence the rate-limiting process under normal plastic deformation conditions. However, the strain-rate sensitivity of BCC metals was found to decrease with the reduction in grain size, especially in the ultrafine-grained and nanocrystalline regimes.[1,11–13] To understand this trend, Wei et al. [11,12] used the $m \sim 1/\sqrt{\tau}$ scaling relation to explain the grain size effect on m based on the Hall–Petch-type elevation of the flow stress (τ) at small grain sizes. More recently, Cheng et al. [2] proposed an alternative explanation. They characterized the type (screw, edge or mixed) and density of the residual dislocations in nanocrystalline Mo prepared through high-pressure torsion: mixed and edge dislocations were observed to be the predominant types for grains with sizes below ~ 100 nm. Based on their observation, Cheng et al. [2] proposed that in such grains the motion of edge and mixed dislocations becomes the rate-controlling defect process in plastic deformation, in lieu of screw dislocations. The plasticity processes mediated by edge/mixed dislocation types are believed to have low rate sensitivity such that the effective m observed in nano-grained BCC metals is small. However, we noted that even in the grain size regime where screw dislocations remain populous, m is already much lower than that in large-grained BCC metals.[2,11] In addition, almost all previous studies of the size dependence of m of BCC metals used polycrystalline materials except the one by Schneider et al.[14] They found that submicron-sized single crystal Mo pillars exhibited strain-rate sensitivity similar to their bulk counterpart. Moreover, compared with the widely studied size-strengthening effect and its underlying mechanism of single crystal BCC metals,[5,6,14–21] much less work have been carried out on m. Recently,

*Corresponding author. Emails: zwshan@mail.xjtu.edu.cn, liju@mit.edu, ema@jhu.edu

we demonstrated through molecular dynamics simulation [19] that very high applied stresses can diminish the velocity difference between screw and edge dislocations in single crystal Mo. As a consequence, BCC Mo behaves more like an FCC metal. This was supported by the fact that when their diameters were reduced below 200 nm or so, single crystal Mo pillars exhibited a higher size-strengthening exponent and mechanical annealing.[19] This prompted us to surmise that the similarities between FCC and BCC metals under high stresses may also manifest in terms of a reduced m, that is, it is the high deformation stress at small sample or grain sizes instead of other factors that cause the reduced m for nanostructured BCC metals, including single crystal BCC metals.

In addition, it has been proposed previously that the sample-size-strengthening exponent α of BCC metals increases with a decrease in the 'critical temperature',[17] defined as the temperature above which the strain rate has little effect on the flow stress of BCC metals because screw and edge dislocations have equal mobility. However, as shown in Table 1, the α for single crystal Mo ranges from 0.29 to 1.0 [5,6,16,18,19] and that for single crystal Nb is from 0.48 to 1.07.[5,6,16] Apparently, just the critical temperature argument cannot rationalize the large scatter of α observed in different BCC metals (Table 1), neither can it explain the size effect on m.

In this work, we choose BCC Fe single crystals as our model material. The sample was acquired from German Tech Co., Ltd. The dimensions of the as-received iron are 3 mm × 2.5 mm × 0.1 mm. A series of [001]-oriented α-Fe pillars with their free-end diameter D ranging from about 200 to 1030 nm and their length-to-diameter ratio of about 3:1 were fabricated using focused ion-beam (FIB) micromachining. More details about the sample preparation procedure are described in [19,22,23]. *In situ* compression tests were carried out in the chamber of the same FIB under SEM mode using a Hysitron PI85 PicoIndenter equipped with flat diamond punches, 1 μm in diameter for pillars with sizes smaller than 500 nm and 7 μm in diameter for pillars larger than 500 nm. All tests were performed at room temperature under displacement rate control. The data acquisition rate for the lowest strain rate has been set to be 100 per second but was adjusted to be higher for a higher strain rate to make sure that enough data points were captured. In order to investigate the residual dislocation structure, we used OmniProbe to lift out deformed samples and then thinned them down gradually with FIB to about 100 nm from both sides. After this, a Model 1040 NanoMill (Fischone Inc.) was employed to minimize the FIB-affected layer with an operation voltage of 800 V, with which the surface damage layer thickness can be reduced down to ~2 nm. The final observation was conducted inside a JEOL 2100F TEM under STEM mode which are known to be insensitive to strain contours and able to image dislocations clearly.

Four groups of samples with their nominal diameter of $D = 200, 300, 475$ and 1000 nm were tested under four different strain rates, ranging from $\dot{\varepsilon} = 10^{-1}$ to 10^{-4} s^{-1}. Regardless of the sample size and the strain rate used, multiple slip bands with obvious shear offsets are always observed after the uniaxial compression test. One typical example is shown in Figure 1. The slip traces which are ~45° inclined to the axial [001] direction suggest a $\{110\}\langle 111\rangle$-type slip system with a Schmid factor of 0.408. The initial free-end diameter of this pillar is about 462 nm (Figure 1(a)). After the compression test, a few large shear offsets can be seen clearly in the postmortem SEM image (Figure 1(b)).

Consistent with the morphology of the deformed pillars, all the engineering stress–strain curves exhibited multiple large load drops (Figure 2(a) and 2(b)), typical of displacement-controlled compression in single crystal metal pillars.[19,24] For $D = 1000$ nm samples, the average flow stress σ increased from about 450 MPa to about 700 MPa when the strain rate $\dot{\varepsilon}$ increased from 3.1×10^{-4} to 2.1×10^{-1} s^{-1} (Figure 2(a)). However, along with the reducing sample size, the strain-rate hardening effect decreases gradually and eventually becomes very small for 200 nm samples (Figure 2(b)). Note that

Table 1. Critical temperature (T_c) of BCC metals; data from Refs. [5,6,16,19,20,25].

Material	W	Mo	Ta	V	Nb	α-Fe
T_c (K)	800	480	450	380	350	**340**
T_{test}/T_c	0.37	0.62	0.66	0.78	0.85	**0.88**
Exponent α size range (nm)	0.21 [16] *5000–200* 0.44 [5] *850–250*	0.38 [16] *5000–350* 0.44 [5] *800–200* 0.29 [19] *1200–200* 1.00 [19] *200–80*	0.41 [16] *3000–340* 0.43 [5] *850–400*	0.79 [20] *2000–200*	0.48 [20] *5000–200* 0.93 [5] *900–250* 1.07 [6] *900–250*	**0.62–0.80** ***1000–200*** ***Our work***

Note: $T_{test} = 298$ K in our experiment.

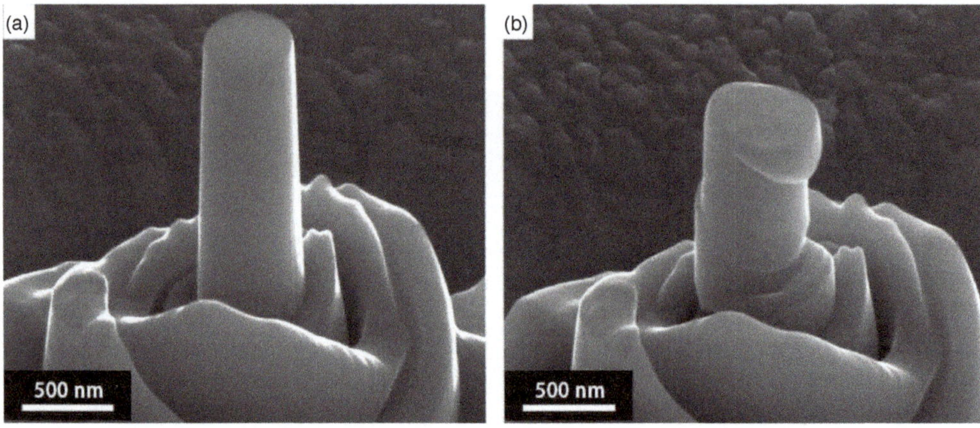

Figure 1. Representative SEM images of [001]-oriented Fe pillar before (a) and after compression (b). The initial free-end diameter is ~ 462 nm.

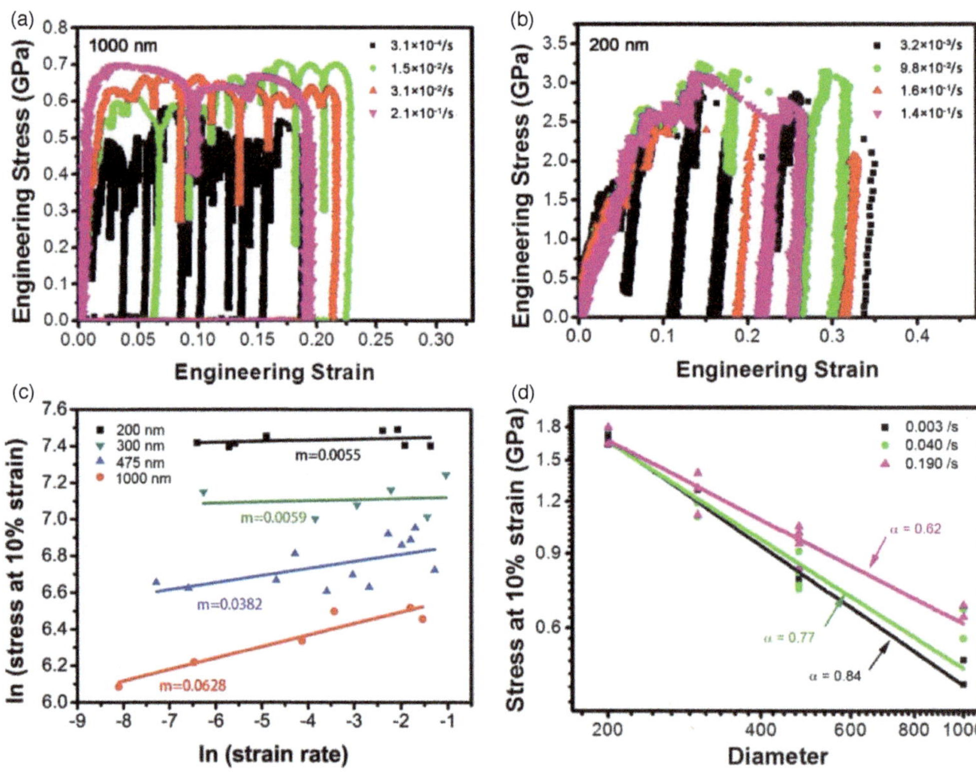

Figure 2. Size dependence of the strain-rate sensitivity of iron pillars with strain rate ranging from 10^{-4} to 10^{-1} s^{-1}. (a) and (b) are representative stress–strain curves of 1000 and 200 nm pillars, respectively. (c) Double logarithmic plot of stress at 10% strain vs. strain rate to determine m for single crystal Fe specimens with their diameters ranging from 200 to 1000 nm. (d) Stress at 10% strain as a function of pillar diameter for [001] oriented Fe pillars. The plot is scaled logarithmically, and shows the strain-rate dependence (three data sets for three strain rates) of the size-strengthening exponent (see linear fit of the data and corresponding exponents in the legend).

200 nm is still much larger than the dislocation-type transition size reported by Cheng et al.[2] In addition, it was interesting to note that regardless of the sample size, the frequency and the amplitude of the load drops (correspond to slip band formation, see Movie S1) decrease with increasing strain rate. The fact that the peak flow stress of $D = 1000$ nm samples (~ 700 MPa, Figure 2(a)) is much lower than that for $D = 200$ nm samples (~ 3000 MPa, Figure 2(b)) is consistent with the well-known tenet of 'smaller is stronger': $\sigma = \sigma_0 + kD^{-\alpha}$, where α is a positive exponent. The critical temperature of α-Fe is 340 K.[25] Compared

with other BCC metals, for example, W,[16] Mo,[16] Ta,[16] V [20] and Nb,[16] the critical temperature of iron is closer to room temperature (for more details, see Table 1). Therefore, it is not surprising to see that the α of iron is almost the largest among the aforementioned BCC metals (see Table 1).

In order to quantify the strain-rate effect, we chose the flow stress at 10% strain as the reference for each sample. The log–log relationship between the strain rate and the flow stress at 10% strain is plotted in Figure 2(c). Despite the scatter of the data, a trend that strain-rate sensitivity decreases with decreasing sample size can be seen clearly from the linear fit for each sample size. The strain-rate sensitivity $m \equiv \partial \ln \sigma / \partial \ln \dot{\varepsilon}$, obtained from the slope of each linear fit in Figure 2(c), decreased more than 10 times, from 0.063 to about 0.0055 when the sample size was reduced from 1000 to 200 nm. The former, that is, $m = 0.063$, is similar to that observed in bulk α-Fe.[26] Therefore, the sample size effect on m starts when the sample dimension was reduced to ~ 1000 nm. This trend remains the same if we choose the stresses at 5% strain. We conclude that BCC iron exhibits a strong 'size' effect on m, and this 'size' D can be either the size of grains in a polycrystalline material, or the size of the specimen for a single crystal.

In addition to 'smaller is stronger' (compare Figure 2(a) and 2(b), for example), we observe that larger samples exhibited a stronger rate dependence (Figure 2(a)) and smaller samples exhibited a much weaker rate dependence (Figure 2(b)). As such, the size-strengthening exponent α measured should be a function of strain rate, rather than remaining constant. Specifically, a lower α is expected at a higher strain rate, and at a fixed lower strain rate the effective α would appear to be higher. This was confirmed by the data plotted in Figure 2(d): when $\dot{\varepsilon}$ increased from 0.003 to 0.19/s, α decreased from 0.84 to 0.62. This change in α due to strain-rate variations at least partly explain the large scatter of α observed in different BCC metals (Table 1) and should be taken into account in the analysis of size-strengthening behavior of materials, but has been rarely paid attention to before.

Because of the strong magnetism of iron, we were not able to carry out the compression tests inside TEM under proper imaging conditions. However, we have managed to check the residual dislocation structure of the deformed iron pillars, as shown in Figure 3. It was found that the residual dislocation structure in the 1000 nm samples is quite different from that of 200 nm samples. For 1000 nm pillar (Figure 3(a)), a high density of dislocations was found in a heavily deformed zone, whereas only a few dislocations are present in other areas. Figure 3(b) is the magnified STEM image of the area framed in Figure 3(a). For 200 nm pillar, however, only one residual dislocation can be found in the heavily deformed zone, as indicated by the black arrows (Figure 3(d)). This agrees well with the expected dislocation escape due to the mechanical annealing phenomena. Because of the single tilt limitation of the TEM holder used, we were not able to identify the type of this dislocation. However, its long and straight feature is the typical characteristics of screw dislocation. In addition, Zhang et al. [27] has carried out *in situ* TEM compression of single crystal Fe–3%Si pillar. They concluded that during the compression of a pillar with its initial diameter of about 120 nm, screw dislocation dominated the deformation process at high stress level. This finding suggests that for the size range we have tested, screw dislocations may still behave quite differently from edge dislocations during deformation. This is also consistent with the observation by Cheng et al. [2] that edge and mixed dislocations take over only at much smaller grain sizes (~ 80 nm). As such, the reduction in m may not be attributed solely to the change in the type of dominant dislocations.

In the following, we attempt to rationalize the observed size effect on m by considering two contributions to the flow stress. Generally in elemental metals when propagation of existing dislocations controls the flow behavior (nucleation is not limiting), the flow stress has two contributors:

$$\sigma(\dot{\varepsilon}) = \sigma_{\text{friction}}(\dot{\varepsilon}) + \sigma_{\text{bowout}}(\dot{\varepsilon}, L), \qquad (1)$$

where σ_{friction} is due to intrinsic lattice friction force on a moving dislocation core. In contrast, σ_{bowout} is due to dislocation–obstacle interactions, where the obstacles may be other dislocations (forest dislocation hardening), free surfaces (source truncation hardening [28]), etc. that impede dislocation line length growth. The magnitude of σ_{bowout} depends on a characteristic distance L between pinning points, which defines a critical curvature through which the dislocations have to squeeze by to propagate. σ_{friction} depends sensitively on $\dot{\varepsilon}$ ($\propto$ dislocation velocity) and temperature because of the small activation volumes of double-kink nucleation and kink migration of the dislocation core. σ_{bowout}, as exemplified by the Orowan equation $\sigma_{\text{bowout}} \propto \mu b/L$, is on the other hand microstructural length scale sensitive but less sensitive to $\dot{\varepsilon}$, because of its usually much larger activation volume involved in bowing out through a critical curvature.[29] According to the source truncation hardening [28] interpretation of 'smaller is stronger' $\sigma = \sigma_0 + kD^{-\alpha}$, $L \propto D^{\alpha}$, and L is the mobile free-arm length [28] between an internal pinning point and the surface. From Equation (1), then, we can deduce that when L is large, for example when a large single crystal of BCC metal is undergoing stage-1 easy glide, $\sigma_{\text{friction}} \gg \sigma_{\text{bowout}}$ dominates the flow stress, so $\sigma \approx \sigma_{\text{friction}}$. Then, the difficulty is in pushing the long screw segments through the crystal, not in bowing out through the sparse microstructural obstacles. In this regime, one expects a larger m ~ 0.06 because of

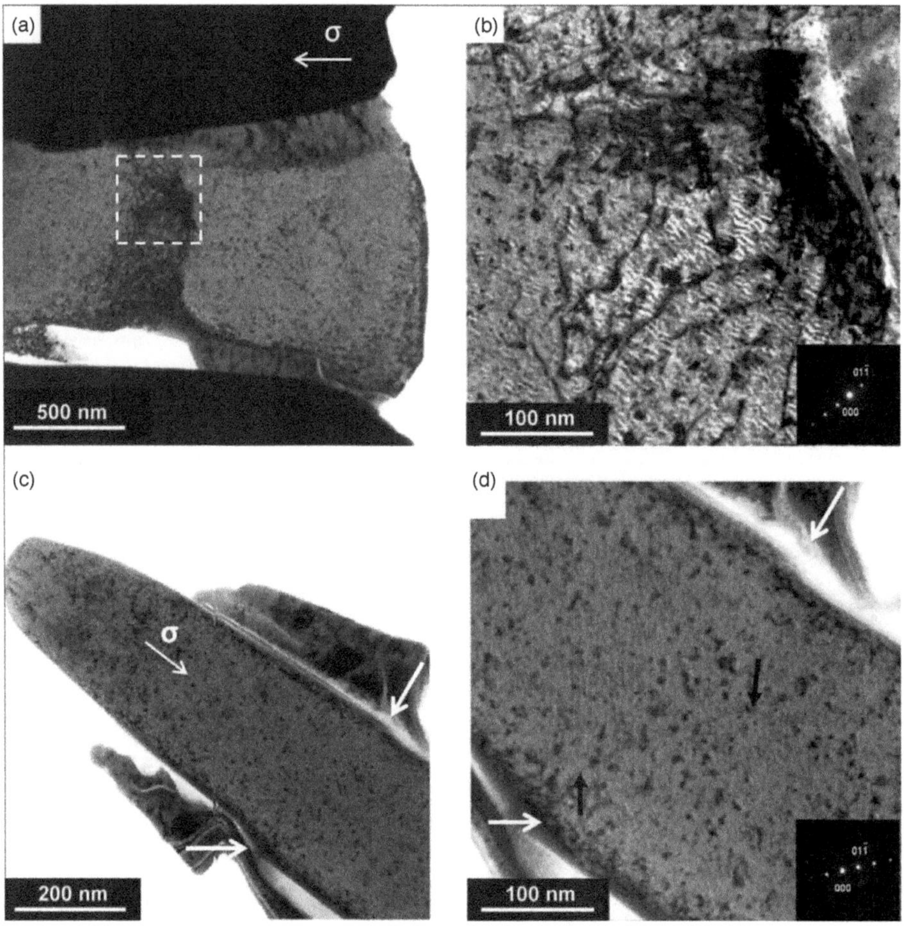

Figure 3. Typical postmortem bright-field STEM micrographs of the pillars after compression tests. (a) 1000 nm; (b) magnified image of the area framed in (a); (c) 200 nm; (d) magnified image of the area indicated by the two long white arrows in (c). Note the dislocation line marked by the two black arrows.

the intrinsically small activation volume of σ_{friction} from atomic-scale double kink nucleation and kink migration activation events. On the other hand, as L is being reduced, for example due to microstructural refinement, sample size reduction, etc., σ_{bowout} starts to overwhelm the intrinsic lattice friction, and $\sigma \approx \sigma_{\text{bowout}} \gg \sigma_{\text{friction}}$. In this regime, the difficulty lies in overcoming the critical curvature for dislocation bowout, not in whether and how the dislocation can move after the bowout. Due to the usually larger activation volume of σ_{bowout}, m will sustain a precipitous drop from ~ 0.06 to ~ 0.005 at this regime transition from friction-to-bowout controlled, as illustrated in Figure 4. In this higher stress, bowout-controlled regime, the physics that controls σ is not that different from that of FCC pillars, as the bowout formula is not very sensitive to screw or edge form. The distinction in mobility may still exist between screw and edge, but they no longer control σ.

Interestingly, after the friction-to-bowout-controlled transition, as L is further reduced, we expect m to gradually rise again, as the activation volume scales with L to some positive power.[29] In the limit of very small L (a few to tens of nanometers) and σ being a significant fraction of the ideal strength, we expect m can rise up to 0.02–0.05, which is the level expected for surface nucleation [30] (instead of propagation)-controlled flow regime. In nanocrystalline materials, m ~ 0.02–0.05 level is expected for slip transmission [31]-controlled mechanism across grain boundaries with large misorientation angle. When D further shrinks, one may trigger surface diffusion induced Coble creep, with m as large as 1,[32,33] as illustrated by Figure 4.

In summary, we have systematically investigated the plastic flow of [001] oriented α-Fe single crystal pillars. Their strength follows the well-established trend of 'smaller is stronger', and the size-strengthening exponent is a function of the strain rate. At the same time, the strain-rate sensitivity decreases by more than 10 times, when the sample diameter decreased from 1000 to 200 nm. Postmortem TEM observation found a high density

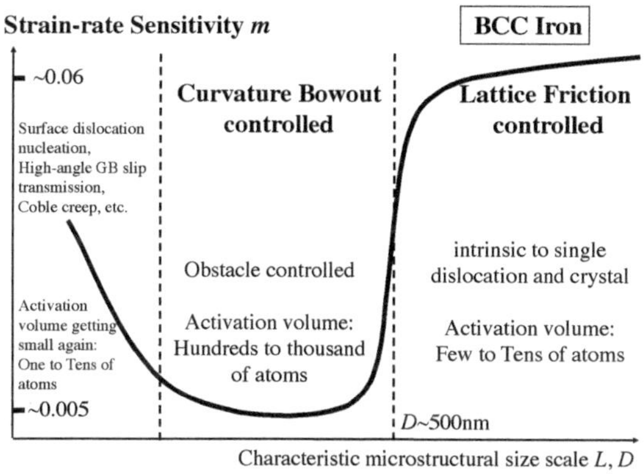

Figure 4. Regimes for strain-rate sensitivity based on characteristic microstructural size scales D. Lattice friction-controlled regime ($D > \sim 500$ nm): lattice friction of screw dislocation sensitively depends on the strain rate because of small activation volume of kink-pair nucleation and migration, thus a high rate sensitivity is typical of bulk BCC metals. Bowout-controlled regime (~ 500 nm $> D > \sim 100$ nm): stress required to bow out a single-arm source overwhelms the lattice friction of screw dislocation, leading to a much smaller rate sensitivity as a result of the large activation volume of single-arm source. Surface/interface-controlled regime ($D < \sim 100$ nm): absence of pre-existing dislocations requires nucleation of dislocations from surface/interface, or non-displacive deformation such as creep, both having small activation volume thus a larger rate sensitivity again.

of residual dislocations in the 1000 nm pillars but few dislocations in the 200 nm pillars. Based on our observations, we proposed that the sample/grain size dependence of m observed in nanostructured BCC metals is mainly a result of the relative magnitudes of dislocation lattice friction and curvature bowout stresses. The latter becomes dominant at small sample (or grain) sizes, entailing a transition of the rate-controlling process (and corresponding activation volume) and hence causing the sample-size-dependent m and the rate-dependent size-strengthening exponent α.

Acknowledgements This research was conducted at Center for Advancing Materials Performance from the Nanoscale (CAMP-Nano) & Hysitron Applied Research Center in China (HARCC), State Key Laboratory for Mechanical Behavior of Materials, Xi'an Jiaotong University, Xi'an 710049, China. Both E.M. and J.L. carried out this work under an adjunct professorship at XJTU.

Disclosure statement No potential conflict of interest was reported by the authors.

Funding This work was supported by the grants from NSFC (51231005, 51321003 and 11132006) and 973 Program of China (2010CB631003, 2012CB619402). We also appreciate the support from the 111 Project of China (B06025). J.L. acknowledges support by NSF DMR-1120901 and DMR-1240933. E.M. was supported at JHU by US-NSF-DMR-0904188.

References

[1] Jia D, Ramesh KT, Ma E. Effects of nanocrystalline and ultrafine grain sizes on constitutive behavior and shear bands in iron. Acta Mater. 2003;51(12):3495–3509.

[2] Cheng GM, Jian WW, Xu WZ, Yuan H, Millett PC, Zhu YT. Grain size effect on deformation mechanisms of Nanocrystalline bcc metals. Mater Res Lett. 2013;1(1):26–31.

[3] Greer JR, Oliver WC, Nix WD. Size dependence of mechanical properties of gold at the micron scale in the absence of strain gradients. Acta Mater. 2005;53(6):1821–1830.

[4] Volkert CA, Lilleodden ET. Size effects in the deformation of sub-micron Au columns. Philos Mag. 2006;86 (33–35):5567–5579.

[5] Kim J-Y, Jang D, Greer JR. Tensile and compressive behavior of tungsten, molybdenum, tantalum and niobium at the nanoscale. Acta Mater. 2010;58(7):2355–2363.

[6] Kim J-Y, Jang D, Greer JR. Insight into the deformation behavior of niobium single crystals under uniaxial compression and tension at the nanoscale. Scripta Mater. 2009;61(3):300–303.

[7] Frick CP, Clark BG, Orso S, Schneider AS, Arzt E. Size effect on strength and strain hardening of small-scale [111] nickel compression pillars. Mater Sci Eng A. 2008;489(1–2):319–329.

[8] Kiener D, Motz C, Rester M, Jenko M, Dehm G. FIB damage of Cu and possible consequences for miniaturized mechanical tests. Mater Sci Eng A. 2007;459 (1–2):262–272.

[9] von Blanckenhagen B, Gumbsch P, Arzt E. Dislocation sources and the flow stress of polycrystalline thin metal films. Philos Mag Lett. 2003;83(1):1–8.

[10] Malow TR, Koch CC, Miraglia PQ, Murty KL. Compressive mechanical behavior of nanocrystalline Fe investigated with an automated ball indentation technique. Mater Sci Eng A. 1998;252(1):36–43.

[11] Wei Q, Cheng S, Ramesh KT, Ma E. Effect of nanocrystalline and ultrafine grain sizes on the strain rate sensitivity and activation volume: fcc versus bcc metals. Mater Sci Eng A. 2004;381(1–2):71–79.

[12] Wei Q, Jiao T, Ramesh KT, Ma E, Kecskes LJ, Magness L, Dowding R, Kazykhanov VU, Valiev RZ. Mechanical behavior and dynamic failure of high-strength ultrafine grained tungsten under uniaxial compression. Acta Mater. 2006;54(1):77–87.

[13] Wei Q, Pan ZL, Wu XL, Schuster BE, Kecskes LJ, Valiev RZ. Microstructure and mechanical properties at different length scales and strain rates of nanocrystalline tantalum produced by high-pressure torsion. Acta Mater. 2011;59(6):2423–2436.

[14] Schneider AS, Clark BG, Frick CP, Gruber PA, Arzt E. Effect of orientation and loading rate on compression behavior of small-scale Mo pillars. Mater Sci Eng A. 2009;508(1–2):241–246.

[15] Weinberger CR, Cai W. Surface-controlled dislocation multiplication in metal micropillars. Proc Natl Acad Sci. 2008;105(38):14304–14307.

[16] Schneider AS, Kaufmann D, Clark BG, Frick CP, Gruber PA, Mönig R, Kraft O, Arzt E. Correlation between critical temperature and strength of small-scale bcc pillars. Phys Rev Lett. 2009;103(10):105501.
[17] Min Han S, Feng G, Young Jung J, Joon Jung H, Groves JR, Nix WD, Cui Y. Critical-temperature/Peierls-stress dependent size effects in body centered cubic nanopillars. Appl Phys Lett. 2013;102(4):041910.
[18] Kim J-Y, Greer JR. Size-dependent mechanical properties of molybdenum nanopillars. Appl Phys Lett. 2008;93(10).
[19] Huang L, Li Q-J, Shan Z-W, Li J, Sun J, Ma E. A new regime for mechanical annealing and strong sample-size strengthening in body centred cubic molybdenum. Nat Commun. 2011;2:547.
[20] Han SM, Bozorg-Grayeli T, Groves JR, Nix WD. Size effects on strength and plasticity of vanadium nanopillars. Scripta Mater. 2010;63(12):1153–1156.
[21] Greer JR, Weinberger CR, Cai W. Comparing the strength of f.c.c. and b.c.c. sub-micrometer pillars: Compression experiments and dislocation dynamics simulations. Mater Sci Eng A. 2008;493(1–2):21–25.
[22] Uchic MD, Dimiduk DM, Florando JN, Nix WD. Sample dimensions influence strength and crystal plasticity. Science. 2004;305(5686):986–989.
[23] Shan ZW, Mishra RK, Syed Asif SA, Warren OL, Minor AM. Mechanical annealing and source-limited deformation in submicrometre-diameter Ni crystals. Nat Mater. 2007;7(2):115–119.
[24] Wang Z-J, Li Q-J, Shan Z-W, Li J, Sun J, Ma E. Sample size effects on the large strain bursts in submicron aluminum pillars. Appl Phys Lett. 2012;100(7): 071906.
[25] Suzuki T, Koizumi H, Kirchner HOK. Plastic flow stress of b.c.c. transition metals and the Peierls potential. Acta Metall Mater. 1995;43(6):2177–2187.
[26] Malow TR, Koch CC. Mechanical properties, ductility, and grain size of nanocrystalline iron produced by mechanical attrition. Metall Mater Trans A. 1998;29(9):2285–2295.
[27] Zhang L, Ohmura T, Sekido K, Hara T, Nakajima K, Tsuzaki K. Dislocation character transition and related mechanical response in a body-centered cubic single crystal. Scripta Mater. 2012;67(4):388–391.
[28] Parthasarathy TA, Rao SI, Dimiduk DM, Uchic MD, Trinkle DR. Contribution to size effect of yield strength from the stochastics of dislocation source lengths in finite samples. Scripta Mater. 2007;56(4):313–316.
[29] Jennings AT, Li J, Greer JR. Emergence of strain-rate sensitivity in Cu nanopillars: transition from dislocation multiplication to dislocation nucleation. Acta Mater. 2011;59(14):5627–5637.
[30] Zhu T, Li J, Samanta A, Leach A, Gall K. Temperature and strain-rate dependence of surface dislocation nucleation. Phys Rev Lett. 2008;100(2):025502.
[31] Zhu T, Li J, Samanta A, Kim HG, Suresh S. Interfacial plasticity governs strain rate sensitivity and ductility in nanostructured metals. Proc Natl Acad Sci. 2007;104(9):3031–3036.
[32] Tian L, Li J, Sun J, Ma E, Shan Z-W. Visualizing size-dependent deformation mechanism transition in Sn. Sci Rep. 2013;3:2113.
[33] Sun J, He L, Lo Y-C, Xu T, Bi H, Sun L, Zhang Z, Mao SX, Li J. Liquid-like pseudoelasticity of sub-10-nm crystalline silver particles. Nat Mater. 2014;13(11): 1007–1012.

5

High-Entropy Alloys: A Critical Review

Ming-Hung Tsai[a]* and Jien-Wei Yeh[b]*

[a] *Department of Materials Science and Engineering, National Chung Hsing University, Taichung 40227, Taiwan;*
[b] *Department of Materials Science and Engineering, National Tsing Hua University, Hsinchu 30013, Taiwan*

High-entropy alloys (HEAs) are alloys with five or more principal elements. Due to the distinct design concept, these alloys often exhibit unusual properties. Thus, there has been significant interest in these materials, leading to an emerging yet exciting new field. This paper briefly reviews some critical aspects of HEAs, including core effects, phases and crystal structures, mechanical properties, high-temperature properties, structural stabilities, and corrosion behaviors. Current challenges and important future directions are also pointed out.

Keywords: High-Entropy Alloys, High-Entropy Materials, Alloy Design, Mechanical Properties, Corrosion Resistance

1. Introduction Most conventional alloys are based on one principal element. Different kinds of alloying elements are added to the principal element to improve its properties, forming an alloy family based on the principal element. For example, steel is based on Fe, and aluminum alloys are based on Al. However, the number of elements in the periodic table is limited, thus the alloy families we can develop are also limited. If we think outside the conventional box and design alloys not from one or two 'base' elements, but from multiple elements altogether, what will we get? This new concept, first proposed in 1995,[1] has been named a high-entropy alloy (HEA).[2]

HEAs are defined as alloys with five or more principal elements. Each principal element should have a concentration between 5 and 35 at.%.[2,3] Besides principal elements, HEAs can contain minor elements, each below 5 at.%. These alloys are named 'HEAs' because their liquid or random solid solution states have significantly higher mixing entropies than those in conventional alloys. Thus, the effect of entropy is much more pronounced in HEAs. Existing physical metallurgy knowledge and binary/ternary phase diagrams suggest that such multi-element alloys may develop several dozen kinds of phases and intermetallic compounds, resulting in complex and brittle microstructures that are difficult to analyze and engineer, and probably have very limited practical value. Opposite to these expectations, however, experimental results show that the higher mixing entropy in these alloys facilitates the formation of solid solution phases with simple structures and thus reduces the number of phases. Such characters, made available by the higher entropy, are of paramount importance to the development and application of these alloys. Therefore, these alloys were named as 'high-entropy' alloys.

Because of the unique multiprincipal element composition, HEAs can possess special properties. These include high strength/hardness, outstanding wear resistance, exceptional high-temperature strength, good structural stability, good corrosion and oxidation resistance. Some of these properties are not seen in conventional alloys, making HEAs attractive in many fields. The fact that it can be used at high temperatures broadens its spectrum of applications even further. Moreover, the fabrication of HEAs does not require special processing techniques or equipment, which indicates that the mass production of HEAs can be easily implemented with existing equipment and technologies. More than 30 elements have been used to prepare more than 300 reported HEAs, forming an exciting new field of metallic materials. This paper reviews some crucial aspects of the field, including core effects, phase formation, mechanical properties, high-temperature properties, and corrosion

*Corresponding author. Emails: mhtsai@nchu.edu.tw; jwyeh@mx.nthu.edu.tw

behaviors. It also points out current challenges and future directions. The physical (magnetic, electrical, and thermal) properties of HEAs have been reviewed elsewhere [4] and are not discussed in this paper.

2. Core Effects [3,5] The multiprincipal-element character of HEAs leads to some important effects that are much less pronounced in conventional alloys. These can be considered as four 'core effects'. This section briefly introduces and discusses the core effects.

2.1. High-Entropy Effect. The high-entropy effect states that the higher mixing entropy (mainly configurational) in HEAs lowers the free energy of solid solution phases and facilitates their formation, particularly at higher temperatures. Due to this enhanced mutual solubility among constituent elements, the number of phases present in HEAs can be evidently reduced. According to G = H − TS (where G is the Gibbs free energy, H is enthalpy, T is temperature, and S is entropy) entropy can stabilize a phase with higher entropy, provided that temperature is sufficiently high. For example, the melting phenomenon of a pure metal is due to the higher entropy of the liquid state relative to the solid state (according to Richard's rule,[6] at melting point the entropy difference between the two states roughly equals the gas constant R). Similarly, among the various types of phases in an alloy, those with higher entropy may also be stabilized at high temperature. In conventional alloys, solid solution phases (including terminal and intermediate solid solutions) have higher mixing entropy than intermetallic compounds do. For HEAs, the difference in entropy between solid solutions and compounds is particularly large owing to the multiprincipal-element design. For example, the configurational entropy of an equimolar quinary random solid solution is $1.61R$. This means that the entropy difference between an equimolar quinary solution and a completely ordered phase (whose entropy is negligible) is about 60% larger than the entropy difference of the melting case mentioned above (i.e. 60% larger than the entropy difference between liquid and solid states of pure metals). Thus, it is highly probable that solid solution phases become the stable phase in HEAs at high temperature. It is expected that with the increase of temperature, the overall degree of order in HEAs decreases. Thus, even those alloys that contain ordered phases in their cast state can transform to random solid solutions at high temperature. However, if the formation enthalpy of an intermetallic compound is high enough to overcome the effect of entropy, that intermetallic compound will still be stable at high temperature.

There is much evidence for the high-entropy effect.[7–13] Here we demonstrate this effect with Figure 1, which shows the XRD patterns of a series of binary to septenary alloys. It is seen that the phases

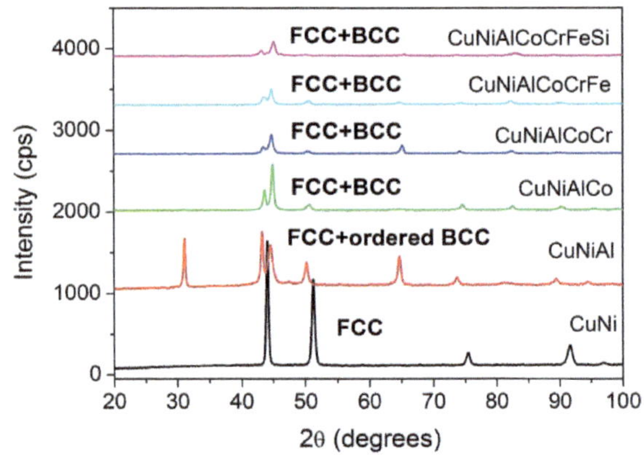

Figure 1. The XRD patterns of a series alloy designed by the sequential addition of one extra element to the previous one. All the alloys have one or two major phases that have simple structures.

in quinary, senary, and septenary alloys remain rather simple: there are only two major phases, and these phases have simple structures such as BCC and FCC (note that the septenary alloy actually contain minor intermetallic phases, although not seen clearly in the XRD pattern). Such simple structures contradict conventional expectation: formation of various kinds of binary/ternary compounds.

Two myths regarding the high-entropy effect are discussed here. The first myth is that the high-entropy effect guarantees the formation of a simple solid solution phase (order/disordered BCC/FCC) at high temperatures. This is not true. Actually, in the very first paper of HEA,[2] the possibility of compound formation was already mentioned. In the same year, the existence of (Cr, Fe)-rich borides in the $AlB_xCoCrCuFeNi$ alloys was observed.[14] As mentioned previously, it is the competition between entropy and enthalpy that determines the phase formation. The large negative formation enthalpy of the borides thus justifies their stability at high temperatures. Similar phenomena have also been observed in other high-temperature annealing experiments.[15–17] The second myth is that only simple solid solution phases (BCC/FCC) are the desired phases for practical applications of HEAs. This is also not true. It is true that these multiprincipal-element simple solid solution phases are unique to HEAs, and they can have outstanding properties. However, as shown later, many HEAs containing intermediate phases also have great potential. Thus, the exploration of HEAs should not be limited to simple solid solution phases.

2.2. Sluggish Diffusion Effect. It was proposed that the diffusion and phase transformation kinetics in HEAs are slower than those in their conventional counterparts.[18] This can be understood from two

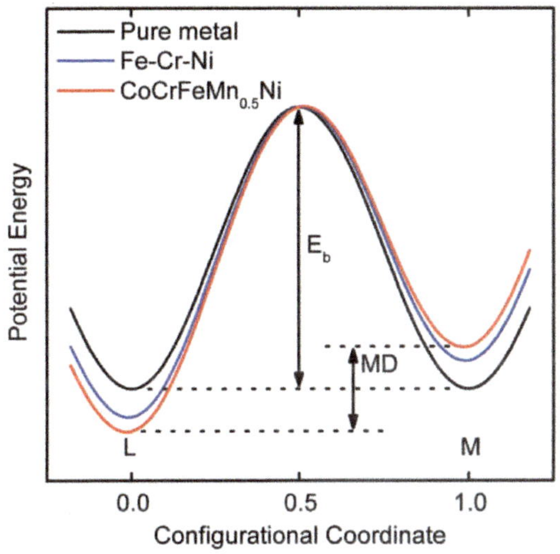

Figure 2. Schematic diagram of the potential energy change during the migration of a Ni atom. The mean difference (MD) in potential energy after each migration for pure metals is zero, whereas that for HEA is the largest.[19]

aspects. Firstly, in HEA the neighboring atoms of each lattice site are somewhat different. Thus, the neighbors before and after an atom jump into a vacancy are different. The difference in local atomic configuration leads to different bonding and therefore different local energies for each site. When an atom jumps into a low-energy site, it becomes 'trapped' and the chance to jump out of that site will be lower. In contrast, if the site is a high-energy site, then the atom has a higher chance to hop back to its original site. Either of these scenarios slows down the diffusion process. Note that in conventional alloys with a low solute concentration, the local atomic configuration before and after jumping into a vacancy is, most of the time, identical. Tsai et al. used a seven-bond model to calculate the effect of local energy fluctuation on diffusion.[19] They showed that for Ni atoms diffusing in Co–Cr–Fe–Mn–Ni alloys (which has a single-phase, FCC structure), the mean potential energy difference between lattice sites is 60.3 meV, which is 50% higher than that in Fe–Cr–Ni alloys (see Figure 2). This energy difference leads to a significantly longer occupation time at low-energy sites (1.73 times longer than that at high-energy sites). Indeed, diffusion couple experiments show that the activation energy of diffusion in the Co–Cr–Fe–Mn–Ni alloys is higher than those in other FCC ternary alloys and pure elements (Table 1).[19] The diffusivity of Ni at the respective melting point of each metal is also calculated (Table 1). The slowest diffusion takes place again in the Co–Cr–Fe–Mn–Ni alloys.

Secondly, the diffusion rate of each element in a HEA is different. Some elements are less active (for example, elements with high melting points) than others so these elements have lower success rates for jumping into vacancies when in competition with other elements. However, phase transformations typically require the coordinated diffusion of many kinds of elements. For example, the nucleation and growth of a new phase require the redistribution of all elements to reach the desired composition. Grain growth also requires the cooperation of all elements so that grain boundaries can successfully migrate. In these scenarios, the slow-moving elements become the rate-limiting factor that impedes the transformation.

The slow kinetics in HEAs allows readily attainable supersaturated state and nano-sized precipitates, even in the cast state.[2,20,21] It also contributes to the excellent performance of HEA coatings as diffusion barriers.[22, 23] Moreover, it allows better high-temperature strength and structural stability, which are discussed in Section 5. For the same reason, HEAs are also expected to have outstanding creep resistance.

2.3. Severe-Lattice-Distortion Effect. The lattice in HEAs is composed of many kinds of elements, each with different size. These size differences inevitably lead to distortion of the lattice. Larger atoms push away their neighbors and small ones have extra space around. The strain energy associated with lattice distortion raises the overall free energy of the HEA lattice. It also affects the properties of HEAs. For example, lattice distortion impedes dislocation movement and leads to pronounced

Table 1. Diffusion parameters for Ni in different FCC matrices. The compositions of Fe-Cr-Ni(-Si) alloys are in wt.%.[19]

Solute	System	D_0 (10^{-4} m^2/s)	Q (kJ/mol)	$T_m(T_s)$ (K)	Q/T_m	D_{T_m} (10^{-13} m^2/s)
Ni	CoCrFeMnNi	19.7	317.5	1607	0.1975	0.95
	FCC Fe	3	314	1812	0.1733	2.66
	Co	0.43	282.2	1768	0.1596	1.98
	Ni	1.77	285.3	1728	0.1651	4.21
	Fe-15Cr-20Ni	1.5	300	1731	0.1733	1.33
	Fe-15Cr-45Ni	1.8	293	1697	0.1727	1.73
	Fe-22Cr-45Ni	1.1	291	1688	0.1724	1.09
	Fe-15Cr-20Ni-Si	4.8	310	1705	0.1818	1.53

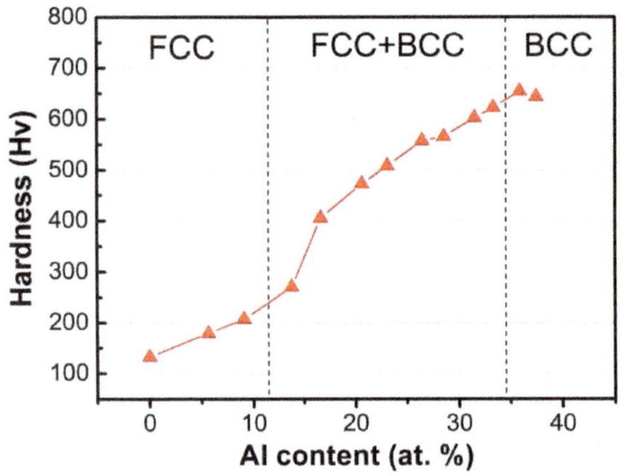

Figure 3. Hardness of the $Al_xCoCrCuFeNi$ alloys as a function of Al content.

solid solution strengthening (see Section 4.4). It also leads to increased scattering of propagating electrons and phonons, which translates to lower electrical and thermal conductivity.[4]

2.4. Cocktail Effect. The properties of HEAs are certainly related to the properties of its composing elements. For example, addition of light elements decreases the density of the alloy. However, besides the properties of the individual composing elements, the *interaction* among the component elements should also be considered. For example, Al is a soft and low-melting-point element. Addition of Al, however, can actually harden HEAs. Figure 3 plots the hardness of the $Al_xCoCrCuFeNi$ alloy as a function of Al content. It is clearly seen that the alloy hardens significantly with the addition of Al. This is partly because of the formation of a hard BCC phase, and partly due to the stronger cohesive bonding between Al and other elements, and its larger atomic size. Thus, the macroscopic properties of HEA not only come from the averaged properties of the component elements, but also include the effects from the excess quantities produced by inter-elemental reactions and lattice distortion.

3. Phase and Crystal Structure In conventional alloys, phases are typically classified into three types: terminal solutions, intermediate solutions, and intermetallic compounds. We can use similar, but expanded concepts to classify phases in HEAs. *Terminal phases* are phases based on one dominating element. The definition of *intermetallic compounds* is not changed: they are stoichiometric and have fixed composition ratio. However, because phases in HEAs typically have a composition range rather than fixed composition ratio, intermetallic compounds are rarely seen in HEAs. *Solution phases* are phases not belonging to the above two categories. This category includes the solid solutions based on both simple (e.g. BCC and FCC) and complex structures (e.g. Laves phase).

In the literature, phases in HEAs are usually classified differently. They are often classified as: random solid solution (e.g. FCC, BCC), ordered solid solution (e.g. B2 and $L1_2$), and intermetallic phases (e.g. Laves phases). This classification may lead to some confusion because by definition, intermetallic phases can also be classified as ordered solid solutions—they have composition ranges and are typically ordered. In view of this, we suggest classifying the phases according to their structure (simple/complex) and ordering (ordered/disordered). A phase is said to be *simple* if its structure is identical to or derived from FCC, BCC or HCP structures. Namely, cI2-W, cF4-Cu, and hP2-Mg structures and their ordered versions (superlattices) such as cP2-CsCl (B2) and cP4-AuCu3 ($L1_2$) structures are simple.[24] If a phase is not simple (e.g. Laves phases), it is said to be *complex*. Therefore, the above three types now becomes: simple disordered phase (SDP), simple ordered phase (SOP), and complex ordered phase (COP).

3.1. Simple Solid Solution or Intermetallics?. One fundamental and important question regarding HEAs is: what kind of phase and crystal structure will form when we mix so many different elements together? To the surprise of most people, simple structures (SDPs and SOPs) are the most frequently seen in as-cast HEAs (see Figure 1). These simple phases originate from the high-entropy effect mentioned previously. Besides simple phases, different kinds of COPs, such as σ, μ, Laves, etc., are also observed in HEAs. Because simple-solution phases with more than five elements are unique to HEAs, researchers have worked intensely to understand the conditions for their formation.

From the classic Hume-Rothery rules, factors that affect the formation of binary solid solutions include atomic size difference, electron concentration, and difference in electronegativity.[25] Besides these factors, enthalpy and entropy of mixing are the most important phase formation parameters for HEAs. Zhang et al. [26] and Guo et al. [27] studied the effect of these parameters on the phase formation of HEAs and obtained similar conclusions: the formation of simple or complex phases depends mainly on the enthalpy of mixing (ΔH_{mix}), entropy of mixing (ΔS_{mix}), and atomic size differences (δ). In order to form sole simple phases (i.e. FCC, BCC, and their mixtures, including both ordered/disordered cases), the following conditions have to be met simultaneously: $-22 \leq \Delta H_{mix} \leq 7\,\mathrm{kJ/mol}$, $\delta \leq 8.5$, and $11 \leq \Delta S_{mix} \leq 19.5\,\mathrm{J/(K\,mol)}$.[27] This is shown in Figure 4, where the boundary of simple phases is shown. These conditions are quite logical: ΔH_{mix} cannot be too large in value because large positive ΔH_{mix} leads to phase separation and large negative ΔH_{mix} typically leads to

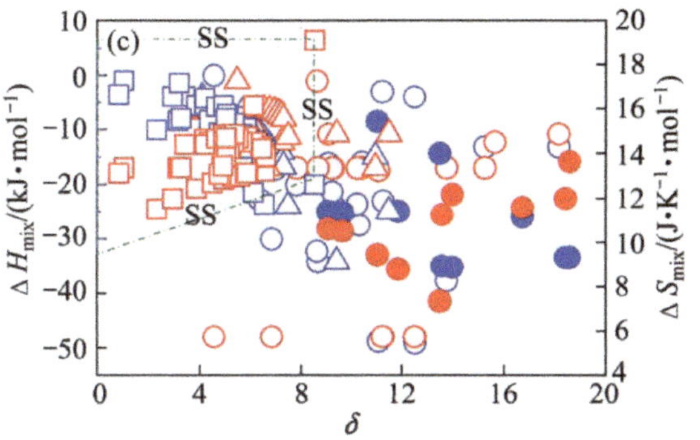

Figure 4. Superimposed effect of ΔH_{mix}, δ, and ΔS_{mix} on phase stability in equiatomic multicomponent alloys and BMGs. Notes: Blue symbols indicate the relationship between ΔH_{mix} and δ, while red ones represents that between ΔS_{mix} and δ. Symbol ○ represents equiatomic amorphous phase-forming alloys; ● represents nonequiatomic amorphous phase-forming alloys; □ represents simple phases and △ represents intermetallic phases. The region delineated by the dash-dotted lines indicates the requirements for simple phases to form.[27]

intermetallic phases. δ has to be small enough since large δ leads to excess strain energy and destabilizes simple structures. ΔS_{mix} has to be large enough because it is the main stabilizing factor for simple phases. If the target of discussion is limited to SDPs only, the conditions are more strict: $-15 \leq \Delta H_{\text{mix}} \leq 5\,\text{kJ/mol}$, $\delta \leq 4.3$, and $12 \leq \Delta S_{\text{mix}} \leq 17.5\,\text{J/(K mol)}$.[26]

The other criterion to judge whether or not sole simple phases will form in an HEA is the parameter $\varepsilon = |T\Delta S_{\text{mix}}/\Delta H_{\text{mix}}|$.[28,29] Based on the concept of entropy–enthalpy competition (see Section 2.1), ε represents the effect of entropy relative to that of enthalpy. Thus, larger ε suggests a higher possibility of forming SDP, which agrees well with the analysis.[28,29] A drawback of the two criterions shown above is that even if an alloy is designed following these criterions, it can still contain intermetallic phases. This can be seen in Figure 4. The 'solid solution' region delineated by the dash-dotted lines still contains triangles, indicating the existence of intermetallic phases. A new thermodynamic parameter has been proposed to solve this problem, and the results will be published soon.[30]

Some points of notice should be mentioned here: Firstly, the above analyses are based on the as-cast state. This is natural because most reported HEAs are in their as-cast state. Secondly, the phases observed macroscopically in as-cast HEAs can contain atomic-scale decompositions/inhomogeneities. This is evidenced by detailed transmission electron microscopy (TEM) and atomic probe analysis.[21,31,32] Thirdly, in HEAs similar disordered and ordered versions of the same base structure often co-exist.[8,33–35] For example, coexistence of BCC and ordered BCC (B2) phases was frequently reported.[8,21,34] Coexistence of FCC and ordered FCC ($L1_2$) phases was also observed.[35] When these phases have nearly identical lattice parameter, their XRD peaks overlap (except the superlattice peaks). Therefore, when the amount of the SOP is small, XRD may reflect only the peaks belonging to SDP, even if SOP exists.[34,35] This indicates that TEM analysis is needed to really confirm the existence of SOP.[34,35]

3.2. *Crystal Structure of Simple Solid Solution Phases.* If an HEA does crystallize into simple phases, what will the structure of the phase be? Virtually all simple solid solution phase observed in HEAs have either BCC or FCC structures.[8,10,36,37] The most critical factor that decides whether an alloy crystallizes into BCC or FCC structure appears to be its VEC (valence electron concentration, the VEC of an alloy is calculated from the

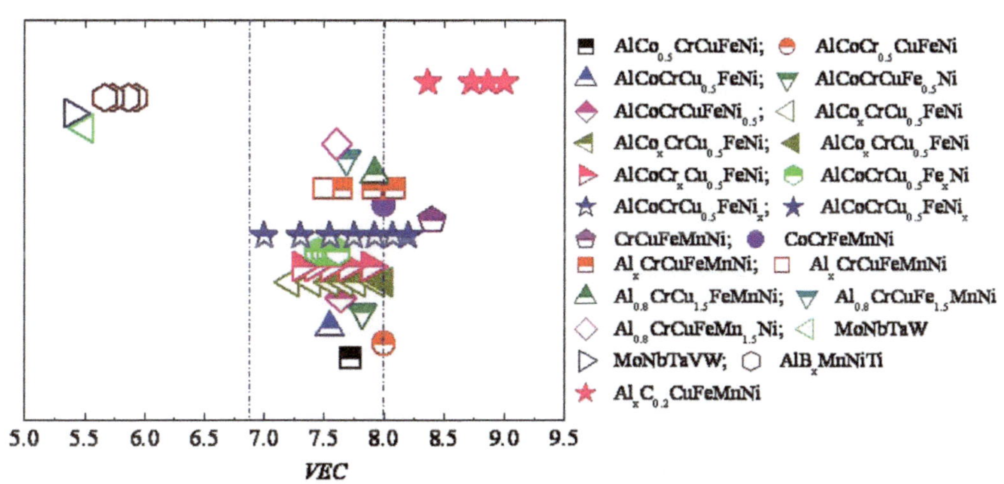

Figure 5. Relationship between VEC and the FCC, BCC phase stability for various HEA systems. Notes: Fully closed symbols for sole FCC phases; fully open symbols for sole BCC phase; top-half closed symbols for mixed FCC and BCC phases.[38]

weighted average VEC of the constituent components). Guo et al. summarized the relationship between VEC and structure of many HEAs (Figure 5).[38] They found that when the VEC of the alloy is larger than 8, the FCC structure is stabilized. When the VEC of the alloy is smaller than 6.87, the BCC structure is stabilized. Co-existence of FCC and BCC phase is observed at VEC values between 6.87 and 8.[38] However, the mechanism behind the effect of VEC on phase formation has not been fully understood so far.

3.3. High-Entropy Bulk Metallic Glass. Although HEAs and bulk metallic glasses (BMGs) are both multi-component and their compositions look similar at first glance, these two classes of materials are based on completely different concepts. For HEAs, more than five principal elements are required. In contrast, most BMGs are based on one or two principal elements. More importantly, BMGs are characterized by their amorphous structure, but there is no structural requirement for HEAs. The concept of HEAs, however, triggered an idea: would there be good glass formers in HEAs? If so, how do we design high-entropy BMGs (HEB-MGs) with good glass-forming ability (GFA) and better properties?

Inoue's empirical rules [39] state that (1) more than three composing element, (2) large atomic size differences, and (3) large negative mixing enthalpies favor the formation of BMGs. The first rule is in line with the design concept of HEAs. Indeed, high configurational entropy in HEAs was found to be beneficial for glass formation. This was evidenced by lower critical cooling rates in BMGs with higher ΔS_{mix}.[27] Another earlier study also reported similar findings.[40] The second and third rules for BMG formation are exactly opposite to the requirements to form simple phases in HEAs (see Section 3.1). This suggests that one should choose HEAs that are predicted to form intermetallic phases rather than simple phases as candidates for HEBMGs. Indeed, very recently Guo et al. analyzed the ΔH_{mix} and δ of some HEAs and BMGs and found that simple-solution-type HEAs and BMGs fall in opposite corners of the δ-ΔH_{mix} plot (shown in Figure 6 [41]). They further pointed out that as long as the competition from intermetallic phases can be ruled out by composition adjustment, HEAs that locate in the amorphous phase region of the δ-ΔH_{mix} plot indeed solidify into metallic glasses.[41] This strategy has been successfully demonstrated in a series of Zr-containing alloys prepared via melt spinning.[41]

Not many HEBMGs have been fabricated till now. Most of these HEBMGs have diameters smaller than 3 mm.[42–46] Takeuchi et al. successfully prepared $Pd_{20}Pt_{20}Cu_{20}Ni_{20}P_{20}$ HEBMG with a maximum diameter of 10 mm.[47] But they suggested that conventional GFA assessment parameters such as T_{rg} (reduced glass

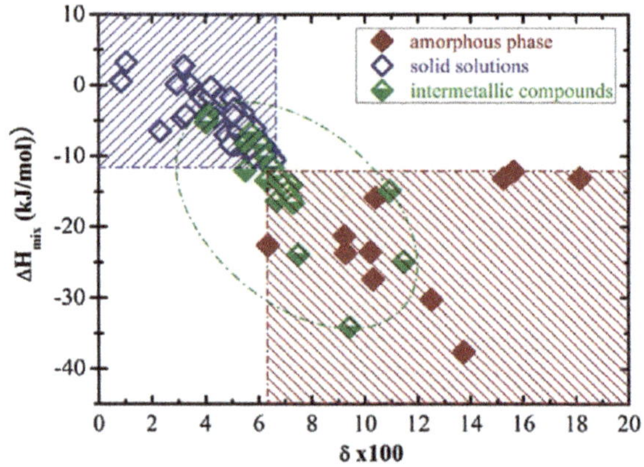

Figure 6. The δ-ΔH_{mix} plot delineating the phase selection in HEAs. Note: The dash-dotted regions highlight the individual region to form simple solid solutions, intermetallic phases and amorphous phases.[41]

transition temperature normalized with liquidus temperature) cannot evaluate the GFA of HEBMG effectively. Other factors, e.g. Gibbs free energy assessments, need to be taken into account.[47] Apparently more study is needed to understand these issues.

The distinct composition design in HEAs brings with their amorphous structure other special properties. For example, an extremely high crystallization temperature (probably the highest among reported BMG) of over 800°C and good performance as diffusion barrier between Cu and Si was observed in 20-nm-thick NbSiTaTiZr alloy film.[23] Additionally, $Zn_{20}Ca_{20}Sr_{20}Yb_{20}(Li_{0.55}Mg_{0.45})_{20}$ shows room temperature homogeneous flow and a plasticity of 25%,[44] which is in sharp contrast to the typical shear banding and brittle behavior of most BMGs. Very recently CaMgZn-SrYb, a high-entropy modification of $Ca_{65}Mg_{15}Zn_{20}$ BMG, has shown improved properties in orthopedic applications.[48] CaMgZnSrYb not only has better mechanical properties, but also promotes osteogenesis and new bone formation after two weeks of implantation. These examples indicate that high-entropy BMGs and amorphous films do have great potential and deserve further exploration—which is good news for both communities.

3.4. Phases at Elevated Temperatures. As mentioned in Section 3.1, our understanding of the phases in HEAs focuses mainly on their as-cast state. However, the cast state is typically not in thermodynamic equilibrium. For alloys that locate in the BCC or FCC phase region suggested by Guo et al. (see Figure 5 [38]), phase transformation may be less significant. For example, no phase formation was found in the Co–Cr–Fe–Mn–Ni FCC alloys during their high-temperature annealing.[14,17] For alloys that locate in the BCC+FCC phase region,

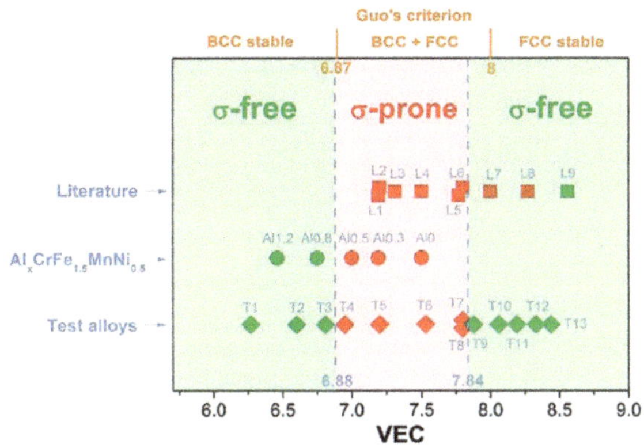

Figure 7. Relationship between the VEC and the presence of σ phase after aging for a number of HEAs. Green and red icons indicate the absence and presence of σ phase after aging, respectively.[52]

however, phase transformation will take place at higher temperatures. In the Al-Co-Cr-Cu-Fe-Ni system, annealing at ~800°C or below increases the fraction of the BCC phase.[49] Annealing at temperatures higher than 800°C increases the fraction of the FCC phase. Because the FCC phase is ductile and the BCC phase is relatively brittle, annealing can be used to tune the mechanical properties.

Some alloys have phases that are stable at intermediate temperatures only. These alloys may contain only simple phases in their as-cast state due to the higher cooling rate of Cu mold casting. Upon annealing, however, the intermediate-temperature phases will appear. For example, η phase [17,50] and σ phase [12,34,51] are known to form in some HEAs. One should pay great attention to the formation of these intermediate-temperature phases, because their formation can lead to significant changes in mechanical properties. Tsai et al. studied the stability of the σ phase.[52] They showed that the formation of the σ phase is predictable: it is directly related to the VEC of the alloy. As shown in Figure 7,[52] there is a σ-phase-forming VEC range for HEAs based on Al, Co, Cr, Cu, Fe, Mn, Ni, Ti, and V. Alloys whose VEC fall in this range develop σ phase upon annealing at 700°C.

3.5. Computer-Aided Phase Prediction. The multi-principal-element nature of HEA translates to an immense amount of possible compositions. It takes huge amounts of time and cost to study all the compositions experimentally. Hence, it is critically important to implement highly efficient, low-cost methods to assist experimental study. For example, phase diagram is one of the most important tools to understand an alloy system. However, developing the phase diagram of just one senary HEA requires tremendous efforts—there are six composition axes! If the phase diagram could be reliably predicted with computational techniques, then experiments are only needed to verify some critical points on the predicted phase diagram, and the work needed is significantly reduced.

The main problem in applying existing phase-prediction techniques (e.g. CALculation of PHAse Diagrams, CALPHAD) on HEAs is probably the lack of a suitable database for multicomponent systems.[53] Thus, Zhang et al. developed a thermodynamic database for the Al–Co–Cr–Fe–Ni alloy system by extrapolating binary and ternary systems to wider composition ranges.[53] This database does not consider quaternary or quinary interaction parameters. Using their database, they obtained phase diagrams of the Al–Co–Cr–Fe–Ni system that agree with available experimental results.

In some cases, it seems that phase diagrams with reasonable accuracy can still be obtained even without the development of new databases. An Ni-based superalloy database was used to predict the equilibrium phases and phase fractions of $Al_{0.5}CoCrCuFeNi$, $Al_{23}Co_{15}Cr_{23}Cu_8Fe_{15}Ni_{16}$ and $Al_8Co_{17}Cr_{17}Cu_8Fe_{17}Ni_{33}$ alloys.[12,31] Except for some minor discrepancies,[12] experiments and predictions agree with each other. These examples demonstrate the effectiveness of computational methods in predicting the phase of HEAs. Although the accuracy of computational phase predictions for other HEA systems (other than Al–Co–Cr–Cu–Fe–Ni) is still not clear (the only result appears to be not as pleasing [54]), it is greatly expected that these techniques will become important tools for the future design and development of HEAs.

3.6. Phase Evolutions in Representative Alloy Systems. The phase evolutions in some important alloy systems are discussed, along with their mechanical properties, in Sections 4.2 and 4.3. This is because the mechanical properties of HEAs are directly related to the phases contained in the alloys (see Section 4.1). Therefore, the best way to understand the mechanical properties is to discuss phase evolution and mechanical properties together.

4. Mechanical Properties

4.1. Basic Concepts and Deformation Behaviors. Because of the wide composition range and the enormous number of alloy systems in HEAs, the mechanical properties of HEA can vary significantly. In terms of hardness/strength, the most critical factors are:

(1) Hardness/strength of each composing phase in the alloy;
(2) Relative volume ratio of each composing phase;
(3) Morphology/distribution of the composing phases.

Table 2. Phases in HEAs as categorized by hardness. Examples and typical hardness range for each category are also listed.

Type	Examples	Typical hardness (HV)
Valence compounds	Carbides, borides, silicides	1000–4000
Intermetallic phases with non-simple structures	σ, Laves, η	650–1300
BCC and derivatives	BCC, B2, Heusler	300–700
FCC and derivatives	FCC, $L1_2$, $L1_0$	100–300

The first factor is largely determined by the crystal structure and bonding of each phase. In our experience, the phases can be roughly classified into four categories each having a different hardness range. This is provided in Table 2. It is important to note that the hardness ranges listed in Table 2 are merely typical ranges and there could be exceptions. Valence compounds are based on very strong covalent bonding and are essentially ceramics. Thus, their hardness is the highest. Non-simple intermetallic phases often lack easily accessible slip systems and dislocation activities are thus largely hindered. BCC and BCC-derivative structures (see the second paragraph of Section 3) are harder than their FCC counterparts because BCC-based structures have stronger directional bonding and lack a truly close-packed slip plane.[55,56]

The general rule to estimate the hardness/strength of an HEA is straight forward: the harder the phase (and the higher the fraction of the hard phase), the harder the alloy. When two HEAs have phases with similar hardness and relative fraction, the distribution of the phases can also play an important role. This is exemplified in later sections. The ductility of HEA is also related to the phase in the alloy. As can be expected, harder phases usually have lower ductility.

The deformation microstructure and mechanism in most HEAs is unclear. However, such information was revealed in a single-phase, FCC CoCrFeMnNi alloy via detailed TEM study.[57] It was found that at small strains ($<2.4\%$), the deformation is governed by planar slip of $1/2\langle 110\rangle$ type dislocations on $\{111\}$ planes. At a strain around 20%, the deformation behavior depends on the deformation temperature. At 77 K, higher strain leads to the prevalence of deformation twinning. At 293 K, however, higher strain leads to the formation of dislocation cell structures. The texture evolution (during cold rolling) and recrystallization behavior in the same alloy have also been studied.[58] The texture after 90% cold rolling was predominantly brass-type, which indicates low stacking fault energy (SFE). Bhattacharjee et al. suggested that the lower SFE in HEAs is due to the higher free energy of the 'perfect crystal' in HEAs. This is explained in further detail in the last paragraph of Section 5. The recrystallization texture in the CoCrFeMnNi alloy retains components from previous cold rolling, which is similar to low-SFE TWIP steels and 316 stainless steel. However, the relative proportions of various texture components between HEA and other low-SFE alloys are different. The main reasons are the slower recrystallization rate in regions with brass texture, and the absence of preferential nucleation of brass-textured grains from shear bands.[58] The low SFE in the CoCrFeMnNi alloy as suggested by Bhattacharjee et al. [58] is indeed confirmed by Zaddach et al.[59] They combined experiments with first-principle calculations and concluded that the SFE of CoCrFeMnNi is between 18.3 and 27.3 mJ/m^2. This value is close to that of conventional low-SFE alloys such as AISI 304L and brass.

The cold deformation and annealing behaviors of another FCC-based HEA, the $Al_{0.5}$CoCrCuFeNi alloy, was also investigated.[60] Deformation twinning is observed again in this alloy. Prevalence of twinning appears to result from the blockage of local slip by the Widmanstätten precipitates. The above results suggest that in FCC HEAs, dislocation slip and twinning are still the main deformation mechanisms.

4.2. The Al-Co-Cr-Cu-Fe-Ni Alloy System. Of all the HEA systems reported, Al-Co-Cr-Cu-Fe-Ni [7,8,21, 35,36,49,60–68] and its Cu-free version Al-Co-Cr-Fe-Ni [69–74] have been studied most comprehensively. Most of the reported HEA systems emerge from them. The main difference between the two is the presence of Cu-rich interdendrite in the former. This is because Cu has positive mixing enthalpies with most of the other elements, and is thus repelled to the interdendrite region. Additionally, remnant Cu in the dendrite region also clusters and forms various kinds of precipitates.[21,35,60] The main phases in cast Al-Co-Cr-Cu-Fe-Ni alloy include FCC, BCC/B2, and the Cu-rich phase (also has an FCC structure). The relative volume ratio of these phases depends on the composition. The relative volume of BCC and FCC phases is related to the VEC of the alloy. Higher VEC typically leads to more of the FCC phase, and vice versa. Al has the strongest effect in this regard,[36] and addition of Al leads to the transition of the FCC phase to the BCC phase. Additionally, when there is sufficient Al, BCC tend to further decompose to a (Al, Ni)-rich B2 phase and a (Cr, Fe)-rich BCC phase that have almost identical lattice parameters.[8,21,71,74] The Cu-rich phase appears clearly when the concentration of Cu is higher than $\sim$10 at.%. Its fraction increases with the content of Cu. In general, hardness of the FCC phase is between HV 100–200.[7,8,70–72] Alloys having a sole FCC phase have ductilities between 20–60%, and usually exhibit significant work hardening.[62,63,69] Hardness of the BCC/B2 phase is typically between HV 500–600. Because addition of Al triggers the formation of hard BCC/B2 phase, and hardness of the alloy increases with Al content [7,8,70–72] (see Figure 3). Alloys having sole BCC/B2 phases probably have ductilities less than 5%.[62] In terms of their plasticity, values of 30% or higher have been reported.[74] Superplasticity was observed in AlCoCrCuFeNi at 800–1000°C, and the elongation was between 405 and 800%.[65]

Effect of annealing on the mechanical properties of the Al-Co-Cr-Cu-Fe-Ni system depends on the phase transformation involved (see Section 3.4). When the selected annealing temperature favors the formation of BCC phase, the alloy becomes harder and more brittle. When the temperature selected increases the fraction of FCC phase, the alloys softens and becomes more ductile. It should be carefully noted that in this alloy system, a long annealing time may lead to the formation of σ phase,[12] which hardens the alloy but remarkably reduces its ductility/plasticity.[12,34,51]

4.3. Derivatives of the Al-Co-Cr-Cu-Fe-Ni Alloy System. This section discusses the mechanical properties of alloys derived from the Al-Co-Cr-Cu-Fe-Ni system. A system is said to be 'derived' from Al-Co-Cr-Cu-Fe-Ni when the number of different principal element between the two systems is less than two. For example, Al-Co-Cr-Cu-Ti is derived from Al-Co-Cr-Cu-Fe-Ni (replace Fe with Ti and remove Ni).

Addition of Ti to Al-Co-Cr-Cu-Fe-Ni usually leads to formation of intermetallic phases such as the Laves phase,[15,75–79] σ phase,[17,75,76] Heusler phase,[17] η-Ni_3Ti,[17,50] and R phase.[76] This is because Ti

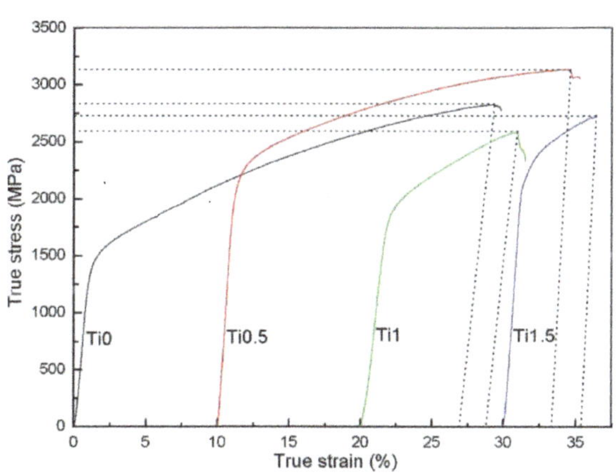

Figure 8. Compressive true stress-strain curves of AlCoCrCuFeNiTi$_x$ alloys with x being 0, 0.5, 1, and 1.5.[77]

has large negative enthalpy of mixing with rest of the composition. These phases strengthen the alloy. For example, $AlCoCrFeNiTi_{0.5}$ has a high yield strength of 2.26 GPa and a plasticity of 23% (Figure 8).[77] The hardness of $Al_{0.5}CoCrCuFeNiTi_x$ alloy can be significantly increased to HV 600 or higher when x is larger than 1.[75] The hardness of Ti-containing alloys may further increase upon annealing owing to the formation of more intermetallic phases.[17,50]

Addition of Mo to the system generally leads to the formation of (Cr, Mo, Co, Fe)-rich σ phase.[80–87] σ phase is a very hard and brittle intermetallic phase. As mentioned previously, it significantly enhances the hardness of the alloy but reduces its ductility/plasticity. For example, addition of 0.5 mole fraction of Mo to AlCoCrFeNi raises its strength from 1,051 to 2,757 MPa.[86]

Addition of Mn leads to two representative systems, the CoCrFeMnNi and $Al_xCrFe_{1.5}MnNi_{0.5}$ alloys. The CoCrFeMnNi alloy is a classic sole-FCC HEA and does not experience phase transformation when annealed. Therefore, it is a good model system to study the behavior of multiprincipal-element solid solutions and has been used to study the grain growth,[88] diffusion,[19] dislocation behavior,[57] and texture evolution [58] of FCC HEAs. This alloy is soft (yield strength ~170 MPa), very ductile (elongation ~60%), and shows high strain hardening capability at room temperature. At 77 K, its strength nearly doubles and its elongation increases to ~80%.[57] The $Al_{0.3}CrFe_{1.5}MnNi_{0.5}$ alloy is a representative age-hardening HEA. When aged between 600 and 900°C, its hardness boosts from about 300 HV to almost 900 HV.[34,51,89] TEM observation shows that the hardening phenomenon is not from precipitation hardening, but from the transformation of the BCC matrix to the σ phase during aging.[34] The hardening phenomenon in this alloy can actually be controlled to occur only near the surface, forming a surface hardening layer that significantly improves wear resistance (see Section 4.5).

Addition of non-metallic elements such as Si and B leads to the formation of valence compounds such as silicides and borides. These compounds also lead to evident hardening. For example, the hardness of AlCoCrFeNi**Si** [90] and Al**B**CoCrCuFeNi [14] are HV 738 and HV 740, respectively. Compound formation also causes loss of toughness. Addition of Si to AlCoCrFeNi by the equimolar ratio reduces its plastic strain from over 25% to ~1%.[90] It should be noted that although valence compounds are typically harder than intermetallic phases, silicide/boride-containing alloys are not necessarily harder than alloys containing intermetallic phase. In the two alloys mentioned above (AlCoCrFeNi**Si** and Al**B**CoCrCuFeNi), the volume fractions of silicides and borides are not high, and the distribution of these compounds is quite localized. Hard phases with such morphology (isolated coarse laths) are less effective in strengthening, particularly when they are embedded in an apparently softer matrix. Besides the above, many other elements have been added to the Al-Co-Cr-Cu-Fe-Ni system (e.g. Zr [91], Nd [91], Nb [92], V [93,94], Y [95], Sn [96,97], Zn [98], and C [99,100]).

4.4. Refractory HEA systems. Some researchers developed HEAs that are based on totally different elements. The most notable system is the refractory HEA system developed by Senkov et al.[10,37,54,101–106] These alloys are composed mainly of refractory elements such as Cr, Hf, Mo, Nb, Ta, V, W, and Zr. The major phase in these alloys is typically BCC. Many of them even have a sole BCC structure. These BCC solid solutions have high strength ranging from 900 to 1,350 MPa.[37,101,105,106] Interestingly, such strength is several times higher than that of the weighted average strength of the composing elements. For example, the rule-of-mixture hardness of MoNbTaVW and HfNbTaTiZr are 1,596 and 1,165 MPa, respectively. However, the hardness measured for these alloys are more than three times higher: 5,250 and 3,826 MPa, respectively. Such high hardness is a result of the solid solution strengthening effect which originated from the difference in atomic size and modulus among the composing elements.[37] Room temperature plasticity of refractory HEA is closely related to composition. The plastic fracture strain is less than 5% for many alloys. However, some alloys can be compressed to 50% without fracture.[37,103,106] It was suggested that the high plasticity in some alloys is owing to the prevalence of deformation twinning.[37,103] It is worth mentioning that refractory HEAs can show superb mechanical properties at elevated temperatures, which is described in Section 5.

4.5. Wear and Fatigue Properties. Wear properties of HEAs, under both abrasive [7,14,51,75,94,107] and adhesive [50,66,80] conditions, have been tested in a number of systems. The adhesive wear properties of tested HEAs are summarized in Figure 9. HEAs composed solely of SDPs typically do not show better wear properties than conventional alloys with similar hardness (e.g. compare $Al_{0.5}CoCrCuFeNi$ and 316SS in Figure 9). Wear resistance is clearly enhanced in the presence of some B2 or COPs (e.g. $Al_2CoCrCuFeNi$ and $AlCoCrFe_{1.5}Mo_{0.5}Ni$). If COPs with high hardness become the main phase, wear resistance is often outstanding. Many HEA systems of this type have comparable or even better performances than conventional wear-resistant alloys such as SKD61 tool steel (AISI H13) and SUJ2 bearing steel (AISI 52100).[50,51,80,107] For example, the wear resistance of the $AlCoCrFeMo_{0.5}Ni$ alloy is comparable to that of SUJ2 (Figure 9). In particular, the $Al_{0.2}Co_{1.5}CrFeNi_{1.5}Ti$ alloy [50] has significantly

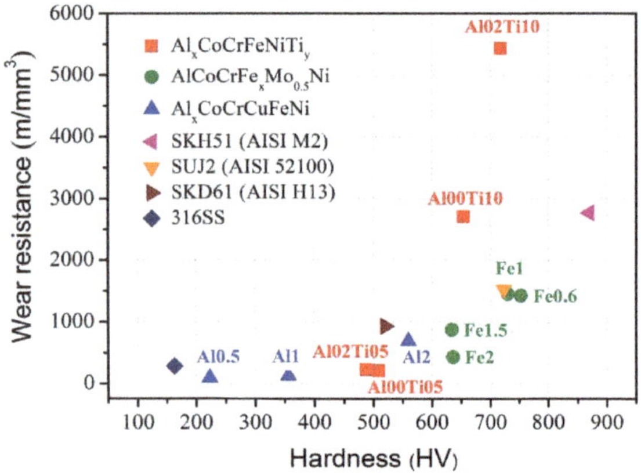

Figure 9. Relationship between hardness and adhesive wear resistances of various HEAs and conventional alloys.[50,66,80]

better wear resistance than SUJ2 and SKH51 (AISI M2, see Figure 9). Its hardness is similar to that of SUJ2, but its wear resistance is 3.6 times that of the latter. The wear resistance of the $Al_{0.2}Co_{1.5}CrFeNi_{1.5}Ti$ alloy is twice that of SKH51, but the hardness of the latter is a lot higher. The outstanding performance is owing to the higher hardness of HEA at high temperatures. Additionally, because the wear mechanism is mild oxidational wear, the better oxidation resistance in the $Al_{0.2}Co_{1.5}CrFeNi_{1.5}Ti$ alloy is also beneficial to its wear resistance.[50]

The $Al_{0.3}CrFe_{1.5}MnNi_{0.5}$ alloy (see also Section 4.3) has a unique surface hardening phenomenon that can increase its wear resistance by ~50%, but only reduce its bending strain by 2%. This is achieved by limiting the formation of hard σ phase at the surface region. Because the heterogeneous nucleation energy and strain energy near the free surface are lower than those in the interior of the alloy, one can carefully select the aging temperature so that the kinetics of the transformation is sufficient high only at the near-surface region. The process is simple (atmospheric aging), applicable to objects with complex shapes, and offers a markedly better combination of mechanical properties than SUJ2 (30% higher wear resistance and 300% larger maximum bending strain).

Preliminary fatigue results are also encouraging. $Al_{0.5}CoCrCuFeNi$ alloy can show long fatigue lives at relatively high stresses, high endurance limits (540–945 MPa) and endurance limit to ultimate tensile strength ratio (0.402–0.703).[67] These values are comparable to or even better than many conventional alloys, such as various steels, Ti alloys, and Zr-based BMGs.[67] Although some degree of scattering in fatigue life was observed, it is probably due to defects introduced during sample preparation. If defects can be carefully controlled, the fatigue resistance characteristics in $Al_{0.5}CoCrCuFeNi$ are very promising.

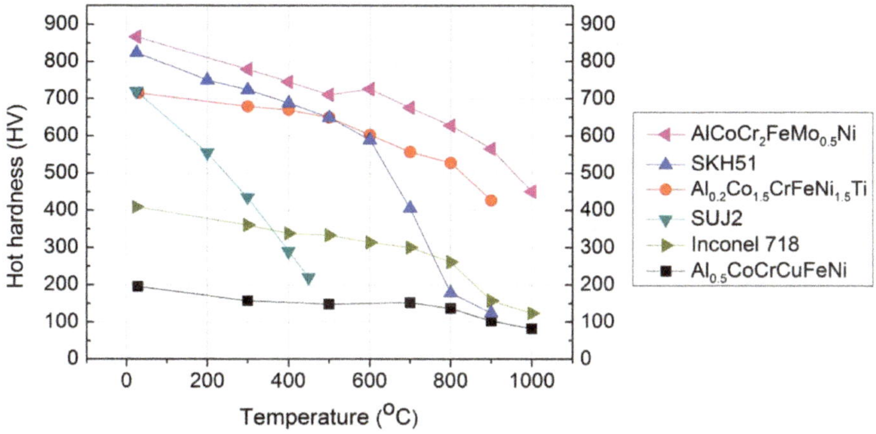

Figure 10. Hardness of various HEAs and conventional alloys as functions of temperature. Some of the data is collected from literature.[50,85]

5. High Temperatures Properties and Structural Stability Many HEAs have exceptional high-temperature strength/hardness, which originates from their excellent resistance to thermal softening. Figure 10 shows the hot hardness of three HEAs and three conventional alloys as functions of temperature.[50,85] The $AlCoCr_2FeMo_{0.5}Ni$ and $Al_{0.2}Co_{1.5}CrFeNi_{1.5}Ti$ alloys have very high hardness at room temperature (more than two times higher than Inconel 718 superalloy). Their hardness decreases slowly with the increase of temperature. SKH51 and SUJ2 also have very high hardness at room temperature (over HV 700). However, in contrast to the gradual softening behavior for the two HEAs, SKH51 and SUJ2 start to soften drastically at 200 and 600°C, respectively. This leads to a significant difference in hardness for the four alloys at 800°C—the two HEAs are still more than two times harder than the Inconel 718 superalloy, while the two conventional alloys become apparently softer than Inconel 718 (Figure 10). At 1000°C, $AlCoCr_2FeMo_{0.5}Ni$ is already more than three times harder than Inconel 718. Also note that even the soft, sole-FCC $Al_{0.5}CoCrCuFeNi$ alloy is quite resistant to thermal softening.

Hsu et al. [85] calculated the high-temperature softening coefficient of the $AlCoCr_xFeMo_{0.5}Ni$ ($x = 0$–2.0) alloys from the hot hardness vs. temperature curves and found that above the transition temperature (onset temperature for rapid softening), the softening coefficient of these HEAs are evidently lower than that of conventional high-temperature alloys such as Inconel 718 and T800 (Table 3).[85] Softening coefficient of some alloys are even less than half that of Inconel 718. The resistance to thermal softening is mainly a result of the sluggish diffusion in HEAs. Above the transition temperature, deformation proceeds via diffusion-related mechanisms. Thus, the slower diffusion rate in HEAs can effectively reduce the degree of thermal softening.

Table 3. Softening coefficient of two $AlCoCr_xFeMo_{0.5}Ni$ alloys and two conventional high-temperature alloys above transition temperature (T_T).[85]

Alloy	Softening coefficient above T_T
Cr-1.5	−2.32E-4
Cr-2	−1.66E-4
T-800	−3.12E-4
In 718	−3.55E-4

Because the transition temperature and diffusion rate of metallic materials are directly related to the melting point of the material, it can be expected that refractory HEAs should have outstanding properties in this regard. Indeed, W-containing refractory HEAs have excellent resistance to thermal softening. This is shown in Figure 11.[101] The yield strengths

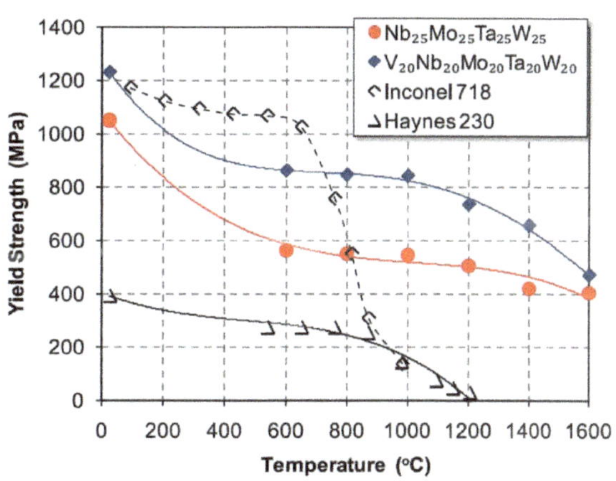

Figure 11. Yield strength as functions of temperature for the $Nb_{25}Mo_{25}Ta_{25}W_{25}$ and $V_{20}Nb_{20}Mo_{20}Ta_{20}W_{20}$ alloys and two superalloys.[101]

of $Nb_{25}Mo_{25}Ta_{25}W_{25}$ and $V_{20}Nb_{20}Mo_{20}Ta_{20}W_{20}$ alloys remain virtually unchanged in the range of 600–1000°C. The two alloys have extremely high yield strength of over 400 MPa at 1600°C. Owing to their good strength at high temperatures, some refractory alloys actually have higher specific yield strength than superalloys such as Inconel 718 and Haynes 230 at higher than 1000°C.[106]

Structural stability is an important issue for high-temperature applications. HEAs generally have good structural stability at high temperatures. This is owing to (1) slower kinetics due to the sluggish diffusion effect (Section 2.2); (2) reduced driving force to eliminate defects (e.g. grain boundaries, interfaces, etc.). The driving force to eliminate defects originates from the free energy difference between defect-containing and defect-free crystals. However, the defect-free lattice in HEAs is still highly strained and full of atomic-scale zigzag, owing to the atomic size differences between the elements. The strain energy of the distorted lattice raises the overall free energy, leading to a reduced energy difference between defect-containing crystal/amorphous structure and the defect-free crystal.[23,108] Thus, the driving force to eliminate defects is reduced. There is much evidence for the good structural stability of HEAs. For example, the activation energy of grain growth for the CoCrFeMnNi alloy (which has a sole FCC structure) is 321.7 kJ/mole^{-1}.[88] This value is twice that of AISI 304LN, indicating a significantly slower kinetics for microstructural coarsening.[88] Growth of precipitates in HEAs is also slow. Precipitates in HEAs treated at moderate temperature (e.g. 700–800°C) often still remain small—diameters on the scale of tens or hundreds of nanometers have been reported.[34,82] Another example is the NbSiTaTiZr amorphous alloy. The crystallization temperature of NbSiTaTiZr is higher than 800°C.[23] This temperature of stability is probably the highest among reported amorphous metals. The excellent structural stability comes not just from the above two reasons. NbSiTaTiZr was designed based on a free energy consideration.[23] Its composition was selected so that the non-crystalline random solid solution state has a low free energy. This lowers its driving force to crystallize into silicides. Additionally, the kinetics in this alloy is further slowed due to the high atomic packing density and high melting points of composing elements.

6. Corrosion Properties The corrosion resistances of some HEAs have been tested in both NaCl and H_2SO_4 solutions.[20,73,109–121] In both solutions, some HEAs show better corrosion properties than 304 SS and even 304L SS and have good resistance to pitting.[109,115–117] Important factors include alloy composition and microstructure, particularly the amount and distribution of corrosion-resistant elements (such as Cr), and the presence of galvanic corrosion. Most of these tested alloys are based on Co, Cr, Fe and Ni. Therefore, the corrosion behavior of the Co–Cr–Fe–Ni alloys is discussed first, followed by the effect of elemental addition to the alloy system.

Corrosion in NaCl Solution: In NaCl solution, the CoCrFeNi single-phase FCC alloy has markedly better corrosion resistance than 304L SS.[111] This is probably owing to its high Cr and Ni content. Addition of Cu to the CoCrFeNi alloy leads to the formation of Cu-rich interdendrite phase, which suffers from galvanic corrosion and severely degrades the corrosion resistance.[111,118] To make things worse, the passive film on the Cu-rich interdendrite regions does not offer good protection, which further narrows down the passivation region.[111] The corrosion property can be improved by reducing the amount of Cu-rich phase via high-temperature annealing.[118] Addition of 0.5 mole fraction of Al to the CoCrFeNi alloy leads to the formation of (Al, Ni)-rich BCC phase and causes galvanic corrosion in NaCl solution.[73] Al is also detrimental to the corrosion resistance of the $Al_xCrFe_{1.5}MnNi_{0.5}$ alloys. Addition of Al significantly reduces the pitting potential and increases the area of localized/pitting corrosion of these alloys in NaCl solution.[114] Although it has higher pitting potential, the passive region is less than that of 304SS. Addition of Mo to the $Co_{1.5}CrFeNi_{1.5}Ti_{0.5}Mo_x$ alloy is beneficial because it evidently raises the pitting potential.[115,119] For example, $Co_{1.5}CrFeNi_{1.5}Ti_{0.5}Mo_{0.1}$ alloy has a wide passivation region of 1.43 V in 1 M NaCl and does not suffer from any pitting.[116,119]

Corrosion in H_2SO_4 Solution: In H_2SO_4 solution, CoCrFeNi also has better corrosion resistance than 304 SS.[117] Addition of Cu to the CoCrFeNi alloy deteriorates its corrosion properties. Similar to the case in NaCl solution, this is due to the formation of the Cu-rich phase. The width of the passivation region is also evidently narrowed.[109,113] Addition of Al is also harmful.[114,117] It leads to the formation of a FCC+BCC or BCC+B2 two-phase structure, in which the Ni,Al-rich BCC/B2 phase is preferentially corroded.[74,117] Additionally, the existence of Al turns the passive film porous and less protective.[117] Replacing Co with Mn in the Co–Cr–Fe–Ni alloy degrades its corrosion resistance in H_2SO_4 solution, and renders its resistance inferior to 304 SS.[114] Addition of B to the $Al_{0.5}CoCrCuFeNi$ alloy leads to the formation of Cr-rich borides and apparently lowers the Cr content in the matrix.[113] This renders the matrix much less corrosion resistant. Addition of Ti to the Co–Cr–Fe–Ni alloy does not affect its outstanding corrosion behavior.[119] Addition of Mo to the Co–Cr–Fe–Ni–Ti alloy is detrimental to the corrosion resistance in H_2SO_4 solution. This is because of the formation of the (Cr, Mo)-rich σ phase, which reduces the concentration of Cr in other phases.[119] The $AlCu_{0.5}CoCrFeNiSi$ alloy has better general corrosion resistance than 304 SS, but its passivation region is smaller.[109]

7. Outstanding Issues and Future Directions HEA is an immense field with a countless number of new alloy systems. Our understanding of these new materials is still very limited and preliminary. As mentioned in Section 4, our knowledge mainly focuses on the Al-Co-Cr-Cu-Fe-Ni system, its derivatives, and the refractory HEAs—just a very tiny fraction of the whole HEA world. More elements and combinations need to be explored to further understand the potentials of HEAs. For example, light HEAs are of great interest.[122–124] Non-metallic elements such as C, B, and N can also be incorporated to achieve even higher strength.[14,99,100,125,126] Additionally, HEAs can also be prepared by other process and/or in other forms. For example, HEAs can be processed by a wrought process including homogenization, hot/cold working, and annealing [62,64] to eliminate casting defects and improve microstructure. HEA powders can be prepared by ball milling.[98,127–131] The powders can subsequently be sintered to form bulk HEAs with ultrafine structure.[98,127,129] Single-crystal of the $Al_{0.3}CoCrFeNi$ HEA has also been prepared successfully by the Bridgman solidification method.[132] Cemented carbides and cermets could be sintered by a proper HEA binder to reduce cost and improve resistance to softening. HEA coatings/claddings can also be prepared by sputtering or other cladding techniques.[22,23,73,133–140]

In terms of the phase and crystal structure of HEAs, most of our knowledge is based on the as-cast state (after vacuum arc melting). Very little is known about the equilibrium phases and phase diagrams, but these are key to the design and development of HEAs. Significantly more combined experimental and computational works are needed in this regard. It is worth mentioning that although equimolar compositions are usually the easiest access to a new alloy system, they probably do not possess the best combination of properties. Therefore, there is probably far more treasure in those non-equimolar alloys with carefully designed composition and tailored microstructures. A good approach is to start from equimolar alloys and then extract the desired non-equimolar composition from the equimolar alloy.

The mechanical properties of HEAs available are mostly about hardness and compressive properties. Because reasonable ductility is crucial to structural applications, more studies on tensile behaviors are definitely needed. Additionally, except for a few examples,[57–59,141] not much is known about the deformation of HEAs, e.g. dislocation behavior, deformation twinning, deformation microstructures, texture evolution, stacking fault energy, etc. This information is the foundation for a thorough understanding of mechanical properties. Microstructural characterization in most works is only conducted by SEM. This is not sufficient for unraveling the structure-property relationship in HEAs because all detailed microstructure observations reveal the existence of nano-sized or even atomic-scale precipitations and/or decompositions.[21,31,32,34,35,98,142] More tests aimed at prolonged service and/or high-temperature applications are also required. These include but are not limited to: fatigue behavior,[67] oxidation resistance,[50,51,104] structural stability, and creep behavior.

HEAs can also be used in other applications. For example, HEAs can be used as hydrogen storage materials,[143,144] radiation resistant materials,[145] diffusion barriers for electronics,[22,23] precision resistors,[4,146] electromagnetic shielding materials, soft magnetic materials,[147] thermoelectric materials, functional coatings, and anti-bacterial materials. Because of the distinct design strategy and unique properties, HEAs and other high-entropy materials (e.g. high-entropy ceramics or even polymers) will probably find applications in fields significantly wider than those listed above. This exciting new virgin field awaits exploration by more scientists.

Acknowledgements MH Tsai gratefully thanks the financial support from the National Science Council of Taiwan under grant NSC 102-2218-E-005-004. The authors acknowledge the valuable discussions with Yih-Farn Kao, Sheng-Chieh Liao, and Dr Kun-Yo Tsai.

References

[1] Huang KH, Yeh JW. A study on multicomponent alloy systems containing equal-mole elements [M.S. thesis]. Hsinchu: National Tsing Hua University; 1996.

[2] Yeh JW, Chen SK, Lin SJ, Gan JY, Chin TS, Shun TT, Tsau CH, Chang SY. Nanostructured high-entropy alloys with multiple principal elements: novel alloy design concepts and outcomes. Adv Eng Mater. 2004;6:299–303.

[3] Yeh JW. Recent progress in high-entropy alloys. Ann Chim-Sci Mat. 2006;31:633–648.

[4] Tsai MH. Physical properties of high entropy alloys. Entropy. 2013;15:5338–5345.

[5] Yeh JW. Alloy design strategies on high-entropy alloys. JOM. 2013.

[6] Gaskell DR. Introduction to the thermodynamics of materials. 3rd ed. Washinton (DC): Taylor & Francis; 1995.

[7] Tong CJ, Chen MR, Chen SK, Yeh JW, Shun TT, Lin SJ, Chang SY. Mechanical performance of the $Al_xCoCrCuFeNi$ high-entropy alloy system with multiprincipal elements. Metall Mater Trans A. 2005;36:1263–1271.

[8] Tong CJ, Chen YL, Chen SK, Yeh JW, Shun TT, Tsau CH, Lin SJ, Chang SY. Microstructure characterization of $Al_xCoCrCuFeNi$ high-entropy alloy system with multiprincipal elements. Metall Mater Trans A. 2005;36:881–893.

[9] Li A, Zhang X. Thermodynamic analysis of the simple microstructure of AlCrFeNiCu high-entropy alloy with multi-principal elements. Acta Metallurgica Sinica (English Letters). 2009;22:219–224.

[10] Senkov ON, Wilks GB, Miracle DB, Chuang CP, Liaw PK. Refractory high-entropy alloys. Intermetallics. 2010;18:1758–1765.

[11] del Grosso MF, Bozzolo G, Mosca HO. Determination of the transition to the high entropy regime for alloys of refractory elements. J Alloys Compd. 2012;534: 25–31.
[12] Ng C, Guo S, Luan J, Shi S, Liu CT. Entropy-driven phase stability and slow diffusion kinetics in an $Al_{0.5}CoCrCuFeNi$ high entropy alloy. Intermetallics. 2012;31:165–172.
[13] Lucas MS, Wilks GB, Mauger L, Munoz JA, Senkov ON, Michel E, Horwath J, Semiatin SL, Stone MB, Abernathy DL, Karapetrova E. Absence of long-range chemical ordering in equimolar FeCoCrNi. Appl Phys Lett. 2012;100:251907.
[14] Hsu CY, Yeh JW, Chen SK, Shun TT. Wear resistance and high-temperature compression strength of Fcc $CuCoNiCrAl_{0.5}Fe$ alloy with boron addition. Metall Mater Trans A. 2004;35:1465–1469.
[15] Zhang KB, Fu ZY. Effects of annealing treatment on phase composition and microstructure of $CoCrFeNiTiAl_x$ high-entropy alloys. Intermetallics. 2012;22:24–32.
[16] Otto F, Yang Y, Bei H, George EP. Relative effects of enthalpy and entropy on the phase stability of equiatomic high-entropy alloys. Acta Mater. 2013;61: 2628–2638.
[17] Lin CW, Tsai MH, Tsai CW, Yeh JW, Chen SK. Microstructure and aging behavior of $Al_5Cr_{32}Fe_{35}Ni_{22}Ti_6$ ternary-like high-entropy alloy. In preparation.
[18] Cheng KH, Lai CH, Lin SJ, Yeh JW. Recent progress in multi-element alloy and nitride coatings sputtered from high-entropy alloy targets. Ann Chim-Sci Mat. 2006;31:723–736.
[19] Tsai KY, Tsai MH, Yeh JW. Sluggish diffusion in Co–Cr–Fe–Mn–Ni high-entropy alloys. Acta Mater. 2013;61:4887–4897.
[20] Chen YY, Duval T, Hong UT, Yeh JW, Shih HC, Wang LH, Oung JC. Corrosion properties of a novel bulk $Cu_{0.5}NiAlCoCrFeSi$ glassy alloy in 288°C high-purity water. Mater Lett. 2007;61:2692–2696.
[21] Singh S, Wanderka N, Murty BS, Glatzel U, Banhart J. Decomposition in multi-component AlCoCrCuFeNi high-entropy alloy. Acta Mater. 2011;59: 182–190.
[22] Tsai MH, Yeh JW, Gan JY. Diffusion barrier properties of AlMoNbSiTaTiVZr high-entropy alloy layer between copper and silicon. Thin Solid Films. 2008;516: 5527–5530.
[23] Tsai MH, Wang CW, Tsai CW, Shen WJ, Yeh JW, Gan JY, Wu WW. Thermal stability and performance of NbSiTaTiZr high-entropy alloy barrier for copper metallization. J Electrochem Soc. 2011;158:H1161–H1165.
[24] Ferro P, Saccone A. Structure of intermetallic compounds and phases. In: Cahn RW, Haasen P, editors. Physical metallurgy. 4th ed. Amsterdam, the Netherlands: Elsevier Science B.V.; 1996.
[25] Cahn RW, Haasen P, editors. Physical Metallurgy. 4th ed. Amsterdam: Elsevier Science B.V.; 1996.
[26] Zhang Y, Zhou YJ, Lin JP, Chen GL, Liaw PK. Solid-solution phase formation rules for multi-component alloys. Adv Eng Mater. 2008;10:534–538.
[27] Guo S, Liu CT. Phase stability in high entropy alloys: formation of solid-solution phase or amorphous phase. Prog Nat Sci: Mater Int. 2011;21:433–446.
[28] Chen ST, Yeh JW. Effect of mixing enthalpy, mixing entropy and atomic size difference on the structure of multicomponent alloys [M.S. thesis]. Hsinchu: National Tsing Hua University; 2009.
[29] Yang X, Zhang Y. Prediction of high-entropy stabilized solid-solution in multi-component alloys. Mater Chem Phys. 2012;132:233–238.
[30] Tsai KY, Tsai MH, Yeh JW. To be submitted.
[31] Manzoni A, Daoud H, Mondal S, van Smaalen S, Völkl R, Glatzel U, Wanderka N. Investigation of phases in $Al_{23}Co_{15}Cr_{23}Cu_8Fe_{15}Ni_{16}$ and $Al_8Co_{17}Cr_{17}Cu_8Fe_{17}Ni_{33}$ high entropy alloys and comparison with equilibrium phases predicted by Thermo-Calc. J Alloys Compd. 2013;552:430–436.
[32] Manzoni A, Daoud H, Völkl R, Glatzel U, Wanderka N. Phase separation in equiatomic AlCoCrFeNi high-entropy alloy. Ultramicroscopy. 2013;132:212–215.
[33] Wang YP, Li BS, Fu HZ. Solid solution or intermetallics in a high-entropy alloy. Adv Eng Mater. 2009;11: 641–644.
[34] Tsai MH, Yuan H, Cheng G, Xu W, Jian WW, Chuang MH, Juan CC, Yeh AC, Lin SJ, Zhu Y. Significant hardening due to the formation of a sigma phase matrix in a high entropy alloy. Intermetallics. 2013;33:81–86.
[35] Tsai MH, Yuan H, Cheng G, Xu W, Tsai KY, Tsai CW, Jian WW, Juan CC, Shen WJ, Chuang MH, Yeh JW, Zhu YT. Morphology, structure and composition of precipitates in $Al_{0.3}CoCrCu_{0.5}FeNi$ high-entropy alloy. Intermetallics. 2013;32:329–336.
[36] Tung CC, Yeh JW, Shun TT, Chen SK, Huang YS, Chen HC. On the elemental effect of AlCoCrCuFeNi high-entropy alloy system. Mater Lett. 2007;61:1–5.
[37] Senkov ON, Scott JM, Senkova SV, Miracle DB, Woodward CF. Microstructure and room temperature properties of a high-entropy TaNbHfZrTi alloy. J Alloys Compd. 2011;509:6043–6048.
[38] Guo S, Ng C, Lu J, Liu CT. Effect of valence electron concentration on stability of fcc or bcc phase in high entropy alloys. J Appl Phys. 2011;109:103505.
[39] Inoue A. Stabilization of metallic supercooled liquid and bulk amorphous alloys. Acta Mater. 2000;48:279–306.
[40] Xia MX, Zhang SG, Ma CL, Li JG. Evaluation of glass-forming ability for metallic glasses based on order-disorder competition. Appl Phys Lett. 2006;89:091917.
[41] Guo S, Hu Q, Ng C, Liu CT. More than entropy in high-entropy alloys: forming solid solutions or amorphous phase. Intermetallics. 2013;41:96–103.
[42] Chen YY, Hong UT, Yeh JW, Shih HC. Mechanical properties of a bulk $Cu_{0.5}NiAlCoCrFeSi$ glassy alloy in 288°C high-purity water. Appl Phys Lett. 2005;87:261918.
[43] Gao XQ, Zhao K, Ke HB, Ding DW, Wang WH, Bai HY. High mixing entropy bulk metallic glasses. J Non-Cryst Solids. 2011;357:3557–3560.
[44] Zhao K, Xia XX, Bai HY, Zhao DQ, Wang WH. Room temperature homogeneous flow in a bulk metallic glass with low glass transition temperature. Appl Phys Lett. 2011;98:141913.
[45] Cunliffe A, Plummer J, Figueroa I, Todd I. Glass formation in a high entropy alloy system by design. Intermetallics. 2012;23:204–207.
[46] Ding HY, Yao KF. High entropy $Ti_{20}Zr_{20}Cu_{20}Ni_{20}Be_{20}$ bulk metallic glass. J Non-Cryst Solids. 2013;364:9–12.
[47] Takeuchi A, Chen N, Wada T, Yokoyama Y, Kato H, Inoue A, Yeh JW. $Pd_{20}Pt_{20}Cu_{20}Ni_{20}P_{20}$ high-entropy alloy as a bulk metallic glass in the centimeter. Intermetallics. 2011;19:1546–1554.
[48] Li HF, Xie XH, Zhao K, Wang YB, Zheng YF, Wang WH, Qin L. In vitro and in vivo studies on biodegradable CaMgZnSrYb high-entropy bulk metallic glass. Acta Biomaterialia. 2013;9:8561–8573.

[49] Wen LH, Kou HC, Li JS, Chang H, Xue XY, Zhou L. Effect of aging temperature on microstructure and properties of AlCoCrCuFeNi high-entropy alloy. Intermetallics. 2009;17:266–269.
[50] Chuang MH, Tsai MH, Wang WR, Lin SJ, Yeh JW. Microstructure and wear behavior of $Al_xCo_{1.5}CrFeNi_{1.5}Ti_y$ high-entropy alloys. Acta Mater. 2011;59: 6308–6317.
[51] Chen ST, Tang WY, Kuo YF, Chen SY, Tsau CH, Shun TT, Yeh JW. Microstructure and properties of age-hardenable $Al_xCrFe_{1.5}MnNi_{0.5}$ alloys. Mater Sci Eng A. 2010;527:5818–5825.
[52] Tsai MH, Tsai KY, Tsai CW, Lee C, Juan CC, Yeh JW. Criterion for sigma phase formation in Cr- and V-containing high-entropy alloys. Mater Res Lett. 2013;1:207–212.
[53] Zhang C, Zhang F, Chen SL, Cao WS. Computational thermodynamics aided high-entropy alloy design. JOM 2012;64:839–845.
[54] Senkov ON, Senkova SV, Woodward C, Miracle DB. Low-density, refractory multi-principal element alloys of the Cr–Nb–Ti–V–Zr system: Microstructure and phase analysis. Acta Mater. 2013;61:1545–1557.
[55] Shigenobu O, Yoshitaka U, Masanori K. First-principles approaches to intrinsic strength and deformation of materials: perfect crystals, nano-structures, surfaces and interfaces. Modelling and Simulation in Materials Science and Engineering. 2009;17:013001.
[56] Reed-Hill RE, Abbaschian R. Physical metallurgy principles. 3rd ed. PWS-KENT; 1991.
[57] Otto F, Dlouhý A, Somsen C, Bei H, Eggeler G, George EP. The influences of temperature and microstructure on the tensile properties of a CoCrFeMnNi high-entropy alloy. Acta Mater. 2013;61:5743–5755.
[58] Bhattacharjee PP, Sathiaraj GD, Zaid M, Gatti JR, Lee C, Tsai C-W, Yeh J-W. Microstructure and texture evolution during annealing of equiatomic CoCrFeMnNi high-entropy alloy. J Alloys Compd. 2014;587:544–552.
[59] Zaddach AJ, Niu C, Koch CC, Irving DL. Mechanical properties and stacking fault energies of NiFeCrCoMn high-entropy alloy. JOM. 2013;65:1780–1789.
[60] Tsai CW, Chen YL, Tsai MH, Yeh JW, Shun TT, Chen SK. Deformation and annealing behaviors of high-entropy alloy $Al_{0.5}CoCrCuFeNi$. J Alloys Compd. 2009;486:427–435.
[61] Zhang KB, Fu ZY, Zhang JY, Shi J, Wang WM, Wang H, Wang YC, Zhang QJ. Annealing on the structure and properties evolution of the CoCrFeNiCuAl high-entropy alloy. J Alloys Compd. 2010;502:295–299.
[62] Tsai CW, Tsai MH, Yeh JW, Yang CC. Effect of temperature on mechanical properties of $Al_{0.5}CoCrCuFeNi$ wrought alloy. J Alloys Compd. 2010;490:160–165.
[63] Wang FJ, Zhang Y, Chen GL, Davies HA. Tensile and compressive mechanical behavior of a CoCrCu $FeNiAl_{0.5}$ high entropy alloy. Int J Mod Phys B. 2009;23:1254–1259.
[64] Kuznetsov AV, Shaysultanov DG, Stepanov ND, Salishchev GA, Senkov ON. Tensile properties of an AlCrCuNiFeCo high-entropy alloy in as-cast and wrought conditions. Mater Sci Eng A. 2012;533: 107–118.
[65] Kuznetsov AV, Shaysultanov DG, Stepanov ND, Salishchev GA, Senkov ON. Superplasticity of AlCoC rCuFeNi high entropy alloy. Mater Sci Forum. 2013;735: 146–151.
[66] Wu JM, Lin SJ, Yeh JW, Chen SK, Huang YS. Adhesive wear behavior of $Al_xCoCrCuFeNi$ high-entropy alloys as a function of aluminum content. Wear. 2006;261: 513–519.
[67] Hemphill MA, Yuan T, Wang GY, Yeh JW, Tsai CW, Chuang A, Liaw PK. Fatigue behavior of $Al_{0.5}CoCrCuFeNi$ high entropy alloys. Acta Mater. 2012;60:5723–5734.
[68] Liu ZY, Guo S, Liu XJ, Ye JC, Yang Y, Wang XL, Yang L, An K, Liu CT. Micromechanical characterization of casting-induced inhomogeneity in an $Al_{0.8}CoCrCuFeNi$ high-entropy alloy. Scr Mater. 2011;64:868–871.
[69] Shun TT, Du YC. Microstructure and tensile behaviors of FCC $Al_{0.3}CoCrFeNi$ high entropy alloy. J Alloys Compd. 2009;479:157–160.
[70] Wang WR, Wang WL, Wang SC, Tsai YC, Lai CH, Yeh JW. Effects of Al addition on the microstructure and mechanical property of $Al_xCoCrFeNi$ high-entropy alloys. Intermetallics. 2012;26:44–51.
[71] Kao YF, Chen TJ, Chen SK, Yeh JW. Microstructure and mechanical property of as-cast, -homogenized, and -deformed $Al_xCoCrFeNi$ ($0 \leq x \leq 2$) high-entropy alloys. J Alloys Compd. 2009;488:57–64.
[72] Li C, Li JC, Zhao M, Jiang Q. Effect of aluminum contents on microstructure and properties of $Al_xCoCrFeNi$ alloys. J Alloys Compd. 2010;504:S515–S518.
[73] Lin CM, Tsai HL. Evolution of microstructure, hardness, and corrosion properties of high-entropy $Al_{0.5}CoCrFeNi$ alloy. Intermetallics. 2011;19:288–294.
[74] Wang YP, Li BS, Ren MX, Yang C, Fu HZ. Microstructure and compressive properties of AlCrFeCoNi high entropy alloy. Mater Sci Eng A. 2008;491: 154–158.
[75] Chen MR, Lin SJ, Yeh JW, Chen SK, Huang YS, Tu CP. Microstructure and properties of $Al_{0.5}CoCrCuFeNiTi_x$ ($x = 0$–2.0) high-entropy alloys. Mater Trans. 2006;47: 1395–1401.
[76] Shun TT, Chang LY, Shiu MH. Microstructures and mechanical properties of multiprincipal component $CoCrFeNiTi_x$ alloys. Mater Sci Eng A. 2012;556: 170–174.
[77] Zhou YJ, Zhang Y, Wang YL, Chen GL. Solid solution alloys of $AlCoCrFeNiTi_x$ with excellent room-temperature mechanical properties. Appl Phys Lett. 2007;90:181904.
[78] Wang XF, Zhang Y, Qiao Y, Chen GL. Novel microstructure and properties of multicomponent $CoCrCuFeNiTi_x$ alloys. Intermetallics. 2007;15:357–362.
[79] Zhang KB, Fu ZY, Zhang JY, Wang WM, Wang H, Wang YC, Zhang QJ, Shi J. Microstructure and mechanical properties of $CoCrFeNiTiAl_x$ high-entropy alloys. Mater Sci Eng A. 2009;508:214–219.
[80] Hsu CY, Sheu TS, Yeh JW, Chen SK. Effect of iron content on wear behavior of $AlCoCrFe_xMo_{0.5}Ni$ high-entropy alloys. Wear. 2010;268:653–659.
[81] Hsu CY, Wang WR, Tang WY, Chen SK, Yeh JW. Microstructure and mechanical properties of new $AlCo_xCrFeMo_{0.5}Ni$ high-entropy alloys. Adv Eng Mater. 2010;12:44–49.
[82] Shun TT, Hung CH, Lee CF. The effects of secondary elemental Mo or Ti addition in $Al_{0.3}CoCrFeNi$ high-entropy alloy on age hardening at 700°C. J Alloys Compd. 2010;495:55–58.
[83] Zhu JM, Zhang HF, Fu HM, Wang AM, Li H, Hu ZQ. Microstructures and compressive properties of multicomponent $AlCoCrCuFeNiMo_x$ under-bar alloys. J Alloys Compd. 2010;497:52–56.
[84] Zhu JM, Fu HM, Zhang HF, Wang AM, Li H, Hu ZQ. Microstructures and compressive properties of

multicomponent $AlCoCrFeNiMo_x$ alloys. Mater Sci Eng A. 2010;527:6975–6979.
[85] Hsu CY, Juan CC, Wang WR, Sheu TS, Yeh JW, Chen SK. On the superior hot hardness and softening resistance of $AlCoCr_xFeMo_{0.5}Ni$ high-entropy alloys. Mater Sci Eng A. 2011;528:3581–3588.
[86] Shun TT, Chang LY, Shiu MH. Microstructure and mechanical properties of multiprincipal component $CoCrFeNiMo_x$ alloys. Mater Charact. 2012;70:63–67.
[87] Juan CC, Hsu CY, Tsai CW, Wang WR, Sheu TS, Yeh JW, Chen SK. On microstructure and mechanical performance of $AlCoCrFeMo_{0.5}Ni_x$ high-entropy alloys. Intermetallics. 2013;32:401–407.
[88] Liu WH, Wu Y, He JY, Nieh TG, Lu ZP. Grain growth and the Hall-Petch relationship in a high-entropy FeCrNiCoMn alloy. Scr Mater. 2013;68:526–529.
[89] Tsao LC, Chen CS, Chu CP. Age hardening reaction of the $Al_{0.3}CrFe_{1.5}MnNi_{0.5}$ high entropy alloy. Mater Des. 2012;36:854–858.
[90] Zhu JM, Fu HM, Zhang HF, Wang AM, Li H, Hu ZQ. Synthesis and properties of multiprincipal component $AlCoCrFeNiSi_x$ alloys. Mater Sci Eng A. 2010;527:7210–7214.
[91] Zhuang YX, Liu WJ, Chen ZY, Xue HD, He JC. Effect of elemental interaction on microstructure and mechanical properties of FeCoNiCuAl alloys. Mater Sci Eng A. 2012;556:395–399.
[92] Ma SG, Zhang Y. Effect of Nb addition on the microstructure and properties of AlCoCrFeNi high-entropy alloy. Mater Sci Eng A. 2012;532:480–486.
[93] Li BS, Wang YR, Ren MX, Yang C, Fu HZ. Effects of Mn, Ti and V on the microstructure and properties of AlCrFeCoNiCu high entropy alloy. Mater Sci Eng A. 2008;498:482–486.
[94] Chen MR, Lin SJ, Yeh JW, Chen SK, Huang YS, Chuang MH. Effect of vanadium addition on the microstructure, hardness, and wear resistance of $Al_{0.5}CoCrCuFeNi$ high-entropy alloy. Metall Mater Trans A. 2006;37:1363–1369.
[95] Hu ZH, Zhan YZ, Zhang GH, She J, Li CH. Effect of rare earth Y addition on the microstructure and mechanical properties of high entropy AlCoCrCuNiTi alloys. Mater Des. 2010;31:1599–1602.
[96] Liu L, Zhu JB, Li L, Li JC, Jiang Q. Microstructure and tensile properties of $FeMnNiCuCoSn_x$ high entropy alloys. Mater Des. 2013;44:223–227.
[97] Liu L, Zhu JB, Zhang C, Li JC, Jiang Q. Microstructure and the properties of $FeCoCuNiSn_x$ high entropy alloys. Mater Sci Eng A. 2012;548:64–68.
[98] Pradeep KG, Wanderka N, Choi P, Banhart J, Murty BS, Raabe D. Atomic-scale compositional characterization of a nanocrystalline AlCrCuFeNiZn high-entropy alloy using atom probe tomography. Acta Mater. 2013;61:4696–4706.
[99] Shun TT, Du YC. Age hardening of the $Al_{0.3}CoCrFeNiC_{0.1}$ high entropy alloy. J Alloys Compd. 2009;478:269–272.
[100] Zhu JM, Fu HM, Zhang HF, Wang AM, Li H, Hu ZQ. Microstructure and compressive properties of multiprincipal component $AlCoCrFeNiC_x$ alloys. J Alloys Compd. 2011;509:3476–3480.
[101] Senkov ON, Wilks GB, Scott JM, Miracle DB. Mechanical properties of $Nb_{25}Mo_{25}Ta_{25}W_{25}$ and $V_{20}Nb_{20}Mo_{20}Ta_{20}W_{20}$ refractory high entropy alloys. Intermetallics. 2011;19:698–706.
[102] Senkov ON, Woodward CF. Microstructure and properties of a refractory $NbCrMo_{0.5}Ta_{0.5}TiZr$ alloy. Mater Sci Eng A. 2011;529:311–320.
[103] Senkov ON, Scott JM, Senkova SV, Meisenkothen F, Miracle DB, Woodward CF. Microstructure and elevated temperature properties of a refractory TaNbHfZrTi alloy. J Mater Sci. 2012;47:4062–4074.
[104] Senkov ON, Senkova SV, Dimiduk DM, Woodward C, Miracle DB. Oxidation behavior of a refractory $NbCrMo_{0.5}Ta_{0.5}TiZr$ alloy. J Mater Sci. 2012;47:6522–6534.
[105] Yang X, Zhang Y, Liaw PK. Microstructure and compressive properties of $NbTiVTaAl_x$ high entropy alloys. Procedia Eng. 2012;36:292–298.
[106] Senkov ON, Senkova SV, Miracle DB, Woodward C. Mechanical properties of low-density, refractory multiprincipal element alloys of the Cr–Nb–Ti–V–Zr system. Mater Sci Eng A. 2013;565:51–62.
[107] Chuang MH, Tsai MH, Tsai CW, Yang NH, Chang SY, Yeh JW, Chen SK, Lin SJ. Intrinsic surface hardening and precipitation kinetics of $Al_{0.3}CrFe_{1.5}MnNi_{0.5}$ multi-component alloy. J Alloys Compd. 2013;551:12–18.
[108] Huang PK, Yeh JW. Inhibition of grain coarsening up to 1000°C in (AlCrNbSiTiV)N superhard coatings. Scr Mater. 2010;62:105–108.
[109] Chen YY, Duval T, Hung UD, Yeh JW, Shih HC. Microstructure and electrochemical properties of high entropy alloys—a comparison with type-304 stainless steel. Corros Sci. 2005;47:2257–2279.
[110] Chen YY, Hong UT, Shih HC, Yeh JW, Duval T. Electrochemical kinetics of the high entropy alloys in aqueous environments—a comparison with type 304 stainless steel. Corros Sci. 2005;47:2679–2699.
[111] Hsu YJ, Chiang WC, Wu JK. Corrosion behavior of $FeCoNiCrCu_x$ high-entropy alloys in 3.5% sodium chloride solution. Mater Chem Phys. 2005;92:112–117.
[112] Chen YY, Hong UT, Yeh JW, Shih HC. Selected corrosion behaviors of a $Cu_{0.5}NiAlCoCrFeSi$ bulk glassy alloy in 288°C high-purity water. Scr Mater. 2006;54:1997–2001.
[113] Lee CP, Chen YY, Hsu CY, Yeh JW, Shih HC. The effect of boron on the corrosion resistance of the high entropy alloys $Al_{0.5}CoCrCuFeNiB_x$. J Electrochem Soc. 2007;154:C424–C430.
[114] Lee CP, Chang CC, Chen YY, Yeh JW, Shih HC. Effect of the aluminium content of $Al_xCrFe_{1.5}MnNi_{0.5}$ high-entropy alloys on. the corrosion behaviour in aqueous environments. Corros Sci. 2008;50:2053–2060.
[115] Chou YL, Wang YC, Yeh JW, Shih HC. Pitting corrosion of the high-entropy alloy $Co_{1.5}CrFeNi_{1.5}Ti_{0.5}Mo_{0.1}$ in chloride-containing sulphate solutions. Corros Sci. 2010;52:3481–3491.
[116] Chou YL, Yeh JW, Shih HC. The effect of molybdenum on the corrosion behaviour of the high-entropy alloys $Co_{1.5}CrFeNi_{1.5}Ti_{0.5}Mo_x$ in aqueous environments. Corros Sci. 2010;52:2571–2581.
[117] Kao YF, Lee TD, Chen SK, Chang YS. Electrochemical passive properties of $Al_xCoCrFeNi$ ($x = 0$, 0.25, 0.50, 1.00) alloys in sulfuric acids. Corros Sci. 2010;52:1026–1034.
[118] Lin CM, Tsai HL, Bor HY. Effect of aging treatment on microstructure and properties of high-entropy $Cu_{0.5}CoCrFeNi$ alloy. Intermetallics. 2010;18:1244–1250.

[119] Chou YL, Yeh JW, Shih HC. Effect of molybdenum on the pitting resistance of $Co_{1.5}CrFeNi_{1.5}Ti_{0.5}Mo_x$ alloys in chloride solutions. Corrosion. 2011;67:085002.

[120] Lin CM, Tsai HL. Effect of annealing treatment on microstructure and properties of high-entropy $FeCoNiCrCu_{0.5}$ alloy. Mater Chem Phys. 2011;128: 50–56.

[121] Yu Y, Xie FQ, Zhang TB, Kou HC, Hu R, Li JS. Microstructure control and corrosion properties of $AlCoCrFeNiTi_{0.5}$ high-entropy alloy. Rare Metal Mater Eng. 2012;41:862–866.

[122] Juan CC, Yeh JW, Chin TS, editors. A novel light high-entropy alloy $Al_{20}Be_{20}Fe_{10}Si_{15}Ti_{35}$. E-MRS Fall Meeting 2009; Warsaw, Poland.

[123] Cotton J, Munitz A, Oliver R, Gomes R, Bourne G, Kaufman M, editors. Search for lower density high entropy alloys. TMS2013; 2013; San Antonio, TX.

[124] Kecskes L, Atwater M, Hammond V, Maupin H, Darling K, editors. Influence of processing parameters on the microstructure and mechanical properties of lightweight high entropy alloys. TMS2013; 2013; San Antonio, TX.

[125] Tang WY, Yeh JW. Effect of aluminum content on plasma-nitrided $Al_xCoCrCuFeNi$ high-entropy alloys. Metall Mater Trans A. 2009;40:1479–1486.

[126] Tang WY, Chuang MH, Lin SJ, Yeh JW. Microstructures and mechanical performance of plasma-nitrided $Al_{0.3}CrFe_{1.5}MnNi_{0.5}$ high-entropy alloys. Metall Mater Trans A. 2012;43:2390–2400.

[127] Varalakshmi S, Kamaraj M, Murty BS. Synthesis and characterization of nanocrystalline AlFeTiCrZnCu high entropy solid solution by mechanical alloying. J Alloys Compd. 2008;460:253–257.

[128] Chen Y-L, Hu Y-H, Tsai C-W, Hsieh C-A, Kao S-W, Yeh J-W, Chin T-S, Chen S-K. Alloying behavior of binary to octonary alloys based on Cu–Ni–Al–Co–Cr–Fe–Ti–Mo during mechanical alloying. J Alloys Compd. 2009;477:696–705.

[129] Varalakshmi S, Rao GA, Kamaraj M, Murty BS. Hot consolidation and mechanical properties of nanocrystalline equiatomic AlFeTiCrZnCu high entropy alloy after mechanical alloying. J Mater Sci. 2010;45: 5158–5163.

[130] Varalakshmi S, Kamaraj M, Murty BS. Formation and stability of equiatomic and nonequiatomic nanocrystalline CuNiCoZnAlTi high-entropy alloys by mechanical alloying. Metall Mater Trans A. 2010;41:2703–2709.

[131] Chen Y-L, Tsai C-W, Juan C-C, Chuang M-H, Yeh J-W, Chin T-S, Chen S-K. Amorphization of equimolar alloys with HCP elements during mechanical alloying. J Alloys Compd. 2010;506:210–215.

[132] Ma SG, Zhang SF, Gao MC, Liaw PK, Zhang Y. A successful synthesis of the CoCrFeNiAl0.3 single-crystal, high-entropy alloy by bridgman solidification. JOM. 2013;65:1751–1758.

[133] Chen JH, Hua PH, Chen PN, Chang CM, Chen MC, Wu W. Characteristics of multi-element alloy cladding produced by TIG process. Mater Lett. 2008;62:2490–2492.

[134] Dolique V, Thomann A-L, Brault P, Tessier Y, Gillon P. Complex structure/composition relationship in thin films of AlCoCrCuFeNi high entropy alloy. Mater Chem Phys. 2009;117:142–147.

[135] Dolique V, Thomann AL, Brault P, Tessier Y, Gillon P. Thermal stability of AlCoCrCuFeNi high entropy alloy thin films studied by in-situ XRD analysis. Surf Coat Technol. 2010;204:1989–1992.

[136] Lin YC, Cho YH. Elucidating the microstructure and wear behavior for multicomponent alloy clad layers by in situ synthesis. Surf Coat Technol. 2008;202:4666–4672.

[137] Chen JH, Chen PN, Hua PH, Chen MC, Chang YY, Wu W. Deposition of multicomponent alloys on low-carbon steel using gas tungsten arc welding (GTAW) cladding process. Mater Trans. 2009;50:689–694.

[138] Chen JH, Chen PN, Lin CM, Chang CM, Chang YY, Wu W. Characterization of multi-element alloy claddings manufactured by the tungsten inert gas process. Surf Coat Technol. 2009;203:2983–2988.

[139] Lin YC, Cho YH. Elucidating the microstructural and tribological characteristics of NiCrAlCoCu and NiCrAl-CoMo multicomponent alloy clad layers synthesized in situ. Surf Coat Technol. 2009;203:1694–1701.

[140] Zhang H, Pan Y, He YZ. Effects of annealing on the microstructure and properties of 6FeNiCoCrAlTiSi high-entropy alloy coating prepared by laser cladding. J Therm Spray Technol. 2011;20:1049–1055.

[141] Zhu C, Lu ZP, Nieh TG. Incipient plasticity and dislocation nucleation of FeCoCrNiMn high-entropy alloy. Acta Mater. 2013;61:2993–3001.

[142] Singh S, Wanderka N, Kiefer K, Siemensmeyer K, Banhart J. Effect of decomposition of the Cr–Fe–Co rich phase of AlCoCrCuFeNi high entropy alloy on magnetic properties. Ultramicroscopy. 2011;111:619–622.

[143] Kao YF, Chen SK, Sheu JH, Lin JT, Lin WE, Yeh JW, Lin SJ, Liou TH, Wang CW. Hydrogen storage properties of multi-principal-component $CoFeMnTi_xV_yZrz$ alloys. Intl J Hydrog Energy. 2010;35:9046–9059.

[144] Kunce I, Polanski M, Bystrzycki J. Structure and hydrogen storage properties of a high entropy ZrTiVCrFeNi alloy synthesized using Laser Engineered Net Shaping (LENS). Int J Hydrog Energy. 2013;38:12180–12189.

[145] Nagase T, Anada S, Rack PD, Noh JH, Yasuda H, Mori H, Egami T. Electron-irradiation-induced structural change in Zr–Hf–Nb alloy. Intermetallics. 2012;26:122–130.

[146] Chen SK, Kao YF. Near-constant resistivity in 4.2-360 K in a B2 $Al_{2.08}CoCrFeNi$. AIP Adv. 2012;2:012111.

[147] Zhang Y, Zuo T, Cheng Y, Liaw PK. High-entropy alloys with high saturation magnetization, electrical resistivity, and malleability. Sci Rep. 2013;3:1455.

6

A Percolative Deformation Process Between Nanograins Promotes Dynamic Shear Localization

Sagi Chen, Daniel Rittel and Dan Mordehai*

Department of Mechanical Engineering, Technion – Israel Institute of Technology, 32000, Haifa, Israel

Shear localization is an important failure mechanism in crystalline solid materials. However, while nanograins (NGs) are universally observed in shear bands, their connection with the formation of the bands is not clear yet. Based on molecular dynamics (MD) simulations, we propose that the shear localization evolves as a percolative process of deformation zones between NGs. The presence of NGs is a necessary condition to relieve some of the elastic energy by dislocation emission and absorption. Moreover, the density and spatial arrangement of those grains are of prime importance to accelerate or delay the formation of a strong localization path.

Keywords: Dynamic Shear Localization, Molecular Dynamics Simulation, Percolation, Nanograins, Dislocations

Dynamic shear localization is a failure mechanism that is a characteristic of high-rate deformation. Narrow strain localization bands, also called adiabatic shear bands (ASBs), tend to develop abruptly, leading to local material softening and catastrophic failure.[1] It has long been argued that the thermo-mechanical coupling effect is responsible for ASB formation in crystalline solids.[2] However, the recent observations of dynamically recrystallized nanograins (NGs) in hexagonal closed packed (HCP) metals,[3,4] which appear locally *long before* the ASB has formed, have led to a new paradigm, namely that dynamic shear localization is driven by microstructural evolutions, rather than by thermal softening. This claim has been supported by recent numerical simulations, which clarified the respective role of microstructural and thermal softening factors.[5] However, the numerical simulations that were aimed at replicating the experimental results did not look into detail at the actual microstructural deformation mechanisms, since a coarse-grained approach was adopted. Moreover, the local matrix softening related to NG formation was surmised based mainly on their low density of dislocations. Therefore, the open issue is whether NGs, which are formed in a crystalline matrix, can actually soften the latter to an extent that strain localization develops abruptly, and if this is the case, how does that happen with respect to the density and spatial layout of those NGs. In other words, does the presence of NGs precipitate dynamic shear localization?

To address this general problem of the heterogeneity of the plastic deformation at its very basic scale, continuum-based simulations are either cumbersome or inadequate. Thus, this study relies on molecular dynamics (MD) simulations, which on the one hand are inherently limited in size, but on the other hand have the great advantage that the dislocation mechanisms manifest themselves naturally; thus, providing a physical picture of the phenomenon at hand. We construct here a model system to show that the presence of NGs by itself does indeed promote a localized deformation mode, while identifying the spatial distribution requirements that accelerate or slow down the process. In the spirit of previous suggestions that shear band formation is a percolative process in amorphous materials,[6,7] we show in this letter that shear band formation can similarly be rationalized as a percolation process [8] leading to deformation zones between NGs in crystalline structures.

The MD simulation code LAMMPS [9] is employed to study the deformation of a HCP lattice with an internal microstructure of NGs. We choose Mg as a representative material using an interatomic interaction described

*Corresponding author. Email: danmord@tx.technion.ac.il

by the embedded atom potential (EAM).[10] This crystallographic structure allows for dislocations on several slip systems to be nucleated when considering a fully periodic slab with a shorter dimension along the $[1\bar{2}10]$ direction. NGs are inserted into the lattice by rotating hexagonal domains of atoms around the $[1\bar{2}10]$ axis by 15° or 37° with respect to the matrix. As we shall further emphasize, the grain boundary formed in these angles gives rise to symmetric dislocation nucleation under tensile stress. The computational cell is first relaxed with the conjugate gradient technique, followed by a relaxation of the stresses via MD time steps, employing the Parrinello–Rahman barostat at $T = 10°$K. Then, the slab is strained along the $[10\bar{1}0]$ direction at a constant strain rate of $10^9\,s^{-1}$. This direction is chosen in order to activate pairs of slip systems (e.g. on $(\bar{1}012)$ and $(10\bar{1}2)$ planes), so that there will be no preferential slip direction which arises solely from the crystal structure. During the deformation, the resulting stress along the tensile direction is calculated. The lattice's microstructure is identified according to the bond order parameter q_6 [11] and visualized using OVITO (for details, see [12]). More information on the simulation setup is given in Section A in the Supplementary Information.

We first consider *homogeneously distributed NGs* on a cubical subset of lattice points (Figure 1(a)). The simulations are repeated with different *sizes of NGs*, while the distance between the center of the grains is kept constant. Characteristic stress–strain curves are shown in Figure 1(b) for a few grain sizes, accompanied by the corresponding atomic configurations during the deformation. Since all the grains are self-similar, their size is characterized by a single parameter, α, which is the largest distance between atoms on the grain boundary of the NG (see the inset in Figure 1(b)). Four distinct stages during the deformation can be identified in the stress–strain curves: the first corresponds to an elastic regime (marked as (I) in Figure 1(b)), which deviates from linearity as the tensile strain increases. In all the simulations, the tensile stress saturates at a value of about 1.9 GPa and remains almost constant during the deformation (second stage marked as (II) in Figure 1(b)). In matrices with the smaller grains (8 nm and below), this stage ends at a tensile strain of about 7%, followed by a third stage during which the material hardens

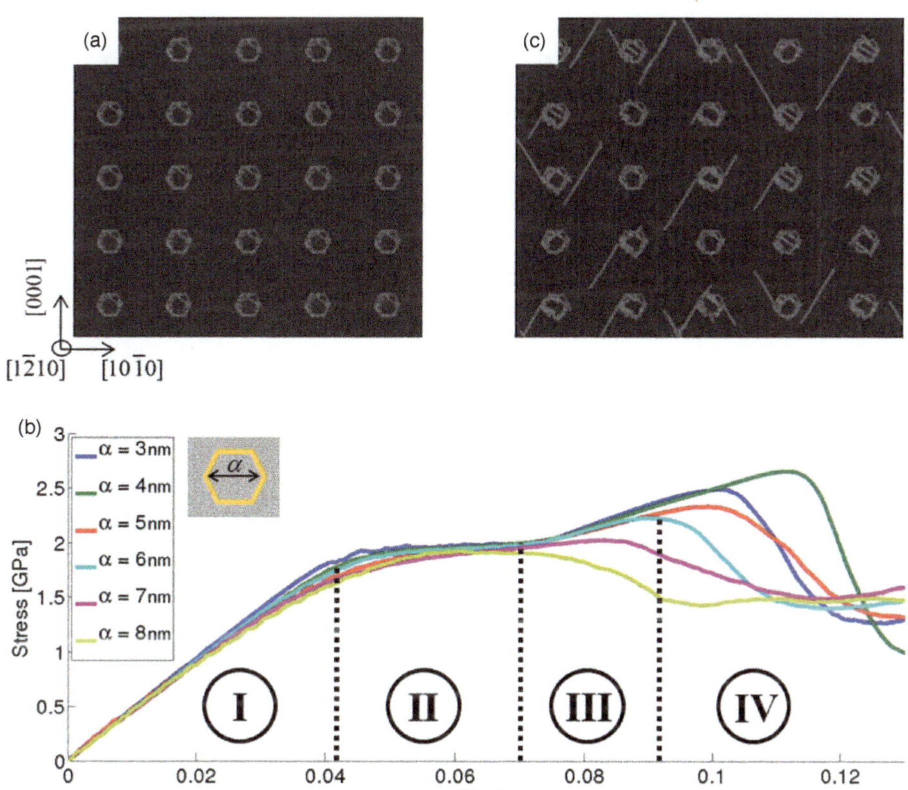

Figure 1. Deformation of lattices with a homogeneous distribution of grains. (a) A representative case of the microstructure after the relaxation (before applying a tensile stress). Grain boundaries, dislocations and stacking faults are shaded in light gray. (b) Stress–strain curves for different grain sizes. (c) The atomic configuration at the onset of the softening stage. A few dislocations that are emitted into the matrix can be observed.

(marked as (III) in Figure 1(b)). Finally, as the strain further increases, noticeable softening occurs (fourth stage, marked as (IV) in Figure 1(b)). The extent of the hardening stage decreases as the grains are made bigger, to an extent that beyond a size of 8 nm, the hardening stage completely disappears and the fourth stage follows directly the second stage. Let us denote here the strain at which softening commences as the *failure strain*.

Microstructural evolutions along the deformation are considered next. During the first stage, no prominent dislocation activity is observed. During the second stage, the main dislocation activity is confined inside the grains and their boundaries. This activity translates into a slight rotation of the grains (through reconstruction of the grain boundaries) but mainly to nucleation of internal dislocations, thereby relieving the internal stresses without plastic deformation in the matrix (outside the NGs). In other words, dislocations' nucleation into the matrix is not observed. Once it becomes harder to deform the material via grain rotation or shearing mechanisms, hardening commences (third stage). As the strain increases, the stresses are relieved by nucleation and emission of partial dislocations on the grain boundaries, which glide into the matrix on two pyramidal planes $(\bar{1}011)$ and $(10\bar{1}1)$ (Figure 1(c)). It is important to mention that the nucleated dislocations are distributed evenly between the two planes and that the rotation angle of the NGs did not impose any directionality due to preferential nucleation into one of these slip planes. The partial dislocations glide away from the grains, and at the onset of the fourth stage, they annihilate on a grain boundary of a different NG, which in some cases results in a persistent faulted plane between two different grain boundaries. This observation is cardinal, in the sense that the failure strain is apparently related to the absorption of dislocations on grain boundaries, which results in a localized deformation around the stacking fault. In addition, when a dislocation is nucleated into the matrix and terminates on another grain, it releases in some cases the elastic energy stored within the grain boundary by triggering emission of new dislocation from the grain boundary, which eventually promotes to softening. We refer the reader to Section B in the Supplementary Information for more details on the dislocation mechanisms. This observation rationalizes the decrease in the failure strain as the grain size is increased, i.e. the dislocations' mean-free path in the matrix decreases as the size increases. Thus, it is suggested that the important parameter that controls the failure strain is the distance between grains.

In that spirit, for a given grain size, one can expect the failure strain to decrease as the *density of NGs* along the slip directions increases. To further examine this point, computational cells with grains aligned along specific directions were constructed and strained as before. The dimensions of the computational cell in each case are determined so that the alignment of the grains will be compatible with the periodic boundary conditions, so that the NGs' area is actually aligned on an infinitely straight line. The characteristics of the mechanical response are similar to the case of a homogeneous distribution of grains, as shown in Figure 2(a). Moreover, the stress–strain curves are indifferent to the grains' distribution up to a tensile strain of 7%, i.e. dislocation activity occurs locally within the grain and on its boundary, independent of the neighboring grains. However, the hardening and softening stages depend on the arrangement of the grains. For instance, the softening stage is postponed when the grains are aligned along the $[10\bar{1}0]$ direction, whereas different arrangements may decrease the failure strain. In particular, the material undergoes relatively low hardening when the center of the NGs is aligned along $[10\bar{1}2]$ and $[\bar{1}012]$ directions (pyramidal planes), which are the preferable directions for dislocation nucleation and glide, found previously with the homogeneous distribution of grains. Examination of the microstructure at failure in each case shows a distinct difference between the case where the grains are aligned along the $[10\bar{1}0]$ and $[10\bar{1}2]$ directions. Here, the major difference is the distance between grains along the activated slip directions. This distance (through the periodic boundary conditions) is large and equal for both activated slip directions in the former case (grain aligned along the $[10\bar{1}0]$ direction). As a result, there is no preferable path for the deformation and the strain is not localized (Figure 2(b)). This contrasts with the case of grains aligned along the $[10\bar{1}2]$ direction, for which this distance becomes the shortest. Thus, dislocations nucleated outside the grain are absorbed in the adjacent grain boundary more rapidly, with only minor interaction within the matrix, leaving a persistent fault between the grains. As the deformation proceeds, more dislocations follow these shorter paths between grains, leading to a localization of the deformation along these paths, as can be seen from Figure 2(c).

These simulations suggest that dislocations are responsible for a two competitive process: hardening due to the interaction between dislocation in the bulk and softening when they terminate on the grain boundaries of NGs. For instance, in the case where the grains are aligned along the $[10\bar{1}0]$ direction, dislocations interact within the bulk before reaching NGs. When grains are aligned along the $[10\bar{1}2]$ direction, dislocations find soft paths between NGs, rather than interacting within the bulk and softening prevails. Interestingly, when the grains are constructed on the second active slip plane, the direction of the soft path follows the grains (see Section C in the Supplementary Information). This emphasizes the importance of the arrangement of the NGs on the directionality of the deformation.

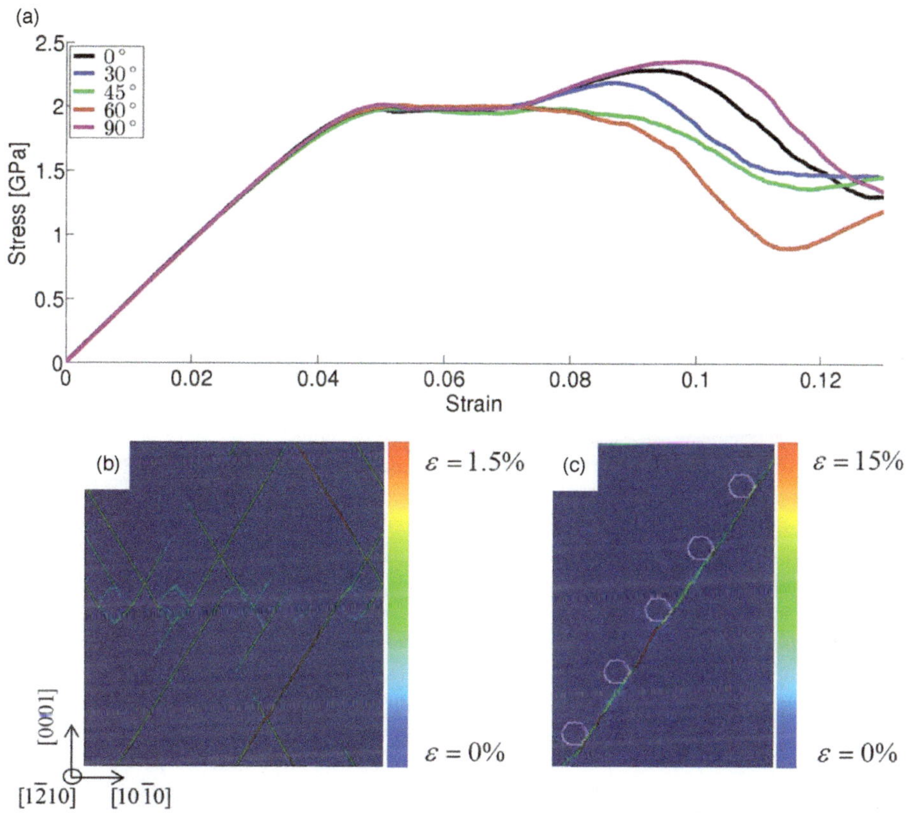

Figure 2. Deformation of lattices with a directional arrangement of grains. (a) Stress–strain curves for different directions. The angle of the grain alignment is measured from the horizontal axis ([10$\bar{1}$0] direction). Von-Mises local shear strain maps at total tensile external strain of 10% for grains aligned along angles of (b) 0° ([10$\bar{1}$0] direction) and (c) 60° (near [10$\bar{1}$2] direction) are shown. A clear strain localization is identified in the latter case.

It is worth mentioning that this behavior is different from that in nano-polycrystalline materials. While in the latter, any morphological change of the NGs involves a coherent deformation of the surrounding NGs,[13] here the NGs are not bounded by other NGs. As a result, upon interaction with dislocations, softening may prevail, while in nano-polycrystalline materials the limitation on grain motion may give rise to hardening.

Based upon these observations, we suggest that recrystallized NGs play a double role in promoting shear localization. Firstly, they release some of the elastic energy stored in the lattice during deformation by nucleating dislocations into the matrix. In addition, dislocations are absorbed on grain boundaries and release some of the elastic energy stored within its grain boundary. In that sense, if dislocations find a short path between the NGs, the deformation will be localized around this path, leading to failure via shear banding. If this path, which we denote as a *soft path*, is not short enough, localization of the deformation may be suppressed. Naturally, quantifying the critical soft path length for shear band formation is beyond the reach of atomistic simulations, and larger simulations that consider also the dislocations' microstructure should be employed (like discrete dislocation dynamics simulations). However, we believe that the typical length of a soft path should correspond to the mean-free path of dislocations in the matrix. If the distance between NGs along the soft path is larger than that, the dislocations will be stored in the matrix and not at the grain boundaries of the NGs. Devincre et al. [14] reported that the mean-free path of dislocations in Cu is of the order of a few micrometers. Adopting this value for other materials, we propose that this order of magnitude is also the minimal distance between recrystallized NGs along the soft path in order to initiate a shear band.

However, the probability of having all NGs aligned along one slip plane is highly unlikely in a real scenario. Can the material find soft paths when NGs are not aligned on a *single* slip plane? It can be surmised that the material is weaker as the *grains are denser*, based on the premise of an increased probability for the dislocations to find a soft path between the grains. To demonstrate that, we next apply a tensile strain on a system with a *distribution of* NGs embedded onto subset of points on a hexagonal sub-grid. The grains' size corresponds to $\alpha = 5\,\text{nm}$. We emphasize that this subset of points is chosen randomly in each simulation. This construction generates different configurations with NGs aligned along different slip planes. Three hundred and twenty configurations were generated in total (two of them are presented in Figure 3).

The average failure strain is calculated from 20 simulations with different arrangements for a given number of grains, and in Figure 4 we plot the averaged values for failure strain as a function of the density of NGs. The error bars indicate the variance of the results around the average value. The failure strain decreases as the density of NGs increases, whereas it is not strongly sensitive to the distribution of NGs for a given density, despite the stochastic nature of this process.

A closer look of the microstructure at the failure strain in each MD simulation rationalizes the failure strain dependency of the NGs density. At the lower density of NGs, the deformation is indeed localized between the grains but there is no clear soft path through the computational cell. For instance, in Figure 3(a) some localized ribbons with high shear strains are observed between NGs. However, the low density of NGs within the matrix allows for dislocation interactions within the

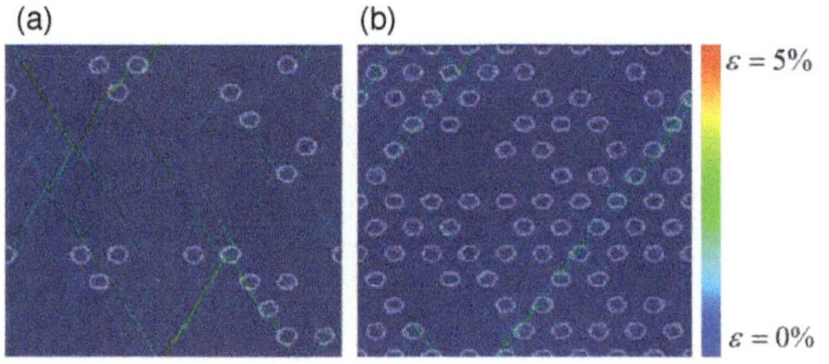

Figure 3. Two representative cases of an inhomogeneous distribution of the NGs with (a) a diffuse and (b) a strong localization. The colors correspond to the Von-Mises local shear strain, at an external tensile strain slightly above the failure strain. For convenience, the atoms within grain boundaries are shaded in light blue on the shear strain maps.

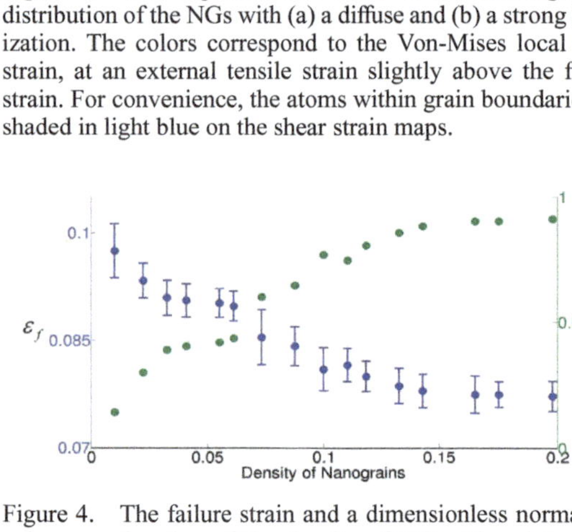

Figure 4. The failure strain and a dimensionless normalized value as a function of the density of NGs. The plot, which summarizes all 320 simulations, shows that failure strain is decreasing with an increase in the density of grains, and above a certain threshold the failure strains become microstructure-independent. This threshold can be viewed as a percolation threshold.

bulk, which promotes hardening and postponed the failure strain. In this case, the localization can be qualified of *diffuse*. For the higher density, the deformation finds a soft path between NGs through the whole computational cell. This is demonstrated in Figure 3(b), where the system finds a soft path along a plane with a high NGs density, which *strongly* localizes the deformation in a specific direction. In other words, when the arial density of NGs exceeds a certain value, here 14%, a soft path prevails, which leads to a strong shear localization. Once a soft path is found, any additional increase in the density of NGs will only add more possible paths and will no longer affect the failure strain, as can be seen from Figure 4.

This behavior is definitely reminiscent of a percolative process, which would adequately describe the catastrophic character of adiabatic shear failure through the attainment of a threshold. Based upon these simulations, we propose the following mechanism. At low densities of NGs, some localized deformation between grains may take place, but its contribution to the total deformation is minor, whereas dislocation interactions within the matrix prevail. For high density of NGs, deformation paths between NG become dominant and shear bands are formed.

We note here that the percolative process proposed here is different from the classical mathematical percolation theories in several aspects, and we name here only a few. While we choose a sub-grid, on which the NGs are formed, not all grid points are occupied; dislocations between NGs intersect also within the bulk and contribute to hardening and not only connecting NGs, the probability to nucleate dislocations changes during the process since the grain boundaries of NGs undergo topological changes after absorbing dislocations. Moreover, in practice, the material is recrystallized dynamically to form the NGs, so that the number of occupied sites is a dynamic process, which is definitely not captured by our MD simulations.

Thus, we do not provide here a quantitative description of the process, rather than identifying that such a process takes place in crystalline materials. Percolation of strain localization zones, as the process toward shear bands formation, was already suggested in other type of structures, e.g. amorphous materials.[6,7] However, the dislocation mechanism reported here has not been previously identified as being responsible for the formation of shear bands in crystalline materials. One should recall the results of Osovski and Rittel,[8] who used the Shannon's entropy of the deformed system to identify a similar kind of percolation threshold, albeit based on continuum calculations.

We conclude from the simulations that there is an aerial (volumetric) density threshold of NGs (recrystallized region) above which a soft path will be found.

We denote this threshold as the shear band threshold. For instance, if we define a parameter $\tilde{\varepsilon}_f = (\varepsilon_{\max} - \varepsilon_f)/(\varepsilon_{\max} - \varepsilon_{\min})$, which is equal to 0 and 1 when the failure strain corresponds to the maximum and minimum values found in our simulations, respectively (see Figure 4), the shear band threshold is the value above which $\tilde{\varepsilon}_f = 1$. In our simulations, this threshold corresponds to a density of 14%.

The picture, as being revealed by our MD simulations, clarifies the process of shear bands formation in crystalline materials. During high-strain rate deformation, small recrystallized NGs are formed *dynamically* within the material, as was previously observed experimentally.[3,4] This *microstructural* heterogeneity gives rise to the localized deformation between NGs by emitting and absorbing dislocations from and into their grain boundaries. However, this process competes with other dislocation mechanisms in the bulk, like hardening and recovery, and below a certain volumetric density of NGs, dislocation interactions within the bulk dominate, i.e. non-localized deformation prevails. As the deformation proceeds, more NGs are formed dynamically and the relative contribution of localized deformation zones between NGs to the overall deformation increases. If the density of NGs overcomes *dynamically* the shear band threshold, these localized zones percolate, and form a soft path on which the deformation is strongly localized, leading to the formation of a shear band. In order to meet this condition, the rate at which the NG's recrystallize dynamically, with respect to the hardening and recovery dislocation mechanisms within the bulk, should be high enough. In that sense, dislocation recovery properties, such as cross-slip, which was shown to be a crucial parameter in controlling dislocation cell structures,[15] may be a key mechanism in determining the failure mode. However, once the NGs are formed dynamically, temperature will not have a crucial effect on the dislocation mechanisms reported here. MD simulations at higher temperatures reveal a similar percolative process to the one reported here. In addition, the process reported here is not necessarily particular to an HCP lattice structure and its extension to other structures is planned in the future. Understanding how to control the failure mode in crystalline material can guide us toward designing new materials and tailoring nanoscale structures with enhanced mechanical properties.[16]

As a final remark, one should note that in this work, the NGs are pre-existing in the matrix, so that the failure strain corresponds to the peak stress situation. By contrast, the previous work of Rittel et al. [3] showed that the NGs can form long before the peak stress is reached and those authors identified the onset of localization with this stage. Since the NG formation process is not addressed in this work, the failure strain is the strain corresponding to the peak stress.

Acknowledgements We thank Dr. S. Osovski for fruitful discussions. This research is partially supported by the Israel Science Foundation (Grant No. 1656/12).

References

[1] Bai Y, Dodd B. Adiabatic shear localization: occurrence, theories, and applications. Oxford: Pergamon Press; 1992.
[2] Zener C, Hollomon JH. Effect of strain rate upon plastic flow of steel. J Appl Phys. 1944;15(1):22–32.
[3] Rittel D, Landau P, Venkert A. Dynamic recrystallization as a potential cause for adiabatic shear failure. Phys Rev Lett. 2008;101(16):165501.
[4] Rittel D, Wang ZG, Merzer M. Adiabatic shear failure and dynamic stored energy of cold work. Phys Rev Lett. 2006;96(7):075502.
[5] Osovski S, Rittel S, Venkert A. The respective influence of microstructural and thermal softening on adiabatic shear localization. Mech Mater. 2013;56:11–22.
[6] Shi Y, Falk ML. Strain localization and percolation of stable structure in amorphous solids. Phys Rev Lett. 2005;95(9):095502.
[7] Liu Y, Liu C, Wang W, Inoue A, Sakurai T, Chen M. Thermodynamic origins of shear band formation and the universal scaling law of metallic glass strength. Phys Rev Lett. 2009;103(6):065504.
[8] Osovski S, Rittel D. Microstructural heterogeneity and dynamic shear localization. Appl Phys Lett. 2012;101(21):211901.
[9] Plimpton S. Fast parallel algorithms for short-range molecular dynamics. J Comp Phys. 1995;117(1): 1–19.
[10] Liu X-Y, Adams JB. Grain-boundary segregation in Al–10%Mg alloys at hot working temperatures. Acta Mat. 1998;46(10):3467–3476.
[11] Steinhardt PJ, Nelson DR, Ronchetti M. Bond-orientational order in liquids and glasses. Phys Rev B. 1983;28(2):784–805.
[12] Stukowski A. Structure identification methods for atomistic simulations of crystalline materials. Model Simul Mater Sci Eng. 2012;20(4):045021.
[13] Wendy XG, Loynachan CN, Wu Z, Zhang Y-W, Srolovitz DJ, Greer JR. Size-dependent deformation of nanocrystalline Pt nanopillars. Nano Lett. 2012;12(12):6385–6392.
[14] Devincre B, Hoc T, Kubin L. Dislocation mean free paths and strain hardening of crystals. Science. 2008;320(5884):1745–1748.
[15] Landau P, Mordehai D, Venkert A, Makov G. Universal strain–temperature dependence of dislocation structures at the nanoscale. Scripta Mater. 2012;66(3–4):135–138.
[16] He G, Eckert J, Loeser W, Schultz L. Novel Ti-base nanostructure–dendrite composite with enhanced plasticity. Nature Mater. 2003;2:33–37.

7

Percolation and Universal Scaling in Composite Infiltration Processing

A. Léger[a*], J. M. Molina-Jordá[b], L. Weber[a] and A. Mortensen[a]

[a]*Laboratory of Mechanical Metallurgy, Institute of Materials, École Polytechnique Fédérale de Lausanne (EPFL), Station 12, 1015 Lausanne, Switzerland;* [b]*Instituto Universitario de Materiales de Alicante, Universidad de Alicante, Apto 99, E-03080 Alicante, Spain*

Composite materials are often made by infiltration, that is, by injecting a liquid matrix between packed solid fibers or particles. We give here direct proof that molten copper, slowly infiltrated into a broad spectrum of preforms, displays in initial phases of the process universal scaling and fractal geometric features that are characteristic signatures of percolation-dominated flow. Implications are that the microstructure of infiltrated composite materials can develop over length scales that far exceed the average preform pore size, and that network models are pertinent in the simulation of composite infiltration processing.

Keywords: Infiltration, Composite Material, Percolation, Universal Scaling, Fractal

Many modern composite materials are made by infiltrating a liquid matrix of polymer, metal or ceramic into a solid porous 'preform' of reinforcing fibers, whiskers or particles; electric vehicle chassis structures,[1] windmill blades [2] and electronic substrates [3] are examples. To eliminate porosity in the final material and to speed the process, pressure is often applied on the infiltrating liquid (Figure 1(a) and 1(b)). Applying pressure is particularly important when the matrix is metallic because metals seldom wet nonmetallic solids typically used as reinforcements;[4] metal meniscuses within the preform are then concave toward the metal, and this creates a back pressure P resisting infiltration.[5,6] With metals P is high: reinforcements are typically micrometric and metal surface tensions are on the order of one joule per square meter, such that P easily exceeds ten atmospheres. Capillarity is thus particularly manifest in metal matrix composite infiltration.

The science underlying capillarity in infiltration is very rich: the complex physics of two-phase flow within porous media have generated a considerable body of knowledge, from which our understanding of composite infiltration processing can draw. To describe the range of pressures P required to force a poorly wetting fluid into a porous body, the well-established semi-empirical relation of Brooks and Corey [6–8] proposes that, when capillary forces and P are equilibrated, the following relation holds:

$$S \equiv \frac{V_m}{1 - V_r} = 1 - \left(\frac{P_b}{P}\right)^{\lambda}. \quad (1)$$

Here, S is the pore saturation in non-wetting fluid (molten metal in the present instance), V_m is the local metal volume fraction per unit volume of material, V_r is the local volume fraction of reinforcement, P is the pressure difference between the two fluids within the preform, while P_b and λ are constants. In composite processing, preforms are initially dry and evacuated; hence, P is here simply the metal pressure.

The Brooks and Corey correlation has repeatedly been shown to agree with the experiment, in soil science, in petroleum extraction and also in composite materials processing.[6,8,9] It is currently used, together with an alternative but closely related relation by van Genuchten,[10] in advanced modeling of composite infiltration processing. In such models, it is typically coupled with continuum transport equations written over representative volume elements (RVEs)

*Corresponding author. Email: alain.leger@epfl.ch

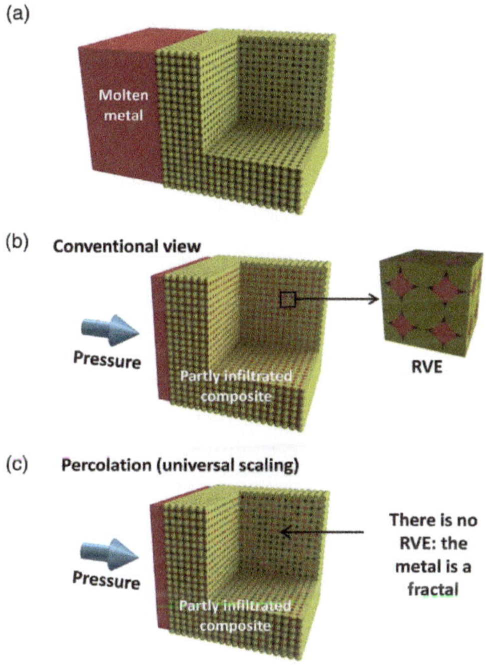

Figure 1. Different views of pressure infiltration. (a) Sketch of metal matrix composite pressure infiltration. (b) In conventional modeling, the process is simulated by integrating local flow laws over a representative volume element (RVE), which is then used as a differential element in a continuum-based simulation of the process across the composite part to be produced. (c) Infiltration as governed by percolation: the metal structure is fractal, invalidating the RVE approach.

encompassing one or a few representative pores [11]; gives a recent example and Figure 1(b) illustrates this approach.

There is, however, a radically different view of infiltration, which has so far been essentially ignored in composite materials processing: infiltration is an archetypical percolation phenomenon.[12–14] Recent summaries of the immense literature on the subject that are specifically focused on two-phase fluid flow through porous media are in Refs. [6,15]. To be specific, infiltration is a variant of percolation known as 'invasion percolation': to be infiltrated with a metal, a pore must not just be 'open' (e.g. conductive in conduction percolation or big enough in the present context); it must also be connected to a stringer of other open pores that extends from the pore in question to the specific location(s) where liquid is fed into the preform.[14,16,17] Invasion percolation in turn comes in several variants. When the infiltrant displaces a liquid, motion of the former around the latter creates isolated blobs of (the displaced, wetting) liquid, which can remain trapped within the porous medium. If, on the other hand, the porous medium is initially evacuated, there is no trapping. Invasion percolation then falls in the same universality class and shows similar spanning cluster fractal characteristics as in ordinary percolation.[6,15,18–20]

This leads to anticipate that universal laws of percolation might govern early phases of the metal pressure infiltration process. Here, we make this connection by giving direct experimental proof that percolation and its manifestations do, indeed, govern the composite pressure infiltration process in its initial phases, as sketched in Figure 1(c). At the same time, we give here one of the clearest extant direct experimental demonstrations of percolation effects in porous media capillarity.

We have designed an apparatus that measures with unique precision capillary forces in metal pressure infiltration under conditions representative of 'real' engineering materials processing. It uses induction heating to melt metal over a porous preform, both being held within an impervious and chemically inert ceramic crucible placed inside an evacuated pressure chamber. Once it is molten, the metal seals the preform from gas in the pressure chamber. The liquid metal can then be pushed into the preform by letting pressurized argon into the closed chamber. Our apparatus is special in that it can measure dynamically the volume of metal displaced [9,21]: as such, it can in fact be viewed as a (very) high-temperature analogue to the mercury porosimeter.[6,22] We achieve this using a floater connected to a rod, the upper end of which exhibits a set of marks, which are visible through a window in the pressure chamber. Movement of the marks is then tracked by a camera. We can work at temperatures up to roughly 1,500 K under applied pressures up to 20 MPa (200 bar); this is the range typical of engineering metal composite processing. Noteworthy features of this apparatus are (i) its precision ($\pm 10\,\mu$m) in tracking dynamically the position of the metal surface above the preform and hence of the metal volume having flown into the preform and (ii) its ability to hold molten metals such as copper at constant (high) temperature and fixed (high) pressure over extended periods of time (hours). A detailed description of the apparatus and its use is given in the Supplementary Online Material.

Precision and stability combined are critical, because P can only be equated with the applied gas pressure if flow of the metal has essentially stopped. Indeed, percolation as described in what precedes is for the ideal case where there are no pressure gradients within the invading liquid and no extraneous destabilizing effects.

If pressure gradients are significant, invasion percolation becomes a variant of what is known as 'gradient percolation' [6,16]: pore filling statistics are then no longer spatially uniform across the preform, and this complicates the problem. Pressure gradients in the liquid arise due to viscous friction; this becomes important if the capillary number, Ca, equals the fluid velocity times its viscosity divided by its surface tension, exceeds unity.[6,23,24] In our experiments, we hold the pressure

Table 1. Solid preforms explored in this work (nature, volume fraction solid and density of the solid) together with measured infiltration characteristics (approximate value of the pressure P_{max} required to fully infiltrate the preform, constants P_c and C in the scaling law describing initial stages of infiltration, Equation (3); constants λ and P_b of the Brooks and Corey relation, Equation (1)).

	Volume fraction, V_r	Bulk density (g/cm^3)	P_{max} ($S = 1$) (MPa)	P_c ($0.6 \geq S \geq 0.2$)	Constant C (MPa$^{-\beta}$)	λ ($0.9 \geq S \geq 0.6$)	P_b ($0.9 \geq S \geq 0.6$)
F320	0.44 ± 0.02	4.0	5	0.25 ± 0.01	2.16 ± 0.17	4.0 ± 0.6	0.23 ± 0.02
AA18	0.54 ± 0.02	4.0	5	0.47 ± 0.02	1.78 ± 0.09	5.0 ± 0.3	0.44 ± 0.02
SAFFIL	0.30 ± 0.01	3.5	8	0.47 ± 0.01	1.08 ± 0.02	2.0 ± 0.1	0.43 ± 0.02
Spherical	0.73 ± 0.02	3.9	10	3.31 ± 0.05	0.89 ± 0.13	3.3 ± 0.6	2.79 ± 0.25
F1000	0.54 ± 0.03	3.9	11	3.52 ± 0.10	0.79 ± 0.03	3.8 ± 0.5	3.15 ± 0.14
Graphite	0.80 ± 0.01	2.3	17	3.38 ± 0.02	0.51 ± 0.02	2.1 ± 0.1	3.09 ± 0.14

constant and wait for the fluid to stabilize: as a result, pores fill gradually while *Ca* falls to a value well below 10^{-4}, reaching at times 10^{-8}; this causes the course of infiltration to be dominated by capillarity.[6,21,23, 25] Gravity also creates pressure gradients within the fluid. These exert an influence if the Bond number, *Bo*, exceeds unity; *Bo* is by definition the fluid density times the gravitational constant *g* times the typical pore diameter squared divided by the liquid surface tension.[6] Here, pores are micrometric and pressures are many atmospheres. Hence, *Bo* is well below 10^{-3} and gravity effects are far weaker than capillary forces, despite the high density of liquid copper.

Another potential source of complexity has to do with the mechanics of fluid transport. Viscous fingering is an instability that appears in infiltration if the ratio of infiltrating to displaced fluid viscosity, *M*, is less than unity [6,23,24]: here, *M* is nearly infinite, so there is no viscous fingering. Also, since wetting angles are large and the metal partial pressure can be neglected, infiltration in present experiments is free of complications observed with a wetting or evaporating liquid, such as pre-wetting film formation or liquid flowing along asperities on the solid.[24] And finally, there is no trapping, and samples of this work are sufficiently large in comparison with the average pore size for finite-size scaling or edge effects to be negligible (these are felt when the sample is on the order of 40 particles across or less [5,6,25,26]).

Summing up, while quite challenging experimentally given the high temperature and high pressures involved, pressure infiltration experiments of this work are not only precise explorations of composite processing; they are also particularly clean and rich experimental embodiments of invasion percolation. Earlier experiments of this type, using for example Wood's metal or mercury, were either affected by gravity, or did not freeze the invading fluid to allow its detailed observation.[13,25,27–31]

We have infiltrated pure molten copper at rates sufficiently slow for pressure gradients to be negligible at 1,423 K into a range of different preforms. These are listed in Table 1 and described in detail in the Supplementary Online Material. Preforms investigated here comprise (i) packed alumina particle preforms, with particles of various origins and hence of diverse shapes and/or sizes (comminuted F320 or F1000, vapor-grown AA18 and atomized spheres; at low oxygen partial pressures the contact angle of Cu on Al_2O_3 at 1,423 K is in the range 120–130° [4]), together with (ii) preforms of pressed (Saffil™) alumina short fibers and (iii) microporous graphite (at low oxygen partial pressures, the contact angle of Cu on carbon at 1,423 K is in the range 140–150° [4]). Figure 2(a) shows the microstructure of fully infiltrated composites together with the associated 'drainage' curves (Figure 2(b)) that plot *S* as a function of *P*. The latter resemble other primary drainage curves found in the literature: invasion of the preform starts at an identifiable threshold pressure P_c, then progresses rapidly as *P* increases further, to then gradually approach full saturation ($S = 1$) in an essentially asymptotic fashion. Typical cross sections through infiltrated samples at low metal saturation are shown in Figure 3(a). Note how, at pressures situated just above the threshold, holes within the metal are much larger than individual preform pores, particles or fibers. This is a sign that pore connectivity plays a role, suggesting in turn that percolation effects might intervene in early phases of pressure infiltration. To show this, we confront the drainage curves, and samples frozen during early phases of infiltration, with predictions of percolation theory.

As mentioned above, invasion percolation without trapping falls in the same universality class and displays similar infinite cluster characteristics as ordinary percolation.[6,15,18–20,32] The fraction *F* of bonds or sites that belong to the infinite percolation cluster then scales with the probability, *p*, that a pore be open to invasion by the metal, according to the scaling law

$$F \alpha (p - p_c)^{\beta}, \tag{2}$$

where p_c is the percolation threshold probability. On a regular bond percolation network p_c roughly equals 1.5 divided by the bond network coordination number

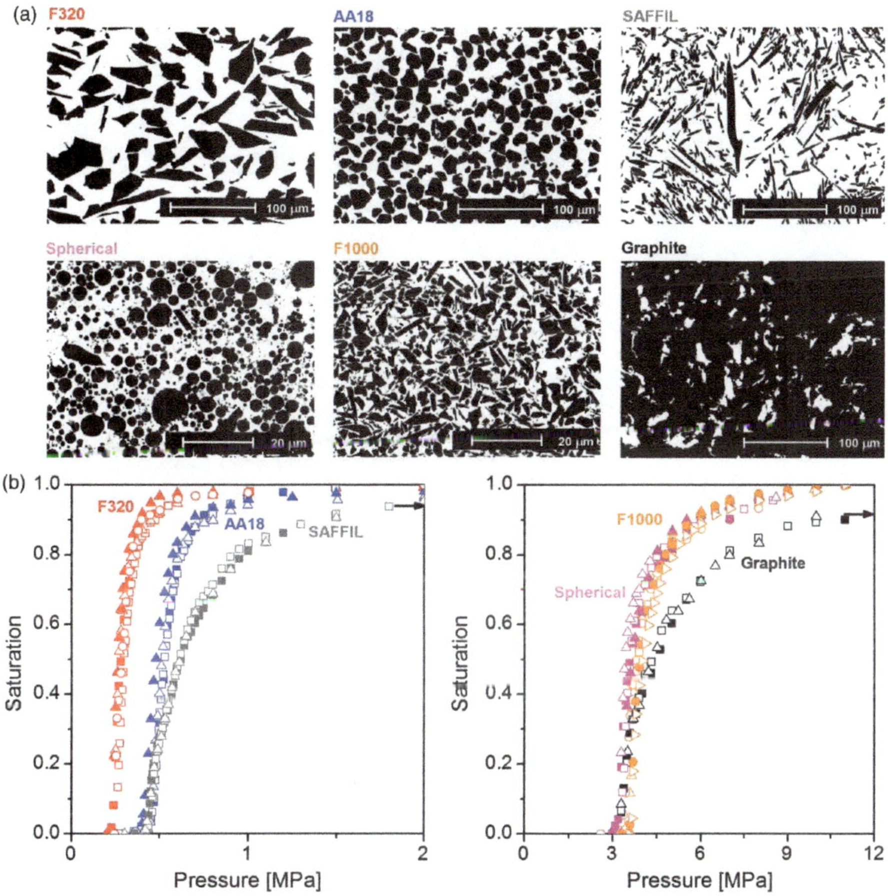

Figure 2. Fully infiltrated samples. (a) Microstructure in each composite investigated here after full infiltration ($S = 1$). (b) Measured drainage curves for each system.

Z.[6,12,14,33,34] The exponent β is universal and is the same for both bond and site percolation; in three dimensional (3-D) $\beta = 0.41$.[6,14] Drainage, defined as the invasion of a porous medium by a non-wetting fluid, is in essence a bond percolation problem.[12,19,35] Indeed, what limits ingress of a non-wetting fluid through the network of open pores is the ability of the liquid to cross, one by one, the narrowest constrictions or 'throats' that join neighboring pores: in the parlance of percolation theory these throats are bonds, while the larger pores that they link are sites. Throat sizes and geometries are statistically distributed across the preform. If we assume that the statistical distribution of pressures, P, needed to push the metal through each throat in the preform is regular, then the fraction of throats traversed, and hence the fraction of pores invaded, will increase just past the threshold pressure P_c according to Equation (2) with the applied pressure P replacing the probability p. If we also assume that the average pore volume that is filled with metal each time a throat is traversed does not vary much as P increases slightly past P_c (which can reasonably be expected if the pore size distribution is also regular), then, for P just above P_c the saturation S will be roughly proportional to the fraction of invaded pores, F. Summing up, one is entitled to expect the (universal) scaling law

$$S = C(P - P_c)^{\beta}, \qquad (3)$$

where $\beta = 0.41$ and C is a system-dependent constant, to describe the drainage curves in the early phases of the pressure infiltration process, that is, when P slightly exceeds P_c. Note that the simple view of drainage that we offer here is not universally adopted (most authors give a finite volume to the throats for example); however, it comes close to that given by Yanuka et al. [36,37] and Equation (3) can also be found in [29].

Figure 3(b) replots drainage curves of Figure 2(b) in coordinates that will show, with a straight line, whether drainage curve data collected for the range of systems in Table 1 agree with Equation (3) for $P \geq P_c$. As seen, within uncertainty, our data indeed do follow straight lines near P_c, proving that Equation (3) is obeyed regardless of the preform. The range of saturations over which the universal scaling law in Equation (3) holds varies

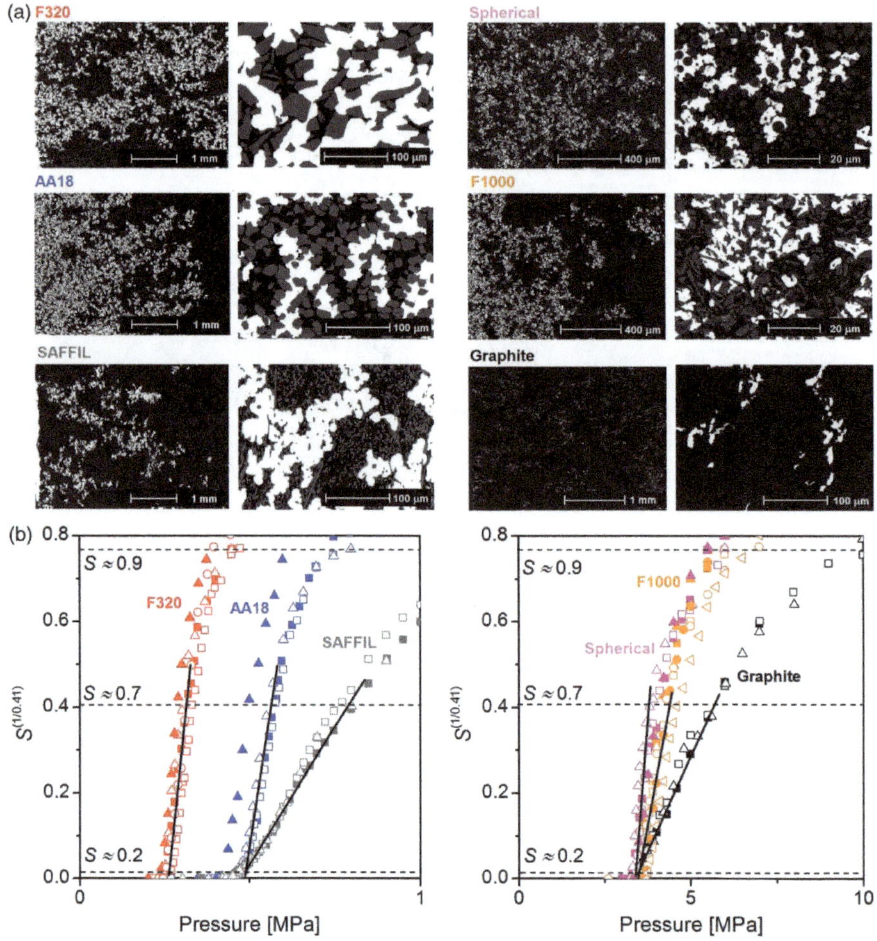

Figure 3. Partially infiltrated samples. (a) Low-magnification (left) and high-magnification (right) view of the microstructure of each composite system, taken at low saturation ($S < 0.5$). (b) Measured drainage curves for each system, plotted to test Equation (1). Average sample saturation for both lower and higher magnification micrographs is 0.1 for Saffil™ short fibers, 0.2 for all other systems.

with the preform; in some systems it stretches all the way to $S = 0.75$, Figure 3(b).

Percolation theory also predicts that, just past the percolation threshold, the cluster of invaded pores is a fractal of dimension $D = 2.53$.[14,16,34] We have applied standard fractal dimension measurement methods [28–31] on two-dimensional (2-D) metallographic cuts through frozen composites at low saturation to measure the 2-D fractal dimension, D_{2D}, of the solidified metal 'tree' formed after partial infiltration when P is just past P_c; Figure 4(a)–(c) illustrates the procedure with an example (a detailed description of the measurements is in the Supplementary Online Material). Results for D_{2D}, and for corresponding 3-D metal cluster fractal dimensions $D \approx 1 + D_{2D}$, are collected in Figure 4(d). As seen, the metal structure is indeed fractal, since measured D_{2D} values are well below 2. The metal cluster fractal dimension increases slightly with the fraction metal and is, for low S values, near that predicted by percolation theory (Figure 4(b) and 4(c)). Microstructures of partially infiltrated composites confirm that first phases of metal pressure infiltration are governed by percolation.

At higher pressures data deviate from universal scaling: there is a crossover to another regime. Replotting the data in Figure 2(b) so as to test the Brooks and Corey relation, Figure 5(a), shows that at higher saturation Equation (1) is well obeyed, while at lower saturation Equation (3) is obeyed, Figures 3(b) and 5(b) (where universal scaling is shown with a more usual log–log plot). What this means from a physical standpoint is that there is a crossover from a dominance of throat size distribution statistics in the early phases of infiltration, to a dominance of the average pore size and

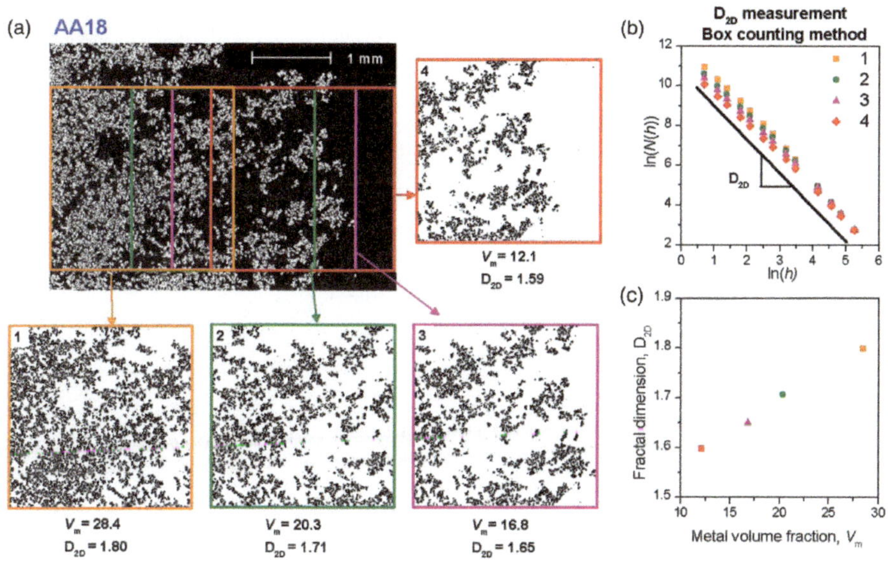

(d)

	Local metal volume fraction, V_m	2-D fractal dimension D_{2D}	3-D fractal dimension $D_{3D, min}$ (D_{2D}+1)
F320	20.6 - 25.8	1.68 1.71	2.7
AA18	12.1 - 28.4	1.59 - 1.80	2.6
SAFFIL	12.2 - 14.7	1.55 - 1.59	2.6
Spherical	9.9 - 15.8	1.55 - 1.65	2.6
F1000	13.3 - 20.3	1.60 - 1.70	2.6
Graphite	3.2 - 3.8	1.33 - 1.38	2.3

Figure 4. Measuring the metal fractal dimension in partly infiltrated composites. (a) Backscattered scanning electron image of a partially infiltrated composite (AA18/Cu at $P = 0.46$ MPa). The colored squares (768 × 768 pixels) indicate the area used to measure the fractal dimension using the box counting method ('ImageJ' function: 'Fractal box counter', size of the boxes: 2, 3, 4, 6, 8, 12, 16, 24, 32, 48, 64, 96, 128 and 192 pixels) and the corresponding thresholded images (in which the metal is turned black and the rest white). (b and c): evolution of the measured two-dimensional fractal dimension D_{2D} computed from the average slope through the range of data (b), showing (c) that it tends to increase somewhat with increasing metal volume fraction V_m (which in turn increases from the preform center to its outer surface in contact with the metal bath); (d): range of variation of D_{2D} with corresponding range for V_m, and final estimated value of $D = (D_{2D} + 1)$, the 3-D fractal dimension of the metal cluster in the most porous (central preform) region for each of the composite systems investigated.

shape in later phases of the metal pressure infiltration process. Percolation theory also predicts this crossover. Indeed, network models of bond-governed percolation evidence a stage, well past the percolation threshold but also well before full saturation, at which the percolation cluster fattens to the point where it contacts essentially all remaining uninfiltrated pores; with $Z = 4$, this happens near $p = 0.63$,[5,12,38,39] that is, well above $p_c = 0.388$.[6,14,34] Past that point infiltration is entirely dictated by the local geometry of each throat and pore in the preform, with essentially no role played by pore connectivity since all empty pores are now accessible to metal. In the literature, this crossover appears in a few simulations of drainage, where effects of connectivity (and hence percolation) are combined with a multiscale, fractal-like, description of the internal porous medium architecture (Figure 4 of [40], Figure 5(c) and 5(d) of [41]). In [15,42] a similar crossover is invoked in connection with predictions of conductivity or global fluid permeability in unsaturated porous media.

Though its existence thus emerges from theory, predicting the crossover location remains a challenge as there are, to our knowledge, no established schemes to predict constant C in Equation (3) for 'real' engineering preforms such as those used here.[15,42] To this question, the present data contribute the observation that the crossover is surprisingly gradual in preforms of the type explored here. Indeed both laws, the universal scaling law expected from percolation

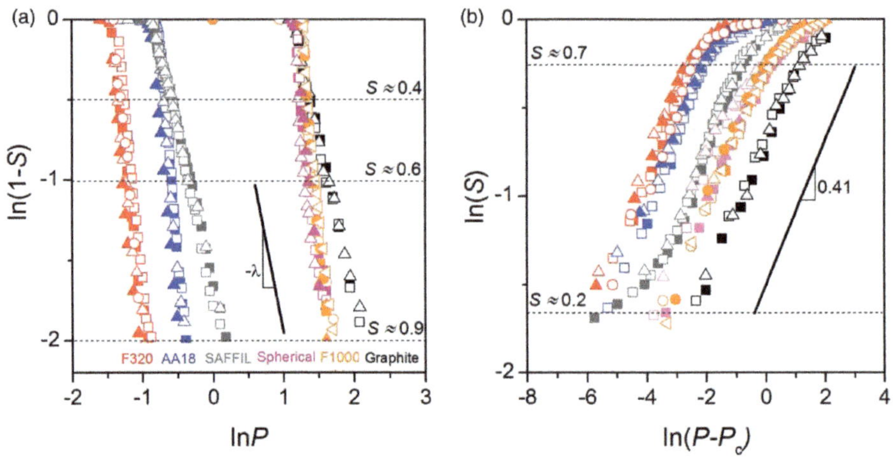

Figure 5. Confrontation of data with Equations (1) and (3) using logarithmic coordinates. Measured drainage curves for all systems explored here (Figure 2(b)) replotted in logarithmic coordinates to test (a) Equation (3), and (b) Equation (1) (using values of P_c obtained by extrapolation of a linear regression of $S^{(1/0.41)}$ over the range $0.02 \leq S^{(1/0.41)} \leq 0.35$, corresponding to $0.2 \leq S \leq 0.65$). Note that portions of the curves in these log-log coordinates are affected in (a) near $S = 1$ by uncertainty in the pore volume (hence data are only plotted up to $S = 0.9$) and in (b) near $S = 0$ by uncertainty in P_c (this does not affect the comparison if data are plotted as in Figure 3(b)).

theory (Equation (3)) and the Brooks–Corey relation (Equation (1)), are simultaneously obeyed by the data over a wide range, which extends (with variations from preform to preform) roughly from $S = 0.2$ to $S = 0.7$, see Figures 3(b) and 5.

Data of this investigation are also interesting as concerns the second, high S, range of infiltration pressures. It is now well established that Equation (1) emerges if one assumes that pore sizes are distributed according to a scaling law, or in other words if one assumes that solid elements in the porous medium, or their surface topography, have a fractal geometry; λ then equals three minus the relevant fractal dimension D.[6,15,43–47] For this reason, power-law saturation expressions such as the Brooks–Corey relation are often called 'fractal' saturation laws; however, the link is neither simple nor univalent.[44] Here, such 'fractal' saturation scaling is observed with short fibers that were spun to a nominally single diameter. In addition, some of the observed λ values also fall sufficiently far above 3, Table 1, that λ cannot equal 3-D. Both observations are at variance with fractal pore models, suggesting in turn that power-law saturation scaling relations such as Equation (1) can have their origin elsewhere.

In summary, we have evidenced that slowly driven copper pressure infiltration is governed by percolation in its initial phases: metal ingress obeys universal scaling laws and the metal first forms a percolation cluster that has the expected fractal characteristics. As such, the present data provide one of the cleanest extant experimental illustrations of the predictions of percolation theory in the context of flow through porous media. The saturation law then crosses over to the Brooks–Corey saturation relation, with both laws valid over roughly half the range of saturations. Present findings have implications concerning fluid flow in materials processing: in initial phases of the infiltration process, universal scaling and fractal geometry govern capillarity, and likely also basic flow parameters such as the relative permeability. Statistics of pore-to-pore connections are then important, such that averaging over a RVE encompassing only a few pores (Figure 1(b)) will go amiss of important features of the infiltration process (Figure 1(c)). The role played by percolation in early phases of infiltration also has implications beyond fluid flow: it will influence heat transfer, and if there are chemical reactions between the metal and the reinforcement, then pore connectivity statistics will influence their rate and the geometrical distribution of reaction products or solute within the final composite (one of the few papers in the literature addressing percolation effects in composite infiltration processing examines this last effect [39]). And finally, implications of the present findings go beyond the processing of composites: an example is with metal ingress into refractories during primary metal production (e.g. in a blast furnace); here too percolation effects are likely to have considerable importance.

Acknowledgements This work was sponsored by the Swiss National Science Foundation, Project No. 200020-137685.

We gratefully acknowledge Mr Willy Dufour, Mr Raphaël Charvet, Mr Cyril Dénéréaz, Mr Claudio Bacciarini and Dr Noelia Rojo Calderon for their contributions in designing, building and perfecting the infiltration apparatus used in this work.

References

[1] Birch S. SAE automotive engineering on-line; 'BMW i3, the inside story: what it's made of, how it's made' [cited 2013 May 9, last accessed July 2014]. Available from: http://www.sae.org/mags/aei/12056/.
[2] Brøndsted P, Lilholt H, Lystrup A. Composite materials for wind power turbine blades. Annu Rev Mater Res. 2005;35:505–538.
[3] CPSTechnologies. Reliability with smart composite products. 2012 [Cited 2014 July]. Available from: http://www.alsic.com/
[4] Eustathopoulos N, Nicholas MG, Drevet B. Wettability at high temperature. Amsterdam: Pergamon-Elsevier Science; 1999.
[5] Dullien FAL. Porous media, fluid transport and pore structure. 2nd ed. New York (NY): Academic Press Inc., Harcourt Brace Jovanovich Publishers, San Diego, USA; 1992.
[6] Sahimi M. Flow and transport in porous media and fractured rock – from classical methods to modern approaches. 2nd, Revised and Enlarged Edition. Weinheim: Wiley-VCH Verlag; 2011.
[7] Brooks RH, Corey AT. Hydraulic properties of porous media. Fort Collins (CO): Civil Engineering Department, Colorado State University; 1964. p. 27.
[8] Bear J, Bachmat Y. Introduction to modeling of transport phenomena in porous media. Dordrecht: Kluwer Academic Publishers; 1990.
[9] Bahraini M, Molina JM, Kida M, Weber L, Narciso J, Mortensen A. Measuring and tailoring capillary forces during liquid metal infiltration. Curr Opin Solid St M. 2005;9:196–201.
[10] van Genuchten MT. A closed-form equation for predicting the hydraulic conductivity of unsaturated soils. Soil Sci Soc Am J. 1980;44:892–898.
[11] Nordlund M, Michaud VJ. Dynamic saturation curve measurement for resin flow in glass fibre reinforcement. Compos Part A-Appl Sci. 2012;43:333–343.
[12] Chatzis I, Dullien FAL. Modeling pore structure by 2-D and 3-D networks with application to sandstones. J Can Petrol Technol. 1977;16:97–108.
[13] deGennes PG, Guyon E. Lois générales pour l'injection d'un fluide dans un milieu poreux aléatoire. J Mécanique. 1978;17:403–432.
[14] Stauffer D, Aharony A. Introduction to percolation theory. Revised 2nd ed. Boca Raton (FL): CRC Press; 1994.
[15] Hunt A, Ewing R. Percolation theory for flow in porous media. Heidelberg: Springer; 2009.
[16] Wilkinson D. Percolation effects in immiscible displacement. Phys Rev A. 1986;34:1380–1391.
[17] Ebrahimi F. Invasion percolation: a computational algorithm for complex phenomena. Comput Sci Eng. 2010;12:84–93.
[18] Wilkinson D, Barsony M. Monte Carlo study of invasion percolation clusters in two and three dimensions. J Phys A-Math Gen. 1984;17:L129–L135.
[19] Berkowitz B, Ewing RP. Percolation theory and network modeling applications in soil physics. Surv Geophys. 1998;19:23–72.
[20] Sheppard AP, Knackstedt MA, Pinczewski WV, Sahimi M. Invasion percolation: new algorithms and universality classes. J Phys A-Math Gen. 1999;32:L521–L529.
[21] Léger A, Calderon N, Charvet R, Dufour W, Bacciarini C, Weber L, Mortensen A. Capillarity in pressure infiltration: improvements in characterization of high-temperature systems. J Mater Sci. 2012;47:8419–8430.
[22] Giesche H. Chapter 2.7, mercury porosimetry. In: Schüth F, Sing KSW, Weitkamp J, editors. Handbook of porous solids. Weinheim: Wiley-VCH; 2002. p. 309–351.
[23] Lenormand R. Flow through porous media: limits of fractal patterns. Proc Roy Soc London. 1989;A423:159–168.
[24] Lenormand R. Liquids in porous media. J Phys-Condens Mat. 1990;2:SA79–SA88.
[25] Thompson AH, Katz AJ, Kronin CE. The microgeometry and transport properties of sedimentary rock. Adv Phys. 1987;36:625–694.
[26] Larson RG, Morrow NR. Effects of sample size on capillary pressures in porous media. Powder Technol. 1981;30:123–138.
[27] Chen JD, Wada N. Visualization of immiscible displacement in a three-dimensional transparent porous medium. Exp Fluids. 1986;4:336–338.
[28] Chen J-D, Dias MM, Patz S, Schwartz LM. Magnetic resonance imaging of immiscible-fluid displacement in porous media. Phys Rev Lett. 1988;61:1489–1492.
[29] Clement E, Baudet C, Hulin JP. Multiple scale structure of non wetting fluid invasion fronts in 3D model porous media. J Phys Lett. 1985;46:L1163–L1171.
[30] Clement E, Baudet C, Guyon E, Hulin JP. Invasion front structure in a 3D model porous medium under a hydrostatic pressure gradient. J Phys D Appl Phys. 1987;20:608–615.
[31] Hulin JP, Clement E, Baudet C, Gouyet JF, Rosso M. Quantitative analysis of an invading-fluid invasion front under gravity. Phys Rev Lett. 1988;61:333–336.
[32] Feder J. Fractals. New York (NY): Plenum Press; 1988.
[33] Vyssotsky VA, Gordon SB, Frisch HL, Hammersley JM. Critical percolation probabilities (bond problem). Phys Rev. 1961;123:1566–1567.
[34] Gouyet JF. Physics and fractal structures. Paris, New York: Masson-Springer; 1996.
[35] Wilkinson D, Willemsen JF. Invasion percolation: a new form of percolation theory. J Phys A-Math Gen. 1983;16:3365–3376.
[36] Yanuka M, Dullien FAL, Elrick DE. Percolation processes and porous media: I. geometrical and topological model of porous media using a three-dimensional joint pore size distribution. J Colloid Interf Sci. 1986;112: 24–41.
[37] Yanuka M. Percolation processes and porous media: II. Computer calculations of percolation probabilities and cluster formation. J Colloid Interf Sci. 1989;127: 35–47.
[38] Frisch HL, Hammersley JM, Welsh DJA. Monte Carlo estimates of percolation probabilities for various lattices. Phys Rev. 1962;126:949–951.
[39] Perham TJ, Chrzan DC, Jonghe LCD. Invasion percolation model of co-interpenetrating ceramic-metal composites. Model Simul Mater Sci. 2002;10:103–119.
[40] Xu K, Daian J-F, Quenard D. Multiscale structures to describe porous media part II: transport properties and application to test materials. Transp Porous Med. 1997;26:319–338.
[41] Bird NRA, Dexter AR. Simulation of soil water retention using random fractal networks. Eur J Soil Sci. 1997;48:633–641.
[42] Berkowitz B, Balberg I. Percolation theory and its application to groundwater hydrology. Water Resour Res. 1993;29:775–794.

[43] deGennes PG. Partial filling of a fractal structure by a wetting fluid. In: Adler D, Fritzsche H, Ovshinsky SR, editors. Physics of disordered media. New York and London: Plenum Press; 1985. p. 227–241.
[44] Bird NRA, Bartoli F, Dexter AR. Water retention models for fractal soil structures. Eur J Soil Sci. 1996;47: 1–6.
[45] Perrier E, Rieu M, Sposito G, de Marsily G. Models of the water retention curve for soils with a fractal pore size distribution. Water Resour Res. 1996;32: 3025–3031.
[46] Gimenez D, Perfect E, Rawls WJ, Pachepsky Y. Fractal models for predicting soil hydraulic properties: a review. Eng Geol. 1997;48:161–183.
[47] Ghanbarian-Alavijeh B, Hunt AG. Comments on, 'More general capillary pressure and relative permeability models from fractal geometry' by Kewen Li. J Contam Hydrol. 2012;140–141:21–23.

8

Hexagonal High-entropy Alloys

Michael Feuerbacher*, Markus Heidelmann and Carsten Thomas

Peter Grünberg Institut PGI-5, Forschungszentrum Jülich GmbH, D-52425 Jülich, Germany

We report on the discovery of a high-entropy alloy (HEA) with a hexagonal crystal structure. Equiatomic samples in the alloy system Ho–Dy–Y–Gd–Tb were found to solidify as homogeneous single-phase HEAs. The results of our electron diffraction investigations and high-resolution scanning transmission electron microscopy are consistent with an Mg-type hexagonal structure. The possibility of hexagonal high-entropy alloys in other alloy systems is discussed.

Keywords: High-entropy Alloy, Structure, Transmission Electron Microscopy

High-entropy alloys (HEAs) constitute a novel field in materials science, currently attracting increasing research interest. The basic idea, pioneered by Yeh in 2004,[1] is to study alloys composed of multiple principal elements solidifying as metallic solid solutions on a simple crystal lattice. Usually, alloys with 4–9, occasionally up to 20 components [2] are considered.

In a multicomponent alloy, the total free energy is largely determined by the free energy of mixing, $\Delta G_{\text{mix}} = \Delta H_{\text{mix}} - T\Delta S_{\text{mix}}$, where ΔH_{mix} is the mixing enthalpy and ΔS_{mix} the mixing entropy. In equimolar or near equimolar compositions, the mixing entropy can become dominant, stabilizing a random solid solution, if the elements are chosen such that the mixing enthalpy is neither too high positive (leading to segregation) nor too high negative (leading to the formation of ordered structures). In the ideal case, the resulting material is a perfectly disordered solid solution with a simple average crystal structure. HEAs are thus related to metallic glasses, but in contrast to the latter, they are thermodynamically stable, in particular at high temperatures.

Up to now numerous alloy systems have been reported in which HEAs exist. Without exception only HEAs with body-centered or face-centered cubic structures have been found. In this paper, we report on the observation of a HEA with a hexagonal crystal structure. We have prepared samples of equiatomic composition in the rare earth (RE)-system Ho–Dy–Y–Gd–Tb and characterized them by means of transmission electron microscopy (TEM), scanning electron microscopy (SEM), energy-dispersive x-ray diffraction (EDX) and scanning transmission electron microscopy (STEM).

We have prepared ingots of equiatomic Ho–Dy–Y–Gd–Tb in a high-frequency levitation furnace under 1 bar Ar atmosphere using 3N high-purity elements. Each ingot was remelted four times to ensure to achieve proper homogenization of the sample. Specimens for microstructural characterization were prepared by means of a dual-beam FEI Helios Nanolab 400S focused ion-beam (FIB) from the centers of the ingots. TEM was carried out using a Philips CM20 transmission electron microscope operated at 200 kV, SEM work using a JEOL 840 equipped with an EDAX energy-dispersive analytical system and STEM work using a probe-corrected FEI Titan 80–300 operated at 300 kV and equipped with a high-angle annular dark-field (HAADF) detector.

Figure 1(a) displays a scanning electron micrograph of the material using a composition-sensitive backscattered-electron detector. Figure 1(b) displays a STEM micrograph of the thin part of a typical FIB sample taken using a Z-contrast sensitive HAADF detector. Both images demonstrate that the sample is perfectly homogeneous, in particular there are no composition variations, no precipitates or secondary phases and no dendrites. We obtain these results consistently with FIB samples taken from various positions in the ingots. Note also that the cooling rate after levitation melting is sufficiently slow to allow for the formation of a homogeneous single phase. The average composition measured by EDX at eight different specimen positions

*Corresponding author. Email: m.feuerbacher@fz-juelich.de

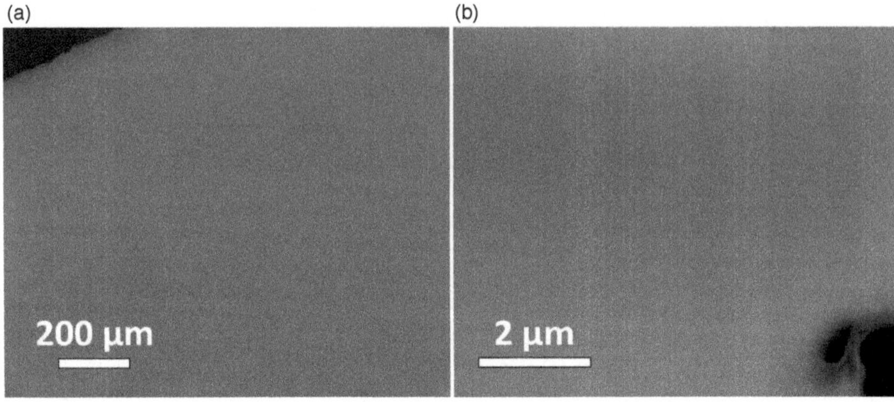

Figure 1. (a) SEM micrograph using a backscattered-electron detector and (b) HAADF-STEM overview of an FIB sample of equiatomic Ho–Dy–Y–Gd–Tb. The material is perfectly homogeneous; no features due to composition variation, precipation, dendrite formation, etc. can be seen. In both micrographs, sample edges (dark) are visible.

is Ho 19.8 at%, Dy 20.0 at%, Y 20.4 at%, Gd 20.1 at%, Tb19.7 at%. Within the experimental uncertainty this corresponds to the nominal equiatomic sample composition, and hence we conclude that the material solidifies congruently.

Figure 2(a) is an electron diffraction pattern taken at the [0 0 0 1] zone axis. The pattern displays perfect hexagonal symmetry, a number of reflections are indexed.

In order to determine the space group we follow a procedure described in [3]. Figure 2(b) is an overlay of two electron diffraction patterns at the $[1\ 1\ \bar{2}\ 1]$ zone axis, a short-time exposure to display zero-order Laue zone reflections and a long-time exposure (inverted to enhance the weak spots) including first-order Laue zone reflections. The whole pattern has an *m* symmetry, which corresponds to the point group 6/*mmm* in a hexagonal system. Figure 2(c)–2(e) shows diffraction patterns at the $[1\ 1\ \bar{2}\ 0]$, $[4\ 1\ \bar{5}\ 0]$ and $[1\ 0\ \bar{1}\ 0]$ zone axes, respectively, which are all perpendicular to [0 0 0 1].

The (0 0 0 1) reflection is visible at the $[1\ 1\ \bar{2}\ 0]$ zone axis, but diminishes when tilting to $[4\ 1\ \bar{5}\ 0]$ and essentially vanishes when tilting further to $[1\ 0\ \bar{1}\ 0]$. This indicates the presence of a 6_3 screw axis parallel to [0 0 0 1]. In the $[1\ 1\ \bar{2}\ 0]$ pattern the $(1\ \bar{1}\ 0\ 1)$ reflections are present, which rules out the $P6_3/mcm$ case.[4] Consequently, the space group must be $P6_3/mmc$. Under stern scrutiny, an extremely weak residual intensity of the (0 0 0 1) reflection in the $[1\ 0\ \bar{1}\ 0]$ zone can be seen. This may be due to a minor deviation from perfect disorder and is neglected in our analysis.

We have carried out tilting diffraction experiments on a single grain to determine the symmetry and angular distance of zone axes around [0 0 0 1]. The results are summarized in Figure 3, which displays the zone axes relevant for structure type determination within reach of the double-tilt goniometer stage. The [0 0 0 1] zone is surrounded by six $\langle\bar{1}\ 1\ 0\ 4\rangle$ and six $\langle\bar{1}\ 2\ \bar{1}\ 6\rangle$ zone axes (only half of which are shown).

The $[\bar{1}\ 1\ 0\ 4]$ zone is tilted by 15.24° with respect to [0 0 0 1] and shows a rectangular pattern that is repeated by five further zone axes rotated by 60° around [0 0 0 1]. The $[\bar{1}\ 2\ \bar{1}\ 6]$ zone is tilted by 17.60° with respect to [0 0 0 1] and shows a pattern with a different aspect ratio of the two reflections spanning the rectangular pattern. Five further equivalent zone axes are arranged around [0 0 0 1] rotated by angles of 60°. The dark lines represent the Kikuchi lines as seen in the TEM. The arrangement of the zone axes, as well as the Kikuchi lines, perfectly reproduces the hexagonal structure. A further zone axis displaying an almost hexagonal pattern, $[\bar{1}\ 2\ \bar{1}\ 3]$, which is tilted by 32° with respect to [0 0 0 1] was recorded. Note that some of the diffraction patterns contain very weak additional reflections, the origin of which is not clear. They may be due to double diffraction, minor deviations from perfect disorder or electron diffraction taking place in neighbored grains. These reflections were not taken into account in our analyses.

All zone axes, Kikuchi lines and reflections can consistently be indexed on the basis of a hexagonal structure. We compared calculated and experimental tilting angles of different zone axes as well as distances of reflections from the direct beam R_{hkil}. For the latter, we compared the ratio of R_{hkil} values of reflections spanning a diffraction pattern to the corresponding calculated d_{hkil} ratio. Due to the relationship $R_{hkil}d_{hkil} =$ const, where the imprecisely known constant is given by the electron wavelength and the camera length of the microscope, these ratios should inversely correspond. Good agreement could be achieved using the lattice parameters $a = 361$ pm and $c = 569$ pm. These lattice

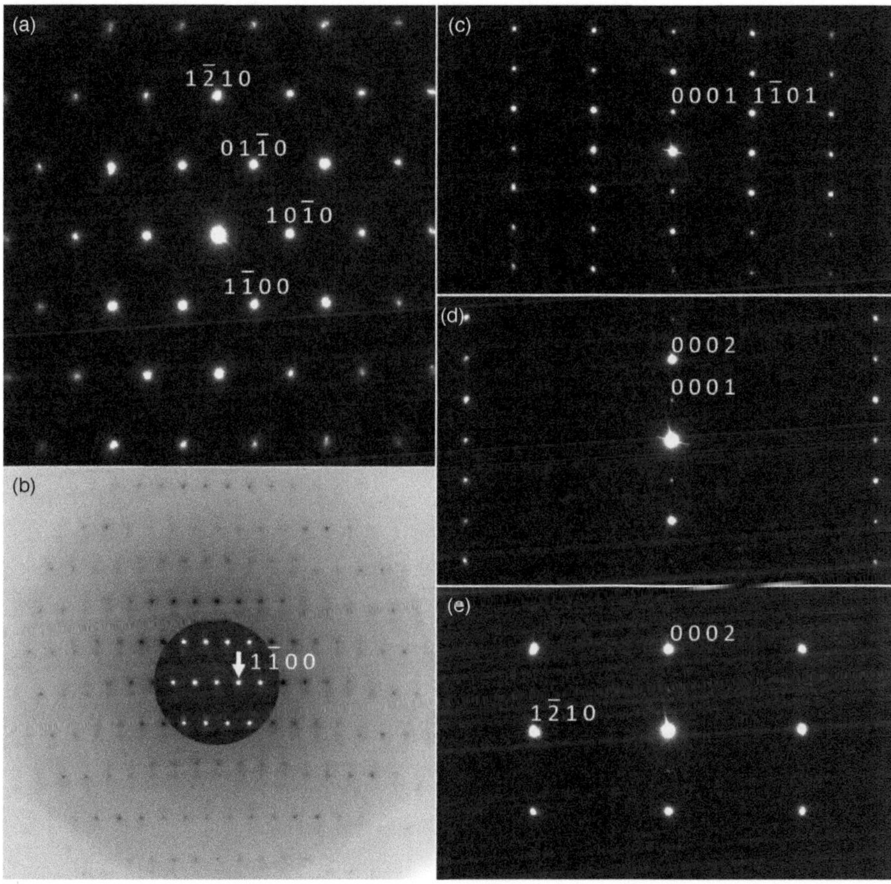

Figure 2. Electron diffraction patterns along different zone axes of the hexagonal structure: (a) [0 0 0 1] zone axis. (b) [1 1 $\bar{2}$ 1] zone axis: overlay of a short-time exposure displaying zero-order Laue zone reflections and a long-time exposure (inverted to enhance the weak spots) including first-order Laue zone reflections. (c) [1 1 $\bar{2}$ 0] zone axis. (d) [4 1 $\bar{5}$ 0] zone axis. (e) [1 0 $\bar{1}$ 0] zone axis.

parameters correspond to the weighted average of the parameters of the constituting elements. A comparison of the experimental and calculated quantities using these parameters is presented in Table 1.

Figure 4 shows STEM micrographs, which were slightly FFT-filtered for high-frequency noise reduction. Figure 4(a) was taken along the [0 0 0 1] direction and in consistency with an Mg-type structure displays a pattern of interconnected hollow hexagons. The inset is an FFT of the image also reflecting the hexagonal structure. Figure 4(b) was taken along [1 1 $\bar{2}$ 0] and the FFT in the inset corresponds to the diffraction pattern in Figure 2(c). The image is also consistent with an Mg-type structure. In particular, the atomic arrangement along this direction excludes a $P6_3/mcm$ space group, which would lead to a pattern of paired atomic rows parallel to the [0 0 0 1] direction. This significantly differs from the almost hexagonal pattern expected for the $P6_3/mmc$ space group, which directly corresponds to the experimentally observed HAADF-STEM image. This observation is consistent with our electron diffraction results. Both images, which were taken with a Z-contrast-sensitive HAADF detector, show that also on this small scale the material is perfectly homogeneous. There is no indication of precipitation or ordering.

The lattice parameters directly measured from the image amount to $a = (363 \pm 30)$ pm and $c = (566 \pm 20)$ pm, which is in agreement with the weighted average of the parameters of the constituting elements. It was observed earlier that the a-lattice constant in cubic HEAs follows a rule of mixtures.[5] Our results indicate that a rule of mixtures also applies to the c-lattice parameter and the c/a ratio.

Our electron diffraction and STEM results are consistent with an Mg-type hexagonal structure. That is, the unit cell has the space group P63/mmc (No. 194) and two atoms per unit cell, on the (1/3, 2/3, 1/4) position and its symmetric equivalent. The structure of the HEA is identical to the structures of the constituting elements, which are also of Mg-type.

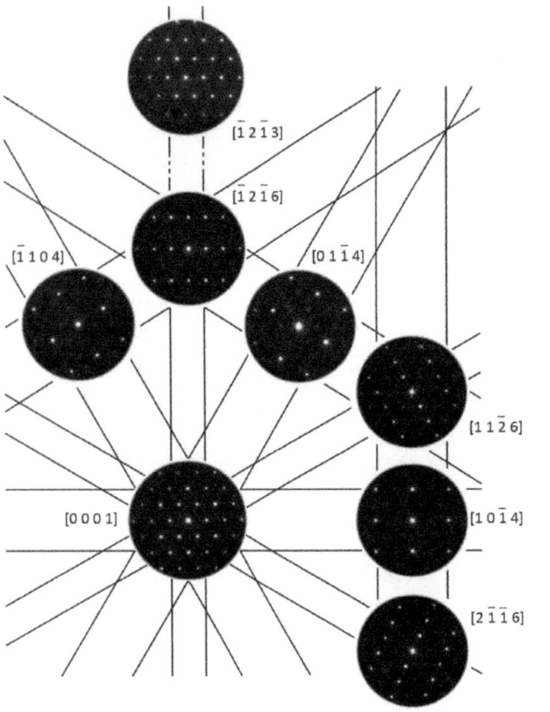

Figure 3. Arrangement of zone axes and corresponding electron diffraction patterns, indexed according to a hexagonal Mg-type structure. The lines correspond to Kikuchi lines as seen in the TEM. Only half of the patterns surrounding [0 0 0 1] are shown. The indicated angle between $[\bar{1}\ 2\ \bar{1}\ 3]$ and $[\bar{1}\ 2\ \bar{1}\ 6]$ is not to scale as indicated by the broken lines.

Table 1. Comparison between experiment and calculations on the basis of a hexagonal Mg-type structure with lattice parameters $a = 361$ pm and $c = 569$ pm.

Zone	Angle to [0 0 0 1] exp. (deg.)	Angle to [0 0 0 1] calc. (deg.)	R_{hkil} ratio exp.	Inverse d_{hkil} ratio calc.
$\bar{1}\ 1\ 0\ 4$	15.24	15.36	1.192[a]	1.198
$\bar{1}\ 2\ \bar{1}\ 6$	16.90	17.60	1.853[b]	1.817
$\bar{1}\ 2\ \bar{1}\ 3$	32.00	32.39	1.151[c]	1.141
$\bar{1}\ 2\ \bar{1}\ 3$	”	”	1.000[d]	1.000

Notes: First and second row: tilting angle of zone with respect to [0 0 0 1]. Third row: ratio of R_{hkil} values of reflections spanning the diffraction patterns. Fourth row: corresponding inverse d_{hkil} ratios.
[a] $2\ \bar{2}\ 0\ 1$ and $1\ 1\ \bar{2}\ 0$ reflections.
[b] $\bar{1}\ 2\ \bar{1}\ \bar{1}$ and $1\ 0\ \bar{1}\ 0$ reflections.
[c] $0\ 1\ \bar{1}\ \bar{1}$ and $1\ 0\ \bar{1}\ 0$ reflections.
[d] $0\ 1\ \bar{1}\ \bar{1}$ and $1\ \bar{1}\ 0\ 1$ reflections.

Criteria for the formation of HEAs are commonly discussed in terms of radius differences and a parameter Ω referring to the thermodynamic properties of the alloy.[6] The radius difference $\delta r = \sqrt{\sum_i c_i(1 - r_i/\bar{r})^2}$, where r_i are the element radii, c_i the mole percentages and $\bar{r}$ the weighted average radius, should be as small as possible. A value below about 6.5% for HEAs in general and below 3.8% for single-phase HEAs [7] is required. In our case $\delta r = 0.77\%$. The thermodynamic parameter $\Omega = T_m \Delta S_{mix}/|\Delta H_{mix}|$, where T_m is the weighted average melting temperature, ΔS_{mix} the mixing entropy and $\Delta H_{mix} = \sum_{i,j,i\neq j} c_i c_j H_{ij}$,[6] should exceed 1.1 (which means that the entropy contribution to the free energy should dominate the enthalpy).

Since the alloy investigated is equiatomic, ΔS_{mix} reaches the maximum value 1.61 R,[1] where $R = 8.314\,\mathrm{J\,K^{-1}\,mol^{-1}}$ is the gas constant. ΔH_{mix} calculated using values from the tables provided by Takeuchi and Inoue [8] amounts to $0\,\mathrm{kJ\,mol^{-1}}$. For Ho–Dy–Y–Gd–Tb not only the sum of the mutual mixing enthalpies is zero,

Figure 4. HAADF-STEM micrographs of equiatomic Ho–Dy–Y–Gd–Tb with insets displaying the corresponding FFTs. (a) Along the [0 0 0 1] direction. (b) Along the $[1\ 1\ \bar{2}\ 0]$ direction.

but each individual component $H_{ij} = 0\,\mathrm{kJ\,mol^{-1}}$ for all i and j. The alloy thus fulfills an even stronger formation criterion. Formally, taking the calculated values for ΔH_{mix} [8] into account, the parameter Ω in our case diverges.

Further criteria supporting HEA formation are equal crystal structures, a high mutual solubility in the binary phase diagrams of all constituting elements and, from a practical point of view, close melting points. All formation criteria are ideally fulfilled for the system investigated.

Note that the same is true for numerous further RE-elements. Similarly positive conditions are present for La, Er, Tm, Lu, Pr, Pm and Nd. If we exclude Pm due to its radioactivity, in total we are left with 11 practically usable elements for which we expect the formation of hexagonal HEAs. Hence, HEAs with higher numbers of components up to 11 and numerous combinations should be possible. The number of possible equiatomic five-element HEAs is 462, and altogether 1486 different HEAs with 5–11 different elements can be combined. The elements Ho, Dy, Y, Gd and Tb used for the alloy investigated in this work, which can be considered a feasibility study, were chosen more or less arbitrarily from the potential candidates due to their immediate availability in our stock. We should point out that some of the elements used undergo a high-temperature transition from hexagonal to cubic structure at 95% (Gd and Tb) and 98% (Dy and Y) of their melting point. The exclusive use of elements that have a hexagonal structure up to their melting point may lead to a larger grained structure. This, however, would limit the system to the four elements Er–Tm–Lu–Ho.

The list can probably be further extended by taking into account elements that have a different crystal structure but also similar radii and low mixing enthalpies with the elements mentioned above. Ce, for example, shows high solubility with all RE-elements (with the presence of a miscibility gap, however) and mixing enthalpies between 0 and $1\,\mathrm{kJ\,mol^{-1}}$. The same seems to hold for Sm, as far as phase diagrams are available.

We have chosen an equiatomic alloy composition for the present work, but we expect that due to the similarity of the constituting elements and hence generous compliance of the formation criteria, the existence ranges of the HEAs are very wide. The production of single-phase ingots with different compositions should therefore also be possible. This expected tolerance to composition variation may open up the possibility to form single-phase HEAs using Mischmetall, which is a more or less undefined alloy of RE-elements in various naturally occurring proportions. Since Mischmetall is considerably cheaper than high-purity single RE-elements, this would eliminate the economical disadvantage of the RE-systems under consideration.

The hexagonal HEAs presented may offer a unique chance to study the magnetism of RE-systems. Mainly related to their 4f electron configurations, RE-elements possess highly complex magnetic behavior displaying a wide variety of properties.[9] The systems presented are unique in the sense that they are topologically ordered but possess random bonds. In the present HEA systems, the included elements and the composition can most freely be varied at constant structure, which may allow for insightful comparative measurements. Furthermore, hexagonal HEAs, including Tm, Yb and Ce, may provide a way to study heavy-fermion behavior in a disordered system.

The hexagonal high-entropy systems also include the possibility to incorporate intentional second phases, that is, for the increase in strength, high-temperature applicability, etc. Potential candidates are Yb, Eu, Zr and Hf, which have a high melting temperature and a high positive mixing enthalpy with the RE-elements considered. The mixing enthalpies with these elements are between 6 and $18\,\mathrm{kJ\,mol^{-1}}$, which will lead to segregation.

We should also point out that recently Gao and Alman [10] predicted the existence of a hexagonal HEA in the system Co–Os–Re–Ru. In this system, the formation criteria are indeed well fulfilled and all constituting elements are mutually soluble. The drawback of this system is that Os reacts with oxygen at room temperature forming the highly toxic osmium tetroxide. Furthermore for the elements proposed, the melting points are strongly different (1768 K (Co) vs. 3459 K (Re)) making the growth procedure practically difficult. Further candidates along these lines are Co–Re–Ru–Tc or the five-element system Co–Re–Ru–Tc–Os, both of which involve the radioactive element Tc.

In conclusion, in this paper we report on the observation of a HEA in the Ho–Dy–Y–Gd–Tb alloy system. Our samples are homogeneous and single-phased, and of hexagonal Mg-type structure.

Acknowledgements The authors thank Ms D. Meertens for focused ion-beam preparation of the TEM samples, and W. Steurer, J. Dolinsek and M. Heggen for the inspiring discussions.

References

[1] Yeh JW, Chen SK, Lin SJ, Gan JY, Chin TS, Shun TT, Tsau CH, Chang SY. Nanostructured high-entropy alloys with multiple principal elements: novel alloy design concepts and outcomes. Adv Eng Mater. 2004;6:299–303.

[2] Cantor B, Chang ITH, Knight P, Vincent AJB. Microstructural development in equiatomic multicomponent alloys. Mat Sci Eng A. 2004;375–377:213–218.

[3] Wei Q, Wanderka, N, Schubert-Bischoff P, Macht MP, Friedrich S. Crystallization phases of the Zr41Ti14Cu12.5

Ni10Be22.5 alloy after slow solidification. J Mater Res. 2000;15:1729–1734.

[4] Morniroli JP, Steeds JW. Microdiffraction as a tool for crystal structure identification and determination. Ultramic. 1992;45:219–239.

[5] Senkov ON, Scott JM, Senkova SV, Miracle DB, Woodward CF. Refractory high-entropy alloys. Intermetallics. 2010;18:1758–1765.

[6] Yang X, Zhang Y. Prediction of high-entropy stabilized solid-solution multicomponent alloys. Mat Chem Phys. 2012;132:233–238.

[7] Miracle DB, Miller JD, Senkov ON, Woodward C, Uchic MD, Tiley J. Exploration and development of high entropy alloys for structural applications. Entropy. 2014;16:494–525.

[8] Takeuchi A, Inoue A. Classification of bulk metallic glasses by atomic size difference, heat of mixing and period of constituent elements and its application to characterization of the main alloying element. Mat Trans. 2005;46:2817–2829.

[9] Jensen J, Mackintosh AR. Rare earth magnetism: structures and excitations. Oxford: Clarendon press; 1991.

[10] Gao MC, Alman DE. Searching for next single-phase high-entropy alloy compositions. Entropy. 2013;15: 4504–4519.

9

Prediction of Spontaneous Plastic Deformation of Irradiated Metallic Glasses due to Thermal Spike-Induced Plasticity

Richard E. Baumer[a,b*] and Michael J. Demkowicz[b]

[a]*Research and Development, The Dow Chemical Company, Building 1776, Midland, MI 48674, USA;* [b]*Department of Materials Science and Engineering, Massachusetts Institute of Technology, Cambridge, MA 02139, USA*

Using large-scale molecular dynamics simulations, we predict that knock-on damage in irradiated metallic glasses yields spontaneous anisotropic deformation, i.e. shape change in the absence of externally applied loads. The root cause of this behavior is anisotropic plastic deformation around non-spherical thermal spikes. Such thermal spike-induced plasticity (TSIP) does not depend on electronic excitations and is distinct from the 'ion hammer' effect. Macroscale TSIP is predicted to occur under unidirectional heavy ion and neutron irradiation. The consequences of TSIP for potential applications of metallic glasses under irradiation are discussed.

Keywords: Metallic Glasses, Radiation Effects, Molecular Dynamics

Metallic glasses have attracted attention due to their good mechanical properties,[1] corrosion resistance,[2] and formability.[3] They also exhibit markedly different responses to radiation than crystalline solids. For example, crystalline alloys swell without limit under continuous irradiation,[4] whereas amorphous materials may expand or contract following irradiation.[5] Irradiated crystalline alloys embrittle [6] while some irradiated amorphous alloys may become more ductile.[7–9]

We performed large-scale molecular dynamics (MD) simulations to better understand the behavior of metallic glasses under irradiation. Our investigation leads us to predict an unexpected new radiation response mechanism of metallic glasses: spontaneous plastic deformation induced by inter-nuclear scattering. This phenomenon is distinct from the well-known 'ion hammer' effect [10] and suggests that some amorphous metals are unsuitable for use as structural materials under neutron irradiation, e.g. in certain nuclear reactor designs.

High-energy neutrons create energetic primary knock-on atoms (PKAs) in solids, which subsequently dissipate their kinetic energy in a cascade of secondary collisions before coming to rest.[11] This process occurs over lengths of nanometers (nm) and times of picoseconds (ps).[12] No experimental methods are currently capable of such spatiotemporal resolution. However, MD simulations are ideally suited to these length and time scales.

We investigate the effect of neutron irradiation on amorphous metals by simulating 475 keV niobium (Nb) PKAs passing through $\sim$1/2 billion-atom models of an amorphous alloy of copper (Cu) and Nb (*a*-CuNb). The PKA energy is at the upper end of the range where inter-nuclear—rather than electronic—scattering dominates its dissipation to the environment in metals.[13] Thus, we carry out our simulation using classical interatomic potentials.[14] Models with differing preparation histories and compositions ranging from 25% Cu to 75% Cu were investigated. The outcomes of these simulations were qualitatively similar. Therefore, we focus on results from a 475 keV PKA in *a*-$Cu_{50}Nb_{50}$ synthesized by rapid quenching from the liquid.

Simulations are conducted in LAMMPS [15] using an embedded atom method potential with Ziegler–Biersack–Littmark short-range interactions.[14] The *a*-$Cu_{50}Nb_{50}$ model is constructed by quenching a 474,353,318-atom, 4,000 K $Cu_{50}Nb_{50}$ liquid at 10^{13} K/s to 300 K in 25 K decrements using a velocity rescaling thermostat, constant number of particles (N), pressure (P), and enthalpy (H) Nosé-Hoover barostat, and periodic boundary conditions. Model properties match those of a previous study on a $\sim$50,000 atom *a*-$Cu_{50}Nb_{50}$

*Corresponding author. Email: rebaumer@dow.com

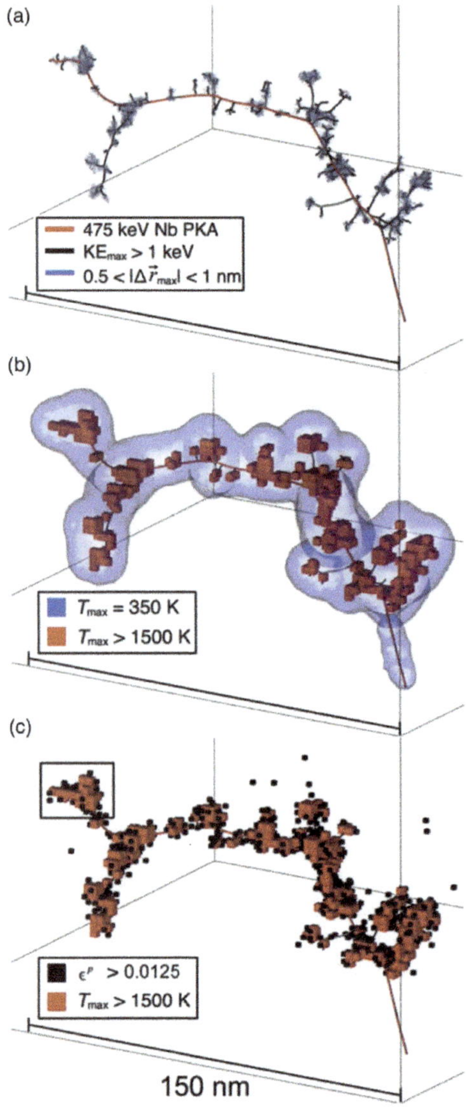

Figure 1. 475 keV Nb PKA passing through a ~1/2 billion-atom model of a-$Cu_{50}Nb_{50}$. (a) Trajectory of PKA plotted in red, >1 keV KAs in black, and atoms displaced by 0.5–1 nm in blue. (b) Temperature fields due to inter-nuclear collisions. Red voxels have $T_{max} > T_G = 1{,}500$ K,[16] blue contour is for $T_{\max} = 350$ K. (c) TS-induced plastic deformation. Black voxels have $\epsilon^{\mathrm{p}} > 0.0125$ at $t = 12$ ps. The boxed TS is analyzed in detail in Figures 2 and 3.

system.[16] During the collision cascade, constant number of particles (N), volume (V), and total energy (E) ensemble equations of motion are integrated using adaptive time steps to ensure energy conservation. The simulation cell size is greater than the projected range of 475 keV Nb ions in a-$Cu_{50}Nb_{50}$ computed using a version of SRIM [17] modified to exclude electronic stopping.[18]

Figure 1(a) shows a 475 keV Nb PKA collision cascade in a-$Cu_{50}Nb_{50}$. The PKA (red trajectory) loses more than 90% of its energy through collisions that transfer more than 1 keV, yielding higher order knock-on atoms (KAs, black trajectories) that terminate in spatially disconnected regions with numerous displacements of 1 nm or less (blue vectors). Rectilinear, collision-free KA trajectories connect these zones, demonstrating that inter-nuclear scattering does not generate ion tracks. Over distances of hundreds of nanometers, the distribution of PKA-induced collisions in Figure 1(a) is comparable to those in crystalline alloys.[19]

To characterize radiation response at lower length scales, we analyze our simulations on a grid of cubic volumes (voxels) with an edge length of 2.6 nm. These voxels are large enough to be representative volume elements of the material. We compute densities, potential energies, and stresses in each voxel by averaging per-atom quantities over voxel constituent atoms;[20] voxel temperatures by fitting the kinetic energy distribution of voxel atoms to the Maxwell–Boltzmann distribution,[21] with temperature as the fitting parameter; time-dependent voxel diffusivities at time t from a linear fit to voxel mean-squared displacement [MSD(t)] in 2 ps time intervals centered at t; and Lagrangian total strains from the best-fit uniform deformation gradient between an initial ($t = 0$ ps) and final voxel atom configuration.[22] Plastic strains are computed by subtracting voxel elastic strain ϵ_{ij}^{e} from total strain, with $\epsilon_{ij}^{e} = S_{ijkl}\sigma_{kl} + \alpha\Delta T\delta_{ij}$, where σ_{kl} is the voxel stress, S_{ijkl} is the compliance tensor, α is thermal expansion, and ΔT is the voxel temperature change. Our model is elastically isotropic with $\lambda = 140.8$ GPa, $\mu = 50.1$ GPa, and $\alpha \approx 9 \times 10^{-6}$ 1/K.

We find that the kinetic energy deposited in regions of dense atomic displacements causes the temperature there to rise rapidly. In some voxels, shown as red cubes in Figure 1(b), the maximum temperature $T_{\max}$ rises above the glass transition temperature of a-$Cu_{50}Nb_{50}$ ($T_{\mathrm{G}} = 1500$ K [16]) within ~1–2 ps and then quenches below T_{G} at rates approaching 10^{14} K/s. Consistent with previous studies,[23,24] structural analysis reveals no crystallization: all voxels return to an amorphous, vitrified state, although the rapid quenching reduces the short-range icosahedral order initially present.[16]

Within voxels where $T_{\max} > T_{\mathrm{G}}$, the temperature dependence of density and diffusivity matches that of uniform liquid Cu–Nb, demonstrating that these voxels are not merely superheated, but rather contain genuine Cu–Nb liquid, albeit only for times of ~10 ps. Connected voxels with $T_{\max} > T_{\mathrm{G}}$ form spatially isolated 'thermal spikes' (TSs) approximately 2–12 nm in diameter. TSs are extensively studied in crystalline solids due to their role in the production of radiation-induced point defects.[12,19,25,26] However, TSs have a distinctly different effect on atomic structure evolution in amorphous metals.[27–29]

By computing voxel-level strains, we find that the material immediately adjacent to TSs undergoes permanent plastic deformation. The tensile work-equivalent

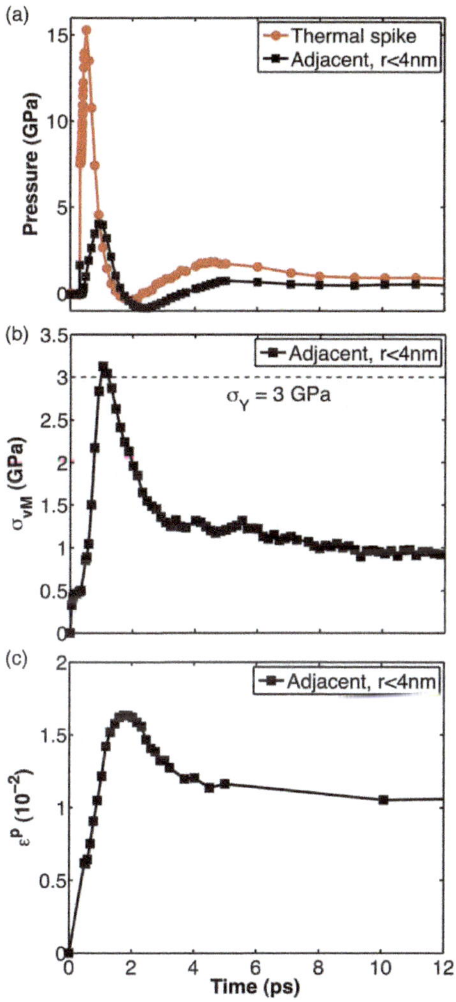

Figure 2. Stresses and strains inside and adjacent to the TS boxed in Figure 1(c). (a) Pressure versus time in the TS and in adjacent material. Von Mises stress (b) and tensile work-equivalent plastic strain (c) versus time, averaged over all voxels within 4 nm of the TS.

plastic strain ϵ^p of black voxels in Figure 1(c) exceeds 1.25%. This plasticity originates with TSs. When the material inside a TS melts, its zero-pressure density decreases to that of an equilibrium liquid. However, the liquid is constrained by the surrounding unmelted material, causing the pressure inside the TS to rise sharply. As in crystalline solids,[19,25] this pressure pulse initiates a stress wave that propagates away from the TS at the longitudinal speed of sound (see supplementary material). Unlike in crystalline solids, close to the TS, the deviatoric component of the stress wave exceeds the yield stress and initiates plastic flow.

We illustrate this process on the TS boxed in Figure 1(c). Figure 2(a) shows that the pressure of the liquid inside the TS shoots up to >15 GPa and drops back to near zero within ~2 ps. A pressure pulse of ~4 GPa is seen with a ~0.5 ps delay in a 4 nm-thick shell of solid immediately adjacent to the TS. By comparing with the analytical solution for stresses around a transiently pressurized inclusion in an isotropic material,[30] we confirmed that the pressure pulse in the liquid initiates the succeeding pulse in the solid. Figure 2(b) shows that the deviatoric stress in the material adjacent to the TS exceeds the yield stress, computed in a uniform a-$Cu_{50}Nb_{50}$ model at the corresponding temperature and strain rate. Concurrently, the deviatoric plastic strain rises to just under 2% and falls back to a steady value of ~1% (likely due to elastic back stresses), as shown in Figure 2(c).

If TSs in irradiated metallic glasses were spherical, then the local plasticity in their vicinity would occur equally in all directions, resulting in no net shape change of the material. However, the TSs in our simulations are typically prolate ellipsoids with major axes along the direction of the KA that initiates them, e.g. the TS boxed in Figure 1(c) and shown close up in Figure 3(a). Under the influence of a high internal pressure, these prolate ellipsoids deform to become more spherical, as illustrated in Figure 3(b), i.e. they expand perpendicular to their major axis and contract parallel to it. Therefore, unlike spherical TSs, plasticity around prolate ellipsoid TSs is not equal in all directions and leads to a net anisotropic strain in the material.

In a cylindrical coordinate system aligned with the TS major axis, the deformation illustrated in Figure 3(b) may be described by a tensile hoop strain $\epsilon_{\theta\theta}$ and compressive radial and axial strains, ϵ_{rr} and ϵ_{zz}, in the material adjacent to the TS. We compute these strains for the TS in Figure 3(a) and plot them as a function of position along its major axis in Figure 3(c), confirming that $\epsilon_{\theta\theta} > 0$ and $\epsilon_{rr}, \epsilon_{zz} < 0$.

We repeat this analysis for the seven largest TSs in our simulation and find that they are also prolate ellipsoids with major axes along KA directions and with cylindrically symmetric plastic strain distributions, similar to those in Figure 3(c). The voxel average plastic strains for each TS, shown in Figure 4(a), satisfy $\bar{\epsilon}_{rr} \leq \bar{\epsilon}_{zz} < \bar{\epsilon}_{\theta\theta}$ and are nearly independent of TS energy. We attribute this fact to the linear dependence of TS volume on deposited knock-on energy, shown in Figure 4(b). The average TS volume per unit knock-on energy is nearly constant at $\langle V_{TS}/E_{TS} \rangle = 14.2 \pm 0.5\,\text{nm}^3/\text{keV}$, giving rise to a narrow distribution of maximum pressures inside TSs (average of 13 GPa) and a similarly narrow distribution of maximum stresses in the adjacent solid material (average of 2.5 GPa).

To enable comparisons of these modeling insights with experiments, we calculate the net plastic strain per unit fluence due to all TSs in our simulation. The voxels that undergo plastic deformation are adjacent to TSs and have maximum temperatures that fall in the range $350 < T_{max} < 1500$ K. We identify these voxels, average the plastic strains in them, and scale the result by the fraction

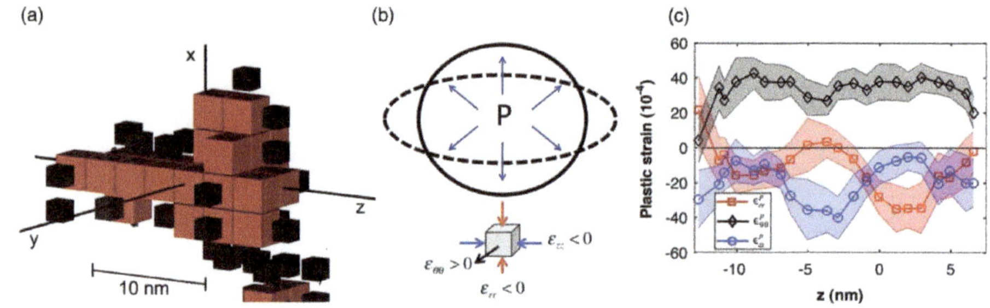

Figure 3. (a) Close-up view of the TS boxed in Figure 1(c) with a cylindrical coordinate system defined along the TS major (z) axis and voxels with $\epsilon^{\mathrm{p}} > 0.0125$ in black. (b) An initially ellipsoidal TS (dashed line) deforms into a more spherical shape (solid line) when internally pressurized, yielding the illustrated plastic strains in surrounding material. (c) Diagonal components of plastic strain computed at $t = 12$ ps in cylindrical coordinates as a function of position along the major axis of the TS in (a). Shaded bands indicate uncertainties.

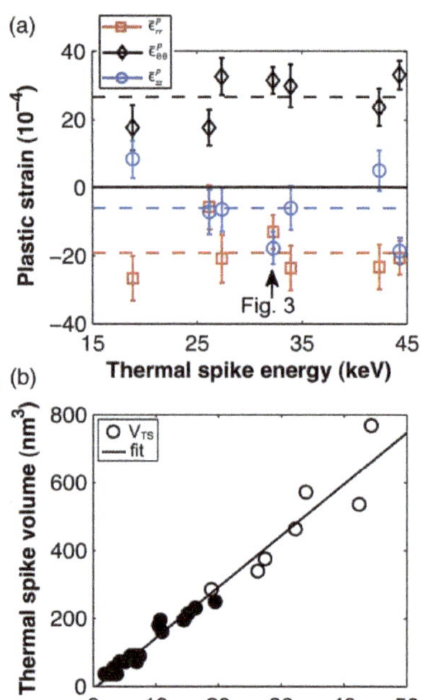

Figure 4. (a) Voxel average plastic strain components in material adjacent to the seven largest TSs as a function of the TS energy. The arrow indicates the TS analyzed in Figures 2 and 3. (b) TS volume V_{TS} versus deposited knock-on energy E_{TS}. Linear fit: $V_{\mathrm{TS}} = (15.1 \pm 0.6\,\mathrm{nm}^3/\mathrm{keV})E_{\mathrm{TS}} - (8.9 \pm 11.2\,\mathrm{nm}^3)$. Open symbols for TSs analyzed in (a).

of the volume of the model that these voxels occupy ($f = 1.12\%$). We find that the PKA causes compressive plastic flow along the cascade direction ($\epsilon_{\mathrm{casc}} = -2.8 \times 10^{-6}$), tensile flow in one lateral direction ($\epsilon_{l1} = 4.4 \times 10^{-6}$), and compressive flow in the other ($\epsilon_{l2} = -0.8 \times 10^{-6}$). In a material subjected to numerous collision cascades, the lateral strains are expected to average to a single value of net expansion in the lateral directions. The deformation includes a hydrostatic component $\epsilon_{\mathrm{H}}^{\mathrm{p}} = 0.2 \times 10^{-6}$, which is small compared with the tensile work-equivalent strain $\epsilon^{\mathrm{p}} = 4.3 \times 10^{-6}$ and consistent with shear-induced dilatation.[1]

Thus, we predict that TS-induced plasticity leads to a net contraction along the incident ion beam direction and expansion perpendicular to it. Barring any recovery or devitrification that may occur over longer times, the radiation-induced plastic strain per fluence along the beam direction is $A = \varepsilon_{\mathrm{casc}}/\phi \approx -10^{-15}\,\mathrm{cm}^2/\mathrm{PKA}$, where $\phi = 1\,\mathrm{PKA/SA} \approx 2.6 \times 10^9\,\mathrm{PKA/cm}^2$ is the effective fluence in our simulation (SA is the model cross-sectional area). To predict the magnitude of TS-induced plasticity under irradiation by a specific high-energy particle, X, A must be further scaled to account for the mean free path between PKAs, λ_{PKA}, and the mean energy of PKAs, E_{PKA}: $A_X = A \times (196\,\mathrm{nm}/\lambda_{\mathrm{PKA}}) \times (E_{\mathrm{PKA}}/475\,\mathrm{keV})$, where 196 nm is the length of our model and 475 keV the PKA energy in our simulation.

As an example, we employ the strain per fluence from MD to compute the radiation-induced plastic strains expected under neutron irradiation. Using a hard-sphere neutron scattering cross section of ~3b (representative of transition metals [31]), the measured fast neutron energy spectrum at the High Flux Isotope Reactor at Oak Ridge National Laboratory,[32] and a neutron energy range of 0.1–20 MeV, we calculate [11] $\lambda_{\mathrm{PKA}} \approx 5.3$ cm and $E_{\mathrm{PKA}} \approx 94$ keV (see supplementary material), yielding a neutron-induced strain per fluence of $A_n \approx -7.3 \times 10^{-22}\,\mathrm{cm}^2/\mathrm{n}$. Thus, for unidirectional neutron beams with fluences of order $10^{20}\,\mathrm{n/cm}^2$, we predict strains of ~7%.

Our findings suggest that TS-induced plasticity (TSIP) will give rise to spontaneous plastic deformation of amorphous metals under neutron irradiation. TSIP is distinct from the 'ion hammer' effect,[10] wherein irradiation of amorphous solids with swift, heavy ions, e.g. 360 MeV Xe ion irradiation, also leads to plasticity. Ion hammering is due to viscoelastic relaxation induced by electronic excitations in continuous ion tracks.[33,34] It is not expected under fission neutron irradiation because maximum PKA energies for neutrons are below the threshold for ion track formation. Whereas ion hammering is due to intense ionization tracks, TSIP arises from plastic deformation around isolated TSs initiated by inter-nuclear collisions and therefore is to be generally expected in amorphous metals under neutron irradiation. A fraction of ion-induced plasticity hitherto attributed to ion hammering may in fact also be due to TSIP, especially under low energy ion bombardment.[35]

We have shown that non-uniform, anisotropic plastic deformation around non-spherical TSs may produce pronounced anisotropic shape changes in metallic glasses irradiated with unidirectional ion or neutron beams. In thermal fission reactors, where the directions of transmutation-induced recoils are close to isotropic, the superposition of all recoil-induced strains is expected to cancel. However, anisotropic neutron fluxes may arise in fast neutron or fusion reactors. TSIP may limit the utility of metallic glasses as structural materials in such cases, despite their other, attractive properties.[7–9,36] TSIP may also enable new applications of amorphous materials in radiation responsive devices, e.g. in sensors where plastic strain is used to monitor dose or in radiation-driven actuation. Moreover, under controlled, unidirectional ion beams, TSIP may provide novel routes for radiation-assisted processing of metallic glasses, such as contactless forming-at-a-distance of amorphous components.

Acknowledgements We thank J. Ziegler for a version of SRIM without electronic stopping and V. V. Bulatov and T. Oppelstrup for assistance with uBGL. We thank W. L. Johnson and S. J. Zinkle for insightful discussions. We acknowledge financial support from the National Science Foundation Graduate Research Fellowship, Grant No. 1122374; the DuPont/MIT Alliance Fellowship; and by the DoE Office of Nuclear Energy, Nuclear Energy Enabling Technologies, Reactor Materials program, contract DE-NE0000533. Computations were performed at Lawrence Livermore National Laboratory through the LLNL 5th Institutional Unclassified Grand Challenge Computing Allocation.

References

[1] Schuh CA, Hufnagel TC, Ramamurty U. Mechanical behavior of amorphous alloys. Acta Mater. 2007;55: 4067–4109.
[2] Inoue A, Takeuchi A. Recent development and application products of bulk glassy alloys. Acta Mater. 2011;59: 2243–2267.
[3] Kumar G, Tang HX, Schroers J. Nanomoulding with amorphous metals. Nature. 2009;457:868–872.
[4] Mansur LK. Theory and experimental background on dimensional changes in irradiated alloys. J Nucl Mater. 1994;216:97–123.
[5] Weber WJ, Ewing RC, Angell CA, Arnold GW, Cormack AN, Delaye JM, Griscom DL, Hobbs LW, Navrotsky A, Price DL, Stoneham AM, Weinberg MC. Radiation effects in glasses used for immobilization of high-level waste and plutonium disposition. J Mater Res. 1997;12:1946–1978.
[6] Odette GR, Lucas GE. Embrittlement of nuclear reactor pressure vessels. JOM. 2001;53:18–22.
[7] Kramer EA, Johnson WL, Cline C. The effects of neutron irradiation on a superconducting metallic glass. Appl Phys Lett. 1979;35:815–818.
[8] Gerling R, Schimansky FP, Wagner R. Restoration of the ductility of thermally embrittled amorphous alloys under neutron-irradiation. Acta Metall. 1987;35:1001–1006.
[9] Magagnosc DJ, Ehrbar R, Kumar G, He MR, Schroers J, Gianola DS. Tunable tensile ductility in metallic glasses. Sci Rep. 2013;3:1096.
[10] Hou M-d, Klaumünzer S, Schumacher G. Dimensional changes of metallic glasses during bombardment with fast heavy ions. Phys Rev B. 1990;41:1144–1157.
[11] Nastasi M, Mayer JW, Hirvonen JK. Ion-solid interactions: fundamentals and applications. Cambridge: Cambridge University Press; 1996.
[12] Averback RS, de la Rubia T. Displacement damage in irradiated metals and semiconductors. Solid State Phys. 1998;51:281 402.
[13] Race CP, Mason DR, Finnis MW, Foulkes WMC, Horsfield AP, Sutton AP. The treatment of electronic excitations in atomistic models of radiation damage in metals. Rep Prog Phys. 2010;73:116501.
[14] Demkowicz MJ, Hoagland RG. Simulations of collision cascades in Cu–Nb layered composites using an EAM interatomic potential. Int J Appl Mech. 2009;1: 421–442.
[15] Plimpton S. Fast parallel algorithms for short-range molecular dynamics. J Comp Phys. 1995;117:1–19.
[16] Baumer RE, Demkowicz MJ. Glass transition by gelation in a phase separating binary alloy. Phys Rev Lett. 2013;110:145502.
[17] Ziegler JF, Ziegler MD, Biersack JP. SRIM-the stopping and range of ions in matter. Nucl Instr Meth B. 2010;268:1818–1823.
[18] Ziegler JF, Biersack JP, Ziegler MD. The stopping and range of ions in matter. Morrisville: LuLu Press; 2008. We used SRIM-2008.3. Available from: http://www.srim.org, which was modified for our use by the authors.
[19] Zarkadoula E, Daraszewicz SL, Duffy DM, Seaton MA, Todorov IT, Nordlund K, Dove MT, Trachenko K. The nature of high-energy radiation damage in iron. J Phys Condens Matter. 2013;25:125402.
[20] Nomura K-i, Chen Y-C, Kalia RK, Nakano A, Vashishta P. Defect migration and recombination in nanoindentation of silica glass. Appl Phys Lett. 2011;99:111906.
[21] Zhu H, Averback RS, Nastasi M. Molecular dynamics simulations of a 10 keV cascade in β-NiAl. Philos Mag A. 1995;71:735–758.
[22] Falk ML, Langer JS. Dynamics of viscoplastic deformation in amorphous solids. Phys Rev E. 1998;57: 7192–7205.
[23] Myers M, Fu EG, Myers M, Wang H, Xie G, Wang X, Chu W-K, Shao, L. An experimental and modeling study on the role of damage cascade formation in nanocrystalization of ion-irradiated $Ni_{52.5}Nb_{10}Zr_{15}Ti_{15}Pt_{7.5}$ metallic glass. Scripta Mater. 2010;63:1045–1048.

[24] Shao L, Chen D, Myers M, Wang J, Tilakaratne B, Wijesundera D, Chu W-K, Xie G, Zare A, Lucca DA. Smoothing metallic glasses without introducing crystallization by gas cluster ion beam. Appl Phys Lett. 2013;102:101604.
[25] Calder AF, Bacon DJ, Barashev AV, Osetsky YN. On the origin of large interstitial clusters in displacement cascades. Philos Mag. 2010;90:863–884.
[26] Skirlo SA, Demkowicz MJ. The role of thermal spike compactness in radiation-induced disordering and Frenkel pair production in Ni_3Al. Scripta Mater. 2012;67: 724–727.
[27] Mayr SG. Impact of ion irradiation on the thermal, structural, and mechanical properties of metallic glasses. Phys Rev B. 2005;71:144109.
[28] Xiao Q, Huang L, Shi Y. Suppression of shear banding in amorphous ZrCuAl nanopillars by irradiation. J Appl Phys. 2013;113:083514.
[29] Avchaciov KA, Ritter Y, Djurabekova F, Nordlund K, Albe K. Controlled softening of $Cu_{64}Zr_{36}$ metallic glass by ion irradiation. Appl Phys Lett. 2013;102: 181910.
[30] Graff KF. Wave motion in elastic solids. Oxford: Clarendon Press; 1975.
[31] Stacey WM. Nuclear reactor physics. Weinheim: Wiley; 2007.
[32] Mahmood ST, Mirzadeh S, Farrell K, Pace III JV, Oliver BM. Neutron dosimetry of the HFIR hydraulic facility. ORNL/TM-12831, Oak Ridge National Laboratory, Oak Ridge, Tennessee 37831; February 1995.
[33] Trinkaus H, Ryazanov AI. Viscoelastic model for the plastic flow of amorphous solids under energetic ion bombardment. Phys Rev Lett. 1995;74: 5072–5075.
[34] van Dillen T, Polman A, Onck PR, van der Giessen E. Anisotropic plastic deformation by viscous flow in ion tracks. Phys Rev B. 2005;71:024103.
[35] Chicoine M, Roorda S, Cliche L, Masut RA. Directional effects during ion implantation: lateral mass transport and anisotropic growth. Phys Rev B. 1997;56: 1551–1560.
[36] Gerling R, Wagner R. Density of neutron irradiated and annealed amorphous $Fe_{40}Ni_{40}B_{20}$. Scripta Metall. 1982;16:963–967.

10

Cracking in a Fe–25Mn–3Si–3Al Steel

Xiaoyang Fang[a], Lihui Zhang[b], Wei Liu[c], Kangying Shu[c], Youtong Fang[b], Yuewu Zeng[a], Liang Meng[a] and Jiabin Liu[a,b]*

[a]*Department of Materials Science and Engineering, Zhejiang University, Hangzhou 310027, People's Republic of China;* [b]*College of Electrical Engineering, Zhejiang University, Hangzhou 310027, People's Republic of China;* [c]*Department of Materials Science and Engineering, China Jiliang University, Hangzhou 310018, People's Republic of China*

Fe–25Mn–3Si–3Al steels obtained much attention because of their excellent mechanical properties. However, the fracture behavior of the steels remains unclear. In this study, Fe–25Mn–3Si–3Al steels were *ex situ* tensile strained and tested by a transmission electron microscope to investigate the relationship between crack and twins. The propagation of crack was examined and related to the dislocation slip and twinning. Cracks initiate related to the moving of dislocations and propagate along the slip bands. Twins parallel and inclined to the crack are produced during crack propagating. The inclined twins notably block and blunt the crack, whereas the parallel twins exhibit an insignificant effect.

Keywords: High Manganese Steels, Deformation, Microstructure, Transmission Electron Microscopy, Twinning

Twinning-induced plasticity (TWIP) steels are expected to be used in the automotive industry due to their high strength and high elongation.[1,2] TWIP steels exhibited no ductile-to-brittle transition even at −196 °C and showed ductile behavior over the whole temperature range.[3,4] The specific energy absorption value of the TWIP steels was about 0.5 J/mm^3, whereas the values of the conventional deep drawing steels were only between 0.16 and 0.25 J/mm^3.[5] The deformation twins were considered to be responsible to the excellent mechanical properties.[6] Plenty of nanotwins were observed in the strained microstructure.[7–10] The formation of twins blocked the movement of dislocations effectively. Several models of interaction of twin boundaries with dislocations were proposed to explain the twinning strengthening mechanism.[11,12] Deformation at a high-strain area would be restrained due to the twinning and the deformation would be transferred to a low-strain area. As a result, the whole specimens could deform homogeneously and the necking would be delayed, leading to the large uniform elongation.

Several *in situ* technologies were adopted to deeply investigate the relationship between excellent mechanical properties and microstructure of TWIP steels.

The evolution of lattice strains was studied via *in situ* neutron diffraction during cyclic loading and a pronounced Bauschinger effect resulting from the dislocation pile-ups at the intersection of stacking faults was found.[13]

TWIP steels presented a dimpled and ductile fracture surface under scanning electron microscopy investigation.[14] The evolution of ductile damage was investigated by three-dimensional X-ray tomography *in situ* tensile test. It was found that the ductile damage process of TWIP steels involved intense nucleation of small voids combined with the significant growth of the biggest cavities.[15] Studies of the fatigue behavior of TWIP steels showed that the crack propagation took place mainly transgranularly, resulting in ductile striations and protrusions on fracture surfaces.[16] Fatigue cracks tended to nucleate on grain and twin boundaries besides slip bands.[16] The fatigue crack passed along different paths, such as grain boundaries, twin boundaries and slip bands. However, the reason for the cracks propagating along the slip bands and twin boundaries still remains unclear. The purpose of this study is to investigate the crack initiation and propagation by a transmission electron microscope (TEM).

*Corresponding author. Email: liujiabin@zju.edu.cn

Table 1. Chemical composition (wt.%) of the alloy determined by ICP-MS.

C	Mn	Si	Al	P	S	Fe
0.018	24.3	3.22	2.89	0.004	0.003	Bal.

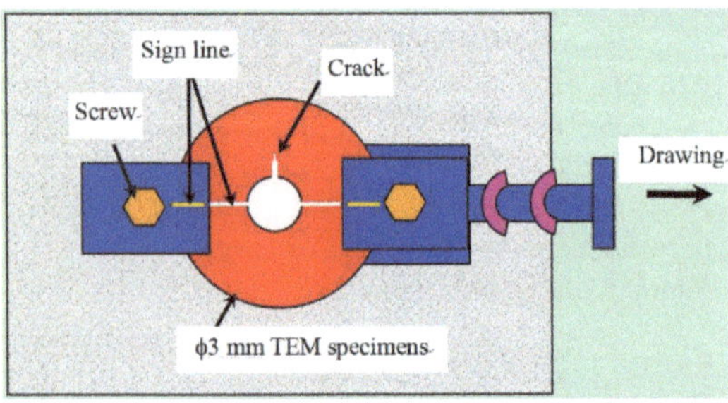

Figure 1. Schematic illustration of the micro-strain holder.

The alloy was melted in a vacuum induction furnace and cast into ingot (200 mm × 200 mm × 400 mm). The chemical composition of the alloy was measured by inductively coupled plasma mass spectrometry (ICP-MS) as given in Table 1. There is a little difference between the measured composition and the designed composition due to the burning loss during the high-temperature melting process. The ingot was homogenized at 1,250 °C for 12 h and then hot-rolled to a thickness of 6 mm at a temperature above 1,000 °C. The hot-rolled plate was cold-rolled at room temperature to a thickness of 2 mm. The cold-rolled sheets were annealed at 1,050 °C for 30 min and then quenched into water.

The microstructure was analyzed by TEM (JEM-2010) operating at 200 kV. The specimens for TEM observation were mechanically thinned to about 100 μm in thickness and then electrolytically thinned by the twin-jet polishing technique using a mixture of 5 vol.% perchloric acid and 95 vol.% alcohol at 20 V. *Ex situ* TEM tensile test was performed on a micro-strain holder as shown in Figure 1. The ϕ3 mm TEM sample was fixed to the holder and tensile strained by drawing the movable arm carefully. The strained sample was transferred into the TEM for observation. Then, the sample was taken out from the TEM and further strained for the next observation. Similar operations were performed to investigate the microstructure evolution with an increasing strain. To repeat the stress condition at each strain, lines were scored

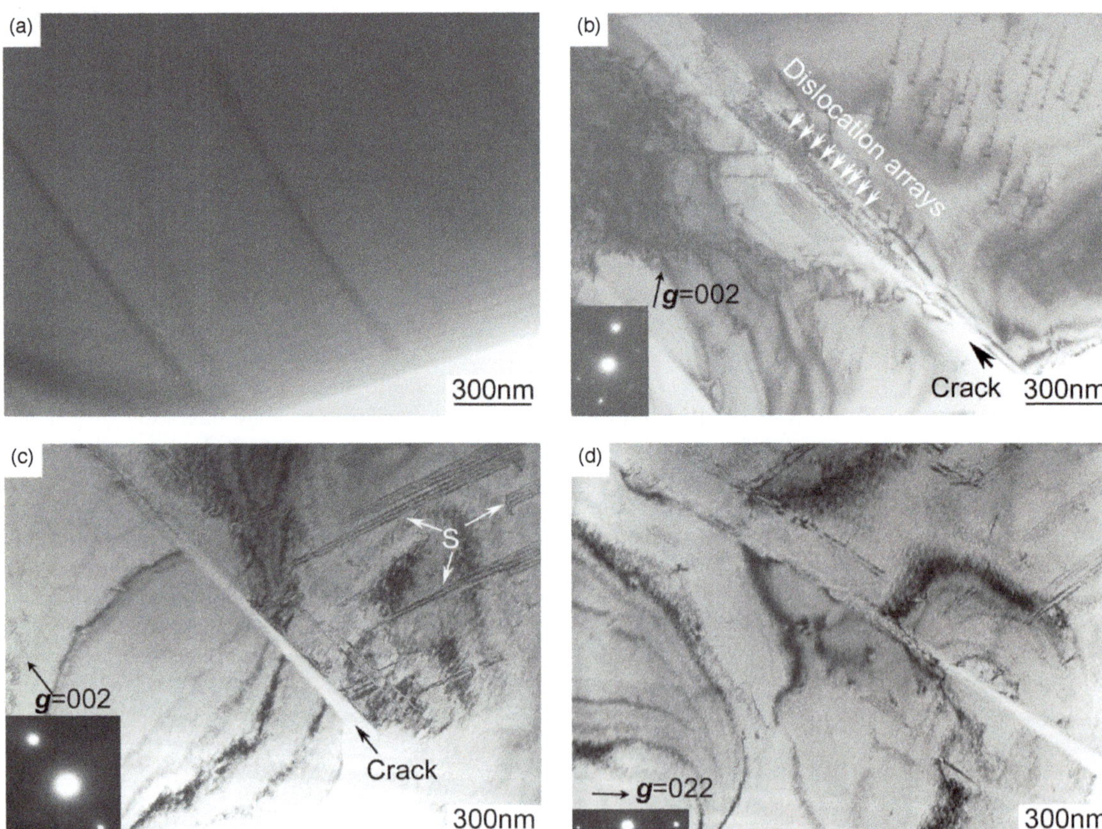

Figure 2. Microstructure of the TEM specimen: (a) before strain, (b) after first tensile strain, (c) after second tensile strain with $g = [002]$ and (d) after second tensile strain with $g = [022]$, insets are the corresponding SAED patterns.

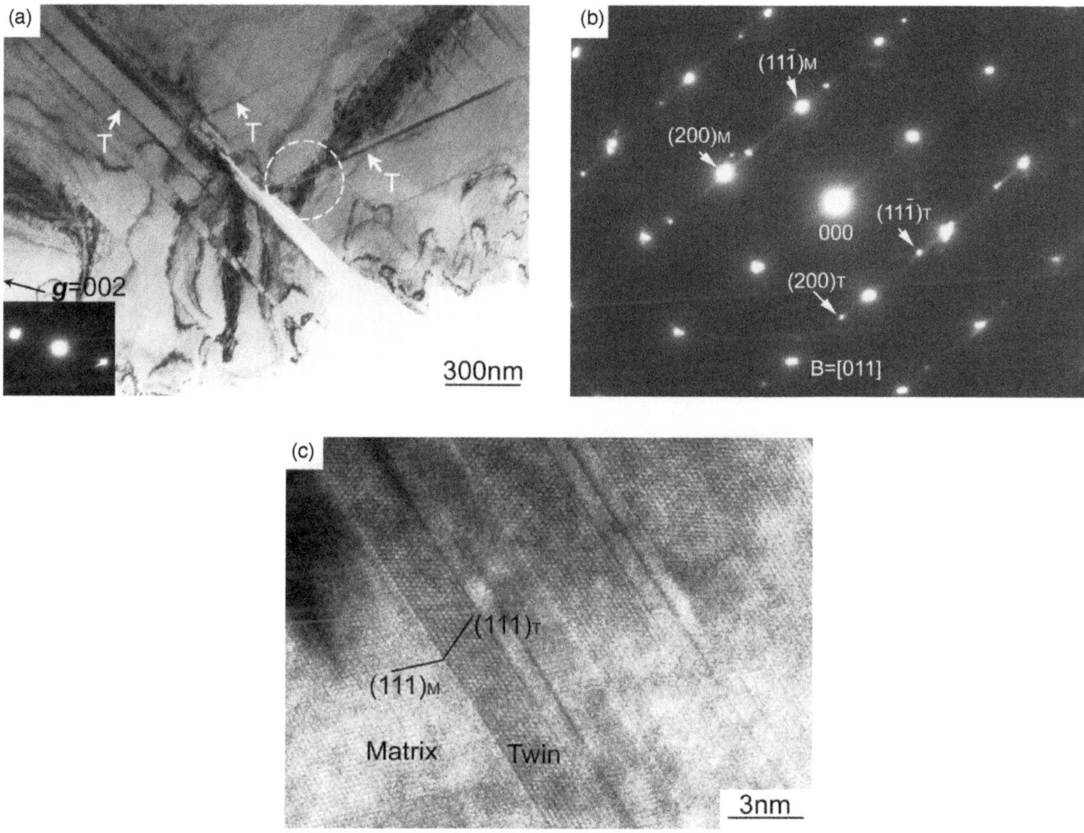

Figure 3. Microstructure of the TEM specimen after third tensile strain at (a) $g = [002]$ (deformation twins are assigned by white arrows), (b) SAED patterns and (c) high-resolution TEM image of the deformation twins corresponding to the white circle in Figure 3(a).

on both the TEM specimen and holder arms to help locate the TEM specimen at the same position on the holder.

There is nearly no dislocation in the specimen before strain as shown in Figure 2(a). After the first tensile strain, a micro-crack is generated from the edge of thin area of TEM specimen (Figure 2(b)). High density of dislocations ahead of the micro-crack is observed. After the second tensile strain, some stacking faults (signed as ‘S’ in Figure 2(c)) nearly inclined to the crack are observed when $g = [002]$, where g is the operating reflection. Most of the stacking faults are invisible except some partial dislocations when $g = [022]$ (Figure 2(d)), which indicates that the stacking faults should be instinct stacking faults.

The microstructure of the TEM specimen after the third tensile strain is shown in Figure 3. The former stacking faults in Figure 2(c) disappear. Some deformation twins (signed as ‘T’ in Figure 3(a)) inclined or parallel to the crack are observed near the crack (Figure 3(a)). The former stacking faults are confirmed to evolve into deformation twins by comparing the microstructure in Figures 2(c) and 3(a) carefully. The deformation twins are further confirmed by the corresponding selected area electron diffraction (SAED) patterns (Figure 3(b)). High-resolution TEM image shows that the twin boundary is flat and the width of the twin is only several nanometers (Figure 3(c)).

The crack in the second and third strained specimens is compared as shown in Figure 4(a). The crack in the second strained specimen is rather sharp. The sharp crack expands obviously at the place where some inclined twins interact with the crack as shown by the red arrow in the third strained specimen. It seems that the inclined twins not only block the crack propagation, but also blunt the crack tip. Similar results that the crack is blunted by the inclined twins are found elsewhere (Figure 4(b)).

Although there are twins parallel to the crack at the head of crack tip (Figure 4(a)), the crack tip still keeps sharp. This result suggests that the twins parallel to the crack have little effect on blocking or blunting the crack. From the investigations of low-temperature fracture of many austenitic steels with high Mn content, transgranular cracking was also observed along $\{111\}$ planes and twin boundaries.[16,17] Those results are consistent with the observation that crack could propagate along the deformation twin boundaries without blunting in this study. Previous studies show extensive evidences

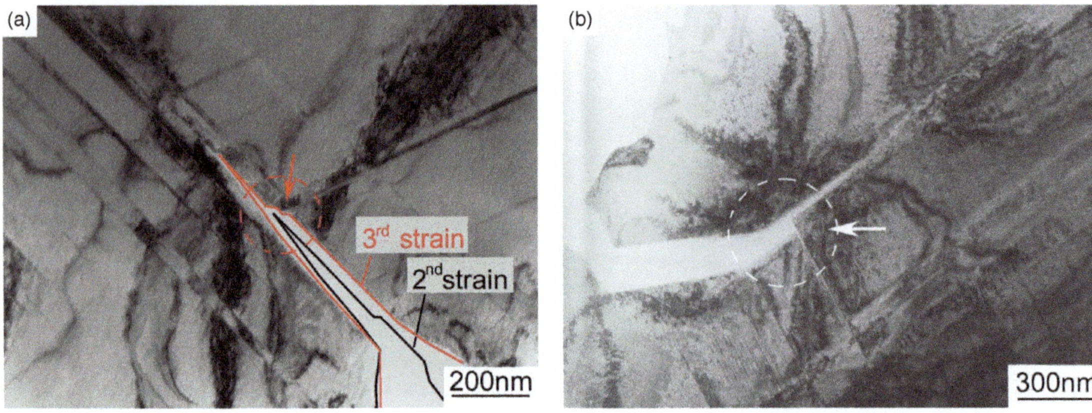

Figure 4. (a) Schematic illustration of the crack propagation during straining and (b) crack interaction with twins.

that twins could serve as favorable sites for the formation of fatigue cracks.[18,19] Twin boundary cracks were the dominating crack events at medium amplitudes in 316 austenitic stainless steels.[19,20] Finite element simulations also confirmed the substantial enhancement of the driving force for the crack formation in a twinned crystal relative to a single crystal.[21] Coherent twin boundaries in face centered cubic (FCC) metals are the ones with the lowest energy among all grain boundaries. In an energetic concept of material damage, such high-coincidence boundaries should be relatively stable and they are generally considered to be a strong barrier to deformation.[22,23] How the crack propagates along the twin boundary easily? This twin boundary crack phenomena has been well explained by Heinz and Neumann.[24] They evaluated the effects of twin boundaries using anisotropic elastic models and concluded that twins induced local stress concentration, which increased the probability of nucleating a fatigue crack at a twin. It has been confirmed that the orientation of twin boundary with the crack played an important role in affecting the crack propagation. An inclined twin boundary orientation is more effective in enhancing the fracture toughness than the parallel orientation.[25] The blunting of crack by the inclined twins could change the propagation way of the crack. However, once the crack turns to parallel with the former inclined twins, it could also propagate along those twin boundaries easily. Although the propagation path of crack is changed and elongated, the crack still keeps sharp along the twin boundary and the driving force for the crack propagation is not released adequately. That may be the reason why the post-uniform elongation is very short in TWIP steels. In other words, although the deformation twins improve the plasticity, they are of little help or even are harmful for the fracture once a crack had originated.

It should be pointed out that the fracture behaviors observed in the thin TEM specimens are not all the same as those in the corresponding bulk materials due to the morphology difference. However, some characterizations of the initiation and propagation of the crack should be hold both in thin specimens and bulk materials.

In conclusion, Fe–25Mn–3Si–3Al TWIP steels were *ex situ* tensile strained and observed using TEM to investigate the initiation and propagation of crack. Cracks initiate related to the moving of dislocations and propagate along the slip bands. Both twins parallel and inclined to the crack are produced during the crack propagation. The twins inclined to the crack notably block and blunt the crack, whereas the parallel twins exhibit no effect on the propagation of the crack.

Acknowledgements The project is financially supported by National Natural Science Foundation of China (No. 11202183) and the National High Technology Research and Development Program (863) of China (No. 2011AA11A101). The authors are indebted to L.N. Wang (Zhejiang Sci-Tech University) for preparing the TEM specimens.

References

[1] Jin JE, Lee YK. Effects of Al on microstructure and tensile properties of C-bearing high Mn TWIP steel. Acta Mater. 2012;60:1680–1688.

[2] Gutierrez-Urrutia I, Raabe D. Grain size effect on strain hardening in twinning-induced plasticity steels. Scripta Mater. 2012;66:992–996.

[3] Grassel O, Kruger L, Frommeyer G, Meyer LW. High strength Fe–Mn–(Al,Si) TRIP/TWIP steels development-properties-application. Int J Plast. 2000;16:1391–1409.

[4] Zhang J, Di H, Mao K, Wang X, Han Z, Ma T. Processing maps for hot deformation of a high-Mn TWIP steel: a comparative study of various criteria based on dynamic materials model. Mater Sci Eng A. 2013;587:110–122.

[5] Frommeyer G, Brux U, Neumann P. Supra-ductile and high-strength manganese-TRIP/TWIP steels for high energy absorption purposes. ISIJ Int. 2003;43:438–446.

[6] Dancette S, Delannay L, Renard K, Melchior MA, Jacques PJ. Crystal plasticity modeling of texture development and hardening in TWIP steels. Acta Mater. 2012;60: 2135–2145.

[7] Liu JB, Liu XH, Liu W, Zeng YW, Shu KY. Microstructure and hardness evolution during isothermal process at 700 °C for Fe–24Mn–0.7Si–1.0Al TWIP steel. Mater Charact. 2010;61:1356–1358.
[8] Allain S, Chateau J-P, Dahmoun D, Bouaziz O. Modeling of mechanical twinning in a high manganese content austenitic steel. Mater Sci Eng A. 2004;387–389:272–276.
[9] Vercammen S, Blanpain B, Cooman BCD, Wollants P. Cold rolling behaviour of an austenitic Fe–30Mn–3Al–3Si TWIP-steel: the importance of deformation twinning. Acta Mater. 2004;52:2005–2012.
[10] Shen YF, Wang YD, Liu XP, Sun X, Lin Peng R, Zhang SY, Zuo L, Liaw PK. Deformation mechanisms of a 20 Mn TWIP steel investigated by in situ neutron diffraction and TEM. Acta Mater. 2013;61:6093–6106.
[11] Zhu YT, Liao XZ, Wu XL. Deformation twinning in nanocrystalline materials. Prog Mater Sci. 2012;57:1–62.
[12] Zhu YT, Wu XL, Liao XZ, Narayan J, Kecskes LJ, Mathaudhu SN. Dislocation twin interactions in nanocrystalline fcc metals. Acta Mater. 2011;59:812–821.
[13] Saleh AA, Pereloma EV, Clausen B, Brown DW, Tomé CN, Gazder AA. On the evolution and modelling of lattice strains during the cyclic loading of TWIP steel. Acta Mater. 2013;61:5247–5262.
[14] Bracke L, Mertens G, Penning J, Cooman BCD, Liebeherr M, Akdut N. Influence of phase transformations on the mechanical properties of high-strength austenitic Fe–Mn–Cr steel. Metall Mater Trans A. 2006;37:307–317.
[15] Fabrègue D, Landron C, Bouaziz O, Maire E. Damage evolution in TWIP and standard austenitic steel by means of 3D X ray tomography. Mater Sci Eng A. 2013;579:92–98.
[16] Hamada AS, Karjalainen LP, Puustinen J. Fatigue behavior of high-Mn TWIP steels. Mater Sci Eng A. 2009;517:68–77.
[17] Liu SC, Hashida T, Takahashi H, Kuwano H, Hamaguchi Y. A study on fractography in the low-temperature brittle fracture of an 18Cr–18Mn–0.7N austenitic steel. Metall Mater Trans A. 1998;29:791–798.
[18] Miao J, Pollock TM, Wayne JJ. Microstructural extremes and the transition from fatigue crack initiation to small crack growth in a polycrystalline nickel-base superalloy. Acta Mater. 2012;60(6–7):2840–2854.
[19] Blochwitz C, Tirschler W. Twin boundaries as crack nucleation sites. Cryst Res Tech. 2005;40(1–2):32–41.
[20] Blochwitz C, Tirschler W. Influence of the crystalline texture on the fatigue damage. Proceedings of the Eighth International Fatigue Congress; 2002; Stockholm. Vol. 3; p. 1625–1632.
[21] Castelluccio GM, McDowell DL. Effect of annealing twins on crack initiation under high cycle fatigue conditions. J Mater Sci. 2013;48(6):2376–2387.
[22] Lu K, Lu L, Suresh S. Strengthening materials by engineering coherent internal boundaries at the nanoscale. Science. 2009;324:349–352.
[23] Jin JE, Lee YK. Strain hardening behavior of a Fe–18Mn–0.6C–1.5Al TWIP steel. Mater Sci Eng A. 2009;527:157–161.
[24] Heinz A, Neumann P. Crack initiation during high cycle fatigue of an austenitic steel. Acta Metall Mater. 1990;38(10):19331940.
[25] Zhou HF, Qu SX, Yang W. Toughening by nano-scaled twin boundaries in nanocrystals. Model Simul Mater Sci Eng. 2010;18:065002.

11

A Novel Low-Density, High-Hardness, High-entropy Alloy with Close-packed Single-phase Nanocrystalline Structures

Khaled M. Youssef[a], Alexander J. Zaddach[b], Changning Niu[b], Douglas L. Irving[b] and Carl C. Koch[b]*

[a]*Department of Materials Science and Technology, Qatar University, PO Box 2713, Doha, Qatar;* [b]*Department of Materials Science and Engineering, North Carolina State University, Raleigh, NC 27695-7907, USA*

A low-density, nanocrystalline high-entropy alloy, $Al_{20}Li_{20}Mg_{10}Sc_{20}Ti_{30}$ was produced by mechanical alloying. It formed a single-phase fcc structure during ball milling and transformed to single-phase hcp upon annealing. The alloy has an estimated strength-to-weight ratio that is significantly higher than other nanocrystalline alloys and is comparable to ceramics. High hardness is retained after annealing.

Keywords: Mechanical Alloying, X-ray Diffraction, Hardness, Nanocrystalline, High-entropy Alloys

High-entropy alloys are a new class of multi-component alloy systems in which the design of the alloys is based not on adding solutes to a single 'base' element, but rather on choosing elements that will form solid solutions when mixed at near-equiatomic concentrations. This behavior has been attributed to the large configurational entropy when five of more elements at near-equiatomic ratios are mixed together.[1,2] These high-entropy alloys (HEAs) have demonstrated the potential for superior properties in many cases. A fundamental understanding of the mechanisms of phase stability of HEAs is still a topic of active research. Guo and Liu demonstrated that compositions likely to form solid solution HEAs have enthalpies of mixing between -22 and 7 kJ/mol and very small differences in atomic radius. The small enthalpy of mixing allows the configurational entropy to dominate the free energy and it has been proposed that the system behaves, in this case, more like an ideal solution.[3] Otto et al. recently published a systematic investigation on the effect of entropy on phase stability of HEAs.[4] The authors demonstrated that high configurational entropy is not the only sufficient criterion to predict whether an equiatomic multi-component alloy will form as a single-phase solid solution. In their study, they found only NiFeCrCoMn to be a single-phase fcc solid solution and HEA. These findings are consistent with the work of Guo and Liu [3] as all the enthalpies of mixing in the respective binary alloys are relatively small and the atoms are similar in size. The authors then replaced individual elements in an equiatomic NiFeCrCoMn alloy one at a time with elements that have the same room temperature crystal structure, similar atomic size, and similar electronegativity as compared with the elements being replaced. All other substitutions to the NiFeCrCoMn alloy led to a multi-phase alloy. Although the replacement elements were 'similar' according to Hume-Rothery rules, the authors found that the substituted elements had a stronger propensity to form secondary phases or intermetallics in their respective binary alloys. This led the system to act less like an ideal solution and resulted in the formation of a multi-phase alloy with a reduced overall entropy.[4] There has also been interest in finding descriptors to predict the likely crystallographic lattice (fcc or bcc) that the solid solution will form upon. One descriptor that has been reasonably successful in predicting the lattice structure is the average valence concentration (or average periodic table group number) (VEC).[5] Fcc phases were determined to be more stable at higher values of VEC (≥ 8) while bcc phases were stable at lower VEC (< 6.87). On the boundary between fcc and bcc (VEC $= 8$), the systems were found to be predominantly

*Corresponding author. Email: cckoch@ncsu.edu

fcc but in some situations the bcc phase has been seen in minute amounts.

A quantitative prediction of whether multi-component systems would form into solid solutions (HEAs), intermetallic containing alloys, or metallic glasses has been presented by Yang and Zhang.[6] They proposed parameters Ω and δ, where Ω is defined as a parameter of the entropy of mixing times the average melting temperature of the elements divided by the enthalpy of mixing. The parameter δ is the mean square deviation of the atomic size of the elements. Their model shows that the high-entropy stabilized solid solution alloys occur for values of $\Omega \geq 1.1$ and $\delta \leq 6.6\%$. These quantitative predictions are mostly consistent with the empirical data in the literature albeit with a few exceptions.[6]

While there is continuing high interest in development of alloys with low densities along with high strength for energy-saving applications such as in transportation and energy, to date there have only been a few reports of studies of low-density HEAs. For purposes of our discussion we define low density as less than 3 g cm^{-3}. Juan et al. [7] reported an $Al_{20}Be_{20}Fe_{10}Si_{15}Ti_{35}$ alloy which had an as-cast microstructure composed of one major and two minor phases. It had a density of 3.91 g cm^{-3}. Li et al. [8,9] studied Mg-containing HEAs. These authors prepared $Mg_x(MnAlZnCu)_{100-x}$ alloys by induction melting and casting.[8] The alloys had multi-phase structures, composed mainly of hcp phases and Al–Mn icosahedral quasicrystalline phases. Depending on the composition, the alloys had densities of 2.20–4.29 g cm^{-3}. They exhibited good compressive strengths (400–500 MPa) at room temperature. However, ductilities were low, 3–5% in compression. Similar structures and mechanical properties were subsequently observed in the equiatomic system cast and cooled in air, water, or brine solutions.[9] Chen et al. used mechanical alloying to prepare BeCoMgTi and BeCoMgTiZn, but the alloys formed only an amorphous phase.[10] To our knowledge, no single-phase low-density high-entropy alloy (LDHEA) has been reported.

In this paper, we report our results on the processing, structure, and mechanical hardness of a low-density HEA.

Phase formation for HEAs can be predicted by calculating the parameters Ω and δ, proposed by Yang et al.[6] Ω represents the competition between enthalpy and entropy, given by

$$\Omega = \frac{T_m \Delta S_{mix}}{|\Delta H_{mix}|}, \quad (1)$$

where $T_m = \sum c_i (T_m)_i$ is the melting temperature of the alloy, $\Delta S_{mix} = -R \sum c_i \ln c_i$ is the entropy of mixing, and $\Delta H_{mix} = \sum_{i<j} 4 c_i c_j \Delta H_{ij}^{mix}$ is the enthalpy of mixing. δ represents the atomic size difference, given by

$$\delta = \sqrt{\sum c_i \left(\frac{1 - r_i}{\bar{r}}\right)^2}. \quad (2)$$

According to ref. [6], $\Omega \geq 1.1$ and $\delta \leq 6.6\%$ should be expected as the criteria for forming high-entropy stabilized solid solution phases. Using data in Table 1,[11] the two parameters of our alloy $Al_{20}Li_{20}Mg_{10}Sc_{20}Ti_{30}$ are $\Omega = 42.6$ and $\delta = 5.2\%$. Therefore, it is expected to be a solid solution.

Gao et al. [12] produced low-density multi-component alloys. $Sr_{20}Ca_{20}Yb_{20}Mg_{20}Zn_{20}$, $Sr_{20}Ca_{20}Yb_{20}Mg_{20}Zn_{10}Cu_{10}$, and $Sr_{20}Ca_{20}Yb_{20}(Li_{0.55}Mg_{.45})_{20}Zn_{20}$. However, the calculated values for Ω and δ fall out of the range [6] that predicts solid solution phases. Indeed, these alloys were all found to be metallic glasses.

Due to the high vapor pressures of several of the elements (Li, Mg), we used mechanical alloying to prepare the alloys instead of melting and casting. The elemental components of better than 99% purity were obtained from Alfa Aesar. In the case of Sc, the as-received pellets were ball milled at liquid nitrogen temperature in order to obtain the starting powder. The powders for the alloy were loaded into a stainless steel vial with stainless steel balls in a high-purity argon atmosphere glove box (oxygen level < 1 ppm). A 20:1 ball-to-powder weight ratio was used. The milling process was performed in a modified SPEX 8000 mixer mill cooled by liquid nitrogen for 2 h, followed by milling at room temperature for 14 h. Liquid nitrogen milling was carried out in order to avoid welding of the powders at the beginning of the mechanical milling process.

The milled powder samples were characterized by X-ray diffraction (XRD) using a Rigaku diffractometer

Table 1. Data from literature to calculate Ω and δ.[2,11]

		Al	Li	Mg	Sc	Ti
r (nm)		0.14317	0.15194	0.16013	0.16410	0.14615
T_m (K)		933	454	922	1814	1943
ΔH^{mix} (kJ mol^{-1})	Li	−4				
	Mg	−2	0			
	Sc	−38	12	−3		
	Ti	−30	34	16	8	

with Cu Kα radiation. Analysis of the crystal structures of the contaminated alloys was done using the PM2 K Whole Powder Pattern Modeling software.[13] Samples were compacted into 6.25 mm diameter disks 3 mm thick under 2 GPa pressure. Microhardness of the milled powder, compacted to full density at room temperature, was measured using a Buehler Micromet microhardness tester with a Vickers indenter at 50 g load and a loading time of 15 s. Under these conditions, the indentation depth is approximately 5.9 μm.

A transmission electron microscopy (TEM) specimen of the as-milled powder was prepared by compacting the powder into a 3 mm diameter disk. Thinning was accomplished using a twin-jet electropolisher with a solution of perchloric acid and methanol. TEM analysis was conducted using a JEOL 2000FX TEM at an accelerating voltage of 200 kV.

The density functional theory (DFT) calculations were performed by the package of the exact muffin-tin orbital method combined with coherent potential approximation or EMTO-CPA.[14] This computational method has been widely used for solid solution systems, for example, stainless steels [15] and NiFeCrCoMn.[16]

The results of our mechanical alloying of the selected alloy, $Al_{20}Li_{20}Mg_{10}Sc_{20}Ti_{30}$, must be divided into two materials—one which contained mainly the five components intentionally added, and another, which besides these components had significant impurity content of nitrogen and oxygen.

The material prepared by mechanical alloying which did not have the high N, O impurity levels was observed to have a single-phase fcc crystal structure in the as-milled condition, as illustrated in Figure 1. It had a nanocrystalline grain size estimated by the Scherrer formula to be about 12 nm. The lattice parameter of this sample determined by the method of Cohen was found to be 0.4323 nm. The mechanical hardness of this sample was very high 5.8 GPa. After annealing this sample at 500°C for 1 h, the crystal structure changed as shown

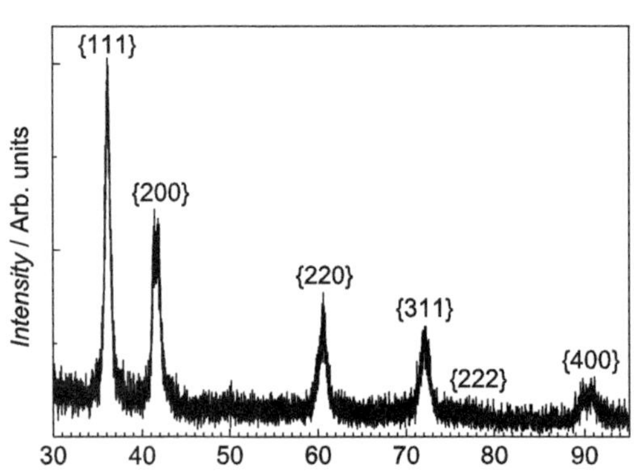

Figure 1. XRD pattern of as-milled alloy.

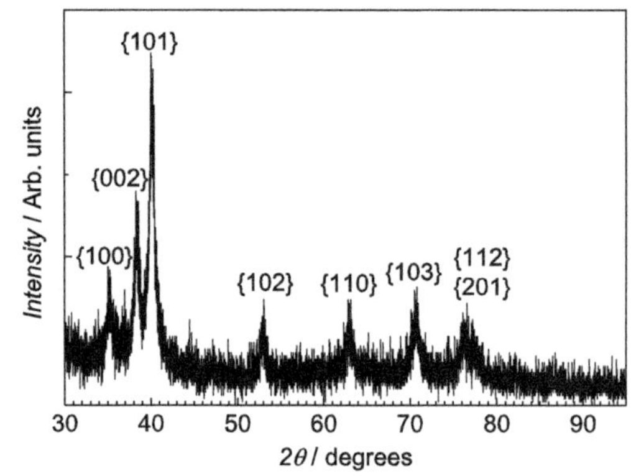

Figure 2. XRD pattern of uncontaminated alloy after annealing at 500°C.

in Figure 2. The new structure was indexed as *hcp* with *c/a* ratio of 1.588. The grain size, again estimated by the Scherrer method, was 26 nm and the mechanical hardness dropped to 4.9 GPa. TEM measurements of the as-milled material, shown in Figure 3, generally agree with the XRD results, giving an average grain size of 20 nm and showing no second phases in the diffraction pattern.

The material with the higher N and O concentrations (0.4 at% N, 1.39 at.% O as determined by chemical analysis) apparently obtained this contamination from a batch of Sc powder which was contaminated during the process of converting the as-received pellets into powder by cryomilling. Subsequent batches did not contain these impurities and were used in the sample described above. The contaminated sample exhibited an as-milled structure essentially identical to that of the un-contaminated sample, that is, single-phase fcc with a grain size of about 12 nm. However, it had a slightly higher mechanical hardness of 6.1 GPa. After annealing at 500°C and 800°C, the structure remained similar to fcc, but some splitting of the diffraction peaks was observed in the XRD results, suggesting either a slight distortion from a perfect cubic lattice or separation into two similar phases with a lattice parameter difference $< 1\%$. Annealing at 500°C decreased the hardness to only 5.9 GPa, while annealing at 800°C resulted in a hardness of 5.75 GPa, consistent with a nanocrystalline grain size. Because of the diffraction peak splitting, the grain size cannot be accurately measured using the Scherrer method, but the broadness of the peaks indicated that the samples are still nanocrystalline after the annealing treatments. The crystal structures, grain size, lattice parameters, and mechanical hardness for these samples are summarized in Table 2.

It has been suggested that one needs very low heats of mixing, approaching ideal behavior, to form solid solutions in these multi-component systems. So it is of high interest that this five component system was found

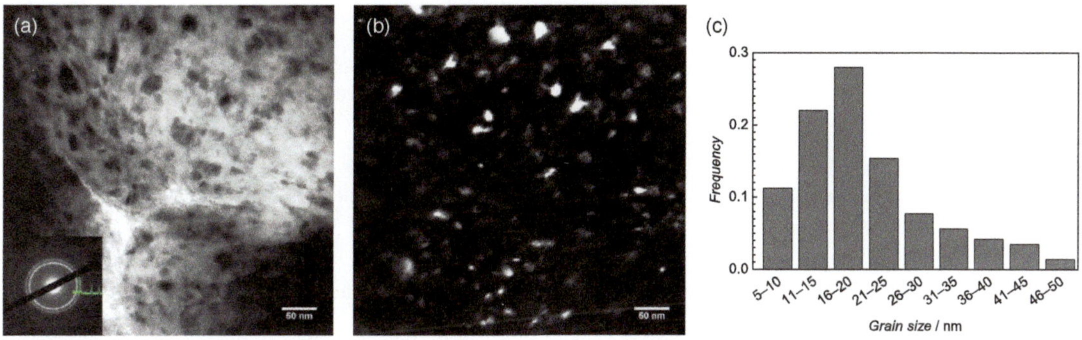

Figure 3. (a) Bright field TEM image and diffraction pattern, (b) dark field image, and (c) grain size distribution of as-milled uncontaminated material.

Table 2. Properties of alloys prepared.

Alloy	Annealing temp. (°C)	Crystal structure	X-ray grain size (nm)	Hardness (GPa)
$Al_{20}Li_{20}Mg_{10}Sc_{20}Ti_{30}$	as milled	fcc	12	5.8
	500	hcp	26	4.9
$Al_{20}Li_{20}Mg_{10}Sc_{20}Ti_{30}$ w/ N, O	as milled	fcc	12	6.1
	500	fcc[a]	nc[a]	5.9
	800	fcc[a]	nc[a]	5.75

[a]The N, O-containing alloys exhibited slight phase separation or distortion upon annealing resulting in a structure similar to fcc. Because the diffraction peaks are split, measurement of the grain size by line profile analysis is inaccurate, but the broadness of the peaks indicates a nanocrystalline structure.

to exhibit a single-phase fcc solid solution in spite of the fact that many of the binary systems in this group exhibit immiscibility (high positive heats of mixing) or form many intermediate phases (high negative heats of mixing). Comparisons of the enthalpy of formation of the solid solution fcc and hcp crystal structures as calculated by DFT show that the hcp phase is the energetically preferred phase, with an enthalpy of formation 0.977 kJ mol^{-1} smaller than the fcc phase.

The sample containing oxygen and nitrogen likely did not transform to hcp upon annealing as the hcp cell contains smaller octahedral interstitial sites. The amount of space for an interstitial atom can be approximated as the distance from the center of the interstitial site to the

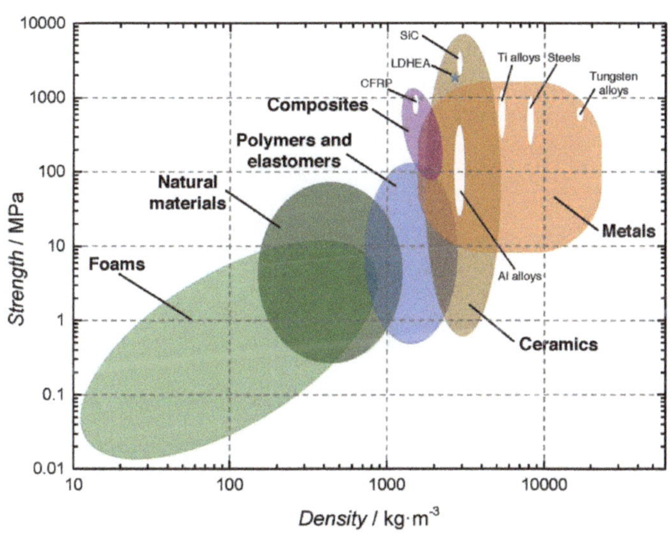

Figure 4. Ashby plot of strength vs. density for engineering materials. (Yield strength for metals and polymers, tear strength for elastomers, compressive strength for ceramics,[19] and tensile strength for composites.) Reproduced with permission from Elsevier 2010.[20]

nearest atomic position minus the average atomic radius of the alloy. Using the lattice parameters measured by XRD, for the fcc octahedral site this distance is 64.4 pm, while for the hcp octahedral site it is 54.7 pm, 15% smaller.

The high mechanical hardness of these alloys can partly be explained by their nanocrystalline grain sizes. However, their hardness values exceed nanocrystalline Al base alloys, for example, by factors of 2–3 times.[17] Because the equilibrium phase is hcp, this suggests that the stacking fault energy of the fcc phase is very low, which may also contribute to the high strength.[18] A comparison of the strength (estimated as hardness/3) and density of these alloys with other structural materials is given in Figure 4. It is clear that these materials are comparable in strength/density values to ceramic materials such as SiC. However, our metallic alloys should exhibit much higher toughness and ductility, which is indicated from the tendency of the powder to weld during milling, the near full density of the compacted powder at room temperature, and the absence of micro cracks around the indents during microhardness indentation.

A low-density (2.67 g cm^{-3}) high-entropy alloy with composition $Al_{20}Li_{20}Mg_{10}Sc_{20}Ti_{30}$ has been prepared by mechanical alloying of elemental powders. The as-milled structure is single-phase fcc, with a nanocrystalline grain size of 12 nm, and a mechanical hardness of 5.9 GPa. The sample without O, N contamination transforms to the hcp structure on annealing at 500°C. The sample containing O, N does not transform, but rather exhibits either a slight phase separation or unit cell distortion upon annealing at 800°C. Calculations of the energies of the competing structures are consistent with the experimental observations. This material exhibits a combination of hardness (strength) and low density that is not equaled by any other metallic material.

These results need to be expanded to a study of this alloy system with various compositions, preparation by processes closer to equilibrium than mechanical alloying, and a detailed study of the mechanical behavior of these unique metallic materials.

Acknowledgements The authors wish to acknowledge the partial support of this research by the National Science Foundation from the Metals and Metallic Nanostructures program under grant DMR-1104930.

Disclosure statement No potential conflict of interest was reported by the author(s).

References

[1] Yeh J-W, Chen S-K, Lin S-J, Gan J-Y, Chin T-S, Shun T-T, Tsau C-H, Chang S-Y. Nanostructured high-entropy alloys with multiple principal elements: novel alloy design concepts and outcomes. Adv Eng Mater. 2004;6:299–303.
[2] Zhang Y, Zuo TT, Tang Z, Gao MC, Dahmen KA, Liaw PK, Lu ZP. Microstructures and properties of high-entropy alloys. Prog Mater Sci. 2014;61:1–93.
[3] Guo S, Liu CT. Phase stability in high entropy alloys: formation of solid-solution phase or amorphous phase. Prog Nat Sci Mater Int. 2011;21:433–446.
[4] Otto F, Yang Y, Bei H, George EP. Relative effects of enthalpy and entropy on the phase stability of equiatomic high-entropy alloys. Acta Mater. 2013;61:2628–2638.
[5] Guo S, Ng C, Lu J, Liu C. Effect of valence electron concentration on stability of fcc or bcc phase in high entropy alloys. J Appl Phys. 2011;109:103505.
[6] Yang X, Zhang Y. Prediction of high-entropy stabilized solid-solution in multi-component alloys. Mater Chem Phys. 2012;132:233–238.
[7] Juan C-C, Yeh J-W, Chin T-S. A novel light high-entropy alloy Al20Be20Fe10Si15Ti35. Paper presented at: E-MRS Fall Meeting, Symposium I; 2009 Sept 14–18; Warsaw, Poland.
[8] Li R, Gao JC, Fan K. Study to microstructure and mechanical properties of Mg containing high entropy alloys. Mater Sci Forum. 2010;650:265–271.
[9] Li R, Gao JC, Fan K. Microstructure and mechanical properties of MgMnAlZnCu high entropy alloy cooling in three conditions. Mater Sci Forum. 2011;686: 235–241.
[10] Chen Y-L, Tsai C-W, Juan C-C, Chuang M-H, Yeh J-W, Chin T-S, Chen S-K. Amorphization of equimolar alloys with HCP elements during mechanical alloying. J Alloys Compd. 2010;506:210–215.
[11] Takeuchi A, Inoue A. Classification of bulk metallic glasses by atomic size difference, heat of mixing and period of constituent elements and its application to characterization of the main alloying element. Mater Trans. 2005;46:2817–2829.
[12] Gao XQ, Zhao K, Ke HB, Ding DW, Wang WH, Bai HY. High mixing entropy bulk metallic glasses. J Non Cryst Solids. 2011;357:3557–3560.
[13] Leoni M, Confente T, Scardi P. PM2K: a flexible program implementing whole powder pattern modelling. Zeitschrift für Kristallographie Suppl. 2006;23:249–254.
[14] Vitos L. Computational quantum mechanics for materials engineers: the EMTO method and applications. London: Springer-Verlag; 2007.
[15] Vitos L, Korzhavyi PA, Johansson B. Stainless steel optimization from quantum mechanical calculations. Nat Mater. 2003;2:25–28.
[16] Zaddach AJ, Niu C, Koch CC, Irving DL. Mechanical properties and stacking fault energies of NiFeCrCoMn high-entropy alloy. JOM. 2013;65:1780–1789.
[17] Youssef KM, Scattergood RO, Murty KL, Koch CC. Nanocrystalline Al–Mg alloy with ultrahigh strength and good ductility. Scr Mater. 2006;54:251–256.
[18] Bahmanpour H, Kauffmann A, Khoshkhoo MS, Youssef KM, Mula S, Freudenberger J, Eckert J, Scattergood RO, Koch CC. Effect of stacking fault energy on deformation behavior of cryo-rolled copper and copper alloys. Mater Sci Eng A Struct Mater. 2011;529:230–236.
[19] Budinski KG, Budinski MK. Engineering materials: properties and selection. 8th ed. Upper Saddle River, NJ: Pearson Prentice Hall; 2005.
[20] Ashby MF. Materials selection in mechanical design. 4th ed. Saint Louis, MO: Elsevier Science & Technology; 2010.

12

Size-Induced Strengthening in Nanostructured Mg Alloy Micropillars

Y.W. Chang, M. Pozuelo* and J.M. Yang

Department of Materials Science and Engineering, University of California Los Angeles, Los Angeles, CA 90095, USA

Size effects on the compressive strength of nanostructured Mg-micropillars were investigated. Mg–10Al alloy micropillars with diameters ranging from 1.5 to 8 μm were prepared by focused-ion-beam and tested under micro-compression. A significant improvement in strength was found by reducing the pillar diameter <3.5 μm. The deformation mechanisms of the compressed pillars were characterized using transmission electron microscopy. We attribute the size-induced strengthening to a less number of dislocation sources along with a higher activity of non-basal deformation mechanisms.

Keywords: Nanostructured Mg Micropillars, Micro-compression Tests, Size Effect, Transmission Electron Microscopy

Studies on size-scale effects i.e. grain size and specimen dimension have attracted considerable attention in recent years because of their influence on the mechanical behavior of materials.[1] In general, it is well accepted that the strength of metals can be improved through refinement of the grain size by the Hall–Petch mechanism. Interestingly, it has also been demonstrated that the strength of the material can be improved by reducing the specimen dimensions, as first reported by Uchic et al. [2] in Ni-based superalloys and by Greer et al. [3] in single crystal Au.

Nowadays, the optimization of the mechanical properties of materials at the micron and submicron scales is crucial for the design of high-performance micro-electro-mechanical devices. Consequently, researchers have focused on investigating both grain size and specimen size effects on the mechanical behavior of micron/submicron-pillars of polycrystalline materials such as Ni,[4,5] Ag,[6] TiAl,[7] Cu,[8] and Pt.[9]

In the case of Mg, many efforts have been dedicated to study the compressive behavior of single crystal Mg micron/submicron-pillars.[10–15] These studies have shown that the compressive strength of single crystal Mg-micropillars mainly depends on the initial dislocation density [11] and the orientation of the basal planes with respect to the loading direction (LD).[10–12] Byer and Ramesh [11] reported a size effect on the yield strength by decreasing the initial dislocation density. Lilleodden [12] observed a remarkable increase in compressive strength along the c-axis for single-crystal Mg-micropillars with diameters <10 μm. It is well known that when the c-axis is nearly parallel or perpendicular to the LD, alternative deformation mechanisms such as non-basal slip and/or twinning can be activated since the Schmid factor for basal plane sliding is very small. For instance, a significant hardening was found on single-crystal Mg-micropillars loaded in compression along the c-axis by the activation of multiple slip systems on pyramidal planes.[10,12] However, it is interesting to note that no sign of twinning is found in those works. It appears that a size effect is also associated with the deformation by twinning. As reported by Yu et al.,[15] the deformation by twinning becomes unfavorable when the specimen size decreases. Ye et al. [13] demonstrated a strong size effect associated with basal plane sliding and extension twinning in single-crystal Mg nano- and micropillars. In fact, it has recently been demonstrated by Prasad et al. [14] that the required stress to nucleate twins in single crystal Mg-micropillars is higher than that in single crystal Mg bulk samples.

While many efforts have been devoted to study the size effect on single-crystal Mg micropillars, the effect of specimen size on polycrystalline Mg micropillars has not been reported yet. In this work, we report the influence of specimen size (diameter) on the compressive strength of nanostructured Mg–10Al micropillars with a

*Corresponding author. Email: pozuelo@ucla.edu

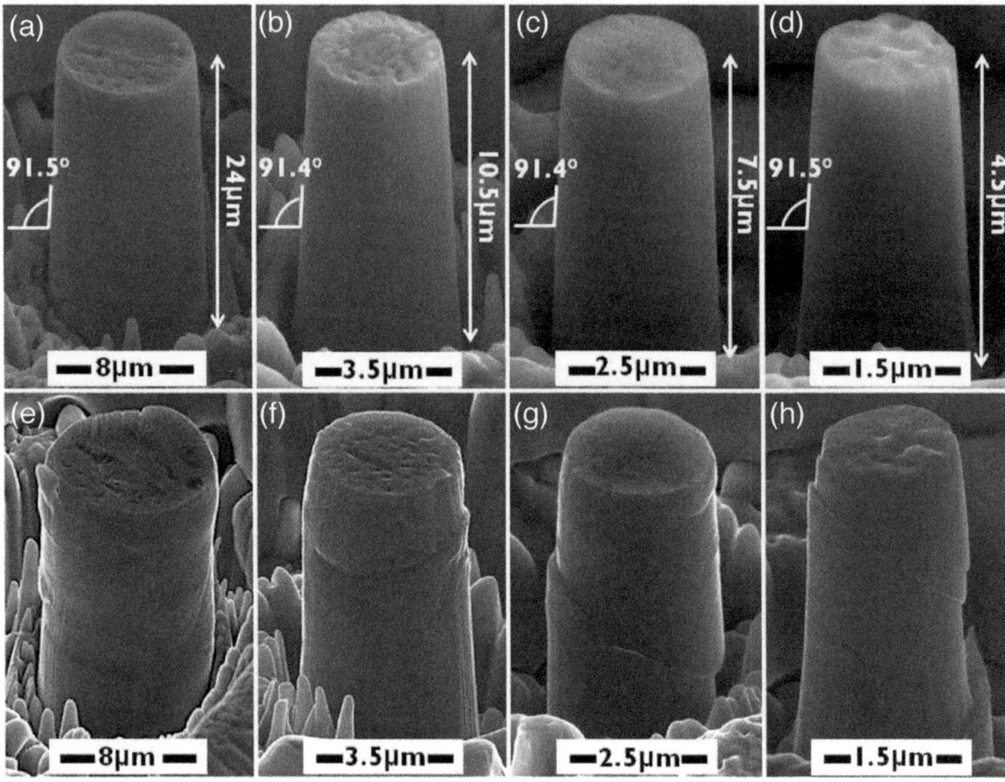

Figure 1. SEM images of nanostructured Mg–10Al micropillars with different diameters (a)–(d) before and (e)–(h) after compression (2% of compressive strain).

bimodal grain size distribution. Using transmission electron microscopy (TEM) we investigate the underlying deformation mechanisms.

Nanocrystalline $Mg_{90}Al_{10}$ (wt.%) powders were prepared by cryomilling for 8 h under liquid nitrogen atmosphere in a Union Process, Szegvari mill at California Nanotechnologies. Cryomilled powders were consolidated using a spark plasma sintering (SPS) system (Syntex Inc., Dr. Sinter Lab TM, model SPS-515S). The sintering process was carried out in vacuum for 5 min at 400°C and under uniaxial pressure of 100 MPa. More details about the material processing can be found in Refs. [16,17] Mg–10Al micropillars with diameters ranging from 1.5 to 8 μm and aspect ratio of 1:3 were fabricated by focused ion beam (FIB). A FEI Nova600 Nanolab DualBeam FIB-scanning electron microscope (SEM) operated at 30 kV was used for this purpose. A series of concentric annular milling patterns with different currents were applied. In order to tailor the pillars into the desired shape and minimize the tapering,[18,19] a low beam current (<1 nA) was used as final milling step. Uniaxial micro-compression tests were performed using a MTS Nanoindenter XP. A diamond indenter with a flat-end tip of 10 μm in diameter was used. The strain rate was controlled in the range of 3×10^{-3} and 5×10^{-4} (s^{-1}). TEM samples of the compressed pillars were prepared by FIB and examined using a FEI-Titan TEM operating at 300 kV.

Four characteristic Mg–10Al micropillars before and after compression are shown in Figure 1. Figure 1(a)–(d) show 52° tilted-view SEM images of those micropillars with top diameters of 8, 3.5, 2.5 and 1.5 μm, respectively, before compression. It is evident to observe that all pillars are slightly tapered as a result of the annular cutting method by FIB. The sidewall taper angle and pillar length are indicated in all images. We measure taper angles ≤1.5°, which are smaller than the values reported in the literature (2–5°) for typical micropillars machined by FIB.[20,21] Figure 1(e)–(h) show 52° tilted-view SEM images of the compressed pillars up to 2% of strain. Shear bands are apparent on the pillars surface, especially for pillars with diameters ≤3.5 μm. In contrast, 8 μm pillars plastically deform showing a barreling shape similar to the bulk samples under compression. This observation suggests that a transition from homogeneous-like to shear-band deformation occurs by reducing the pillar diameter.

The results of the micro-compression tests are shown in Figure 2 and presented in Table 1. At least five samples were tested for each pillar diameter as shown in Table 1. The average yield strength and the standard deviation have also been included. In order to avoid any

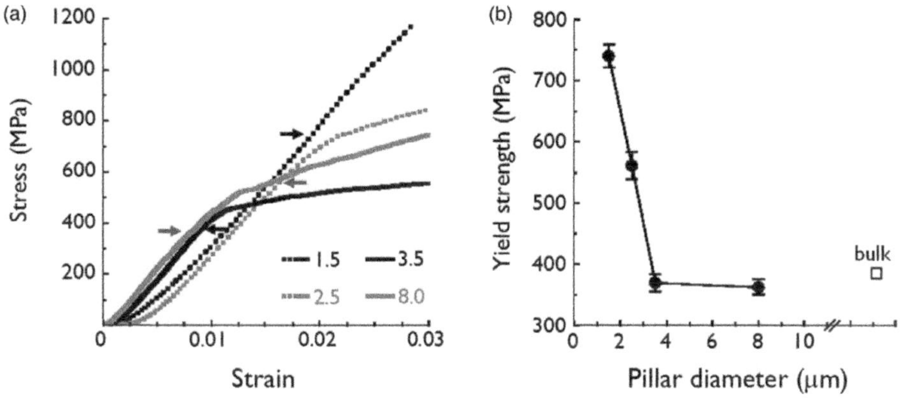

Figure 2. (a) Compressive stress–strain curves of nanostructured Mg–10Al micropillars. (b) Compressive yield strength as a function of pillar diameter.

Table 1. Compressive properties of nanostructured Mg–10Al micropillars with different diameters.

Pillar diameter (μm)	Yield strength (MPa)	Strain at failure (%)	Average yield strength (MPa)	Average strain at failure (%)
1.5	742	2.4	746 ± 19	2.6 ± 0.2
	729	2.7		
	778	2.7		
	746	2.8		
	734	2.5		
2.5	577	3.6	554 ± 21	3.9 ± 0.3
	536	3.7		
	549	4.4		
	575	3.9		
	533	3.7		
3.5	347	4.2	364 ± 15	4.5 ± 0.2
	374	4.6		
	373	4.8		
	379	4.3		
	349	4.7		
8.0	373	4.8	362 ± 12	4.7 ± 0.3
	362	5.5		
	350	4.1		
	373	4.3		
	350	4.7		

overestimation of the measured stress due to the tapering, the stress was obtained using the load divided by the middle cross-section area of the pillars. Figure 2(a) shows representative engineering stress vs. strain curves for nanostructured Mg–10Al pillars with different diameters. From the linear slope of the stress–strain curve, we estimate the elastic modulus, E, ~50 GPa, which is in good agreement with that measured by nanoindentation with a Berkovich tip (52 GPa). The yield strength, which is determined as the end of the proportional limit of each curve is highlighted by an arrow.

Pillars with 8 and 3.5 μm diameters present similar yield strength values i.e. 373 and 374 MPa, respectively. It is interesting to note that the yield strength increases significantly when decreasing the pillar diameter. In particular, the yield strength values for 2.5 and 1.5 μm pillars are 549 and 746 MPa, respectively. Therefore, an increase of around 50% and 100% is obtained when decreasing the pillar size to 2.5 and 1.5 μm, respectively. Additionally, the stress–strain curves of Figure 2(a) indicate a less ductile behavior for the smallest pillars as confirmed by their values of strain at failure in Table 1. Thus, in addition to a significant increase in strength, a reduction in ductility is observed for pillars with smaller diameter.

The trend of the yield strength as a function of pillar diameter can be better seen in Figure 2(b). These results suggest a strong size effect on the strength of nanostructured Mg micropillars with diameters smaller than 3.5 μm. In contrast, no size effect was observed when the diameters are higher than 3.5 μm. In fact, the compressive strength of micropillars with diameters higher than 3.5 μm is similar to that of the bulk sample (397 MPa).[17]

TEM analysis was performed in order to study the underlying deformation mechanisms during microcompression. Figure 3(a) is a bright-field TEM image of a compressed 2.5 μm micropillar showing the characteristic bimodal grain size distribution after SPS sintering. As described in our previous work,[17] the bimodal grain size distribution consists of fine-grains ~35 nm and coarse-grains ~400 nm. No apparent texture was found. See Ref. [17] for more microstructural details. We identify at least three shear bands forming an angle with respect to the LD of ~70° (labeled 1 and 2 at the top of the pillar) and ~60° (labeled 3 at the bottom of the pillar). High-resolution TEM (HRTEM) image from the shear band 1 (Figure 3(b)) reveals extension twins on the characteristic $\{10\bar{1}2\}$ planes (highlighted by yellow solid lines) in a grain with the c-axis almost perpendicular to the LD. In fact, nanotwins of <1 nm widths can be observed. This has been confirmed by indexing its fast Fourier transform (FFT) (inset). Interplanar distances measurements indicate the misorientation relation between the parent

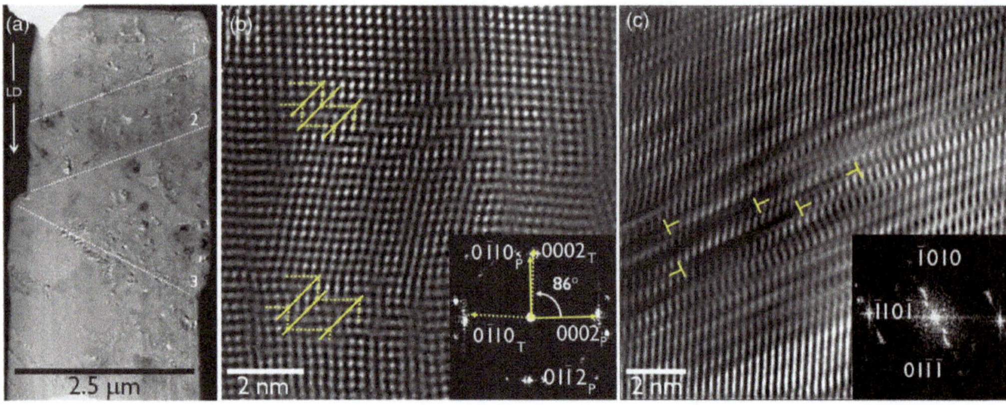

Figure 3. (a) Bright-field TEM image of a compressed 2.5 μm micropillar. (b) HRTEM image showing extension twins in a grain oriented to the $\langle 2\bar{1}\bar{1}0\rangle$ zone axis. Inset is its FFT. (c) HRTEM image showing prismatic dislocations terminated at SFs in a grain oriented to the $\langle 1\bar{2}1\bar{3}\rangle$ zone axis. Inset is its FFT.

(solid lines) and the twin (dashed lines) close to the expected angle of 86° for a grain oriented to the $\langle 2\bar{1}\bar{1}0\rangle$ zone axis. Note that $\pm(0002)_T$ (basal planes from the twin) and $\pm(0002)_P$ (basal planes from the parent) are almost perpendicular to each other.

In addition to the twinning mechanism, we find prismatic dislocations (highlighted by ⊥) terminated at stacking faults (SFs) in a grain close to the shear band 2 (Figure 3(c)). Inset is the indexed FFT of the image indicating the $\{1\bar{1}00\}$ planes where dislocations occur for a grain oriented to the $\langle 1\bar{2}1\bar{3}\rangle$ one axis. The characteristic streaks parallel to the $\langle 1\bar{1}00\rangle$ direction as a result of prismatic SFs are also apparent.

From the TEM analysis, we find that extension twinning and prismatic slip can be activated during compression of 2.5 μm nanostructured Mg-micropillars. While extension twinning is a common deformation mechanism seen in Mg bulk samples, higher stress is required to nucleate extension twins in micropillars.[14] In addition, it is well known that the critical resolved shear stress for prismatic slip and twinning is always higher than that for the basal slip.[22] Therefore, we believe that the enhanced strength observed in nanostructured Mg–10Al micropillars with 2.5 μm diameter is mainly due to the activation of the aforementioned deformation mechanisms.

From our previous work based on *in situ* SEM micro-compression studies of nanostructured Mg–10Al micropillars with 4 μm diameter,[23] we found in addition to basal plane sliding, contraction twins in a grain oriented with the *c*-axis almost parallel to the LD.

These results highlight the important role that the grain orientation plays in the strengthening of nanostructured Mg micropillars. As demonstrated by the TEM analysis of 2.5 μm micropillars, alternative deformation mechanisms such as non-basal slip and/or twinning can be activated in grains with the *c*-axis nearly parallel or perpendicular to the LD, since the Schmid factor for basal plane sliding is very small. These observations suggest that the contribution from grains favorably oriented for basal slip (within a bimodal grain distribution randomly oriented without any preferential texture [17]) to the pillar strength decreases with decreasing pillar diameter. Additionally, by reducing pillar diameter the number of grain boundaries also decreases,[24] which significantly reduces the number of potential dislocation sources.[4] Hence, the applied stress required for nucleation or activation of dislocations, i.e. yield strength also increases. Based on this, we conclude that the size-induced strengthening is due to a smaller number of dislocation sources along with a higher activity of non-basal dislocation sliding and twinning.

In summary, we have demonstrated a strong specimen size effect on the compressive strength of nanostructured Mg–10Al alloy micropillars machined by FIB. We find that the yield strength increases significantly when the pillar diameter is <3.5 μm. An increase of around 50% and 100% is obtained when the pillar diameter decreases to 2.5 and 1.5 μm, respectively. In contrast, no size effect is observed for pillar diameter ≥3.5 μm. We attribute the size-induced strengthening to a less number of dislocation sources along with a higher activity of non-basal dislocation sliding and twinning.

Acknowledgements This work was funded by the Army Research Office (ARO) under contract No. W911NF-09-1-0558. The authors are particularly grateful to the ARO program manager, Dr Larry Russell for his supervision and to Christopher Melnyk (California Nanotechnologies Inc.) for his technical assistance. They are also grateful to Noah Bodzin for his assistance with the TEM sample preparation.

References

[1] Greer JR, De Hosson JTM. Plasticity in small-sized metallic systems: intrinsic versus extrinsic size effect. Prog Mater Sci. 2011;56:654–724.

[2] Uchic MD, Dimiduk DM, Florando JN, Nix WD. Sample dimensions influence strength and crystal plasticity. Science. 2004;305:986–989.
[3] Greer JR, Oliver WC, Nix WD. Size dependence of mechanical properties of gold at the micron scale in the absence of strain gradients. Acta Mater. 2005;53:1821–1830.
[4] Rinaldi A, Peralta P, Friesen C, Sieradzki K. Sample-size effects in the yield behavior of nanocrystalline nickel. Acta Mater. 2008;56:511–517.
[5] Jang D, Greer JR. Size-induced weakening and grain boundary-assisted deformation in 60 nm grained Ni nanopillars. Scripta Mater. 2011;64:77–80.
[6] Chen XX, Ngan AHW. Specimen size and grain size effects on tensile strength of Ag microwires. Scripta Mater. 2011;64:717–720.
[7] Edalati K, Toh S, Iwaoka H, Watanabe M, Horita Z, Kashioka D, Kishida K, Inui H. Ultrahigh strength and high plasticity in TiAl intermetallics with bimodal grain structure and nanotwins. Scripta Mater. 2012;67:814–817.
[8] Jang D, Cai C, Greer JR. Influence of homogeneous interfaces on the strength of 500 nm diameter Cu nanopillars. Nano Lett. 2011;11:1743–1746.
[9] Gu XW, Loynachan CN, Wu Z, Zhang Y, Srolovitz DJ, Greer JR. Size-dependent deformation of nanocrystalline Pt nanopillars. Nano Lett. 2012;12:6385–6392.
[10] Byer CM, Li B, Cao B, Ramesh KT. Microcompression of single-crystal magnesium. Scripta Mater. 2010;62:536–539.
[11] Byer CM, Ramesh KT. Effects of the initial dislocation density on size effects in single-crystal magnesium. Acta Mater. 2013;61:3808–3818.
[12] Lilleodden E. Microcompression study of Mg (0001) single crystal. Scripta Mater. 2010;62:532–535.
[13] Ye J, Mishra RK, Sachdev AK, Minor AM. *In situ* TEM compression testing of Mg and Mg–0.2 wt.% Ce single crystals. Scripta Mater. 2011;64:292–295.
[14] Prasad KE, Rajesh K, Ramamurty U. Micropillar and macropillar compression responses of magnesium single crystals oriented for single slip or extension twinning. Acta Mater. 2014;65:316–325.
[15] Yu Q, Mishra RK, Minor AM. The effect of size on the deformation twinning behavior in HCP Ti and Mg. JOM. 2012;64:1235–1240.
[16] Pozuelo M, Melnyk C, Kao WH, Yang JM. Cryomilling and spark plasma sintering of nanocrystalline magnesium-based alloy. J Mater Res. 2011;26:904–911.
[17] Pozuelo M, Chang YW, Yang JM, Enhanced compressive strength of an extruded nanostructured Mg–10Al alloy. Mater Sci Eng A. 2014;594:203–211.
[18] Zhang H, Schuster BE, Wei Q, Ramesh KT. The design of accurate micro-compression experiments. Scripta Mater. 2006;54:181–186.
[19] Shim S, Bei H, Miller MK, Pharr GM, George EP. Effects of focused ion beam milling on the compressive behavior of directionally solidified micropillars and the nanoindentation response of an electropolished surface. Acta Mater. 2009;57:503–510.
[20] Kiener D, Motz C, Dehm G. Micro-compression testing: a critical discussion of experimental constraints. Mater Sci Eng A. 2009;505:79–87.
[21] Volkert CA, Lilleodden ET. Size effects in the deformation of sub-micron Au columns. Phil Mag. 2006;86:5567–5579.
[22] Wang YN, Huang JC. Texture analysis in hexagonal materials. Mater Chem Phys. 2003;81:11–26.
[23] Pozuelo M, Chang YW, Yang JM. In-situ microcompression study of nanostructured Mg alloy micropillars. Mater Lett. 2013;108:320–323.
[24] Alshehri O, Yavuz M, Tsui T. Manifestation of external size reduction effects on the yield point of nanocrystalline rhodium using nanopillars approach. Acta Mater. 2013;61:40–50.

13

Current-Activated, Pressure-Assisted Infiltration: A Novel, Versatile Route for Producing Interpenetrating Ceramic–Metal Composites

Liangfa Hu[a], Ankush Kothalkar[a], Morgan O'Neil[b], Ibrahim Karaman[a,b] and Miladin Radovic[a,b]*

[a]*Department of Materials Science and Engineering, Texas A&M University, College Station, TX 77843, USA;*
[b]*Department of Mechanical Engineering, Texas A&M University, College Station, TX 77843, USA*

A novel technique is proposed and demonstrated for processing interpenetrating ceramic–metal composites by the infiltration of molten metals/alloys into ceramic foams utilizing a current-activated, pressure-assisted technique. Ti_2AlC foams were prepared and used as the infiltration preform. Aluminum alloy 6061 (AA6061) has been infiltrated into the Ti_2AlC foams, resulting in AA6061/Ti_2AlC composites. The microstructure, composition, and distribution of phases in the composites were studied. The AA6061/Ti_2AlC composites were studied under compression testing, and the results were compared with those of the pure components of the composites. Advantages of this method in comparison to other methods for processing metal–ceramic composites are discussed.

Keywords: MAX Phase, Infiltration, Ceramic–Metal Composites, Aluminum Composites

1. Introduction Ceramic–metal composites have been under investigation for more than 40 years. On the one hand, adding a ceramic phase in light metals such as aluminum and magnesium results in composites with higher strength, elastic moduli, hardness, wear resistance and reduced coefficient of thermal expansion.[1] On the other hand, metal phases are incorporated in ceramics mainly to increase their toughness.[2] Among various methods for producing the ceramic–metal composites, such as powder metallurgy, stir casting, diffusion bonding, and deposition,[3] the infiltration of molten metals into ceramic preforms containing continuous open porosity is one of the most popular methods.[4] The processes for infiltration can be divided into two categories: (i) pressureless infiltration,[5] where the molten metals or alloys wet the ceramic preforms sufficiently to be spontaneously drawn into the preform, and (ii) pressure infiltration,[6] in which an external pressure is used to actively drive the molten metals or alloys into the preform.

Poor wettability of ceramic substrate with molten metal makes the pressureless infiltration process very slow or even in some instances impossible.[7] In addition, poor wetting can also yield weak bonding between the metal and ceramic phases, resulting in inferior mechanical properties of the composites.[8] Therefore, numerous methods have been used to enhance the wettability. These methods include modifying the ceramic surface,[9] tailoring the chemistry of the metals and alloys,[9,10] the use of controlled atmospheres [10,11] and much higher temperatures than the melting points of the metals/alloys.[12] However, the components of the composites may be degraded due to chemical attack or introduction of new reaction phases at high temperatures.[13] In fact, the majority of infiltration-based composites are made by pressure infiltration processes. The external pressure can be mechanical as in squeeze infiltration,[14] or an applied gas pressure.[15] With the apparatus being made of steel, the squeeze infiltration is normally carried out at temperatures lower than the melting point of steel, while the gas pressure infiltration process normally requires a highly sealed container.[16] Moreover, both methods take relatively long time and may not be practical in many cases where the molten metals are highly reactive with the ceramic preforms.

Our interest concerns developing a rapid infiltration method based on a current-activated, pressure-assisted technique to minimize the two-phase reactions. The current-activated and pressure-assisted powder

*Corresponding author. Email: mradovic@tamu.edu

densification method utilizes simultaneously very large electric currents and applied loads to achieve rapid consolidation of ceramic or metallic powders.[17] In fact, it has been clearly shown that this method is viable to co-sinter highly reactive ceramic–metal systems.[18]

Aluminum (Al) and their alloys have been used extensively in various industries – such as automotive and aerospace – due to their lightweight, excellent corrosion resistance, good castability and machinability.[19,20] Two ternary carbides, namely Ti_2AlC and Ti_3AlC_2, belonging to the family of MAX phases, where M is an early transition metal, A is an A-group element (from groups 13 to 16), and X is C and/or N, have high stiffness, high strength, good resistance to thermal shock and oxidation, and high thermal and electrical conductivities; they are also lightweight and easily machinable.[21,22] Such unique properties make them promising candidate materials for use as reinforcements in Al and its alloys and extend the application temperatures of Al and its alloys.[23] As a matter of fact, Ti_3AlC_2/Al composites have been fabricated using powder mixture co-sintering.[23] The yield stress of the Ti_3AlC_2/Al composite has been found to be twice as high as that of the Al matrix in the 20–500°C temperature range.[23] However, the Ti_3AlC_2 particles were barely sintered in the Ti_3AlC_2/Al composites prepared by powder mixture co-sintering, because the co-sintering temperature was kept low to avoid substantial and harmful reactions. Therefore, to process interpenetrating Al/MAX phase composites with marginal two-phase reactions infiltration of molten metals/alloys into ceramic foams has to be considered. This method has been already used to process the Ti_2AlC/Mg composite that exhibits higher strength in tension and compression than other Mg composites, as well as high damping capacity.[24] However, pressureless infiltration is not an easy task because of the poor wettability of MAX phase preform with molten Al. The challenge for processing interpenetrating Al/MAX phase composites is therefore to achieve high density while controlling the reaction between MAX phase and Al.

The present paper reports on the current-activated, pressure-assisted infiltration (CAPAI) method for producing interpenetrating ceramic–metal composites where poor wettability and severe reactions prohibit application of conventional processing methods. Ti_2AlC foams were infiltrated by molten aluminum alloy 6061 (AA6061), resulting in AA6061/Ti_2AlC composites. The composite microstructures were studied. The mechanical properties of the AA6061/Ti_2AlC composites were characterized under compression testing, and the results were compared with those of the pure components of the composites, namely solution heat treated AA6061 and Ti_2AlC foam.

2. Experimental Details Ti_2AlC powders (MAXthal 211, Sandvik Heating Technology AB, Sörkvarnsvägen, Sweden) with the particle size in the 45–90 μm range, and NaCl powders (Sigma-Aldrich, St. Louis, MO, USA) with the particle size in the 355–500 μm range were used to process porous preforms following procedures described elsewhere in more detail.[25] The preparation of the Ti_2AlC foams involves three steps: (i) cold pressing a NaCl–Ti_2AlC powder mixture in 40/60 volume ratio; (ii) dissolving the NaCl in distilled water, and (iii) pressureless sintering of the porous green body under flowing argon at 1400°C for 4 h.

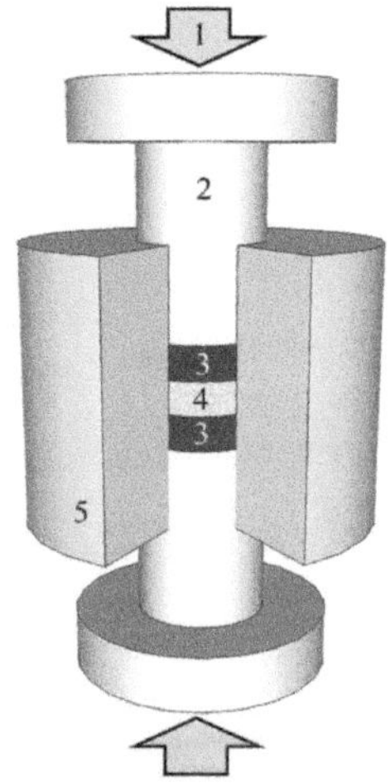

Item	Description
1	Coupled current / pressure
2	Graphite punch
3	Metal or alloy
4	Porous ceramic preform
5	Graphite die

Figure 1. Schematic of the melt infiltration approach showing the coupled current/pressure (item 1), the graphite die set (items 2 and 5), and the 'sandwich'-like set-up of the metal/alloy discs (item 3) and the porous preform disc (item 4).

AA6061 discs (McMaster-Carr, GA, USA) with a diameter of 20 mm and a thickness of 4 mm were used for the infiltration process. As shown in Figure 1, a disc (20 mm in diameter and 4 mm in thickness) of Ti_2AlC foam was 'sandwiched' in between two AA6061 discs, and placed in a graphite die, followed by infiltration (SPS 25-10, GT Advanced Technologies, CA) at 750°C under 5 MPa uniaxial pressure for 1 min. A high dc current was applied, increasing from 0 to 1250 A in 4 min and then stabilized at 860 A for 1 min at the sintering temperature of 750°C. The pulse cycle was 10 ms on and 10 ms off. The 'sandwich' set-up enables more uniform infiltration of molten AA6061. The chamber was evacuated and held at 10^{-6} torr for 10 min before heating. The heating rate was 200°C per minute. It takes less than 10 min to process the samples, including heating, soaking, and cooling. Graphite foils were applied between the samples and graphite die before infiltration. The temperature was measured at the bottom graphite punch, which was immediately below the surface that was in contact with the sample.

The porosity of all samples was determined by the alcohol immersion method based on Archimedes' principle, as specified in ASTM C20-00.[26] The theoretical

Figure 2. (a) A photograph of a porous Ti_2AlC preform, and SEM images of (b) the preform showing open pores and struts, (c) enlarged view on the strut showing both closed pores and open pores. (d) A photograph of an as-infiltrated AA6061/Ti_2AlC composite, and SEM images of (e) the composite showing AA6061 (dark gray) infiltrated into the open pores, (f) enlarged view on the strut showing Ti_2AlC (1, light), titanium aluminide (2, gray), and Al_2O_3 (3, dark). The composition of phases 1, 2, and 3 are according to the EDS results (not shown here).

density values of 4.11 (g/cm^3) [27] and 2.70 (g/cm^3) [28] for fully dense Ti_2AlC and AA6061, respectively, were used to calculate the theoretical density of composites using the rule of mixture and assuming that the effect of the eventual reaction phases on the theoretical density is negligible. The relative density is thus the measured density value divided by the corresponding theoretical rule-of-mixture density value, i.e. $3.55\ g/cm^3$.

The phase composition of starting powders and as-infiltrated composites was determined using an X-ray diffractometer, XRD (D8 Discover, Bruker, WI, USA) with Cu $K\alpha$ radiation (wavelength $= 1.542$ Å) at 40 kV and 30 mA. The two theta range varied from 8° to 80° with a step size of 0.04° and a step time of 1.5 s. The XRD results were analyzed utilizing the Inorganic Crystal Structure Database (ICSD). The microstructure, phase composition, and distribution were characterized using a field emission scanning electron microscopy, SEM (JSM-7500F, JEOL, Tokyo, Japan) equipped with energy-dispersive spectroscopy (EDS). The accelerating voltage and emission current were 15 kV and 20 mA, respectively. The duration of spot scan of EDS was 60 s per spectrum. The compressive strength of the samples with dimensions of 3.5 mm × 3.5 mm × 7 mm was measured by using a universal testing machine (MTS810, MTS, MN, USA) with a strain rate of $1.4 \times 10^{-4}\ s^{-1}$. All samples for compressive testing were cut by electrical discharge machining.

3. Results and Discussions

3.1. Microstructure, Density, and Phase Analysis.

Figure 2(a)–(c) shows the Ti_2AlC foam prepared with 40 vol.% NaCl pore former with particle size in the 355–500 µm range. The open and closed porosity of the preform are 35.7 ± 1.0 vol.% and 5.1 ± 0.4 vol.%, respectively, which means that 87.5% of overall porosity (sum of the open and closed porosity) are open. Both SEM and optical images in Figure 2(a)–(c) show that the open pores are homogeneously dispersed in the Ti_2AlC, while some closed pores can be found in the struts of the foam due to the incomplete sintering of Ti_2AlC particles during the pressureless sintering. Meanwhile, Figure 2(c) clearly illustrates the strong connectivity of Ti_2AlC grains on the struts of the foam. The strong network of grains is critical to obtain high mechanical strength of porous preform, which turns into the struts of the composite after infiltration. As a component of the infiltrated composite, the strong Ti_2AlC networks play an important role to reach high compressive strengths in the composites.

For infiltration processes, ceramic preforms have to be made before the metal is infiltrated. The preform structures are normally packed particles or fibers,[29] and they can in some instances be solid foams.[30] The mechanical behavior of the preform is important, not only during handling and placement of the preform, but also because the preform may be stressed during the pressure infiltration. MAX phases are kinking-nonlinear elastic solids which can dissipate significant amount of energy during cyclic loading due to the formation and annihilation of fully reversible incipient kink bands.[31] Previous investigations have shown exceptional damage tolerance of MAX phase foams.[25] Specifically, the compressive strength of Ti_2AlC decreases almost linearly with increasing porosity and less rapidly when compared with common ceramics, such as Al_2O_3, Si_3N_4, and ZrO_2, whose compressive strength drops dramatically with increasing porosity.[32] In addition to relatively high strength of Ti_2AlC foams, another advantage of using them as preforms is the easy control of pore size and porosity which allows producing interpenetrating ceramic–metal composites with customized structures.

Figure 2(d)–(f) shows a photograph and the microstructure of the AA6061/Ti_2AlC composite processed by the CAPAI method. The open and closed porosity are 1.0 ± 0.9 vol.% and 4.6 ± 0.2 vol.%, respectively. More than 97% of all open pores are successfully infiltrated with AA6061 in the composites. Although the open porosity dramatically decreased due to infiltration with AA6061, the closed porosity hardly changed after infiltration. Therefore, the majority of the residual porosity in the composites comes from closed pores that are of micrometer size and isolated in the Ti_2AlC area. In contrast, pores or voids can hardly be found in the AA6061 areas and in the two-phase boundaries. This suggests that the AA6061 infiltration occurred dominantly in the open pores, and volume fractions of the components in the final composites, e.g. ceramic phase, metallic phase, open and closed pores, are tunable by adjusting the initial porosities in the ceramic preform. Note that the amount of reaction phase in the AA6061/Ti_2AlC composite processed in this work is much smaller than that in the previously reported Ti_3AlC_2/Al composites prepared by powder mixture co-sintering.[23] Indeed, Ti-Al-C phase diagram [33] clearly shows that Al reacts with Ti_2AlC at 727°C to form stable phases like titanium aluminide intermetallic phases. Even when the Ti_3AlC_2/Al composites were processed at a lower temperature by the co-sintering method, that method takes much longer time than the infiltration process and thus results in more reaction between two phases. This suggests that the CAPAI method effectively reduces the reactions in highly reactive ceramic–metal systems.

Figure 3 shows the XRD results of the starting materials, AA6061 and Ti_2AlC foam, and the as-infiltrated AA6061/Ti_2AlC composites. The results clearly show that the Ti_2AlC foam contains, in addition to Ti_2AlC, some Ti_3AlC_2 and a very small amount of Al_2O_3. These two impurity phases come from the starting commercial MAXtal211 powder. Ti_2AlC, Ti_3AlC_2 and Al are three major phases in the AA6061/Ti_2AlC composite in the presence of a small amount of Al_2O_3. All phases identified in the composites were also found in the starting materials, namely AA6061 and Ti_2AlC foam, and no new phases were detected by the XRD. Although a new

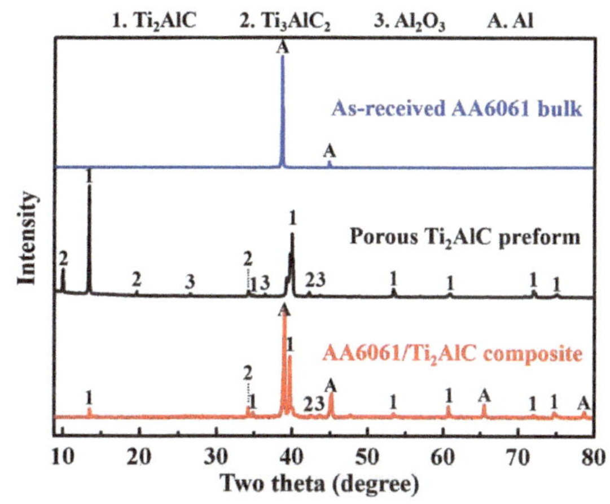

Figure 3. XRD results of the AA6061/Ti_2AlC composite and its pure components, namely AA6061 and Ti_2AlC foam. All XRD data were collected at room temperature. The identification of phases was done according to the ICSD collection code 165460 for Ti_2AlC, 153266 for Ti_3AlC_2, 10425 for Al_2O_3, 43423 for Al.

phase, intermetallic titanium aluminide (Figure 2(f)), was identified by the EDS in the AA6061/Ti_2AlC composite, its amount is so small that they are not detectable in the XRD.

The CAPAI technique is utilized here for the first time for the infiltration process to take advantage of its three inherent benefits of CAPAI. First, the coupled current/pressure provides simultaneous heating and pressing that actively drive the molten metals into the preform. Second, the fast and controllable heating rate, as high as 500°C/min, offers an easy and efficient way to avoid the reactions between the ceramic preforms and the metal/alloy melts during infiltration. Third, the controllable atmosphere, i.e. either vacuum or inert gas, makes it easy to investigate the effects of atmosphere on the infiltration mechanism and to control possible oxidation and reactions between ceramic and metallic phases.

Particular advantages of this technique arise when one compares the CAPAI method to conventional infiltration methods. First, unlike the time-consuming pressureless infiltration processes, the CAPAI process can be very fast and efficient, e.g. 10 min for a single infiltration process, including heating/melting, infiltrating and cooling/solidification. Second, in contrast to the pressure infiltration where temperature is limited below the melting point of steel and high cost is associated with sealing, the infiltration temperature for the CAPAI can be as high as 2200°C and sealing for highly pressurized gas is not required. Note that the current or electric field could not be excluded as a factor of enhancing the wettability of 6061 Al onto the MAX phase, even though the mechanism is not clear at this point.

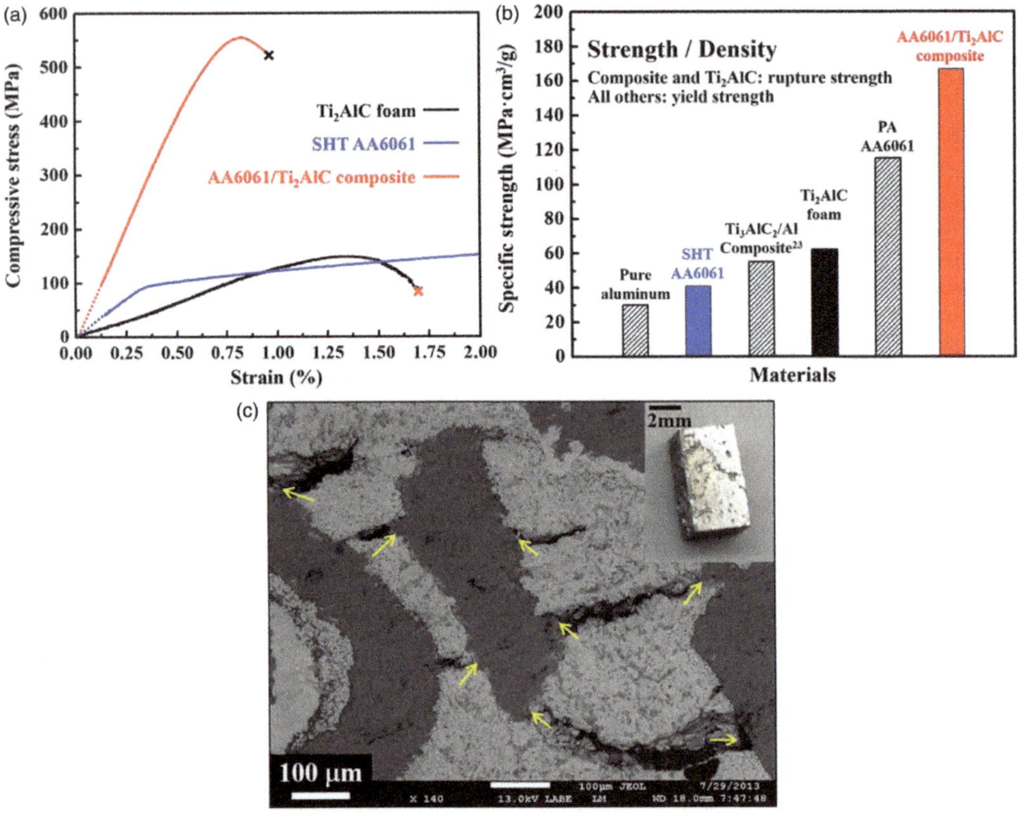

Figure 4. (a) Compressive stress–strain curves of the AA6061/Ti_2AlC composites, SHT AA6061, and Ti_2AlC foam. (b) Specific strength of the AA6061/Ti_2AlC composites, peak aged (PA) AA6061, and Ti_2AlC foam, the Ti_3AlC_2/Al composites,[23] SHT AA6061, and pure aluminum. (c) Surface morphology of the post-compression composite. The inset shows that the sample failed by cracking in the direction of approximately 45° relative to the loading direction with a tortuous crack path. The cracks propagate from bottom right to top left. The arrow indicates that cracks in the ceramic phase were effectively stopped and deflected by metallic phase.

3.2. Mechanical Properties. It has been demonstrated up to this point that the CAPAI method allows processing novel ceramic–metal composites which could not otherwise be obtained using powder metallurgy and conventional infiltration techniques. Furthermore, the section below also demonstrates that the CAPAI method can also result in superior properties in the composites when compared to their pure components.

Figure 4(a) shows the compressive stress–strain curves of the AA6061/Ti_2AlC composites, solution heat treated (SHT) AA6061 (to be more comparable with its condition in the composite), and Ti_2AlC foam. The compressive strength of the AA6061/Ti_2AlC composite is 560 MPa, which is about four times as high as that of the Ti_2AlC foam (150 MPa) and about five times as high as the yield strength of the SHT AA6061 (110 MPa [34]). More importantly, Figure 4(b) shows that the specific strength, i.e. compressive/yield strength divided by density, of the AA6061/Ti_2AlC composite is 167 MPa cm^3/g, which is about 3 times as high as that of the Ti_2AlC foam (62 MPa cm^3/g), 1.5 times as high as that of the peak aged AA6061 (115 MPa cm^3/g [34]), 3 times as high as that of previously reported value of the Ti_3AlC_2/Al composite (55 MPa cm^3/g [23]), and 4 times as high as the yield strength of the SHT AA6061 (41 MPa cm^3/g). Interestingly enough, post-testing analysis of fracture surfaces (Figure 4(c)) shows distinctive evidence of crack propagation resistance of the AA6061/Ti_2AlC composite.

As the inset in Figure 4(c) shows, all tested samples failed by cracking in the direction of approximately 45° relative to the loading direction with a tortuous crack path. This observation suggests that the composite shows a graceful failure, and does not crush and fragment under compression. Higher magnification SEM image in Figure 4(c) of the area around the fracture surface clearly shows that cracks in the ceramic phase were effectively stopped and deflected by the metallic phase. More importantly, no cracks in the interfacial areas have been observed after compression testing, indicating exceptionally strong bonding between two phases in the composite.

4. Conclusions In summary, the CAPAI method is proposed and demonstrated as a novel, efficient and rapid technique for producing interpenetrating ceramic–metal composites by the infiltration of molten metals and alloys into porous ceramic preforms. AA6061/Ti_2AlC composites were successfully processed using this method. The major findings are summarized as follows:

(1) More than 97% of the open porosity was infiltrated with molten metal even after a very short processing time. Most of the residual pores in the composites are closed pores and are found in the Ti_2AlC area as a result of incomplete sintering during preparation of the preform. This method offers an efficient route for producing interpenetrating ceramic–metal composites with customizable structures by controlling the structure of the ceramic preform.
(2) Little reaction is observed in the AA6061/Ti_2AlC system, which suggests that the CAPAI method allows processing novel, highly reactive ceramic–metal composite systems which could not otherwise be obtained using powder metallurgy and conventional infiltration techniques.
(3) This method can result in composites with superior properties: the specific compressive strength of the AA6061/Ti_2AlC composites is about 1.5 times of the specific yield strength of peak aged AA6061.

Acknowledgements This work was supported by the US Air Force Office of Scientific Research, MURI Program (FA9550-09-1-0686) to Texas A&M University. The authors would like to thank the program manager Dr. David Stargel for his support.

References

[1] Mortensen A, Llorca J. Metal matrix composites. In: Clarke DR, Ruhle M, Zok F, editors. Annual review of materials research, Vol 40. Palo Alto (CA): Annual Reviews; 2010. p. 243–270.
[2] Evans AG. Perspective on the development of high-toughness ceramics. J Am Ceram Soc. 1990;73:187–206.
[3] Evans A, Marchi CS, Mortensen A. Processing of metal matrix composites. Metal matrix composites in industry, an introduction and a survey. Boston (MA): Kluwer Academic; 2003.
[4] Mortensen A. Melt infiltration of metal matrix composite. In: Clyne TW, editor. Comprehensive composite materials, vol 3: metal matrix composites. Oxford: Pergamon; 2000.
[5] Aghajanian MK, Rocazella MA, Burke JT, Keck SD. The fabrication of metal matrix composites by a pressureless infiltration technique. J Mater Sci. 1991;26:447–454.
[6] Garcia-Cordovilla C, Louis E, Narciso J. Pressure infiltration of packed ceramic particulates by liquid metals. Acta Mater. 1999;47:4461–4479.
[7] Li JG. Wetting of ceramic materials by liquid silicon, aluminum and metallic melts containing titanium and other reactive elements-a review. Ceram Int. 1994;20:391–412.
[8] Liu J, Sinner J, Higginson R, Zhou ZX. Interfacial reactions and wetting in Al-Mg/oxide ceramic interpenetrating composites made by a pressureless infiltration technique. Compos Sci Technol. 2012;72:886–893.
[9] Sobczak N, Asthana R, Ksiazek M, Radziwill W, Mikulowski B. The effect of temperature, matrix alloying and substrate coatings on wettability and shear strength of Al/Al_2O_3 couples. Metall Mater Trans A. 2004;35A: 911–923.
[10] Sercombe TB, Schaffer GB. On the role of magnesium and nitrogen in the infiltration of aluminium by aluminium for rapid prototyping applications. Acta Mater. 2004;52:3019–3025.
[11] Rao BS, Jayaram V. New technique for pressureless infiltration of Al alloys into Al_2O_3 preforms. J Mater Res. 2001;16:2906–2913.
[12] Toy C, Scott WD. Ceramic metal composite produced by melt infiltration. J Am Ceram Soc. 1990;73:97–101.
[13] Warrier SG, Blue CA, Lin RY. Control of interfaces in Al-C fiber composites. J Mater Sci. 1993;28:760–768.
[14] Lange FF, Velamakanni BV, Evans AG. Method for processing metal-reinforced ceramic composites. J Am Ceram Soc. 1990;73:388–393.
[15] Prielipp H, Knechtel M, Claussen N, Streiffer SK, Mullejans H, Ruhle M, Rodel J. Strength and fracture toughness of aluminium/alumina composites with interpenetrating networks. Mater Sci Eng A. 1995;197:19–30.
[16] Lemster K, Graule T, Kuebler J. Processing and microstructure of metal matrix composites prepared by pressureless Ti-activated infiltration using Fe-base and Ni-base alloys. Mater Sci Eng A. 2005;393:229–238.
[17] Munir ZA, Anselmi-Tamburini U, Ohyanagi M. The effect of electric field and pressure on the synthesis and consolidation of materials: a review of the spark plasma sintering method. J Mater Sci. 2006;41:763–777.
[18] Hu L, Kothalkar A, Karaman I, Proust G, Radovic M. Morphology and phase analysis in spark plasma sintered shape memory alloy/MAX phase composites. J Alloys Comp. Under review.
[19] Kulekci MK. Magnesium and its alloys applications in automotive industry. Int J Adv Manuf Technol. 2008;39:851–865.
[20] Miller WS, Zhuang L, Bottema J, Wittebrood A, De Smet P, Haszler A, Vieregge A. Recent development in aluminium alloys for the automotive industry. Mater Sci Eng A. 2000;280:37–49.
[21] Barsoum MW, Radovic M. Elastic and mechanical properties of the MAX phases. In: Clarke DR, Fratzl P, editors. Annual review of materials research, Vol 41. Palo Alto (CA): Annual Reviews; 2011. p. 195–227.
[22] Radovic M, Barsoum MW. MAX phases: bridging the gap between metals and ceramics. Am Ceram Soc Bull. 2013;92:20–27.
[23] Wang WJ, Gauthier-Brunet V, Bei GP, Laplanche G, Bonneville J, Joulain A, Dubois S. Powder metallurgy processing and compressive properties of Ti_3AlC_2/Al composites. Mater Sci Eng A. 2011;530:168–173.
[24] Amini S, Ni CY, Barsoum MW. Processing, microstructural characterization and mechanical properties of a Ti_2AlC/nanocrystalline Mg-matrix composite. Compos Sci Technol. 2009;69:414–420.
[25] Hu L, Benitez R, Basu S, Karaman I, Radovic M. Processing and characterization of porous Ti_2AlC with controlled porosity and pore size. Acta Mater. 2012;60:6266–6277.
[26] ASTM C20-00. Standard test methods for apparent porosity, water absorption, apparent specific gravity, and bulk

density of burned refractory brick and shapes by boiling water. West Conshohocken: ASTM international; 2010.

[27] Barsoum MW. The $M_{(N+1)}AX_{(N)}$ phases: a new class of solids; thermodynamically stable nanolaminates. Prog Solid State Chem. 2000;28:201–281.

[28] Ross RB. Metallic materials specification handbook. 4th ed. London: Chapman & Hall; 1992.

[29] Gil R, Jinnapat A, Kennedy AR. Pressure-assisted infiltration of molten aluminium into open cell ceramic foams: experimental observations and infiltration modelling. Compos Pt A. 2012;43:880–884.

[30] Mattern A, Huchler B, Staudenecker D, Oberacker R, Nagel A, Hoffmann MJ. Preparation of interpenetrating ceramic-metal composites. J Eur Ceram Soc. 2004;24: 3399–3408.

[31] Barsoum MW, Zhen T, Kalidindi SR, Radovic M, Murugaiah A. Fully reversible, dislocation-based compressive deformation of Ti_3SiC_2 to 1 GPa. Nat Mater. 2003;2: 107–111.

[32] Hu L, Wang CA. Effect of sintering temperature on compressive strength of porous yttria-stabilized zirconia ceramics. Ceram Int. 2010;36:1697–1701.

[33] Viala JC, Vincent C, Vincent H, Bouix J. Thermodynamic approach of the chemical interaction between al and tic. Mater Res Bull. 1990;25:457–464.

[34] Davis JR. Metals handbook. Materials Park (OH): ASM International; 1998.

14

An Ideal Ultrafine-Grained Structure for High Strength and High Ductility

C. X. Huang[a]*, W. P. Hu[b], Q. Y. Wang[c]*, C. Wang[d], G. Yang[d] and Y. T. Zhu[e,f]

[a]School of Aeronautics and Astronautics, Sichuan University, Chengdu 610065, People's Republic of China; [b]Institute of Physical Metallurgy and Metal Physics, RWTH Aachen University, Aachen 52056, Germany; [c]Institute for Mechanics of Advanced Materials and Structures, Chengdu University, Chengdu 610106, People's Republic of China; [d]Central Iron and Steel Research Institute, Beijing 100081, People's Republic of China; [e]Department of Materials Science and Engineering, North Carolina State University, Raleigh, NC 27695, USA; [f]School of Materials Science and Engineering, Nanjing University of Science and Technology, Nanjing 210094, People's Republic of China

An ideal ultrafine-grained (UFG) microstructure for high strength and high ductility should have short dislocation-slip path to impede dislocation slip and very low dislocation density to ensure more room for dislocation accumulation. Such a microstructure is hard to produce, especially for UFG metals produced by severe plastic deformation techniques. Here, we report an ideal UFG structure produced by reverse transformation of deformation-induced martensite in 304 L austenitic stainless steel. It produced small grains and a high density of nanotwins for both high strength and high ductility. This approach is applicable to face-centered cubic metals with low stacking fault energy.

Keywords: Ultrafine Grains, Dislocations, Twins, Stacking Faults, Strength and Ductility

What is an ideal ultrafine-grained (UFG) structure for high strength and high ductility? First, to produce high strength in a pure metal or alloy, the path for free slip of dislocations needs to be short to make it hard for dislocation to slip. In other words, high spatial density of barriers to block dislocation slip is needed. Second, to produce high tensile ductility, effective accumulation of crystalline defects such as dislocations, stacking faults (SFs) and/or twin boundaries (TBs) is needed to produce a high strain-hardening rate to deter necking under tensile strain.[1–3] This requires microstructures that can effectively accumulate crystalline defects such as dislocations and have initial low dislocation density to leave enough room for dislocation accumulation.

It is a challenge to produce UFG microstructures with the ideal characteristics described above, especially in large work pieces. UFG metals and alloys are usually produced by a top-down approach named severe plastic deformation (SPD) because it can produce bulk UFG work pieces that are large enough for real-world structural applications at reasonably low cost.[4] SPD techniques refine grains by plastically deforming a metal to extremely large plastic strain, which results in dislocation rearrangement and grain subdivision.[5–7] Consequently, high density of dislocations often exists in the as-processed UFG metals,[8,9] which leaves little room for further dislocation accumulation. This leads to low strain hardening and low ductility, which has been reported in UFG materials produced by the SPD methods such as equal channel angular pressing (ECAP),[10] accumulative roll bonding [11–13] and high pressing torsion.[14] Annealing UFG metals to lower dislocation density can recover some of the ductility, but often at the cost of significant reduction in strength.[15,16] In addition, some dislocation structures produced by SPD could be relatively stable and hard to be fully annealed out without significant grain growth, which dramatically decreases the strength.[17]

Here, we report an ideal UFG microstructure in 304 L austenitic stainless steel (304 L SS) that was

*Corresponding author. Email: chxhuang@scu.edu.cn, wangqy@scu.edu.cn

produced by reverse transformation of deformation-induced martensite. This UFG microstructure is composed of fine grains and high density of twins to ensure high strength, while its low dislocation density leaves more room for strain hardening. Furthermore, the low stacking fault energy (SFE) of 304 L SS leads to simultaneous accumulation of both dislocations and SFs, making it more effective in strain hardening. As a result, a combination of high yield strength (810 MPa) and high uniform elongation (30%) is obtained in tensile tests.

The 304 L SS has a composition (wt%) of 0.007 C, 18.46 Cr, 11.82 Ni, 1.61 Si, 0.008 S, 0.018 P, 0.29 Mn and the balance Fe. The UFG microstructure was produced by means of phase-reversion annealing of heavily cold-deformed austenite.[18,19] Specifically, the coarse-grained (CG) samples were first processed by ECAP at room temperature (RT) for six passes, which produced a high volume fraction ($\sim 70\%$) of deformation-induced martensite with a mean grain size of ~ 90 nm.[20] This was followed by annealing at 625°C for 60 min, which reversely transformed all martensite back to austenite with ultrafine grain sizes and high density of nanotwins. Figure 1(a) shows the typical transmission electron microscopy (TEM) image of the phase-reverted microstructure with a mean grain size of 660 nm. It reveals equiaxed grains with sharp grain boundaries (GBs), annealing nanotwins, and very low dislocation density. Approximately 75% of grains were observed containing annealing nanotwins in the TEM micrograph. The actual fraction of grains that contain annealing twins should be higher since some twins without appropriate orientation may be not observed under TEM.[21] Figure 1(b) is an electron back scattered diffraction (EBSD) image, which shows a high fraction of the $\Sigma 3$ TBs, as marked with red lines. The average thickness of the twins is ~ 110 nm, as shown in Figure 1(c). The GB misorientation distribution is shown in Figure 1(d). Ignoring the boundaries with misorientation less than 2°, the high-angle GBs ($> 15°$) comprise 92% of the total GBs. The highest peak at $\sim 60°$ results from $\Sigma 3$ TBs, which is 23.1% of the total boundaries.

By uniaxial tensile testing at RT, the yield strength and ultimate tensile strength of the UFG samples were measured to be 810 and 970 MPa, respectively. The high yield strength can be attributed to fine grain sizes, according to the Hall–Petch relationship. In addition,

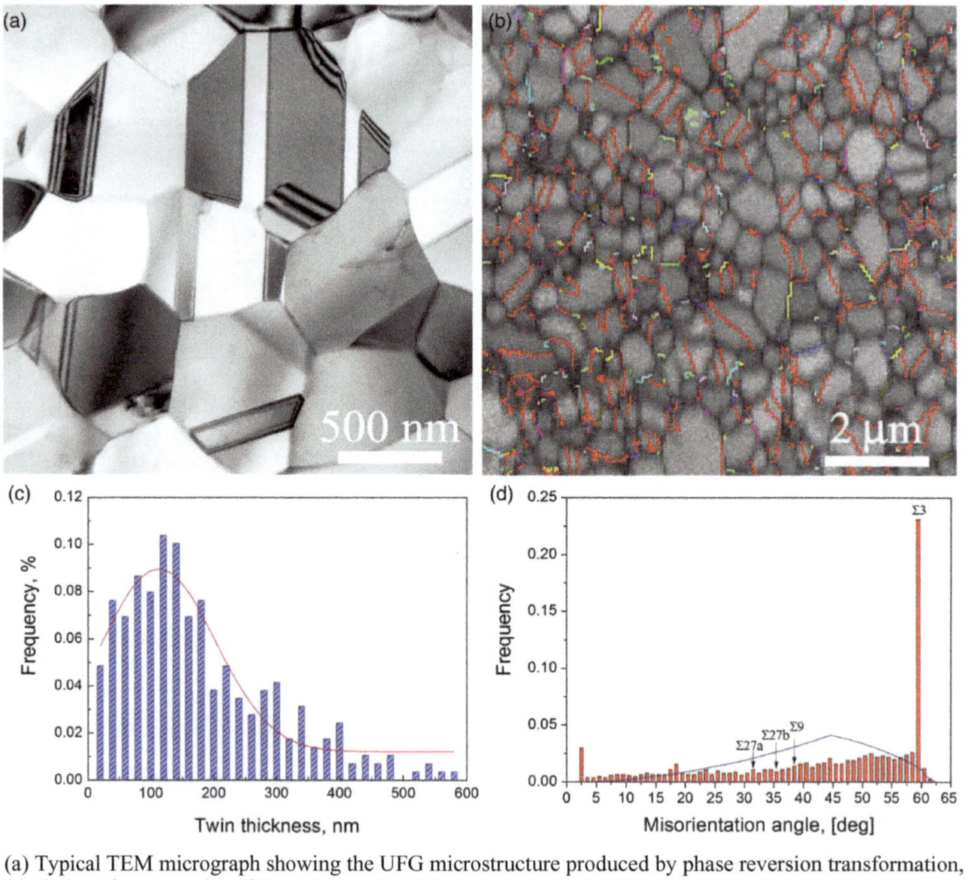

Figure 1. (a) Typical TEM micrograph showing the UFG microstructure produced by phase reversion transformation, (b) EBSD image. The black lines denote random GBs and the colored lines represent coincident-site lattice boundaries, of which the $\Sigma 3$ TBs are represented by red lines, (c) distribution of twin thickness, (d) distribution of GB misorientation angles.

the high density of nanotwins should also have contributed to the observed high yield strength, because TBs play a similar role to that of GBs.[22,23] A high uniform elongation of 30% was obtained, which is much higher than that of most UFG materials.[8–14] Figure 2(a) shows the representative true stress–strain curves of the current ideal UFG 304 L sample and its CG counterpart. The samples used for tensile tests are in dog-bone shape with gauge dimensions of $1 \times 2 \times 8$ mm^3 (see Figure S1 in the Supplementary material). A yield drop in stress is observed in the UFG sample, suggesting a strong pinning of mobile dislocations, which underpins the high yield stress. After yielding, the Lüders deformation is activated for a few percent of strain with low strain-hardening rate (see Figure S1).With further straining, the UFG sample undergoes high strain hardening, which produces a large uniform plastic deformation. X-ray diffraction analysis at tensile strains of 10.9%, 19.2% and 29.3% revealed no strain-induced martensite (see Figure S2), indicating that no martensitic transformation occurred during tensile deformation. It has been reported that decreasing austenite grain size can enhance the austenite stability to impede strain-induced martensite transformation.[24,25] In addition, the relatively high nickel and very low carbon contents in this 304 L steel also improved the austenite stability. Therefore, the high strain-hardening rate of the current UFG sample was caused by the accumulation of dislocations and/or other crystalline defects.

Figure 2(b) compares the yield strength and uniform elongation of the current sample with those reported in the literature for 304 L SSs with Ni concentration in the range of 10–12%.[26–30] As shown, for most reported data points, the yield strength usually decreases with increasing uniform elongation. However, the strengths-uniform elongations of the current UFG samples are clearly superior to those reported in literature. The only other data point (the pink triangle [30]) that is comparable to the current one is from a sample that had strain-induced martensite transformation during tensile deformation. It has been reported that the strain-induced martensite causes undesired hydrogen embrittlement [31–33] and decreases corrosion resistance of austenitic SSs.[34–36] These problems are avoided in the current ideal UFG sample, which has more stable austenite due to ultrafine grain sizes.

In order to identify the deformation mechanism, the UFG samples after tensile testing were characterized by TEM and high-resolution TEM (HRTEM). Figure 3(a) shows the typical microstructure at low magnification after ~28% tensile strain. It reveals very high dislocation density, in sharp contrast with the initial microstructure before testing (Figure 1(a)). Figure 3(b) further shows the defect structure in the interior of a single grain at higher magnification by tilting the incident beam to parallel to the [011] zone axis. It reveals a high density

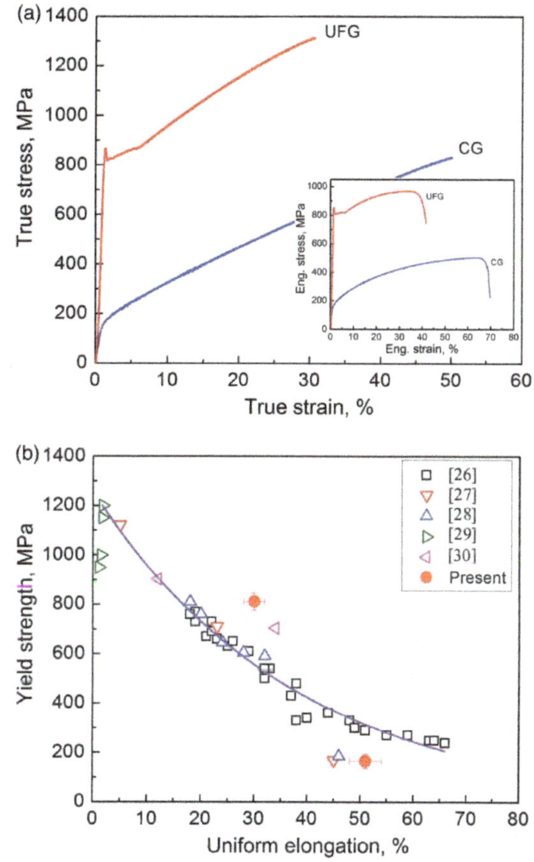

Figure 2. (a) Typical tensile true stress–strain curves for UFG and CG 304 L SSs at a quasi-static strain rate of 5×10^{-4} s^{-1} and RT. The inset is engineering stress–strain curves, (b) excellent strength–ductility combination of the current ideal UFG SS in comparison with other 304 L SSs (with Ni content of 10 ~ 12%).

of straight lines lying on the $(1\bar{1}1)$ plane, with lengths of several tens to hundreds of nanometers and inter-line spacing of 4 ~ 60 nm. The selected-area diffraction (SAD) pattern shows obvious streaks along the $[1\bar{1}1]$ direction, indicating that these lines are edge-on SFs. The formation of high density of SFs is due to the low SFE of this 304 L SS, ~34 mJ/m^2.[37] Low SFE promotes the dissociation of a full dislocation into two partial dislocations during plastic deformation. In UFG and nanocrystalline materials, the leading Shockley partial dislocations can originate from GBs and glide into grain interiors, leaving behind wide SFs.[38–40] In low SFE materials, the SFs can be very wide, even across the entire grains, as shown in Figure 3(b).

Figure 3(c) and 3(d) show annealing nanotwins in two ultrafine grains. High densities of dislocations are clearly shown near some TBs, suggesting that TBs are effective in blocking and storing dislocations. It is also seen that the interaction between dislocations and TBs made TBs curved and deviate from the original straight

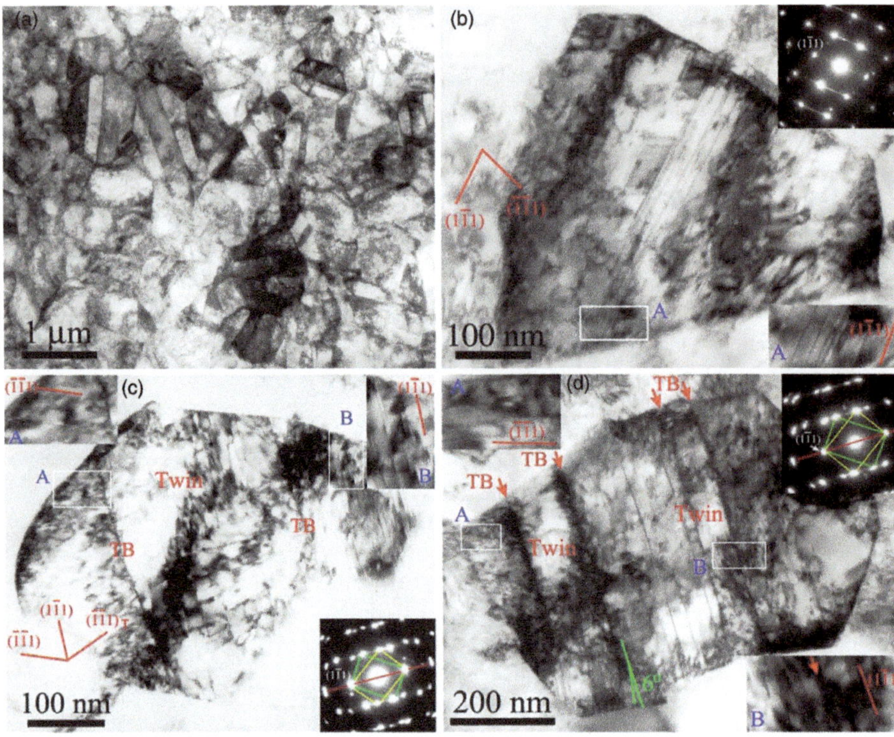

Figure 3. Typical TEM micrographs showing the microstructures and defects in ultrafine austenitic grains after ~28% tensile strain: (a) overview of the microstructure, (b) defect structure in a 520 nm grain, (c) a 400 × 570 nm grain with an annealing twin, (d) a 500 × 700 nm grain with two nanotwins. The corresponding SAD patterns with [011] zone axis are shown in the insets. The insets denoted by capital letters A and B are enlarged from the corresponding white frame regions, which reveal different slip planes activated in different regions.

line, which is consistent with a previous report.[41] In addition, the nanotwins lead to activation of more slip systems for dislocations/SFs. As seen in the inset images enlarged from the corresponding white frame regions in Figure 3(c) and 3(d) (denoted by capital letters A and B), the dislocations/SFs on the left part of matrix glided on the $(\bar{1}\bar{1}1)$ plane, while on the right side they glided on the $(1\bar{1}1)$ plane. Within twin lamellae, the dislocations can glide on both $(1\bar{1}1)$ and $(\bar{1}\bar{1}1)_T$ slip planes, as evidenced by the rhombic grids in the center of Figure 3(c). It is expected that the activation of multiple slip systems produces high strain-hardening rate, because the simultaneous activation of different slip systems needs a higher shear stress,[42] and the dislocations from different slip systems may entangle and react with each other.[43] The dislocation density in deformed grains is measured to be about 1.8×10^{15} m^2.

The generation of SFs was also observed within nanotwin lamellae, as demonstrated by the straight lines indicated by arrows in Figure 4. These SFs either traveled across twin lamellae or terminated inside the twin. The lower-left inset is an HRTEM image viewed from the [011] zone axis, confirming that these lines are SFs. The SFs can originate from TBs via dislocation-TB reactions. As revealed by molecular dynamic simulation [44] and analytical model,[45] when a perfect 60° dislocation meets a TB, it can be dissociated into a 30° partial and a 90° partial. The partials can transmit across the TB to slip in the twin to produce SFs, depending on the orientation and magnitude of local stress. As shown in Figure 4, most of the SFs are originated from TBs, which is consistent with the above twin-dislocation interaction theory.

The above microstructure and defect observations indicate that the ideal UFG structure produced by phase-reversion annealing can effectively accumulate a high density of dislocations during uniaxial tensile test, which explains its high strain-hardening rate. The contribution of dislocation accumulation to the plastic flow stress can be evaluated by the Taylor dislocation hardening relation, $\sigma = M\alpha Gb\sqrt{\rho}$, where σ is the true flow stress, M is the Taylor factor, α is the material constant, G is the shear modulus, b is the Burgers vector, and ρ is the dislocation density accumulated in grain interior during plastic strain. In this study, the dislocation density after tensile test is as high as 1.8×10^{15} m^2, which can be considered accumulated during tensile deformation because the initial dislocation density is extremely low (see Figure 1(a)). Using the typical parameters for 304 L SS ($M = \sqrt{3}$, $\alpha = 0.28$,[46] $G = 78$ GPa and

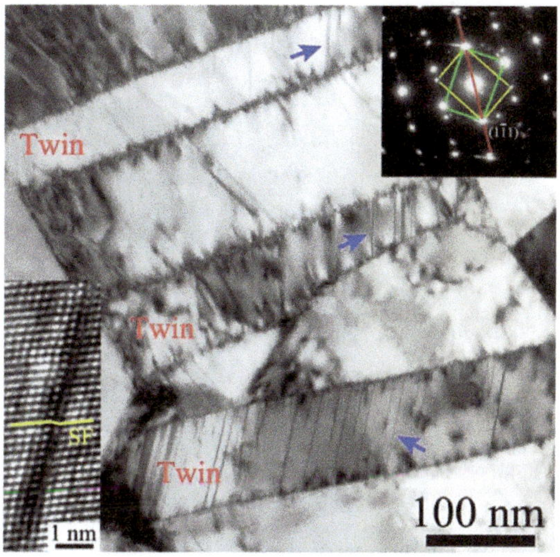

Figure 4. High density of SFs (indicated by arrows) formed inside nanotwin lamellae. The upper-right inset is corresponding SAD pattern with the [011] zone axis. The lower-left inset is an HRTEM image showing a stacking fault formed within a nanotwin.

$b = 0.2535$ nm), the increase in flow stress is calculated to be about 407 MPa. This value is close to the increase (470 MPa) of flow stress from yield point (810 MPa) to the true stress at 28% strain (1,280 MPa), as shown in Figure 2(a).

It should be noted that due to the low SFE of 304LSS, most dislocations are dissociated with SF ribbons, and some SFs have one or both ends pinned at GBs or TBs, as seen in Figures 3 and 4. This makes dislocation recovery via cross-slip and climb very difficult, because these processes require the wide SF ribbon to constrict into a full dislocation first. In addition, the SFs are effective barriers for blocking and accumulating dislocations.[47,48] When a moving dislocation encounters with a SF, it can either interact with the SF or accumulate near the SF if the applied stress is not high enough to activate the reaction.[45] It has been reported that SFs are very effective in promoting dislocation accumulation in UFG Cu–Zn alloy.[40] Therefore, lowering SFE has been found to enhance dislocation density and hence improve the strain-hardening rate of these alloys.[40,49–51]

Another important feature of the current UFG sample is the presence of a high density of nanotwins. These nanotwins must have contributed to the high strain-hardening rate. It has been reported that TBs are strong obstacles for blocking gliding dislocations and effective sites for storing dislocations,[44,45,51–53] which is also clearly shown in Figure 3(c) and 3(d). Lu et al. [54] proposed four types of TB-related dislocations to explain the strain hardening of nanotwinned Cu. They are glissile Shockley partials, sessile dislocation locks and extended SFs at TBs, as well as thread dislocations in the twin/matrix lamellae. All of these dislocations can be found in our samples, as shown in Figures 3 and 4. However, fabricating large bulk UFG materials with high density of nanotwins for structural applications is still a challenge to date. The current work provides a new approach to produce bulk low SFE UFG materials, which can be engineered with high density of annealing nanotwins for high strength and high ductility.

Other approaches have also been reported to improve the strength and ductility of UFG metals. For example, introducing nano-precipitates into the UFG matrix can recover some strain hardening. However, the increase in uniform elongation is still very limited, usually less than 10% in UFG Al alloys, for instance.[2,55] One approach for regaining high strain hardening is phase transformation, which can lead to high strength and good ductility in some austenitic steels.[18,30,56,57] For example, bulk UFG 301 grade austenitic SSs exhibited very high tensile yield strength (900–1,000 MPa) and high total elongation (30–40%) via strain-induced martensitic transformation.[18,19] However, the strain-induced martensite results in significant hydrogen embrittlement [31–33] and decreased corrosion resistance [34–36] in austenitic SSs, both of which are potential hazards for some applications such as nuclear reactor structural materials.[58,59] To avoid such problems, strain-induced martensite transformation has to be avoided during the plastic deformation of austenitic SSs. The ideal UFG structure produced by phase reversion in the current work avoids this problem and may have potential applications where resistance to hydrogen embrittlement and corrosion is critical.

In summary, we have produced an ideal UFG microstructure with fine grains, high density of annealing twins and low dislocation density by phase-reversion annealing of deformation-induced martensite in 304 L SS. It exhibited a superior combination of strength and ductility. The high strength resulted from the small grain size and high density of annealing twins, while the high ductility resulted from the high density of twins and initial low dislocation density, which were not only effective in blocking and storing dislocations, but also provided more room for dislocation accumulation. The low SFE of 304 L SS also helped with increasing the strain-hardening rate by producing SFs and dissociated dislocations with wide SF ribbons. These structural and deformation features ensured a high strain-hardening rate for high ductility. Finally, the ideal microstructure presented here can only be produced in fcc metals with low SFEs. Any other approaches that can produce fine grain size with high dislocation barriers and low dislocation density are expected to produce similar superior mechanical properties.

Acknowledgements This work was supported from National Natural Science Foundation of China (11172187 and 11172188). C.X. Huang acknowledges the Program for New Century Excellent Talents in University (NCET-12–0372) and Outstanding Young Scholars of Sichuan University (2012SCU04A05).

References

[1] Zhu YT, Liao XZ. Nanostructured metals-retaining ductility. Nature Mater. 2004;3:351–352.
[2] Zhao YH, Liao XZ, Cheng S, Ma E, Zhu YT. Simultaneously increasing the ductility and strength of nanostructured alloys. Adv Mater. 2006;18:2280–2283.
[3] Zhao YH, Bingert JF, Liao XZ, Cui BZ, Han K, Sergueeva AV, Mukherjee AK, Valiev RZ, Langdon TG, Zhu YT. Simultaneously increasing the ductility and strength of ultrafine-grained pure copper. Adv Mater. 2006;18:2949–2953.
[4] Valiev RZ, Estrin Y, Horita Z, Langdon TG, Zechetbauer MJ, Zhu YT. Producing bulk ultrafine-grained materials by severe plastic deformation. JOM. 2006;58:33–39.
[5] Bay B, Hansen N, Hughes DA, Kuhlmann-Wilsdore D. Overview no. 96: evolution of fcc deformation structures in polyslip. Acta Metal Mater. 1992;40:205–219.
[6] Bay B, Hansen N, Kuhlmann-Wilsdore D. Deformation structures in light rolled pure aluminum. Mater Sci Eng A. 1989;113:385–397.
[7] Huang JY, Zhu YT, Jiang H, Lowe TC. Microstructures and dislocation configurations in nanostructured Cu processed by repetitive corrugation and straightening. Acta Mater. 2001;49:1497–1505.
[8] Meyers MA, Mishra A, Benson DJ. Mechanical properties of nanocrystalline materials. Prog Mater Sci. 2006;51:427–556.
[9] Witkin DB, Lavernia EJ. Synthesis and mechanical behavior of nanostructured materials via cryomilling. Prog Mater Sci. 2006;51:1–60.
[10] Valiev RZ, Langdon TG. Principles of equal-channel angular pressing as a processing tool for grain refinement. Prog Mater Sci. 2006;51:881–981.
[11] Saito Y, Tsuji N, Utsunomiya, Sakai T, Hong RG. Ultra-fine grained bulk aluminum produced by accumulative roll-bonding (ARB) process. Scripta Mater. 1998;39:1221–1227.
[12] Hausol T, Hoppel HW, Goken M. Tailoring materials properties of UFG aluminum alloys by accumulative roll bonded sandwich-lick sheets. J Mater Sci. 2010;45:4733–4738.
[13] Beyerlein IJ, Mara NA, Carpenter JS, Nizolek T, Mook WM, Wynn TA, McCabe RJ, Mayeur JR, Kang K, Zheng S, Wang J, Pollock TM. Interface-driven microstructure development and ultra high strength of bulk nanostructured Cu-Nb multilayers fabricated by severe plastic deformation. J Mater Sci. 2013;28:1799–1812.
[14] Zhilyaev AP, Langdon TG. Using high-pressure torsion for metal processing: fundamentals and applications. Prog Mater Sci. 2008;53:893–979.
[15] Park KT, Kim YS, Lee JG, Shin DH. Thermal stability and mechanical properties of ultrafine grained low carbon steel. Mater Sci Eng A. 2000;293:165–172.
[16] Yu YC, Kao PW, Chang CP. Transition of tensile deformation behaviors in ultrafine-grained aluminum. Acta Mater. 2005;53:4019–4028.
[17] Stolyarov VV, Zhu YT, Lowe TC, Valiev RZ. Microstructure and properties of pure Ti processed by ECAP and cold extrusion. Mater Sci Eng A. 2001;303:82–89.
[18] Somani MC, Juntunen P, Karjalainen LP, Misra RDK, Kyrolainen A. Enhanced mechanical properties through reversion in metastable austenitic stainless steels. Metall Mater Trans A. 2009;40:729–744.
[19] Misra RDK, Kumar BR, Somani M, KarJalainen P. Deformation processes during tensile straining of ultrafine/nanograined structures formed by reversion in metastable austenitic steels. Scripta Mater. 2008;59:79–82.
[20] Huang CX, Yang G, Deng B, Wu SD, Li SX, Zhang ZF. Formation mechanism of nanostructures in austenitic stainless steel during equal channel angular pressing. Phil Mag. 2007;87:4949–4971.
[21] Wu XL, Zhu YT. Inverse grain-size effect on twinning in nanocrystalline Ni. Phys Rev Lett. 2008;101:025503.
[22] Lu L, Shen YF, Chen XH, Qian LH, Lu K. Ultrahigh strength and high electrical conductivity in copper. Science. 2004;304:422–426.
[23] Lu K, Lu L, Suresh S. Strengthening materials by engineering coherent internal boundaries at the nanoscale. Science. 2009;324:349–352.
[24] Yoo CS, Park YM, Jung YS, Lee YK. Effect of grain size on transformation-induced plasticity in an ultrafine-grained metastable austenitic steel. Scripta Mater. 2008;59:71–74.
[25] Turtletaub S, Suiker ASJ. Grain size effects in multiphase steels assisted by transformation-induced plasticity. Inter J Solids Mech. 2006;43:7322–7336.
[26] Mataya MC, Brown EL, Riendeau MP. Effect of hot-working on structure and strength of type 304 L austenitic stainless steel. Metall Trans A. 1990;21:1969–1987.
[27] Qu S, Huang CX, Gao YL, Yang G, Wu SD, Zang QS, Zhang ZF. Tensile and compressive properties of AISI 304 L stainless steel subjected to equal channel angular pressing. Mater Sci Eng A. 2008;475:207–216.
[28] Huang CX, Yang G, Gao YL, Wu SD, Zhang ZF.: Influence of processing temperature on the microstructures and tensile properties of 304 L stainless steel by ECAP. Mater Sci Eng A. 2008;485:643–650.
[29] Ravi Kumar B, Mahato B, Sharma S, Sahu JK. Effect of cyclic thermal process on ultrafine grain formation in AISI 304 L austenitic stainless steel. Metall Mater Trans A. 2009;40:3226–3234.
[30] Ravi Kumar B, Raabe D. Tensile deformation characteristics of bulk ultrafine-grained austenitic stainless steel produced by thermal cycling. Scripta Mater. 2012;66:634–637.
[31] Eliezer D, Chakrapani DG, Altstetter CJ, Puch EN. The influence of austenite stability on the hydrogen embrittlement and stress-corrosion cracking of stainless steel. Metall Trans A. 1979;10:935–941.
[32] Han G, He J, Fukuyama S, Yokogawa K. Effect of strain-induced martensite on hydrogen environment embrittlement of sensitized austenitic stainless steels at low temperatures. Acta Mater. 1998;46:4559–4570.
[33] Zhang L, Wen M, Imade M, Fukuyama S, Yokogawa K. Effect of nickel equivalent on hydrogen gas embrittlement of austenitic stainless steels based on type 316 at low temperatures. Acta Mater. 2008;56:3414–3421.
[34] Hamada AS, Karjalainen LP, Somani MC. Electrochemical corrosion behavior of a novel submicron-grained

austenitic steel in an acidic NaCl solution. Mater Sci Eng A. 2006;431:211–217.
[35] Alvarez SM, Bautista A, Velasco F.: Influence of strain-induced martensite in the anodic dissolution of austenitic stainless steel in acid medium. Corros Sci. 2013;69: 130–138.
[36] Balusamy T, Sankara Narayanan TSN, Ravichandran K, Song park II, Lee MH. Influence of surface mechanical attrition treatment (SMAT) on the corrosion behavior of AISI 304 stainless steel. Corros Sci. 2013;74:332–344.
[37] Schramm RE, Reed RP. Stacking-fault energies of 7 commercial austenitic stainless-steels. Metall Trans A. 1975;6:1345–1351.
[38] Zhu YT, Liao XZ, Wu XL. Deformation twinning in nanocrystalline materials. Prog Mater Sci. 2012;57:1–62.
[39] Huang CX, Wang K, Wu SD, Zhang ZF, Li GY, Li SX. Deformation twinning in polycrystalline copper at room temperature and low strain rate. Acta Mater. 2006;54:655–665.
[40] Wang ZW, Wang YB, Liao XZ, Zhao YH, Lavernia EJ, Zhu YT, Horita Z, Langdon TG. Influence of stacking fault energy on deformation mechanism and dislocation storage capacity in ultrafine-grained materials. Scripta Mater. 2009;60:52–55.
[41] Cao Y, Wang YB, Chen ZB, Liao XZ, Kawasaki M, Ringer SP, Langdon TG, Zhu YT. De-twinning via secondary twinning in face-centered cubic alloys. Mater Sci Eng A. 2013;578:110–114.
[42] Seeger A, Diehl J, Mader S, Rebstock H. Work-hardening and work softening of face-centred cubic metal crystals. Phil Mag. 1957;2:323–350.
[43] Wu XL, Zhu YT, Wei GY, Wei Q. Strong strain hardening in nanocrystalline nickel. Phys Rev Lett. 2009;103:205504.
[44] Jin ZH, Gumbsch P, Albe K, Ma E, Lu K, Gleiter H, Hahn H. Interactions between non-screw lattice dislocations and coherent twin boundaries in face-centered cubic metals. Acta Mater. 2008;56:1126–1135.
[45] Zhu YT, Wu XL, Liao XZ, Narayan J, Kecskes LJ, Mathauhu SN. Dislocation-twin interactions in nanocrystalline fcc metals. Acta Mater. 2011;59:812–821.
[46] Kassner ME. Taylor hardening in five-power-law creep of metals and Class M alloys. Acta Mater. 2004;52:1–9.
[47] Jian WW, Cheng GM, Xu WZ, Koch CC, Wang QD, Zhu YT, Mathaudhu SN. Physics and modeling of strengthening of metals by parallel stacking faults. Appl Phys Lett. 2013;103:133108.
[48] Jian WW, Cheng GM, Xu WZ, Yuan H, Tsai MH, Wang QD, Koch CC, Zhu YT, Mathaudhu SN. Ultrastrong Mg alloy via nano-spaced stacking faults. Mater Res Lett. 2013;1:61–66.
[49] Zhao YH, Zhu YT, Liao XZ, Horita Z, Langdon TG. Tailoring stacking fault energy for high ductility and high strength in ultrafine grained Cu and its alloy. Appl Phys Lett. 2006;89:121906.
[50] Huang CX, Hu W, Yang G, Zhang ZF, Wu SD, Wang QY, Gottstein G. The effect of stacking fault energy on equilibrium grain size and tensile properties of nanostructured copper and copper-aluminum alloys processed by equal channel angular pressing. Mater Sci Eng A. 2012;556:638–647.
[51] Huang CX, Hu WP, Wang QY. Strain-rate sensitivity, activation volume and mobile dislocations exhaustion rate in nanocrystalline Cu-11.1at%Al alloy with low stacking fault energy. Mater Sci Eng A. 2014;611:274–279.
[52] Zhu T, Gao HJ. Plastic deformation mechanism in nanotwinned metals: an insight from molecular dynamics and mechanistic modeling. Scripta Mater. 2012;66:843–848.
[53] Li N, Wang J, Misra A, Zhang X, Huang JY, Hirth JP. Twinning dislocation multiplication at a coherent twin boundary. Acta Mater. 2011;59:5989–5996.
[54] Lu L, You ZS, Lu K. Work hardening of polycrystalline Cu with nanoscale twins. Scripta Mater. 2012;66:837–842.
[55] Horita Z, Ohashi K, Fujita T, Kaneko K, Langdon TG. Achieving high strength and high ductility in precipitation-hardened alloys. Adv Mater. 2005;17:1599–1602.
[56] Ma YQ, Jin JE, Lee YK. A repetitive thermomechanical process to produce nano-crystalline in a metastable austenitic steel. Scripta Mater. 2005;52:1311–1315.
[57] Huang CX, Yang G, Wang C, Zhang ZF, Wu SD. Mechanical behaviors of ultrafine-grained 301 austenitic stainless steel produced by equal-channel angular pressing. Metall Mater Trans A. 2011;42:2061–2071.
[58] Scott P. A review of irradiation assisted stress corrosion cracking. J Nucl Mater. 1994;211:101–122.
[59] Zinkle SJ, Was GS. Materials challenges in nuclear energy. Acta Mater. 2013;61:735–758.

15

Improved Fatigue Strengths of Nanocrystalline Cu and Cu–Al Alloys

Xianghai An*, Qingyun Lin, Shiding Wu and Zhefeng Zhang*

Shenyang National Laboratory for Materials Science, Institute of Metal Research, Chinese Academy of Sciences, 72 Wenhua Road, Shenyang 110016, People's Republic of China

Increasing the fatigue strengths of high-strength materials, especially their endurance limits is essentially challenging. The fatigue strengths of nanocrystalline (NC) Cu and Cu–Al alloys processed by high-pressure torsion were prominently enhanced with a decrease in stacking fault energy. This remarkable escalation can be attributed not only to their significantly increased tensile strengths in macro-scale, but also to the essentially improved microstructure stability and reduced strain localization within shear bands in microscale. Owing to the limitation of their intrinsic fatigue mechanisms, the fatigue limits of NC Cu and Cu–Al alloys cannot always acquire appreciable improvement with their monotonic strengths.

Keywords: Nanocrystalline, Stacking Fault Energy, Fatigue Strength, Fatigue Mechanism, Severe Plastic Deformation

Superior strength and hardness furnished by the nanocrystalline (NC) metallic materials provide them great potentials as prospective engineering materials, while the low ductility makes materials scientists endeavor devotedly to manipulate the nanostructures architecturally for improving their overall properties.[1–5] Regarding the long-term performance of metallic structures, apart from the monotonic properties of NC metals, their fatigue strength is of critical and crucial concern, with an essential impact on their practical applications.[6–8] In comparison to the coarse-grained (CG) counterparts, NC materials present enhanced fatigue strengths due to the high tensile strengths. Nevertheless, this increase, particularly of their endurance limits, is smaller than that of the tensile strengths, which may be attributed to the change of their fatigue damage mechanisms.[7,8] Thus, improving their fatigue strengths is significant scientifically and challenging technologically to enhance the integrity and reliability of NC metallic materials as structure components.

It is generally accepted that upgrading the tensile strengths of materials will prominently increase their fatigue strengths.[9,10] In fact, the fatigue damage mechanisms also play a crucial role in their fatigue strengths.[7–10] For NC materials, the extensive thermally activated microstructure instability in terms of grain coarsening and shear banding is the predominant cyclic deformation mechanism, driving their final failure.[10–14] Recent results revealed that decreasing the stacking fault energy (SFE) via alloying design can significantly increase the fatigue strengths of the NC Cu–Al and Cu–Zn alloys processed by equal channel annular pressing (ECAP) through the enhancement of their tensile strengths and stabilization of the nanostructure via critically restricting the relevant mechanisms.[15,16]

As well known, severe plastic deformation (SPD) processes remarkably affect the formation of nanostructures and corresponding mechanical properties.[17,18] Compared with ECAP, high-pressure torsion (HPT), another popular SPD technique, can produce remarkably finer grains and a higher fraction of high-angle grain boundaries (GBs) and then harvest higher tensile strengths.[18,19] This indicates that the fatigue strengths of the NC materials processed by ECAP may be further escalated when prepared by HPT. However, for the high-strength materials, the increase in their fatigue limits is extremely difficult and the fatigue strengths may remain constant or even decrease with a further increase in the tensile strengths, while the detailed mechanisms are still mysterious.[9,10,20] In this investigation, NC Cu and Cu–Al alloys prepared by HPT were cyclically deformed to critically explore several questions. Do the fatigue strengths still increase with lowering of the SFE

*Corresponding authors. Emails: anxianghai@gmail.com, zhfzhang@imr.ac.cn

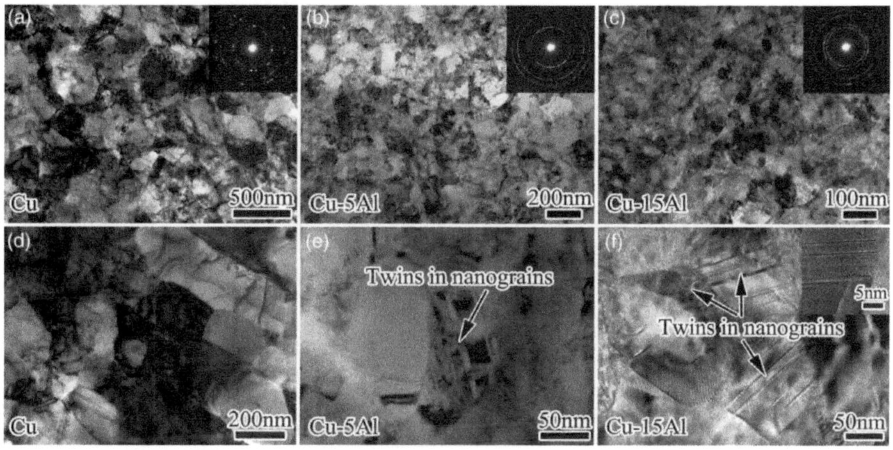

Figure 1. Bright-field TEM images of typical microstructures of NC Cu, Cu–5 at% Al alloy and Cu–15 at% Al alloy.

in NC Cu–Al with higher strength obtained by HPT? Due to the upgraded tensile strength, are the fatigue endurance limits significantly enhanced when compared with those produced by ECAP? What is the fundamental relationship between tensile strength, fatigue strength and fatigue damage mechanism of NC materials? These essential issues will be meticulously elucidated and uncovered through systematic investigations and detailed analysis.

Pure Cu of 99.998% purity, Cu–5 at% Al alloy and Cu–15 at%Al alloy were selected and the detailed microstructures of CG materials have been reported elsewhere.[18,21] In order to be suitable for fatigue experiments, disks for HPT processing were shaped with a diameter of 30 mm and a thickness of 2.5 mm (Figure S1). These disks were processed using HPT facilities at the Erich Schmid Institute, Austrian Academy of Science, Austria, through five revolutions under an imposed pressure of 5.0 GPa at room temperature (RT) under quasi-constrained conditions.[22,23] Following HPT, the specimens with a gauge section of 8 mm × 2 mm × 1 mm and 4 mm × 2 mm × 1.5 mm were taken from the off-center positions within the disks at positions of ~5 mm from the centers for tensile test and symmetrical cyclic push–pull (stress ratio $R = -1$) experiments, respectively, and then any problem associated with microstructural inhomogeneities in the center of the disks can be avoided. All mechanical tests were carried out at RT using an Instron 8,871 servohydraulic testing machine. At least four tensile tests were performed on each sample condition to substantiate the reproducibility of the stress–strain curves at a strain rate of 5×10^{-4} s^{-1}, while the fatigue experiments were conducted by using a triangular wave form with a frequency of 30 Hz under constant stress control. To characterize the fatigue damage mechanism, the fine microstructural features and post-fatigued microstructures of these samples were characterized using an FEI Tecnai F20 transmission electron microscope (TEM) operating at 200 kV. Their surface damage morphology was examined by a field emission gun scanning electron microscope (LEO SUPRA35 SEM).

The characteristic microstructures of NC Cu and Cu–Al alloys exhibit the homogeneously distributed nanograins, as revealed in Figure 1. Extensive measurements revealed that these nanograins are much finer than those processed by ECAP and the average grain sizes decrease with lowering of the SFE by values of ~140, ~65, ~40 nm for NC Cu, Cu–5 at%Al alloy and Cu–15 at% alloy, respectively (Figure S2 and Table 1). This means that no matter which kinds of SPD methods are applied, the sizes of the nanograins are notably decreased with the reduction of the SFE due to the restricted cross-slip and the transformation of grain refinement mechanisms from dislocation subdivision to twins fragmentation.[14,21,24,25] In addition, deformation twins or stacking faults can be detected more frequently in the nanograins with a decrease in the SFE, while the high-resolution TEM substantiates that the deformation twins may be formed via the successive emission of Shockley partials from GBs,[26] as shown in the inset of Figure 1(f). Therefore, SFE is still a crucial factor governing the deformation mechanism in the materials with nanosacle grains, which will significantly affect the mechanical properties.

Typical tensile engineering stress–strain curves and related tensile properties of NC Cu and Cu–Al alloys are shown in Figure 2(a) and 2(b), respectively. Originating from solid solution strengthening and finer grains with lowering of the SFE, the strengths of NC Cu and Cu–Al alloys can be drastically enhanced with increasing Al contents.[16,18,21,27] Compared with those prepared by ECAP, these NC materials exhibit substantially a higher strength due to the much more extensive grain

Table 1. The grain size, tensile and fatigue properties of NC Cu and NC Cu–Al alloys processed by ECAP [16] and HPT.

	Metals	Grain size (nm)	Ultimate tensile strength (MPa)	Uniform elongation (%)	Fatigue strength (MPa)	σ_f' (MPa)	b
ECAP [16]	Cu	~200	422 ± 4	2.2 ± 0.1	100	1,737	− 0.168
	Cu–5 at% Al	~100	573 ± 5	2.9 ± 0.2	150	1,878	− 0.147
	Cu–11 at% Al	~70	770 ± 10	3.4 ± 0.3	190	1,962	− 0.140
HPT	Cu	~140	481 ± 4	3.3 ± 0.2	100	2,238	− 0.186
	Cu–5 at% Al	~65	776 ± 7	4.2 ± 0.1	170	2,754	− 0.170
	Cu–15 at% Al	~40	942 ± 15	3.4 ± 0.2	200	3,032	− 0.164

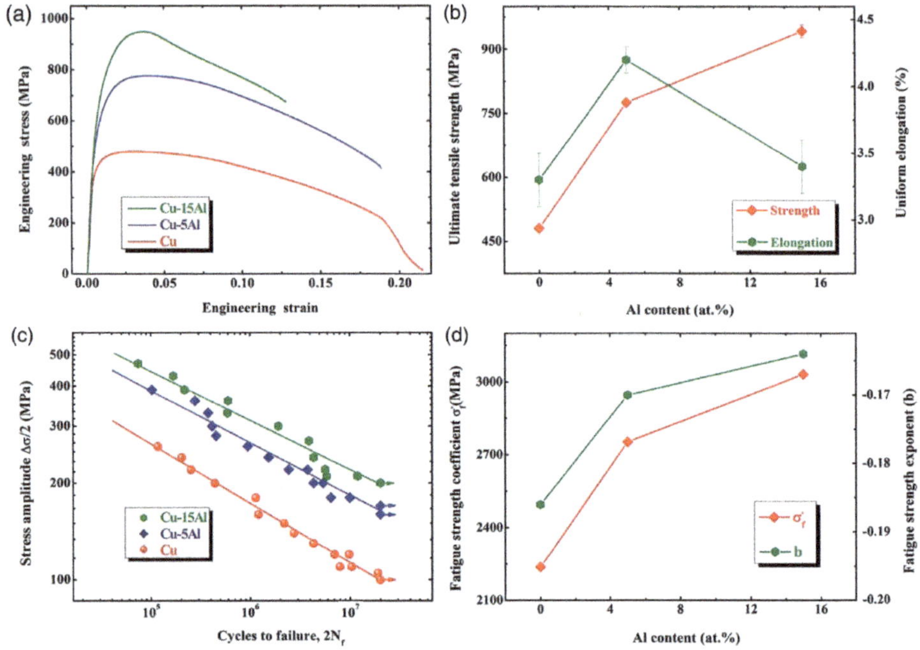

Figure 2. (a) Tensile engineering stress–strain curves of NC Cu and Cu–Al alloys; (b) the evolution of ultimate tensile strength and uniform elongation with Al content; (c) dependence of fatigue lives ($2N_f$) on the stress amplitudes of NC Cu and Cu–Al alloys and (d) the evolution of fatigue strength coefficient σ_f' and fatigue strength exponent b in the Basquin equations with the Al content in NC Cu and Cu–Al alloys.

refinement (Table 1).[16,18,21] However, different from the trends that the uniform elongation of NC Cu and Cu–Al alloys processed by ECAP presents a mild improvement with lowering of the SFE,[16,21] the ductility of the materials processed by HPT increases with the reduced SFE, but ultimately there is a reversal at the lowest SFE where the ductility decreases. Similar phenomena were also reported earlier for NC Cu and Cu–Zn alloys processed by HPT.[28,29] The counterturn evolution of ductility with SFE could originate from the extremely fine grains not able to store crystalline defects and the easy initiation of the plastic instability stemming from the GB activities during subsequent tensile testing,[18,28,29] both of which impair the ductility of NC metals with extremely low SFE.

The dependence of the reversal fatigue life ($2N_f$) on the stress amplitudes of NC Cu and Cu–Al alloys is exhibited on a log–log scale in Figure 2(c). It is apparent that the fatigue endurance limits remarkably increase with a decrease in the SFE in the NC Cu and Cu–Al alloys processed by HPT. This implies that SFE does play an intrinsic role in the fatigue strength, regardless of the kinds of SPD techniques applied, as revealed in Table 1.[15,16] Based on the well-known Basquin law, the reversal fatigue life is closely related to the stress amplitude in a fully reversed, constant-amplitude fatigue test as expressed in the following form [10]:

$$\frac{\Delta\sigma}{2} = \sigma_f'(2N_f)^b \text{ or } \mathrm{Ln}\left(\frac{\Delta\sigma}{2}\right) = \mathrm{Ln}(\sigma_f') + bLn(2N_f), \quad (1)$$

where σ_f' is the fatigue strength coefficient and b is known as the fatigue strength exponent or Basquin exponent. This predicates that the fatigue strength at certain cycles is mainly dominated by two parameters, that is, σ_f'

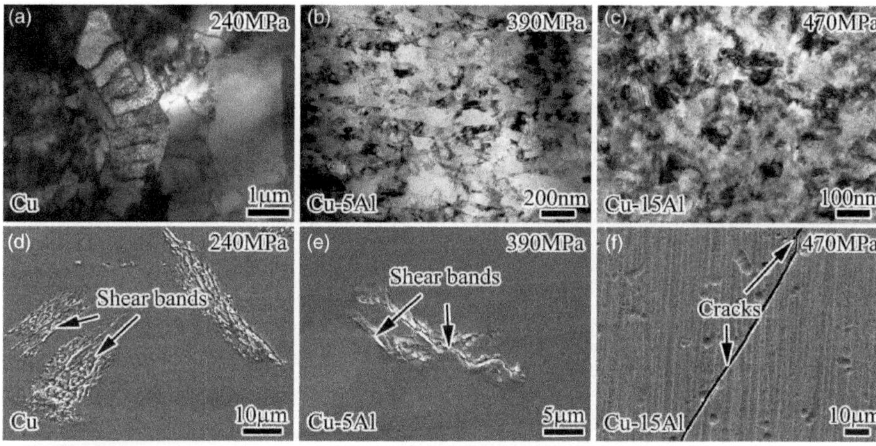

Figure 3. (a–c) The TEM micrographs of post-deformation microstructures and (d–f) SEM images of surface damage morphology of the NC Cu and Cu–Al alloys fatigued at the constant stress amplitudes of $\Delta\sigma/2 \approx 0.5\sigma_{UTS}$.

and b. Therefore, two kinds of strategies can be developed to improve the fatigue strength of materials (Figure S3), that is, increasing σ_f' or upgrading b. In the present investigation, the calculated values of σ_f' and b are apparently increased at the same time with the reduction of SFE in NC Cu and Cu–Al alloys as exhibited in Figure 2(d) and Table 1. The enhancement of σ_f' with lowering of the SFE can be reasonably expected due to its affinitive relationship with the monotonic strength in macroscale.[9,10] In fact, upgrading b essentially improves the slopes of the Basquin plots and then enhances the fatigue strength of the materials (Figure S3). Although the physical fundament of b was rarely exploited,[9,10] it is intimately pertinent to the microstructures and intrinsic fatigue damage mechanism in microscale.[30]

The representative post-deformation microstructures and surface damage morphology of the NC Cu and Cu–Al alloys fatigued at the constant stress amplitude of $\Delta\sigma/2 \approx 0.5\sigma_{UTS}$ are displayed in Figure 3. Apparently, extensive grain growth and strain localization in shear bands with striking protrusions occurred in the NC Cu after cyclic deformation, as revealed in Figure 3(a) and 3(d). However, these microstructural damages are exceptionally ameliorated with a decrease in the SFE of NC Cu–Al alloys. As exhibited in Figure 3(b) and 3(c), some grains slightly increased to $\sim$100–200 nm in the NC Cu–5 at%Al alloy, while there were no detectable changes of microstructures in the NC Cu–15 at%Al alloy. Similarly, the density of the shear bands and the extent of protrusions are remarkably decreased in the NC Cu–5 at%Al alloy, and only the primary cracks are prominently evident in the NC Cu–15 at%Al alloy in which there are no shear bands, as illustrated in Figure 3(e) and 3(f), respectively. In addition, these enhanced microstructural stability and lessened surface damage can also be conspicuous at a lower stress amplitude (Figure S4). This signals that lowering the SFE of NC materials can essentially reduce the intensity of microstructural damages induced by cyclic deformation.

It is well documented that grain growth via GB migration and highly localized shear bands, both of which are ubiquitous and greatly impair the fatigue strengths in the NC materials with high or medium SFE, are the primary fatigue damage mechanisms.[6–8,11–16] As grain coarsening is a thermally activated process,[7,8,11,14] lowering the SFE by alloying can essentially restrict the closely relevant mechanisms, decrease the GB energy and then confine the GB migration.[16,31,32] Besides, the formation of twins with low excess energy, the interaction between Al atoms with GBs and the formation of short-range order (SRO) are also beneficial to the improvement of microstructures stability.[16,33,34] Although the formation mechanism of the shear bands during cyclic deformation is not well understood, this kind of surface damage was significantly lessened with lowering of the SFE as well, which could be attributed to the changed GB natures stemming from the lowered SFE.[35] Additionally, the nature of planar slip in the materials with low SFE and SRO will facilitate reducing the strain localizations as well.[36] In light of the causality between the fatigue cracks and the shear bands, prominently improved shear banding behavior can conduce to strengthen the fatigue cracking resistance. As mentioned above, b in the Basquin equation, another vital ingredient in principle to dominate the fatigue strength, plays a crucial role in bridging fatigue strength, microstructures and fatigue damage mechanisms in microscale. Therefore, the extraordinarily meliorated microstructural damage processes can be intrinsically reflected by the increase in b with lowering of the SFE in Basquin plots.

Based on the foregoing strategies of improving the fatigue strength, it is rational that the Basquin plots can be polished up advantageously with a decrease in the

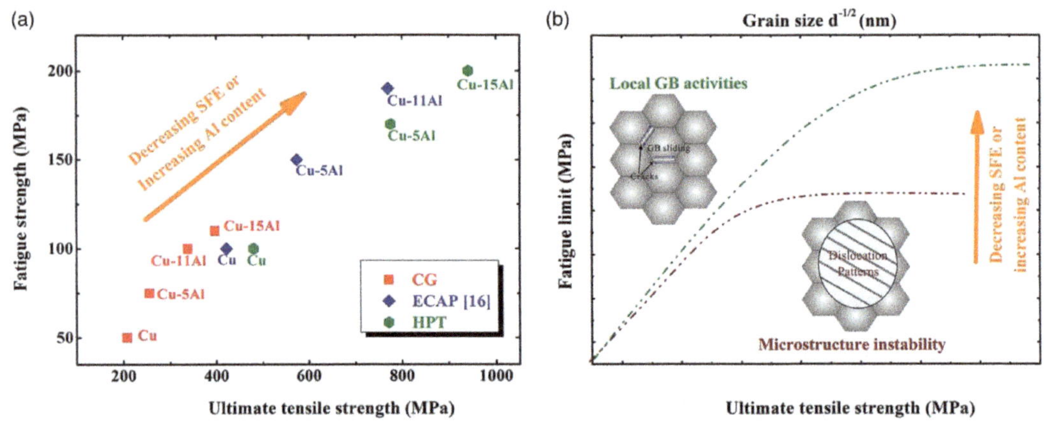

Figure 4. (a) Relationship between ultimate tensile strength and fatigue strength of pure Cu and Cu–Al alloys, showing that the tensile strength and fatigue limits can be simultaneously increased with lowering of the SFE or increasing the Al content in Cu and Cu–Al alloys; (b) schematic of relationship between ultimate tensile strength and fatigue strength in Cu and Cu–Al alloys showing that the fatigue strength cannot constantly increase with ultimate tensile strength due to different fatigue damage micromechanisms.

SFE in NC Cu and Cu–Al alloys, as exhibited in Figure S5, since the σ_f' is remarkably enhanced due to the rather finer grain size and solid solution hardening, and b is simultaneously improved deriving from the increased microstructure stability and lessened surface damage. Therefore, independent of SPD-processing methods, the fatigue strength, especially the fatigue endurance limits, can be essentially upgraded with lowering of the SFE in NC Cu and Cu–Al alloys, which is of significance in guiding the structural design of NC metallic materials with superior mechanical properties.

Referring to the closely linked relationship between σ_f' and monotonic strength, boosting the monotonic strength is the most reasonably crucial approach to increase the fatigue strength, as revealed by the general trend in Figure 4(a). However, present results can seriously remind us to realize that if b decreases drastically, the fatigue strength cannot be prospectively improved even though the tensile strength is significantly increased via enhancing the severity of the SPD methods, as displayed in Table 1 and Figure 4. Based on the significance of b, this limitation should be closely related to their microstructures and easy activation of their intrinsic fatigue damage mechanisms.

In the case of NC Cu, although finer grain sizes obtained by HPT can render a higher monotonic strength when compared with that obtained by ECAP, the fatigue strength is still around 100 MPa, which is analogous to those of NC Cu with high purity.[13,37] The identical characteristic in these materials during cyclic deformation is their disappointing low microstructure stability due to the non-equilibrium structures of GBs produced by SPD.[38,39] It is well known that high density of extrinsic dislocations resides in or near these GBs and can furnish the GBs to have higher energies,[38] which makes them thermodynamically unstable and readily moveable during long-termed cyclic deformation.[39] Apart from the non-equilibrium nature of GB structures, high purity of samples is another essential prerequisite resulting in the GB migration and grain growth.[40] The stress for the moving extra dislocations to activate the GB migration in NC Cu has been calculated roughly to be $\sim$160 MPa.[16,41] Such a stress can be readily achieved in the local regions of NC Cu processed by SPD. Therefore, the grain growth of NC Cu with high purity still happened during cyclic deformation even at the stress amplitudes of 100 and 90 MPa.[13,37] The formation of mechanically driven annealing twins in several coarsened grains can further authenticate the easy occurrence of the grain growth in NC Cu during fatigue.[42] As a result, large regional stress or strain incompatibility due to the difference in microstructures will accumulate and facilitate fatigue damage, triggering the final fatigue fracture. Therefore, although higher density of dislocations and finer grains in the NC Cu processed by HPT can advance the increase in the tensile strength and σ_f', these also gear up the local grain growth due to the higher mechanical energy stored in the materials,[4,43] and then significantly decrease b, as given in Table 1. This deterioration essentially compromises the benefits of increasing σ_f' for the further betterment of fatigue strength of NC Cu. As revealed in Figure 4(b), microstructure instability becomes a vital bottleneck inhibiting the additional improvement of the endurance limit of NC Cu with high purity.

With respect to the fatigue strength of NC Cu–Al alloys with high Al concentration, it cannot also be upgraded as much as the monotonic strength, as shown in Figure 4(a) and 4(b). In comparison with that of NC Cu–11 at%Al alloy processed by ECAP, the monotonic strength of NC Cu–15 at%Al alloy by HPT has more than 20% enhancement, whereas the fatigue strength is only increased by approximately 5%. Since most of mechanical energy produced during fatigue is deposited in the GBs of NC materials,[43] the fatigue damage mechanism of NC Cu–Al alloys with low SFE is mainly pertinent to GB activities. Due to the significant suppression of the GB migration in NC Cu–Al alloys with low SFE, other GB activities, such as atom shuffling or local GB sliding/rotation, which produce the insignificant microstructural variation, become crucial in accommodation of the cyclic plasticity.[16,44] Although the calculation of the stress to activate these GB activities during fatigue is challengingly difficult, these processes can be more readily initiated in the materials with extremely small grain sizes.[1,27,43,44] Moreover, the dislocations lodged in GBs will facilitate the grain sliding and rotations across the GBs.[44] This is coincidently consistent with the result that the reversal ductility of NC Cu–Al and Cu–Zn alloys prepared by HPT at the lowest SFE originated from these GBs activities.[18,28,29] During these processes, nanovoids were formed and then merged to nucleate the small cracks.[16,45] Besides, the profuse deformation twins in the nanograins of NC Cu–Al alloys also promote the generation of nanocracks during deformation because of the high local stresses originating from the intersection between twins and GBs.[46,47] Thus, as exhibited in Figure 4(b), these local GB activities in the NC Cu–Al alloys with low SFE with extremely fine grains will inevitably impair and impede the further improvement of their fatigue strengths.

To summarize, the prominent enhancement of their fatigue strengths is essentially achieved with a decrease in the SFE in NC Cu and Cu–Al alloys, irrespective of SPD techniques applied. Based on the Basquin law, the remarkable advancement of the fatigue strengths with lowering of the SFE can be attributed to the significantly increased fatigue strength coefficient σ_f' closely related to the boosted tensile strengths and the extraordinarily upgraded Basquin exponent b, originating from the ameliorated microstructure stability and reduced strain localization. Although augmenting the severity of SPD can apparently escalate the monotonic tensile strengths in the NC Cu and Cu–Al alloys, the corresponding promotion of their fatigue strengths was less than hoped for, as illustrated in Figure 4(b). The principal choke points are their inherent fatigue micromechanisms, that is, the microstructure instability and readily activated GB activities, respectively. Hence, for the NC materials with a high monotonic strength, appropriately tuning the microstructures and intrinsic fatigue mechanism in microscale (the value of b) via different strategies is inherently crucial to increase the fatigue strength in lieu of the mandatory improvement of σ_f'. These discoveries can enliven and enrich our understanding on the cyclic deformation behaviors of NC materials processed by SPD.

Acknowledgements The authors would like to thank Prof. H. Mughrabi for stimulating discussions and revising the manuscript and thank Prof. Pippan R for the NC sample preparations. This work was supported by the National Natural Science Foundation of China under Grants Nos. 50931005, 51101162 and 51331007.

Disclosure statement No potential conflict of interest was reported by the authors.

References

[1] Meyers MA, Mishra A, Benson DJ. Mechanical properties of nanocrystalline materials. Prog Mater Sci. 2006;51:427–556.
[2] Zhu YT, Liao XZ. Nanostructured metals—retaining ductility. Nature Mater. 2004;3:351–352.
[3] Lu K, Lu L, Suresh S. Strengthening materials by engineering coherent internal boundaries at the nanoscale. Science. 2009;324:349–352.
[4] Fang TH, Li WL, Tao NR, Lu K. Revealing extraordinary intrinsic tensile plasticity in gradient nano-grained copper. Science. 2011;331:1587–1590.
[5] Wu XL, Jiang P, Chen L, Zhang JF, Yuan FP, Zhu YT. Synergetic strengthening by gradient structure. Mater Res Lett. 2014;2:185–191.
[6] Vinogradov A, Hashimoto, S. Fatigue of severely deformed metals. Adv Eng Mater. 2003;5:351–358.
[7] Mughrabi H, Höppel HW, Kautz M. Fatigue and microstructure of ultrafine-grained metals produced by severe plastic deformation. Scr Mater. 2004;51:807–812.
[8] Mughrabi H, Höppel HW. Cyclic deformation and fatigue properties of very fine-grained metals and alloys. Int J Fat. 2010;32:1413–1427.
[9] Forrest PG. Fatigue of metals. Oxford: Pergamon Press; 1962.
[10] Suresh S. Fatigue of materials. 2nd ed. Cambridge: Cambrige Univeristy Press; 1998.
[11] Höppel HW, Zhou ZM, Mughrabi H, Valiev RZ. Microstructural study of the parameters governing coarsening and cyclic softening in fatigued ultrafine-grained copper. Phil Mag A. 2009;82:1781–1794.
[12] Wu SD, Wang ZG, Jiang CB, Li GY, Alexandrov IV, Valiev RZ. The formation of PSB-like shear bands in cyclically deformed ultrafine grained copper processed by ECAP. Scr Mater. 2003;48:1605–1609.
[13] Goto M, Han SZ, Euh K, Kang JH, Kim SS, Kawagoishi N. Formation of a high-cycle fatigue fracture surface and a crack growth mechanism of ultrafine-grained copper with different stages of microstructural evolution. Acta Mater. 2010;58:6294–6305.
[14] Malekjani S, Hodgson PD, Cizek P, Hilditch TB. Cyclic deformation response of ultrafine pure Al. Acta Mater. 2011;59:5358–5367.
[15] Zhang ZJ, An XH, Zhang P, Yang MX, Yang G, Wu SD, Zhang ZF. Effects of dislocation slip mode on high-cycle fatigue behaviors of ultrafine-grained Cu–Zn alloy processed by equal-channel angular pressing. Scr Mater. 2013;68:389–392.

[16] An XH, Wu SD, Wang ZG, Zhang ZF. Enhanced cyclic deformation responses of ultrafine-grained Cu and nanocrystalline Cu–Al alloys. Acta Mater. 2014;74:200–214.

[17] Valiev RZ, Islamgaliev RK, Alexandrov IV. Bulk nanostructured materials from severe plastic deformation. Prog Mater Sci. 2000;45:103–189.

[18] An XH, Lin QY, Wu SD, Zhang ZF, Figueiredo RB, Gao N, Langdon TG. The influence of stacking fault energy on the mechanical properties of nanostructured Cu and Cu–Al alloys processed by high-pressure torsion. Scr Mater. 2011;64:954–957.

[19] Zhilyaev AP, Langdon TG. Using high-pressure torsion for metal processing: fundamentals and applications. Prog Mater Sci. 2008;53:893–979.

[20] Pang JC, Li SX, Wang ZG, Zhang ZF. Relationship between tensile and fatigue strengths of metallic materials. Mater Sci Eng A. 2013;564:331–341.

[21] Qu S, An XH, Yang HJ, Huang CX, Yang G, Zang QS, Wang ZG, Wu SD, Zhang ZF. Microstructural evolution and mechanical properties of Cu–Al alloys subjected to equal channel angular pressing. Acta Mater. 2009;57:1586–1601.

[22] Hohenwarter A, Pippan R. Fracture toughness evaluation of ultrafine-grained nickel. Scr Mater. 2011;64:982–985.

[23] Figueiredo RB, Pereira PHR, Aguilar MTP, Cetlin PR, Langdon TG. Using finite element modeling to examine the temperature distribution in quasi-constrained high-pressure torsion. Acta Mater. 2012;60:3190–3198.

[24] Balogh L, Ungár T, Zhao Y, Zhu YT, Horita Z, Xu C, Langdon TG. Influence of stacking-fault energy on microstructural characteristics of ultrafine-grain copper and copper-zinc alloys. Acta Mater. 2008;56: 809–820.

[25] An XH, Lin QY, Wu SD, Zhang ZF, Figueiredo RB, Gao N, Langdon TG. Significance of stacking fault energy on microstructural evolution in Cu and Cu–Al alloys processed by high-pressure torsion. Phil Mag. 2011;91:3307–3326.

[26] Zhu YT, Liao XZ, Wu XL. Deformation twinning in nanocrystalline materials. Prog Mater Sci. 2012;57:1–62.

[27] Zhang Y, Tao NR, Lu K. Effects of stacking fault energy, strain rate and temperature on microstructure and strength of nanostructured Cu-Al alloys subjected to plastic deformation. Acta Mater. 2011;59:6048–6058.

[28] Zhao YH, Liao XZ, Horita Z, Langdon TG, Zhu, YT. Determining the optimal stacking fault energy for achieving high ductility in ultrafine-grained Cu–Zn alloys. Mater Sci Eng A. 2008;493:123–129.

[29] Zhang P, An XH, Zhang ZJ, Wu SD, Li SX, Zhang ZF, Figueiredo RB, Gao N, Langdon TG. Optimizing strength and ductility of Cu–Zn alloys through severe plastic deformation. Scr Mater. 2012;67:871–874.

[30] Kun F, Carmona HA, Andrade JS Jr, Herrmann HJ. Universality behind Basquin's law of fatigue. Phys Rev Lett. 2008;100:094301-1–3.

[31] Berbenni S, Bhasker PB, Cherkaoui M. A micromechan ics-based model for shear-coupled grain boundary migration in bicrystals. Int J Plasticity. 2013;44:68–94.

[32] Brandl C, Germann TC, Perez-Bergquist AG, Cerreta EK. Grain boundary motion under dynamic loading: mechanism and large-scale molecular dynamics simulations. Mater Res Lett. 2013;1:220–227.

[33] Anderoglu O, Misra A, Wang J, Hoagland RG, Hirth JP, Zhang X. Plastic flow stability of nanotwinned Cu foils. Int J Plast. 2010;26:875–886.

[34] Pan QS, Lu QH, Lu L. Fatigue behavior of columnar-grained Cu with preferentially oriented nanoscale twins. Acta Mater. 2013;61:1383–1393.

[35] Sangid MD, Maier HJ, Sehitoglu H. The role of grain boundaries on fatigue crack initiation—an energy approach. Int J Plast. 2011;27:801–821.

[36] Li P, Li SX, Wang ZG, Zhang ZF. Fundamental factors on formation mechanism of dislocation arrangements in cyclically deformed fcc single crystals. Prog Mater Sci. 2011;56:328–377.

[37] Khatibi G, Horky J, Weiss B, Zehetbauer MJ. High cycle fatigue behaviour of copper deformed by high pressure torsion. Int J Fat. 2010;32:269–278.

[38] Huang JY, Zhu YT, Jiang H, Lowe, TC. Microstructures and dislocation configurations in nanostructured Cu processed by repetitive corrugation and straightening. Acta Mater. 2001;49:1497–1505.

[39] Tucker, GJ, McDowell, DL. Non-equilibrium grain boundary structure and inelastic deformation using atomistic simulations. Int J Plast. 2011;27:841–857.

[40] Kunz L, Lukàš P, Svoboda M. Fatigue strength, microstructural stability and strain localization in ultrafine-grained copper. Mater Sci Eng A. 2006;424:97–104.

[41] Li JC. Mechanical grain growth in nanocrystalline copper. Phys Rev Lett. 2006;96:215506-1–4.

[42] An XH, Lin QY, Wu SD, Zhang ZF. Mechanically driven annealing twinning induced by cyclic deformation in nanocrystalline Cu. Scr Mater. 2013;68:988–991.

[43] Schiøtz J. Strain-induced coarsening in nanocrystalline metals under cyclic deformation. Mater Sci Eng A. 2004;375–377:975–979.

[44] Cheng S, Zhao Y, Wang Y, Li Y, Wang XL, Liaw PK, Lavernia EJ. Structure modulation driven by cyclic deformation in nanocrystalline NiFe. Phys Rev Lett. 2010;104:255501-1–4.

[45] Farkas D, Willemann M, Hyde B. Atomistic mechanisms of fatigue in nanocrystalline metals. Phys Rev Lett. 2005;94:165502-1–4.

[46] Ueno H, Kakihata K, Kaneko Y, Hashimoto S, Vinogradov A. Enhanced fatigue properties of nanostructured austenitic SUS 316 L stainless steel. Acta Mater. 2011;59:7060–7069.

[47] Ovid'ko IA, Sheinerman AG. Generation of nanocracks at deformation Twins in nanomaterials. Mater Res Lett. 2013;1:168–173.

16

Computational Design and Synthesis of Nitrogen-Substituted Carbon and Silicon Clathrates

Kwai S. Chan[a]*, Michael A. Miller[a], Wuwei Liang[a,b] and Carol Ellis-Terrell[a]

[a]*Southwest Research Institute, 6220 Culebra Road, San Antonio, TX 78238, USA;* [b]*Math Works, 3 Apple Hill Drive, Natick, MA 01760, USA*

A computational approach was utilized to design the framework and to identify potential guest atoms for stabilizing nitrogen-substituted carbon and silicon clathrates. Two new series of N-substituted carbon and silicon clathrates compounds were discovered as potential candidate materials. One of the hybrid C−N clathrates was successfully synthesized using an industrial arc-melting technique. The theoretical bulk moduli of these N-substituted carbon and silicon clathrates were computed and they are comparable to those of C_3N_4, Si_3N_4, and SiC. Some N-substituted carbon clathrates may be suitable for application as hard materials.

Keywords: Hybrid C−N Clathrates, Hybrid Si−N Clathrates, Hybrid Cage Structure, Superhard Materials

Carbon clathrates, which can be considered as analogs of carbon fullerene materials, are composed of C_{20}, C_{24}, and C_{28} cages linked through SP^3-covalent bonds. Theoretical computations have shown that both Type I carbon clathrate (C_{46}) and Type II carbon clathrate (C_{136}) may exist as metastable phases under high pressures.[1–3] Like silicon clathrates (Si_{46}),[4,5] Type I carbon clathrates, C_{46}, is predicted to exist in a simple cubic structure that belongs to the space group $Pm\bar{3}n$, Number 223, and exhibits a lattice parameter of 6.696 Å.[3] The cage structure of Type I carbon clathrates (C_{46}) is shown in Figure 1. Insertion of guest atoms such as Li, Na, or Ba into the cage structures has been predicted to be feasible under high pressures.[1–3] However, the energies of the formation for the Type I and Type II carbon clathrates are extremely high and neither Type I nor Type II carbon clathrates have been synthesized.

In general, the structure of Type I intermetallic clathrates based on Si, Ge, or Sn can be characterized by a framework of X atoms forming a 3D cage structure.[6–9] Some of the framework atoms can be substituted by elements such as Al, Ga, Cu, and Ag to form a framework that can be represented as M_yX_{46-y}. The empty space within the cage structure can serve as host sites for guest atoms, A. There are two small cages that can host two guest atoms (2a sites) and there are six large cages that can host six atoms (6d sites). Type I clathrates can generally be described by the formula: $A_xM_yX_{46-y}$, where A represents the guest atoms and x is the number of guest atoms. M represents the substitution atoms on the framework, y is the number of the substitution atoms, and X represents the framework atoms. It is also possible for vacancies to exist on the framework. Under this circumstance, Type I clathrates are formally represented as $A_xM_yX_{46-y-z}\square_z$ for non-stoichiometric compounds, where $\square$ represents a vacancy and z is the number of vacancies on the framework. For simplicity, $A_xC_yX_{46-y}$ is used in this paper to signify both stoichiometric and non-stoichiometric compounds. For non-stoichiometric Type I clathrates, the number of framework atoms is less than 46 and vacancies occur at the site positions of the missing atoms.

The substitution of Type I silicon framework with carbon atoms to form a series of hybrid carbon–silicon clathrates ($A_xC_ySi_{46-y}$) has been investigated theoretically by Chan et al.[10] The energy of formation of $A_xC_ySi_{46-y}$ has been found to decrease with increasing lattice parameter of the unit cell. Some of these newly discovered Type I hybrid carbon–silicon clathrates ($Ba_8C_6Si_{40}$, $Ba_8C_{20}Si_{26}$, and $Ba_8C_{23}Si_{23}$), which were predicted to exhibit small positive energies of formation, were successfully synthesized utilizing a vacuum arc-melting technique.[10] One implication of the findings of this investigation is that the carbon or silicon

*Corresponding author. Email: kchan@swri.edu, kwai.chan@swri.org

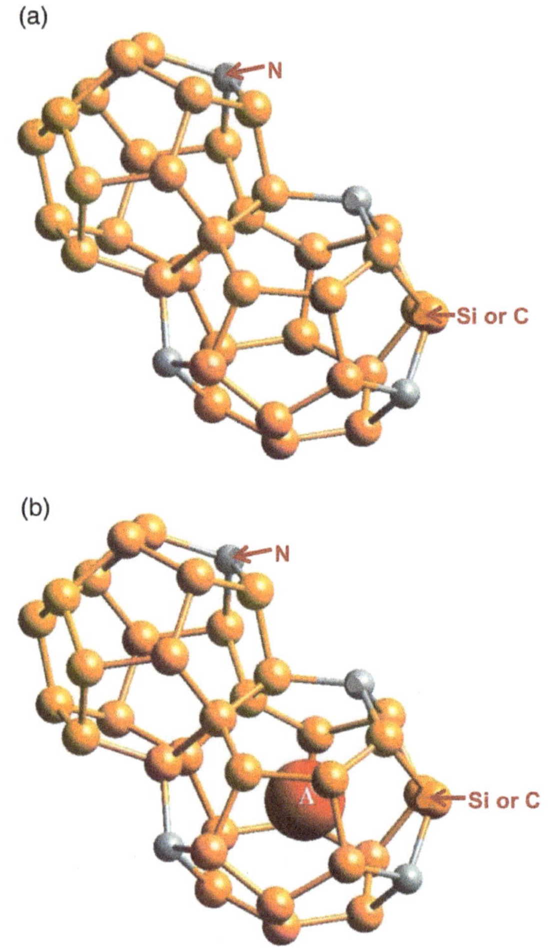

Figure 1. (a) Cage structure of Type I nitrogen–carbon clathrate, N_yC_{46-y}, or nitrogen–silicon clathrate, N_ySi_{46-y}, without guest atoms. (b) Cage structure of Type I guest atom stabilized nitrogen–carbon clathrate ($A_xN_yC_{46-y}$) or nitrogen–silicon clathrate ($A_xN_ySi_{46-y}$) with x number of guest atom A resided within the cage.

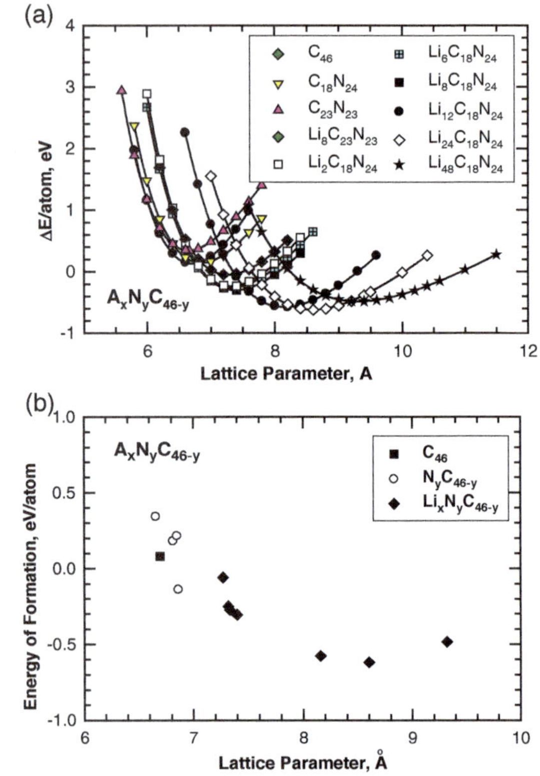

Figure 2. (a) Computed energy curves for Type I $A_xN_yC_{46-y}$ with various amounts of Li guest atoms. (b) A comparison of the energy of formation for C_{46}, N_yC_{46-y}, and $Li_xN_yC_{46-y}$ as a function of lattice parameter.

framework of Type I clathrates may be reduced and stabilized by substitution of C or Si framework atoms with small-sized atoms.

In this study, Type I hybrid nitrogen–carbon and nitrogen–silicon clathrates were designed by substituting some of the C atoms on the C_{46} framework and some of the Si atoms on the Si_{46} framework with N atoms, respectively. First-principles computational studies using the Car–Perrinello Molecular Dynamics (CPMD) code [11,12] indicate that the carbon atoms in the C_{46} framework can be partially substituted by nitrogen atoms to form a hybrid carbon–nitrogen clathrate, which can be represented by N_yC_{46-y}. Figure 1(a) shows a representation of the Type I N_yC_{46-y} clathrates. Furthermore, guest atoms can be inserted into the cage structure to stabilize the hybrid nitrogen–carbon clathrate by reducing the energy of formation to form a class of nitrogen-substituted carbon clathrates, represented as $A_xN_yC_{46-y}$. These hybrid structures do not exist in nature and thus represent a novel structure of matter neither known in the open literature nor covered in existing patents on clathrate compounds. Figure 1(b) shows a structural representation of the Type I $A_xN_yC_{46-y}$ clathrate compounds. The influence of Li guest atom (A = Li) on the structural stability of the hybrid C−N framework was also considered. Figure 2(a) shows that insertion of Li atoms results in a lowering of the energy formation accompanied by an expansion of the framework for Li atoms ranging from 2 to 48. Table 1 presents a summary of the computed lattice constants and energy of formations for individual Type I clathrate compounds. The computed values of the energy of formation per atom for C_{46}, N_yC_{46-y-z}, and $Li_xN_yC_{46-y}$ are presented as a function of the lattice parameter in Figure 2(b), which compares the energy of formation for C_{46} and N_yC_{46-y} without and with Li guest atoms. The results in Figure 2(b) show that nitrogen substitution can lower the energy of formation of N_yC_{46-y} compared with that for C_{46}. Furthermore, Li

Table 1. Summary of computed lattice constant, a_o, and energy of formation, E_f, for Type I clathrate compounds.

Compound	a_o (Å)	E_f (eV/atom)
C_{46}	6.6973	.0815
$C_{18}N_{24}$	6.8622	−.1377
$C_{19}N_{27}$	6.8148	.1803
$C_{20}N_{26}$	6.8502	.2144
$C_{23}N_{23}$	6.6596	.3416
$Li_2C_{18}N_{24}$	7.3192	−.25
$Li_6C_{18}N_{24}$	7.3336	−.2742
$Li_8C_{18}N_{23}$	7.3979	−.3051
$Li_8C_{23}N_{23}$	7.2716	−.0578
$Li_{12}C_{18}N_{24}$	8.1592	−.5765
$Li_{24}C_{18}N_{24}$	8.6003	−.6192
$Li_{48}C_{18}N_{28}$	9.3206	−.4832
Si_{46}	10.238	.0571
$Si_{18}N_{24}$	9.0215	−.6338
$Si_{23}N_{23}$	9.0546	−.5417
$Li_8Si_{18}N_{24}$	9.8924	−.6092

insertion reduces the energy of formation to negative values, which indicates that $Li_xN_yC_{46-y}$ clathrates are stable compounds.

The hybrid nitrogen and silicon clathrate was obtained by substituting some silicon atoms on the Type I silicon clathrate framework, Si_{46}, with nitrogen atoms to produce a nitrogen–silicon framework. As illustrated in Figure 1(a), the hybrid N−Si clathrates framework is represented by N_ySi_{46-y} and consists of y nitrogen atoms and 46-y silicon atoms with a regular arrangement of 20-atom and 24-atom cages fused together through 5-atom pentagonal rings (Type I clathrate). This compound can be inserted with guest atoms to form $A_xN_ySi_{46-y}$, Figure 1(b). This series of compounds was designed by substituting some of the Si atoms on the Si_{46} framework with N atoms in order to form a Si_3N_4-like compound with Li guest atoms. The values of the energy of formation for N_ySi_{46-y} are compared against those of C_{46}, Si_{46}, and $Li_xN_yC_{46-y}$ in Figure 3. For the compositions considered, the N_ySi_{46-y} compounds are stable compounds as their energy of formation values are negative. However, the insertion of Li into the framework causes the lattice to expand, as shown in Figure 3.

Figure 4 is a summary plot that depicts the energy of formation for nitrogen-substituted clathrate compounds, which include $A_xN_yC_{46-y}$ and $A_xN_ySi_{46-y}$, against those of C_{46}, Si_{46}, and $A_xC_ySi_{46-y}$ from Chan et al.[10] Nitrogen substitution is predicted to lower the energy of formation and produces stable carbon and silicon clathrates. These first-principles computations have led to the identification of two new classes of nitrogen-substituted Type I clathrates based on the carbon or silicon framework: (1) N-substituted carbon clathrates without (N_yC_{46-y}) and with guest atoms ($A_xN_yC_{46-y}$) and (2) N-substituted silicon clathrates without (N_ySi_{46-y}) and with guest atoms ($A_xN_ySi_{46-y}$).

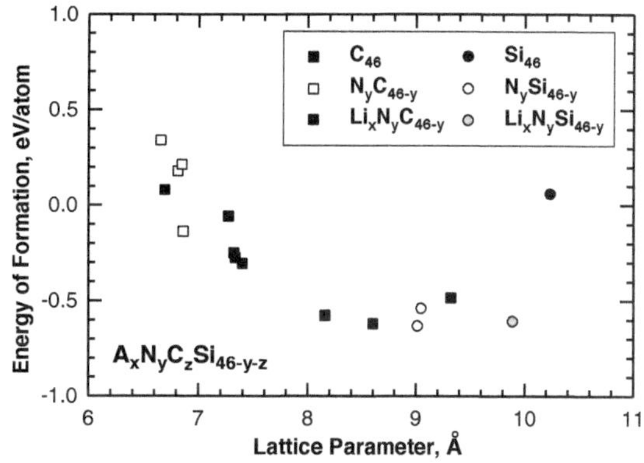

Figure 3. A comparison of the energy of formation for C_{46}, N_yC_{46-y}, $Li_xN_yC_{46-y}$, Si_{46}, N_ySi_{46-y}, and $Li_xN_ySi_{46-y}$ as a function of lattice parameter.

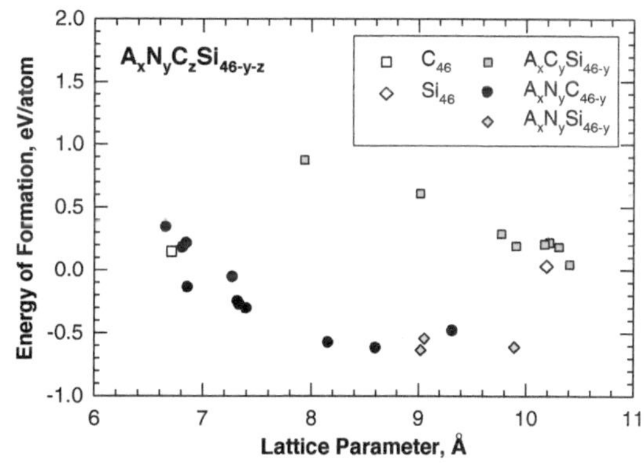

Figure 4. A comparison of the energy of formation for $A_xC_ySi_{46-y}$, $A_xN_yC_{46-y}$, and $A_xN_ySi_{46-y}$ against those of C_{46} and Si_{46} as a function of lattice parameter showing that nitrogen substitution of the framework atoms on C_{46} or Si_{46} results in stable carbon or silicon clathrate compounds. The results for $A_xC_ySi_{46-y}$ are from Chan et al.[10]

The results in Figure 4 also suggest the existence of nitrogen-substituted carbon and silicon clathrates without ($N_yC_zSi_{46-y-z}$) and with guest atoms ($A_xN_ySi_{46-y-z}$) on the basis that nitrogen substitution may lower the energy of formation in both carbon clathrates and silicon clathrates. In a recent study,[10] Ba-stabilized carbon silicon clathrates ($Ba_8C_6Si_{40}$, $Ba_8C_{20}Si_{26}$, and $Ba_8C_{23}Si_{23}$) have been synthesized and shown to exist as metastable phases at ambient temperature. Since Figure 4 shows that the energies of formation for N-substituted carbon and silicon clathrates are lower than those of $A_xC_ySi_{46-y}$, it is likely that nitrogen substitution of the hybrid carbon–silicon framework may lower the energy of formation further to turn them into more stable compounds. It is anticipated that N, C, and Si may form a ternary N−C−Si framework represented by the formula $N_yC_zSi_{46-y-z}$, with $1 \leq y \leq 44$ and $1 \leq z \leq 45 - y$. Furthermore, this new class of new Type I clathrates with mixed nitrogen, silicon, and carbon atoms on the framework of the cage structure may be stabilized by inserting guest atoms

Table 2. Candidate compositions and elemental or alloy starting components for nitrogen-substituted carbon clathrates synthesis by vacuum arc-melting.

Composition	Target clathrate compound	D (g/cc) FW (g/mol)	Ba	Mg	g-$C_3N_{4+x}H_y$	Total (g/mol) Batch (g)	Estimated density (g/cc) Estimated volume (cc)
		D (g/cc)	3.51	1.74	2.34		
		FW (g/mol)	137.327	24.305	40.0962		
1	$Mg_8C_{18}N_{24}$	Atom ratio		8	6		
		Weight (g)	0.00	194.44	240.58	435.02	2.07
		Batch	**0.00**	**11.17**	**13.83**	25.00	12.08
2	$Ba_8C_{18}N_{24}$	Atom ratio	8		6		
		Weight (g)	1,098.62	0.0	240.58	1,339.19	3.30
		Batch	**20/60**	**0.0**	**4.51**	25.11	7.61
	Total element/batch		**20.60**	**11.17**	**18.34**		

Table 3. Summary of targeted compounds and actual compounds produced by arc-melting.

Composition	Target compound	Actual compounds produced	Type I clathrate
1	$Mg_8N_{18}C_{28}$	MgO, g-C_2N_4, and $MgCN_2$	No
2	$Ba_8N_{18}C_{24}$	Ba, g-C_3N_4, and $Ba_8C_{18}N_{24}$	Yes ($Ba_8C_{18}N_{24}$)

A inside the cage structure wherein the composition of this series of Type I clathrates is qualitatively represented by the formula $A_xN_yC_zSi_{46-y-z}$, with $1 \leq y \leq 44$ and $1 \leq z \leq 45 - y$, and A can be guest atoms such as Ba, Li, and other similar alkaline and rare-earth metals. Some of the potential compounds in this series of ternary nitrogen, carbon, and silicon clathrates may include $Ba_xN_8C_{10}Si_{23}$, $Ba_xN_8C_{10}Si_{28}$, $Ba_xN_8C_{10}Si_{24}$, $Ba_xN_8C_{14}Si_{24}$, $Li_xN_8C_{10}Si_{23}$, $Li_xN_8C_{10}Si_{28}$, $Li_xN_8C_{10}Si_{24}$, $Li_xN_8C_{14}Si_{24}$, and other similar permutations of N, C, and Si. It is noted that vacancies are assumed to be present on the framework of non-stoichiometric compounds when the number of total framework atoms is less than 46. Although not investigated in this study, the presence of vacancies on the framework can significantly affect the electronic properties of these hybrid clathrate compounds.

To validate the first-principles computations, syntheses of $Mg_8C_{18}N_{24}$ and $Ba_8C_{18}N_{24}$ were attempted by vacuum arc-melting of admixture of graphitic carbon nitride (g-$C_3N_{4+y}H_y$) and appropriate metal powders, with detailed compositions shown in Table 2. Synthesis of $Li_8C_{18}N_{24}$ was not attempted because of difficulties of handling Li metal during arc-melting and the lack of lithium compounds that can be combined to give the desired compositions of $Li_8C_{18}N_{24}$. In addition, previous first-principles calculations on $Li_8C_6Si_{40}$ and $Ba_8C_6Si_{40}$ have also shown that Ba atoms are more effective than the Li atoms in stabilizing the hybrid silicon–carbon framework.[10] Based on prior experience in synthesizing hybrid silicon–carbon clathrates,[10] it was anticipated that Ba would be more effective in stabilizing the hybrid carbon–nitrogen framework and offer a greater chance for a successful synthesis of nitrogen-substituted carbon clathrates. The rationale was borne out

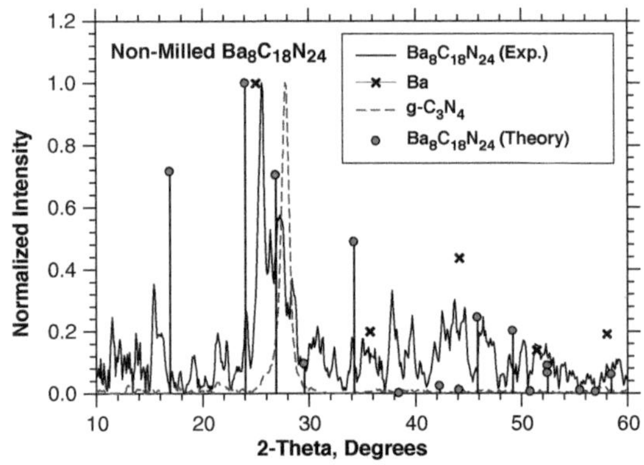

Figure 5. PXRD patterns measured for Composition (2) subsequent to arc-melting of powdered admixture of Ba and g-$C_3N_{4+x}H_y$.

by attempting to synthesize $Mg_8C_{18}N_{24}$, which was not successful. Powder x-ray diffraction (PXRD) of the non-milled $Mg_8C_{18}N_{24}$ powders indicate the absence of any peaks that can be attributed to a Type I clathrate structure and all of the peaks in the PXRD pattern can be identified and attributed to those of the graphitic carbon nitride, MgO, and $MgCN_2$, as summarized in Table 3.

In contrast, synthesis of $Ba_8C_{18}N_{24}$ by the vacuum arc-melting technique was successful. In particular, $Ba_8C_{18}N_{24}$ was synthesized by arc-melting appropriate amounts of Ba and graphitic carbon nitride (g-$C_3N_{4+x}H_y$) as the starting materials. Admixtures of Ba and g-$C_3N_{4+x}H_y$ (in the proportion of 20.6 g Ba and 4.51 g of g-$C_3N_{4+x}H_y$ powders) were arc-melted to make about 25.11 g of the product, consisting of $Ba_8C_{18}N_{24}$ plus some amounts of unreacted starting materials. PXRD data of the arc-melted product (i.e. not purified) is presented in Figure 5. Some of the reflection peaks in the x-ray diffraction (XRD) spectra correspond to unreacted Ba and g-$C_3N_{4+x}H_y$ starting materials. However, the remaining reflections in the XRD spectra do not belong to Ba and g-$C_3N_{4+x}H_y$, and some have been attributed to the Type I clathrate structure in question. The theoretically computed reflections for $Ba_8C_{18}N_{24}$ were obtained by first optimizing the Type I clathrate structures using first-principles CPMD computations to obtain the equilibrium crystallographic parameters, followed by computing the corresponding reflections in the XRD spectrum using the Diamond analysis and visualization software.[13] The comparison indicates that small quantities of Type I clathrate compound are present in the arc-melted $Ba_8C_{18}N_{24}$ product. The crystal structure of this clathrate compound is close to those of $Ba_8N_{18}N_{24}$ based on the characteristic reflections at 2θ of 16.9°, 24.0°, 26.9°, 34.3°, 44.9°, and 49.2° for the nitrogen-substituted carbon clathrate. The products identified by XRD are summarized in Table 3.

The bulk modulus of the various intermetallic clathrate compounds was computed using the first-principles approach according to the expression given by [14]

$$\frac{\Delta E}{V} = \frac{9}{2} B\delta^2, \tag{1}$$

where $\Delta E/V$ is the energy change per unit cell volume, B is the bulk modulus, and δ is the normal strain in the three principal directions of the unit cell. A plot of $\Delta E/V$ versus δ was obtained for each unit cell of individual clathrate compounds and the data were fitted to Equation (1). The regression coefficient was then used to obtain the bulk modulus, B. A summary of the theoretical bulk modulus for various intermetallic clathrate compounds is represented in Table 4.

Also shown in Table 4 for comparison are theoretical or experimental data (indicated by an asterisk) of bulk modulus for diamond, carbon nitride, silicon nitride, and silicon carbide from the literature.[1–3,15–21] The results in Table 4 indicate that a wide range of bulk modulus can be obtained from Type I hybrid C−N, N−Si, and C−Si clathrates, depending on the framework atoms. Some of the carbon clathrate compounds exhibit bulk moduli that are in the range of 245–374 GPa, which are not as high as diamond C or β-C_3N_4, but are comparable to silicon nitride (C-Si_3N_4) and carbon nitride. Thus, carbon clathrates may be considered for potential applications as hard materials.

Table 4. Computed values of the bulk modulus of selected Type I clathrates compared with theoretical and experimental values of the bulk modulus of diamond, carbon nitride, silicon nitride, and silicon carbide from the literature.

Compound	Structure	B (GPa)
C_{46}	Type I clathrate (Simple cubic)	373.7 409 [1] 363.7 [2] 371 [3]
$C_{18}N_{24}$	Type I clathrate	257.35
$C_{23}N_{23}$	Type I clathrate	311.6
$Li_8C_{23}N_{23}$	Type I clathrate	245.18
$Si_{23}C_{23}$	Type I clathrate	124.03
Si_6C_{40}	Type I clathrate	196.8
$Si_{18}N_{24}$	Type I clathrate	96.08
$Si_{23}N_{23}$	Type I clathrate	114.13
C_6Si_{40}	Type I clathrate	62.9
C (Diamond)	Diamond cubic	438.8 [15] 442* [16]
α-C_3N_4	Hexagonal	378.7 [15]
β-C_3N_4	Cubic spinel	419.1 [15]
c-Si_3N_4	Cubic spinel	300 [17] 300* [18]
SiC	Cubic (Zinc blend)	225* [19–21]

Note: Experimental values are indicated by an asterisk (*).

First-Principles Computational Method An *ab initio* molecular dynamics code based on the CPMD method [11,12] was utilized to investigate theoretically and systematically the effects of small-atom substitution on the framework and insertion into the empty space inside the cage structure on the energy formation and lattice constant of selected intermetallic clathrate compounds. The CPMD code [11] is a plane wave implementation of the Density Functional Theory.[12] It uses an approximation frozen-core approach where only the chemically active valence electrons are dealt with explicitly and the inert core electrons are considered frozen together with the nuclei as rigid non-polarizable ion cores. It is capable of both first-principles wave function optimization (static calculation) and *ab initio* molecular dynamics calculations. The PBE functional [22] and projector-augmented wave [23,24] potentials were used along with the plane wave basis sets for the geometry optimization and self-consistent total energy calculations. The energy cutoff for the plane wave basis set was 2,041 eV. The convergence energy criterion was set to be 3×10^{-5} eV. Reciprocal space was sampled using $3 \times 3 \times 3$ Monkhorst Pack meshes centered at Gamma. The number of k-point was 64 ($4 \times 4 \times 4$). For crystalline solids, the energy and equilibrium lattice constant of a unit cell at ground state (0 K) were computed by calculating the energies for unit cells with different lattice constants by performing wave function optimization with CPMD. A high-order (e.g. 4th order) polynomial was fitted to the energy points. The equilibrium lattice constant was determined by finding the lattice constant corresponding to the minimum energy on the polynomial curve. The formation energies were calculated by subtracting the total energies of the elements from the energy of the structure, then dividing by the total number of atoms. For example, the formation energy, ΔE_{form}, for $A_xN_yC_{46-y-z}\Box_z$ was calculated using the equation given by

$$\Delta E_{\text{form}} = \frac{E(A_xN_yC_{46-y-z}\Box_z) - xE(A) - yE(N) - (46-y-z)E(C)}{x+46-z}, \quad (2)$$

where $E(N)$, $E(C)$, and $E(A)$ are the energies per atom for nitrogen, C (diamond), and A metal, respectively. Similarly, the same approach was utilized to compute the energy of formation for $A_xN_ySi_{46-y}$ by replacing C with Si for all pertinent energy parameters.

Synthesis Method Table 2 provides a list of the candidate series and the elemental or alloy components constituting the admixture of starting materials. Magnesium powder (>99.87%, 32 mesh) was obtained from Noah Technology (San Antonio, TX). Barium (99.2*, < 0.8 % Sr, in pieces) was purchased from Alfa Aesar (Ward Hill, MA) and graphitic carbon nitride powders (99.5%, 730 µm) were obtained from Carbodeon, Ltd. (Vantaa, Finland). The targeted compositions were mixed together inside an argon-filled glove box at the appropriate stoichiometric ratios as indicated in Table 3, each admixed composition was individually packaged in a stainless steel tubular container and sealed with Swagelok fittings before removing it from the argon-filled glove box. Both packaged compositions were then shipped to Sophisticated Alloys, Inc. (Butler, PA) for vacuum arc-melting in an industrial vacuum arc-melter. The arc-melting process was conducted in an argon atmosphere under a sub-atmospheric pressure.

Acknowledgements This work was supported by The Internal Research Program of Southwest Research Institute. We acknowledge the Texas Advanced Computing Center of the TerraGrid Network and the Extreme Science and Engineering Discovery Environment (XSEDE) High Performance Computing Facilities for providing the computational resources for the CPMD calculations. Clerical assistance by Ms L. Salas, SwRI, in the preparation of this manuscript is acknowledged.

References

[1] Perottoni CA, da Jornada JAH. The carbon analogues of type-I silicon clathrates. J Phys: Condens Matter. 2001;13:5981–5998.

[2] Wang J-T, Chen C, Wang D-S, Mizuseki H, Kawazoe Y. Phase stability of carbon clathrates at high pressure. J Appl Phys. 2010;107:063507, 1–4.

[3] Rey N, Muñoz A, Rodríquez-Hernández P, San Miguel A. First-principles study of lithium-doped carbon clathrates under pressure. J Phys: Condens Matter. 2008;20:215218, 1–7.

[4] Adams, GB, O'Keeffe M, Kemkov AA, Sankey OF, Huang Y-M. Wide-band-gap Si in opne four-fold-coordinated clathrate structures. Phys Rev B. 1994;49: 8084–8053.

[5] Melinon P, Keghelian P, Perez A, Champagnon B, Guyot Y, Saviot L, Reny E, Cros C, Pouchard M, Dianoux AJ. Phonon density of states of silicon clathrates: characteristic width narrowing effect with respect to the diamond phase. Phys Rev B. 1999;59: 10099–10103.

[6] Rogl P. Intermetallic clathrates: a challenge for thermoelectric. Austria: Institute of Physical Chemistry, University of Vienna; 2006.

[7] San-Miguel A, Toulemonde P. High-pressure properties of group IV clathrates. High Press Res. 2005;25(3): 159–185.

[8] Cros C, Pouchard M. The clathrate phases of silicon and related elements (C, Ge, Sn): a historical approach. C R Chim. 2009;12:1014–1056.

[9] Nolas GS, Schujman SB, Slack GA. Chapter 6, Semiconductor clathrates: a phonon glass electron crystal material with potential for thermoelectric applications. In: Tritt TM, editor. Semiconductors and semimetals. Vol 69. San Diego: Academic Press; 2001. p. 255–300.

[10] Chan KS, Miller MA, Liang W, Ellis-Terrell C, Peng X. First-principles computational design and synthesis of hybrid carbon-silicon clathrates. J Mater Sci. 2013.

[11] Car R, Parrinello M. Molecular dynamics: an *ab initio* electronic structure and molecular dynamics program. Version 3.13.1. The CPMD consortium; Stuttgart: IBM and Max Planck Institute. 2008 [cited 2008 Jun 27]. Available from: http://www.cpmd.org

[12] Car R, Parrinello M. Unified approach for molecular dynamics and density functional theory. Phys Rev Lett. 1985;55(22):2471–2474.

[13] Diamond Version 3.2. Crystal Impact. Bonn, Germany; 2013. Available from: http://www.crystalimpact.com

[14] Wang SQ, Ye HQ. *Ab initio* elastic constants for the lonsdaleite phases of C, Si and Ge. J Phys: Condens Matter. 2003;15:5307–5134.

[15] He J, Guo L, Guo X, Liu R, Tian Y, Wang H, Gao C. Predicting hardness of dense C_3N_4 polymorphs. Appl Phys Lett. 2006;88:101906-1–3.

[16] Teter DM. Computational alchemy: the search for new superhard materials. MRS Bull. 1998;23:22–27.

[17] He H, Sekine T, Kobayashi T, Hirosaki H, Suzuki I. Shock-induced phase transition of β-Si_3N_4 to c-Si_3N_4. Phys Rev B. 2000;62:11412–11417.

[18] Jiang JZ, Lindelov H, Gerward I, Stahl K, Reeio JM, Mori-Sanchez P, Carlson S, Mezouar M, Dooryhee E, Fitch A, Frost DJ. Compressibility and thermal expansion of cubic silicon nitride. Phys Rev B. 2002;65:161302(R)-1–4.

[19] Carnahan RD. Computational properties of silicon carbide. J Am Ceram Soc. 1968;51:223–224.

[20] Yean DH, Ritter JR. Estimates of isothermal bulk moduli for group IVA crystals with the zincblende structure. J Phys Chem Solids. 1971;32:653–655.

[21] Aguado F, Baonza VG. Prediction of bulk modulus at high temperatures from longitudinal phonon frequencies: application to diamond, c-BN, and 3C-SiC. Phys Rev B. 2006;73:024111-1–5.

[22] Perdew P, Burke K, Ernzerhof M. Generalized gradient approximation made simple. Phys Rev Lett. 1996;77(18): 3865–3868.

[23] Blochl PE. Projector augmented-wave method. Phys Rev B. 1994;50(24):17953–17979.

[24] Kresse G, Joubert D. From ultrasoft pseudopotentials to the projector augmented-wave method. Phys Rev B. 1999;59(3):1758–1775.

17

Thermally Induced Interdiffusion and Precipitation in a Ni/Ni_3Al System

C. Sun[a]*, E. Martínez[a], J.A. Aguiar[b], A. Caro[a], J.A. Valdez[a], K. Baldwin[a], Y. Xu[a], B.P. Uberuaga[a], O. Anderoglu[a] and S.A. Maloy[a]

[a]*Los Alamos National Laboratory, Los Alamos, NM 87545, USA;* [b]*National Renewable Energy Laboratory, Golden, CO 80401, USA*

Ordered Ni_3Al intermetallic precipitates constitute the main hardening sources of Ni-based superalloys. Here, we report the interdiffusion and precipitation behavior in a Ni/Ni_3Al model system. The deposition of Ni_3Al on a pure Ni layer at 500°C generated $L1_2$-structured γ' (Ni_3Al) precipitates, preferentially at the interface. After annealing at 800°C for 1 h, interdiffusion between Ni and Ni_3Al layers occurred, and the γ' precipitates that grew near the parent Ni/Ni_3Al interface are $\sim$2.8 times larger in size than those formed in the matrix. Monte Carlo simulations indicate that vacancies preferentially diffuse along the Ni/Ni_3Al interface, increasing the probability of precipitation.

Keywords: Ni-based Superalloy, Interdiffusion, Precipitation, Ni/Ni_3Al Interface, Thermal Annealing

1. Introduction Ni-based superalloys exhibit high strength, long-time rupture strength, and excellent resistance to oxidation at elevated temperatures.[1–3] In Ni-based superalloys, the $L1_2$-structured γ' precipitates ($Ni_3(Al,Ti)$) contribute significantly to the high-temperature strength and the fcc-structured γ matrix can accommodate extensive plastic deformation, resulting in a good combination of high strength and excellent ductility at high temperatures.[4,5] Diffusion-driven mass transport between γ' precipitates and γ matrix affects the ordering and dissolution behavior of the γ' precipitates and thus the mechanical properties. Fundamental studies on the solute redistribution and microstructure evolution of the γ/γ' system under heat treatments are crucial to understand their behavior for the high-temperature applications.

Diffusion of alloy components is significantly influenced by the state of ordering.[6] In disordered solid solution alloys, diffusion typically occurs through the random walk of thermal vacancies. In ordered intermetallic, the random walk of vacancies can disrupt the equilibrium ordering arrangement of atoms. However, diffusion through nonrandom motion of vacancies allows the atom–vacancy exchanges to take place without sacrificing long-range ordering.[7] In $L1_2$-structured Ni_3Al, there are two major self-diffusion mechanisms proposed. One view holds that Ni atoms diffuse through site exchanges with a neighboring vacancy on its own sublattice without concomitant disordering, where a six-jump vacancy sequence may be operative with Ni and Al moving simultaneously during diffusion.[8] Another perspective is that the minor element, Al, occupying lattice sites of the sublattice of the major atoms (i.e. Ni) can diffuse through the Ni-sublattice.[9] Watanabe et al. studied the interdiffusion in a Ni/Ni_3Al diffusion couple and pointed out that the interdiffusivities follow the modified Darken's equation.[10]

The nucleation of γ' precipitates is normally considered as a diffusional transformation and occurs through classic nucleation and growth mechanism. In the theory of nucleation, the precipitation of a second phase can result in the reduction of total free energy of the system and the rate of precipitation significantly depends on the temperature of undercooling at which the precipitation initiates.[11] The coarsening of γ' precipitates in the γ matrix is a diffusion-driven Ostwald ripening process.[12–16] During heat treatment, the interfacial energy can provide a driving force for

*Corresponding author. Emails: csun@lanl.gov, chengsun2008@gmail.com

coarsening and morphological changes of γ' precipitates, which in turn results in a reduction of total interfacial energy.[12,17] The morphological change of γ' precipitates is significantly influenced by the elastic energy associated with the lattice parameter misfit between the matrix and the precipitate.[18] In misfit-free (matrix and precipitate) Ni–Cr–Al alloys, γ' precipitates grow as spheres as predicted by the Lifshitz–Slyozov–Wagner (LSW) theory.[19–21] With increasing misfit between the matrix and the precipitate, by the addition of molybdenum for instance, the morphology of γ' precipitates in a Ni–Cr–Al–Mo alloy changed from spheroidal to cuboidal.[22]

Pristine defects in the components have an important impact on the coarsening kinetics of γ' precipitates.[23,24] High-angle grain boundaries are considered as solute diffusion paths contributing to the growing of precipitates that are located on grain boundaries. Dislocations intersecting the precipitate surface can generate fast pipe diffusion and thus serve as paths for precipitate coarsening.[23] Here, we used Ni/Ni_3Al as a γ/γ' model system to explore the role of interfaces on precipitation and the competing tendencies to order and dissolve the supersaturated Ni_3Al layer. Interface promotes the formation of γ' precipitates due to the preferential vacancy diffusion along the interface.

2. Experiments and Modeling Methodologies Ni/Ni_3Al multilayers were synthesized by magnetron sputtering at 500°C under vacuum at a pressure of 3×10^{-3} torr. The base pressure of the system was 1×10^{-8} torr. The first Ni layer with a thickness of 100 nm was deposited on an MgO substrate, the second Ni_3Al layer with a thickness of 200 nm was cosputtered from Al and Ni targets, and the third layer was the Ni–Al cap layer (~ 100 nm in thickness). The DC power used on the targets was varied as necessary to achieve the required stoichiometry. After deposition, Ni/Ni_3Al films were annealed at 800°C for 1 h with gettered Ar flowing through the tube and cooled down to room temperature at a controlled rate of 5°C/min. Cross-sectional samples were prepared for transmission electron microscopy (TEM) observation by mechanical polishing followed by subsequent ion-milling.

TEM was performed on an image-corrected FEI Titan at Los Alamos National Laboratory (LANL), operated in TEM mode at 300 keV. To acquire structural high-angle annular dark- and bright-field images, the 200 keV JEOL ARM at Arizona State University was utilized to perform scanning transmission electron microscopy (STEM). Spectral imaging was utilized to acquire the Ni-L and Al-K near-edge fine structures with the best achievable spatial and energy resolution for each microscope. Particular attention was paid to minimize the effects of electron irradiation on the sample materials, to assure that all the reported results are not a consequence of electron-beam-induced damage, by utilizing lower dose imaging techniques.

An off-lattice kinetic Monte Carlo algorithm [25,26] was used to study the Ni_3Al precipitation process in the presence of misfit strain. One vacancy was introduced in the system and its exchange rates can be calculated by using Glauber dynamics, that is,

$$r_j = \frac{x_j}{1 + x_j}, \tag{1}$$

$$x_j = \exp\left\{\frac{-\Delta E_j}{kT}\right\}, \tag{2}$$

where ΔE_j is the change in energy between the final and initial configurations. The pressure of the sample was relaxed every 100 Monte Carlo steps to take into account the internal strain fields due to lattice mismatch. Periodic boundary conditions were applied in every dimension. Once the rates were calculated, the traditional Bortz–Kalos–Lebowitz (BKL) algorithm [27] was used to study the evolution of the microstructure using the rates as defined in Equation (2). The embedded-atom method (EAM) potential developed by Mishin [28] was used to calculate ΔE_{j}, as, in our experience, it best reproduces the experimental phase diagram. This algorithm does not capture the dynamics of the process, but elucidates the most important effects in the competition between ordering and interdiffusion. In particular, because prefactors and barriers for each rate are not calculated, the 'time scale' is not representative of the experiment, but the trajectories should be reasonable.

3. Results and Discussion Figure 1 shows the microstructure of the as-deposited Ni/Ni_3Al film. Figure 1(a) shows the bright-field TEM image and the inset selected area electron diffraction (SAED) pattern obtained along the [100] zone axis, containing MgO substrate, Ni, and Ni_3Al layers. The inset SAED pattern shows that the Ni_3Al layer was a chemically partially ordered structure due to the presence of intense diffraction maxima at the $\langle 110 \rangle$ orientation. Complementary energy dispersive X-ray spectroscopy (EDS) line scans were performed across the film (Figure 1(b)). A Ni layer with the thickness of ~ 100 nm and a $Ni_{0.73}Al_{0.27}$ layer of ~ 200 nm were resolved. Figure 1(c) shows the bright-field TEM image under kinematic imaging conditions with a diffraction vector $g = [110]$. To image the precipitates, dark-field TEM images were taken as shown in Figure 1(d). The bright columnar structure in the Ni_3Al layer in Figure 1(d) was identified as the γ' (Ni_3Al in $L1_2$ structure) phase. Interestingly, some isolated nanometric γ' particles were formed on the Ni/Ni_3Al interface during the deposition at 500°C, as marked by the arrows in Figure 1(d).

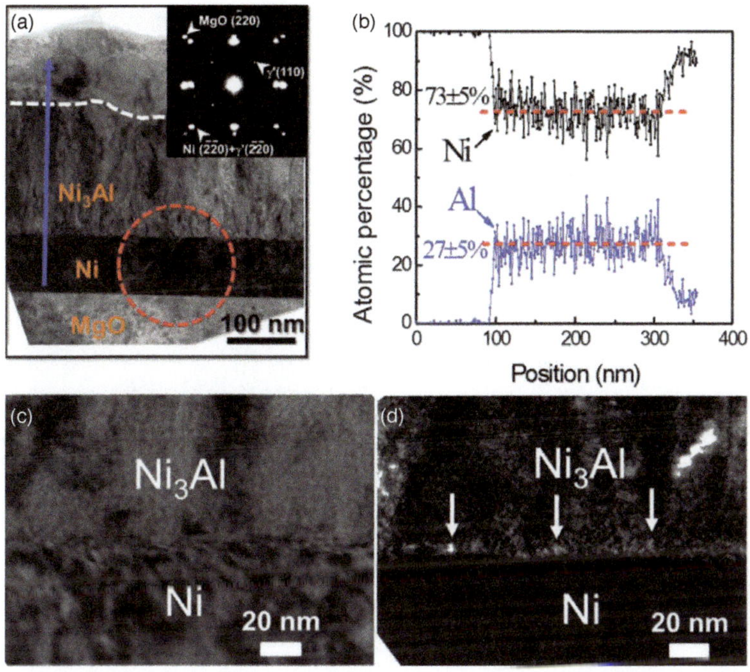

Figure 1. (a) TEM images obtained from the as-deposited pristine Ni/Ni_3Al thin-film multilayer on an MgO substrate. SAED pattern was taken from the circled area, including the MgO substrate, Ni, and Ni_3Al layers. Superlattice reflections, as indicated in the inset, suggest an ordered structure in the Ni_3Al layer. (b) EDS line scan across the film. The average ratio of Ni to Al concentration is measured to be 73%:27%. (c) Bright-field TEM image of the as-deposited Ni/Ni_3Al film. (d) Dark-field image with diffraction vector $g = [110]$, showing columnar γ' (Ni_3Al) precipitate phase in the Ni_3Al layer and some isolated nanoscaled γ' (Ni_3Al) particles formed on the Ni/Ni_3Al interface.

After annealing at 800°C for 1 h, the sample homogenized and no clear interfaces were observed (Figure 2(a)). The SAED pattern obtained from the annealed film suggests long-range ordering. A complementary EDS line scan performed across the film in Figure 2(b) shows a slight gradient in the Ni and Al compositions. Supplementary Figure S1 shows the microstructure of an as-annealed sample. The γ' precipitates appear as bright spots and these precipitates were formed throughout the film. A high-resolution STEM image obtained from the annealed film, shown in Figure 2(c), highlighted the Z-contrast difference between the matrix and the precipitates. Fast Fourier Transformed (FFT) diffraction patterns from selected regions of the STEM image suggested that these regions were composed of a disordered fcc-structured γ matrix and an ordered $L1_2$-structured γ' precipitates. Figure 2(d) shows a bright-field TEM image of a local area in the annealed film. A magnified annular bright-field (ABF) image of the area outlined by the box is shown in Figure 2(e), revealing a high density of precipitates (dark contrast features). Figure 2(f) and 2(g) shows STEM-based EDS chemical mappings of Ni and Al; interpretation of the EDS maps suggests the enrichment of Al in γ' precipitates. X-ray powder diffraction (XRD) profiles of the Ni/Ni_3Al film before and after annealing are shown in Figure S2. In the as-deposited state, the lattice parameters (a) of Ni and Ni_3Al were measured to be 3.552 and 3.561 Å, respectively, and thus the misfit strain (ε) was $\sim$0.25%, estimated by $\varepsilon = \Delta a/a$. After annealing, due to the intermixing of Ni and Al across the film and the greater amount of $L1_2$ precipitation, the lattice parameter of the matrix ($Ni_{80}Al_{20}$) and the precipitates (Ni_3Al) became 3.542 and 3.583 Å, respectively, and the misfit strain was $\sim$1.16%. Comparing the XRD profiles in Figure S2, more superlattice peaks, such as Ni_3Al (111) and Ni_3Al (210), were observed in the as-annealed state, which implies the evolution of texture, where grains with different orientations are formed.

The above observations lead to the following picture for the evolution of precipitates in the γ/γ' system under thermal annealing. In the Ni layer, the diffusion of Al from the Ni_3Al layer first leads to the formation of a Ni–Al solid solution. As the interdiffusion extends, γ' precipitates are nucleated due to the supersaturation of Al. In the Ni_3Al layer, the initial γ' phase evolves into nanometer-sized γ' precipitates under annealing. The partitioning of the elements Ni and Al causes changes in the lattice parameter of the matrix, and therefore changes the lattice mismatch between the matrix and the precipitates. The loss of Al and the gain of Ni in the Ni_3Al layer result in the increase of mismatch strain between the γ'

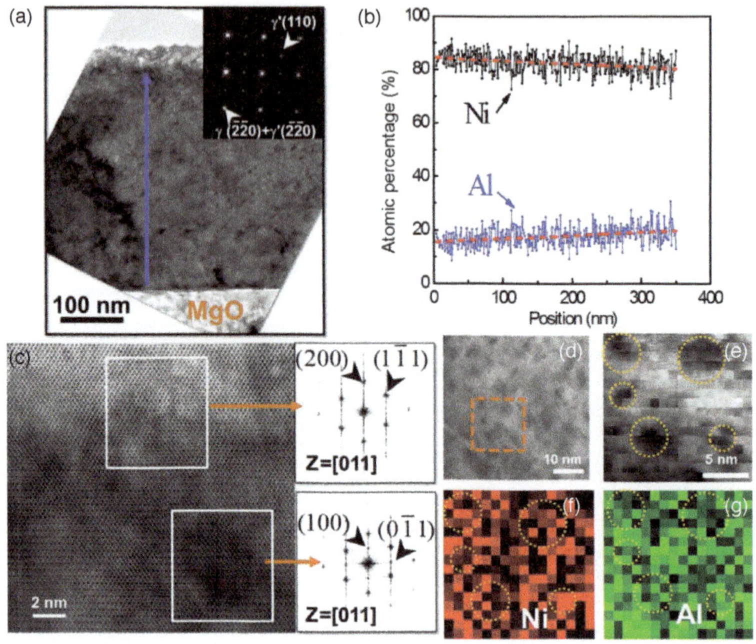

Figure 2. (a) Bright-field TEM image of the as-annealed film, showing no clear Ni/Ni_3Al interfaces. The inset SAED pattern shows an ordered structure of the film. (b) Measured chemical profile of the film after annealing. Thermally induced chemical intermixing over the entire film occurs under annealing. (c) High-resolution STEM image showing atomic (Z) contrast in a local area of the annealed sample. FFT patterns show clear evidence of a disordered matrix and ordered precipitates. (d) Bright-field TEM image in a local area, showing a high density of precipitates. (e) High-resolution STEM annual bright-field (STEM-ABF) image obtained from the area within the box in (d). Precipitates are marked with circles. (f and g) High-resolution EDS mapping of Ni and Al distribution in (e).

precipitates and the matrix. The increase in the lattice mismatch can also induce changes in the morphology of the γ' precipitates. For the small γ' precipitates, the morphology tends to be spherical in order to minimize the interfacial energy, while for the large γ' precipitates, as the strain energy increases, the shape of the precipitates evolves into platelets or facets.[18] In our current study, both spherical and plate-like precipitates form in the matrix after thermal annealing, as shown in Figure 2(d).

Bright-field TEM images were collected near the Ni/Ni_3Al interface of the as-annealed sample, as shown in Figure 3(a). Compared with the precipitates in the matrix, the size of the precipitates became larger and the geometry of these precipitates became faceted, as shown in Figure 3(b). STEM images of the as-deposited and as-annealed samples are shown in Figure S3, and the comparison reveals that the precipitates formed at the interface exhibited a lower mass contrast. A high-resolution EDS line scan across the precipitate is plotted in Figure 3(c). From the EDS line scan it can be seen that the precipitates had a higher concentration of Al and lower concentration of Ni compared with the background matrix; the atomic concentration ratio of Ni to Al in the precipitates was close to 3:1. Figure 3(d) shows a high-resolution STEM image of the phase boundary between the matrix and a precipitate. The Z-contrast imaging near the phase boundary suggests aggregation of Al in the precipitate, which is consistent with the measured chemical profile of γ' precipitates studied with EDS. Statistical analysis was performed on the measured sizes of precipitates in the annealed sample. The results of this analysis are presented in Figure 3(e) and 3(f) and show that the average precipitate size was $\sim$11 nm in the vicinity of the original parent interface, which was $\sim$2.8 times larger than that of the precipitates in the matrix. Annealing twins observed between Ni/Ni_3Al and Ni/MgO interfaces, as shown in Figure S4, might be the result of interface relaxation during heat treatment.

To better understand the role of the interface in enhancing the precipitation of γ', we used kinetic Monte Carlo to model the evolution of a fully coherent Ni/Ni_3Al interface under thermal annealing. One sample oriented in the direction [100], [010], and [001] was generated with dimensions 4.3 nm $\times$ 4.3 nm $\times$ 14.2 nm containing 23,040 atoms (Figure S5). The figure shows only Ni atoms that have eight Ni and four Al first-nearest neighbors and Al atoms with 12 Ni first-nearest neighbors; these are the environments that belong to an $L1_2$ structure. A vacancy was added randomly to the sample and evolved as described in the Methods section. Figure 4 shows the microstructural evolution of the system at

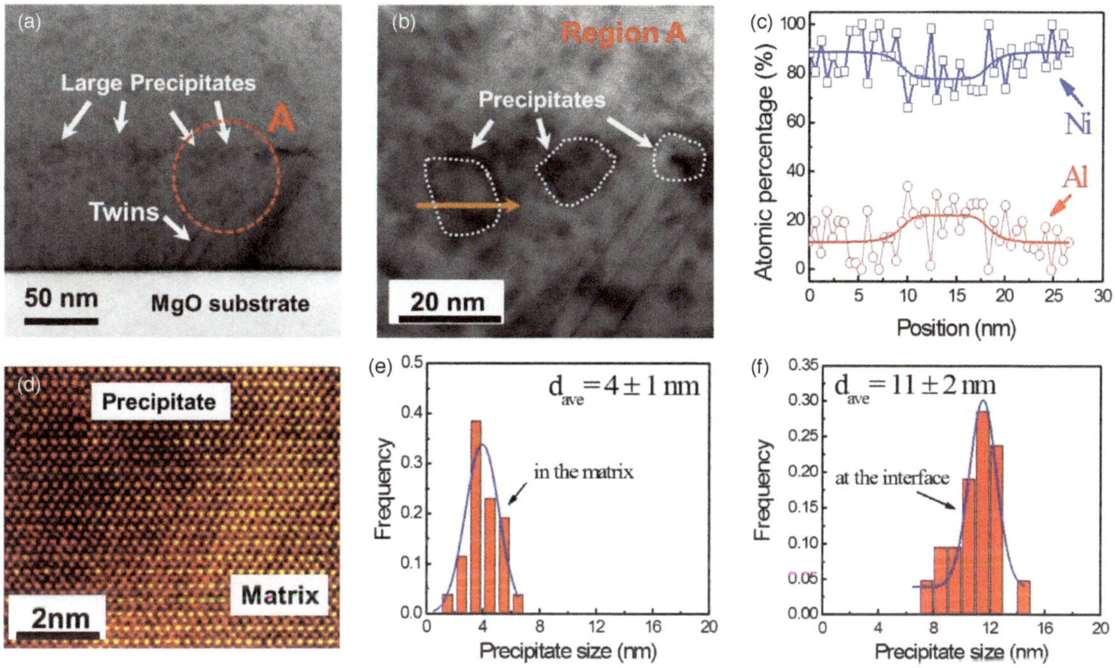

Figure 3. (a) Bright-field STEM image of the precipitates within the annealed sample located near the parent Ni/Ni_3Al interface. Larger sized precipitates and annealing twins were formed. (b) Magnified STEM micrograph of region 'A' showing large faceted precipitates. (c) High-resolution line scan across the precipitate marked by the orange arrow in (b). The atomic percent ratio of Ni to Al within the precipitate is ~3:1. (d) High-resolution STEM image of the precipitate/matrix showing a coherent interface was formed. (e) The average size of the precipitates in the matrix is measured to be ~4 nm. (f) The average size of the precipitates near the parent Ni/Ni_3Al interface is ~11 nm.

500°C, mimicking the deposition conditions, as given by the Monte Carlo simulations at different time steps, 3.60×10^5, 7.00×10^5, and 1.17×10^6. Figure 4(a), 4(d), and 4(g) shows the atomic configuration of the system at different stages. It is clear that ordering occurs in the system with the nucleation and growth of $L1_2$ precipitates. We also analyzed the distribution of Al in the sample and the average of the vacancy position along 10^5 Monte Carlo steps with the average steps. Figure 4(b), 4(e), and 4(h) shows these results, with the blue dashed curve displaying the Al distribution and the black curve showing the vacancy probability. Two main results can be extracted from these calculations: (1) The vacancy spends most of the time close to the interface (within the Ni-rich region). (2) There is Al intermixing into the Ni-rich layer that does not lead to the formation of $L1_2$ (over the length of the simulations studied) while ordering occurs in the Al supersaturated region. Figure 4(c), 4(f), and 4(i) describe the Al content belonging to an $L1_2$ structure. We observe the development of a peak close to one of the interfaces, where the vacancy spends most of the time (note that periodic boundary conditions were applied). Shaded regions are used as guides to the eye that show the Al undersaturated regions. These results show that the interface plays an important role in the formation of γ' precipitates, increasing the propensity for the formation of these precipitates along the Ni/Ni_3Al interface. The longer time that the vacancy spends at the interface enhances the nucleation of γ' precipitates. This is particularly interesting as, in this model, there are no misfit dislocations at the interface—it is fully coherent. Thus, even a coherent interface leads to trapping of the vacancy and enhanced precipitation. We expect the effect to be even more dramatic when dislocations are present.

Based on the aforementioned experimental and simulation studies, we can conclude that the γ' particles are favorably formed on the Ni/Ni_3Al interface during the deposition at 500°C as a result of the enhanced vacancy diffusion along the interface. Under thermal annealing at 800°C, the interface gradually becomes invisible due to the interdiffusion of Ni and Al across the interface, and the γ' precipitates in the vicinity of the parent interface grow much faster compared to those nucleated in the matrix because of the trapping of vacancies by the interface. In addition, the Monte Carlo simulation implies that the interface between the γ' precipitates and the γ matrix may also serve as the vacancy diffusion path and promote the coarsening of the γ' precipitates.

4. Conclusions Interdiffusion and precipitation behavior in a model γ/γ' system were studied by an

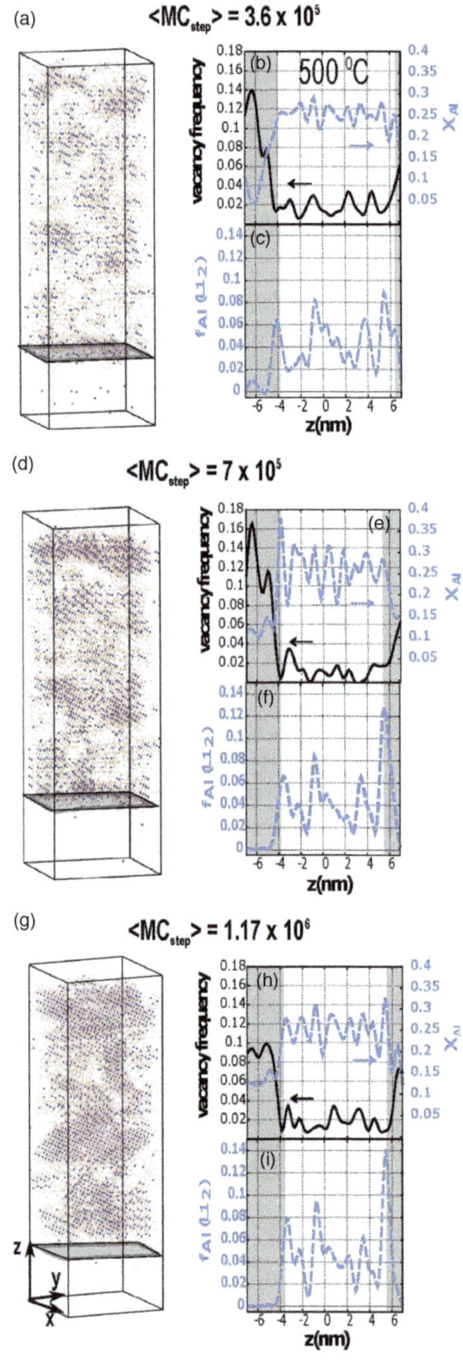

Figure 4. (a), (d), and (g) Distribution of Ni and Al atoms (brown and blue, respectively) in $L1_2$ structure at 500°C after 3.60×10^5, 7.00×10^5, and 1.17×10^6 steps, respectively. The original structure has two sharp Ni/Ni_3Al interfaces, one at the top boundary of the cell, and one at about one-fifth of the cell dimension from the bottom. (b), (e), and (h) show the average vacancy frequency in the z direction (black) and the total Al distribution (blue) for the different Monte Carlo steps. (c), (f), and (g) show the fraction of Al in the $L1_2$ structure for the different steps. Shaded areas show the approximate width of the Al undersaturated region.

experiment and Monte Carlo simulations. Interdiffusion, induced by thermal annealing, of the elements Ni and Al across the Ni/Ni_3Al interface resulted in the disappearance of the parent interface. Under thermal annealing at 800°C, ordered γ' precipitates with an average size of $\sim$4 nm were formed throughout the film sample and larger γ' precipitates with an average size of $\sim$11 nm were created near the parent Ni/Ni_3Al interface. The pristine Ni/Ni_3Al interface promotes the formation of the γ' precipitates. Vacancies preferentially diffuse along the interface, accelerating the precipitation of Ni_3Al particles. Our current study shows that the Ni/Ni_3Al interface has pronounced impact on the γ' precipitation during thermal annealing, suggesting the importance of an interface design in Ni-based superalloys for high-temperature applications.

Acknowledgements We acknowledge the financial support by LDRD program in Los Alamos National Laboratory. Los Alamos National Laboratory is operated by Los Alamos National Security, LLC, for the National Nuclear Security Administration of the U.S. DOE under contract DE-AC52-06NA25396. We thank Dr Robert Dickerson for helpful discussions. We also acknowledge that parts of the TEM work were performed at the LeRoy Eyring Center for High Resolution Electron Microscopy at Arizona State University (ASU) in collaboration with Toshihiro Aoki, and some of the high-resolution STEM work was conducted at the Center for Nanophase Materials Science, which is a DOE Office of Science User Facility.

Disclosure Statement No potential conflict of interest was reported by the authors.

References

[1] Pollock TM, Tin S. Nickel-based superalloys for advanced turbine engines: chemistry, microstructure and properties. J Propul Power. 2006;22:361–374.
[2] Furrer D, Fecht H. Ni-based superalloys for turbine discs. JOM. 1999;51:14–17.
[3] Unocic R, Viswanathan G, Sarosi P, Karthikeyan S, Li J, Mills M. Mechanisms of creep deformation in polycrystalline Ni-base disk superalloys. Mater Sci Eng A. 2008;483–484:25–32.
[4] Caron P. High γ' solvus new generation nickel-based superalloys for single crystal turbine blade applications. In: Pollock TM, Kissinger RD, Bowman RR, Green KA, McLean M, Olson S, Schirra JJ, editors. Superalloys. Warrendale, PA: The Minerals, Metals & Materials Society; 2000. p. 737–746.
[5] Payton EJ, Phillips PJ, Mills MJ. Semi-automated characterization of the phase in Ni-based superalloys via high-resolution backscatter imaging. Mater Sci Eng A. 2010;527:2684–2692.
[6] Mehrer H. Diffusion in intermetallics. Mater Trans JIM. 1996;37:1259–1280.
[7] Yasuda HY, Nakajima H, Koiwa M. Diffusion in L12-type intermetallic compounds. Defect Diffus Forum. 1993;95–98:823–830.

[8] Nakajima H, Nonaka K, Sprengel W, Koiwa M. Self-diffusion and interdiffusion in intermetallic compounds. Mater Sci Eng A. 1997;239–240:819–827.
[9] Koiwa M, Numakura H, Ishioka S. Diffusion in L12 type intermetallic compounds. Defect Diffus Forum. 1997;143–147:209–222.
[10] Watanabe M, Horita Z, Sano T, Nemoto M. Electron microscopy study of Ni/Ni_3Al diffusion-couple interface—II. Diffusivity measurement. Acta Metall. 1994;42:3389–3396.
[11] Svoboda J, Fischer FD, Fratzl P, Kozeschnik E. Modelling of kinetics in multi-component multi-phase systems with spherical precipitates: I: theory. Mater Sci Eng A. 2004;385:166–174.
[12] Baldan A. Review progress in Ostwald ripening theories and their applications to nickel-base superalloys Part I: Ostwald ripening theories. J Mater Sci. 2002;37:2171–2202.
[13] Wendt h, Haasen P. Nucleation and growth of γ'-Precipitates in Ni-14 at.% Al. Acta Metall. 1983;31:1649–1659.
[14] Hirata T, Kirkwood DH. The prediction and measurement of precipitate number densities in a nickel-6.05 wt.% aluminium alloy. Acta Metall. 1977;25:1425–1434.
[15] Ges A, Fornaro O, Palacio H. Long term coarsening of γ' precipitates in a Ni-base superalloy. J Mater Sci. 1997;32:3687–3691.
[16] Ma Y, Ardell AJ. Coarsening of γ (Ni–Al solid solution) precipitates in a γ' (Ni_3Al) matrix. Acta Mater. 2007;55:4419–4427.
[17] Maebashi T, Doi M. Coarsening behaviours of coherent γ' and γ precipitates in elastically constrained Ni–Al–Ti alloys. Mater Sci Eng A. 2004;373:72–79.
[18] Gleiter H, Hornbogen E. Precipitation hardening by coherent particles. Mater Sci Eng. 1968;2:285–302.
[19] Lifshitz IM, Slyozov VV. The kinetics of precipitation from supersaturated solid solutions. J Phys Chem Solids. 1961;19:35–50.
[20] Greenwood GW. The growth of dispersed precipitates in solutions. Acta Metall. 1956;4:243–248.
[21] Wagner C. The determination of small deviations from the ideal stoichiometric composition of ionic crystals and other binary compounds. Prog Solid State Chem. 1971;6:1–15.
[22] Loomis W, Freeman J, Sponseller D. The influence of molybdenum on the γ' phase in experimental nickel-base superalloys. Metall Trans. 1972;3:989–1000.
[23] Jayanth CS, Nash P. Factors affecting particle-coarsening kinetics and size distribution. J Mater Sci. 1989;24:3041–3052.
[24] Hin C. Kinetics of heterogeneous grain boundary precipitation of Ni 3 Al in nickel alloy. J Phys D Appl Phys. 2009;42:225309.
[25] Sadigh B, Erhart P, Stukowski A, Caro A, Martinez E, Zepeda-Ruiz L. Scalable parallel Monte Carlo algorithm for atomistic simulations of precipitation in alloys. Phys Rev B. 2012;85:184203.
[26] Martínez E, Caro A. Atomistic modeling of long-term evolution of twist boundaries under vacancy supersaturation. Phys Rev B, 2012;86:214109.
[27] Bortz AB, Kalos MH, Lebowitz JL. A new algorithm for Monte Carlo simulation of Ising spin systems. J Comput Phys. 1975;17:10–18.
[28] Mishin Y. Atomistic modeling of the γ and γ'-phases of the Ni–Al system. Acta Mater. 2004;52:1451–1467.

18

Atomic-Scale Comparison Between $\{\bar{1}101\}$ and $\{\bar{1}102\}$ Twin Growth Mechanisms in Magnesium

L. Leclercq[a], L. Capolungo[a*] and D. Rodney[b,c]

[a]*George Woodruff School of Mechanical Engineering, Georgia Institute of Technology Lorraine, UMI 2958 Georgia Tech-CNRS, 57070 Metz, France;* [b]*Laboratoire de Science et Ingénierie des Matériaux et des Procédés, Grenoble Institute of Technology, CNRS, UJF, 38402 Saint Martin d'Hères, France;* [c]*Institut Lumière Matière, Université Lyon 1, CNRS, UMR 5306, 69622 Villeurbanne, France*

The two most frequently observed twins in hexagonal close-packed (HCP) Mg, $\{\bar{1}101\}$ and $\{\bar{1}102\}$ twins, have surprisingly different properties and morphologies, with $\{\bar{1}101\}$ twins appearing under higher stresses and being much thinner than $\{\bar{1}102\}$ twins. By considering the atomic-scale elementary properties of the twin interfaces and their disconnections, we show that (1) the transverse propagation of $\{\bar{1}101\}$ twins is hindered by the absence of low-energy mobile interfaces, whereas $\{\bar{1}102\}$ twins benefit from prismatic-basal interfaces and (2) the thickening of $\{\bar{1}101\}$ twins is slowed by higher energy barriers against both the nucleation and propagation of disconnections along their interfaces.

Keywords: Pure Magnesium, Twinning, Disconnections, Molecular Statics

1. Introduction Twinning is a deformation mechanism particularly important in hexagonal close-packed (hcp) crystals, providing together with pyramidal slip, a means to accommodate deformation along the *c*-axis.[1–3] A significant body of experimental and theoretical studies, dating back to the early 1950s,[4–6] has been dedicated to the subject. Of particular interest is the question of the activation and the effect of twinning on the plastic response of materials such as magnesium and its alloys.[7–10]

Amongst the three twinning modes experimentally reported in pure and alloyed magnesium, two are of particular interest: $\{\bar{1}102\}$ and $\{\bar{1}101\}$ twins.[8,11–15] The former induces an extension of the *c*-axis while the latter leads to a compression of the *c*-axis. Provided loads are suitably oriented with respect to the texture, these two twin types are readily activated, yet present distinct characteristics. $\{\bar{1}102\}$ Tension twins, an example of which is shown from an atomistic simulation in Figure 1, are commonly observed in several hcp metals (i.e. Ti,Be,Zr,Hf), including pure Mg and Mg alloys AZ31 and ZK60A.[16–18] In the case of Mg, tensile twins are reported to be activated under low critical resolved shear stresses (CRSS) between 2 and 8 MPa.[19,20] These twins can consume close to the entire parent grains.[8,11,21] The twin morphology depends on the total accumulated shear strain in the twinning shear direction and results from multiple nucleation events followed by growth and coalescence of the twin domains.

On the contrary, $\{\bar{1}101\}$ compression twins, which have been observed in pure Mg as well as in Mg AZ31,[16,17] are activated under relatively high CRSS, between 30 and 100 MPa,[19,23–25] comparable to that required to activate pyramidal slip (30–80 MPa [23,26–29]). Compressive twinning is thus in competition with pyramidal slip. Moreover, even if no quantification of the thickness of $\{\bar{1}101\}$ twins is available in the literature, they typically appear to be narrow. For example, in extruded AZ31, following compressive loading up to 10% strain in the extrusion direction, $\{\bar{1}101\}$ twins were observed to be 2.5 μm thick.[14]

Interestingly, the rational for these marked differences remains unclear to this date. In the absence of spinoidal mechanisms, understanding the mechanism of twinning requires the study of three phenomena: nucleation, transverse propagation (i.e. propagation of the twin nucleus across the parent phase) and thickening. Twin nucleation has been described, theoretically, in

*Corresponding author. Email: laurent.capolungo@me.gatech.edu

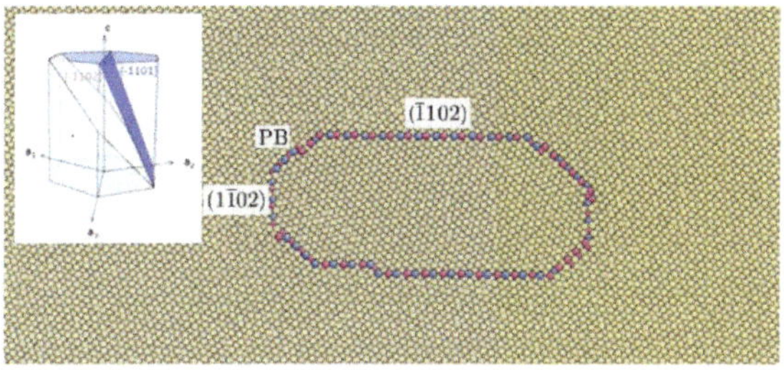

Figure 1. Nucleus of a $\{\bar{1}102\}$ twin from an atomistic simulation. The inset illustrates the mirror planes of the $(\bar{1}101)$ and $(\bar{1}102)$ twins in hcp crystals. The atoms are coloured according to their coordination number using the Atomeye software.[22]

terms of both homogeneous and heterogeneous processes; the latter being more probable from experimental observations.[17,30,31] The first heterogeneous nucleation model proposed is the pole mechanism of Thompson and Millard,[32] which considers the dissociation of $\langle c \rangle$ dislocations into twinning dislocations. Mendelson [33,34] then developed a similar but more detailed model in which twinning dislocations result from the dissociation of $\langle a \rangle$, $\langle c \mid a \rangle$ or $\langle c \rangle$ dislocations. Recently, Ghazisaeidi and Curtin [35] studied the dissociation of $\langle c \rangle$ and $\langle c + a \rangle$ dislocations to nucleate $\{\bar{1}102\}$ twins after Capolungo and Beyerlein [36] had studied the conditions to nucleate stable twin loops by dissociating a single $\langle a \rangle$ dislocation. Other potential structures and conditions for generating twin nuclei have also been identified in recent atomistic studies, which are based either on a shuffling mechanism from an existing grain boundary [37] or on the decomposition of particular dislocations, such as the $\{\bar{1}102\}$ twin nucleus resulting from the relaxation of a dipole of partial dislocations suggested by Wang et al.[38] An evaluation of the stresses necessary to nucleate twins was provided by Luque et al. [39] who used molecular dynamics simulations on Mg single-crystal nanopillars.

To date, transverse twin propagation has attracted much less attention. To the authors' knowledge, Wang et al. [38] and Xu et al. [40] provided the only studies of this phenomenon. In both works, only $\{\bar{1}102\}$ twins were considered. In [40], a specific low-energy interface, which was systematically found at the front of the growing twin, is observed to drive transverse propagation. This interface places face-to-face a basal plane of the matrix and a prismatic plane of the twin (or vice versa) and is consequently referred to as prismatic-basal (PB) interface. This interface was also identified in [37] and observed experimentally in cobalt in [41]. One open question is whether a similar interface is involved in either the transverse propagation or the thickening of $\{\bar{1}101\}$ twins.

Twin thickening has been extensively studied and occurs through the propagation of disconnections along the twinning plane.[42–45] Disconnections are line defects characterized by a Burgers vector in the twin interface and a step normal to the twinning plane. In the case of $\{\bar{1}101\}$ twins, Christian and Mahajan,[1] Serra et al.,[45] Serra and Bacon [46] and Wang et al. [47,48] reported the disconnections to be two or four crystallographic planes high. Similarly in the case of tensile twins, disconnections were fully characterized in work of Serra et al.,[45] Serra and Bacon [46] and Pond et al.[49]

The purpose of the present article is to rationalize the experimentally observed structural differences between $\{\bar{1}101\}$ and $\{\bar{1}102\}$ twins. To this end, molecular statics simulations were performed, using the embedded atom method interatomic potential developed by Liu et al.[50] First, twin transverse propagation is studied by extending the results of Xu et al. [40] to $\{\bar{1}101\}$ twins. In particular, we investigate whether a specific interface, analogous to the PB interface, exists for $\{\bar{1}101\}$ twins. The second part of this article focuses on thickening. Energies necessary to propagate two-layer and four-layer twinning disconnections are investigated on the two types of twin interfaces (i.e. $\{\bar{1}102\}$ and $\{\bar{1}101\}$). Our work points to an extrinsic growth mechanism for $\{\bar{1}101\}$ twins mediated by plasticity, while the growth mechanism of $\{\bar{1}102\}$ twins is compatible with an intrinsic mechanism (i.e. not mediated by the interaction with other defects).

2. Transverse propagation As stated above, our investigations start from the results obtained by Xu et al.[40] Since the real nature of twin nuclei is still unknown, twin transverse propagation was studied from two hypothetical initial nuclei. In this part, we first recall the results found in [40] for $\{\bar{1}102\}$ and apply a similar methodology to $\{\bar{1}101\}$ twins.

The first method to generate a nucleus consists in introducing a dipole of partial dislocations in a perfect crystal and letting the system relax under an applied shear strain via energy minimization. The Burgers vector of the partials, $\pm b_n$, matches the minimum of the γ-line calculated on the twinning plane along the $[1\bar{1}01]$ direction, as suggested by Wang et al.[38] Upon imposing a sufficiently large shear strain parallel to the twin interface, growth of the nucleus can be activated. The resulting twinned structure is bounded by $(\bar{1}102)$ and $(1\bar{1}02)$ planes and PB interfaces, as shown in Figure 1. The second type of nucleus created in [40] was inspired by Eshelby's inclusion problem.[51] The nucleus is obtained by cutting a sub-domain, applying mirror symmetry to the atoms inside the sub-domain and placing the atoms back in the original space with an additional strain applied to recover the initial shape of the sub-domain. Such nucleus was also found to grow under sufficiently large shear strains parallel to the twin plane. These two methods revealed the importance of PB interfaces to transverse propagation of $\{\bar{1}102\}$ twins in [40]. One particularly important feature is that PB interfaces readily form regardless of the method used to build the initial nucleus.

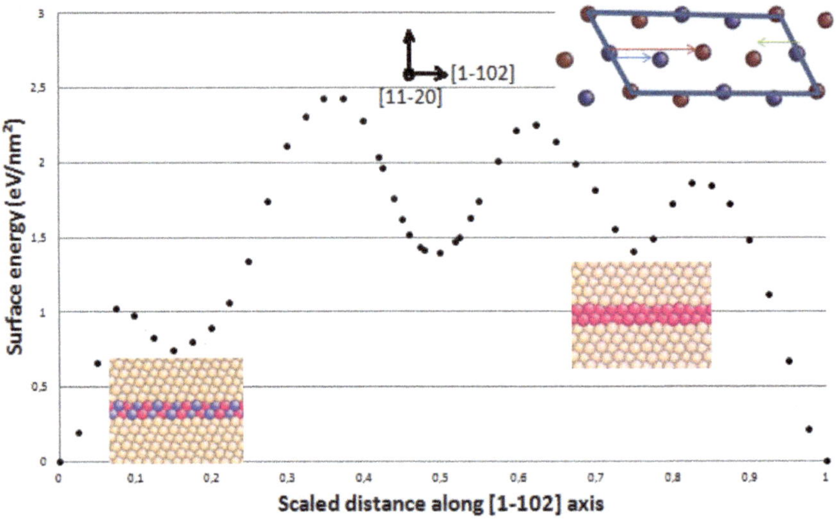

Figure 2. γ-line calculated in the $\{\bar{1}101\}$ twin interface along the $[1\bar{1}02]$ direction. Distances are given as a fraction of the length of the $[1\bar{1}02]$ vector. The atomic configurations correspond to the two stable stacking faults. The unit cell is given as an inset, arrows corresponding to the fault vector of the minima.

In the present letter, the two above methods were applied to $\{\bar{1}101\}$ twins. We first calculated the γ-line on the twinning plane along the $[1\bar{1}02]$ direction in order to apply the dipole method. The result is shown in Figure 2. The γ-line has three local minima, generically noted b_m, located at 0.17*$[1\bar{1}02]$, 0.5*$[1\bar{1}02]$ and 0.75*$[1\bar{1}02]$, resulting in five potential Burgers vectors: 0.17*$[1\bar{1}02]$ (0.17–1)*$[1\bar{1}02]$, 0.5*$[1\bar{1}02]$, 0.75*$[1\bar{1}02]$ and (0.75 − 1)*$[1\bar{1}02]$. For each minimum, we chose the lowest absolute value between b_m and $(b_m - 1)$ since the smaller b_m has a lower energy.

Referring to the unit cell in Figure 2, one can notice that all three minima correspond to displacements essentially positioning atoms at previously occupied sites, accordingly to the topological approach of Pond et al.[49] We then introduced the corresponding dipoles of partial dislocations into periodic cells of dimensions $58.8 \times 24.4 \times 0.32\,\mathrm{nm}^3$, containing approximately 20,000 atoms. Unlike $\{\bar{1}102\}$ twins where there is a minimal length for the dipole that results in the growth of a twin, all dipole lengths provide the same result for the $\{\bar{1}101\}$ twin. The 0.17*$[1\bar{1}02]$ Burgers vector results in a stable stacking fault of low energy, given by the absolute minimum of the γ-line. This Burgers vector is represented by the blue arrow in the unit cell given in Figure 2. The 0.5*$[1\bar{1}02]$ Burgers vector, represented by the red arrow, yields an unstable structure, relaxing into the previously identified stable stacking fault outside of the initial dipole. The (0.75 − 1)*$[1\bar{1}02]$ Burgers vector, represented by the green arrow, results in another stable configuration but with a higher stacking fault energy. No dipole of partial dislocations thus leads to the formation of an incipient twin, even under applied shear strains up to 10%. It is consequently not possible to use the dipole method to make a $\{\bar{1}101\}$ twin grow.

One may also wonder if a mechanism similar to the propagation of PB interfaces in $\{\bar{1}102\}$ twins could result in $\{\bar{1}101\}$ twin propagation. In order to search for potential low-energy interfaces that would be comparable to PB interfaces, we computed the energy of the twin interface as a function of its orientation with respect to the $\{\bar{1}101\}$ mirror plane, as done in [40] for $\{\bar{1}102\}$ twins. To do so, the first step is to be able to produce a twinned domain of arbitrary shape. For that, we noticed that a twinned domain can be obtained starting from a perfect crystal by propagating a screw dislocation of Burgers vector $[11\bar{2}0]/6$ every other dense pyramidal plane and letting the system relax. This is equivalent to propagating two-layer disconnections on the twinning plane, since the latter have the same $[11\bar{2}0]/6$ screw component and the edge component (approximately 0.16*a in Mg [46]) that is necessary to recover the entire twinning dislocation forms during the energy minimization. We then created a cylindrical volume in which the interface between the matrix and the twin corresponds to a diameter of the cylinder (Figure 3). The energy of each interface was calculated by summing the energies of all atoms in the volume and subtracting their cohesive energy. Boundary effects were minimized by including the cylinder of interest (blue) in a larger one (red). The angle between the interface and the twinning plane is referred to as θ and we chose to set $\theta = 0$ when the interface coincides with the twinning plane. One obtains the energies of the interfaces as a function of their orientation with respect to the $\{\bar{1}101\}$ mirror plane by making θ vary between 0° and 180° (Figure 3(a)).

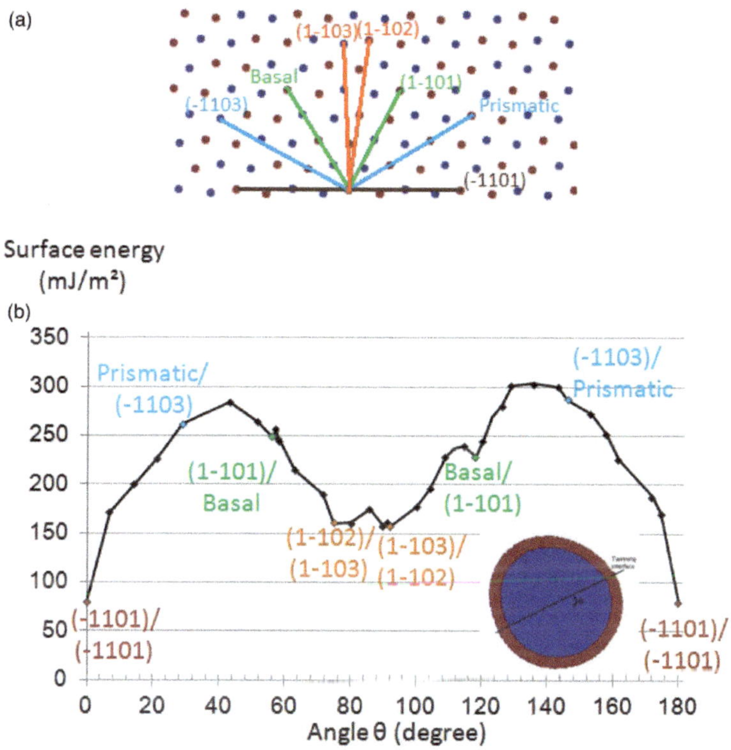

Figure 3. Planes with symmetry properties with respect to the normal to the ($\bar{1}101$) twinning plane (a) and energies of twinned interfaces as a function of their angle θ with respect to the horizontal [$1\bar{1}02$] direction (b).

Several low-index planes are identified in Figure 3(a) that are close to symmetrical with respect to the normal to the twinning plane and result in particular interfaces. Such relationship between symmetrical planes is purely geometrical since symmetry operations with respect to the twinning plane or with respect to its normal result in the same crystal reorientation. The ($\bar{1}101$) mirror plane is the orientation of minimum energy. Three other pairs of particular conjugated planes were identified from Figure 3(a) and used as boundaries to build hexagonal twin nuclei with the method inspired from Eshelby's inclusion problem. ($1\bar{1}02$)/($1\bar{1}03$) interfaces were studied first since they are the second lowest energy interfaces. They are reminiscent of the PB interfaces for {$\bar{1}102$} twins since they correspond to marked minima in the energy curve. Prismatic/($\bar{1}103$) interfaces were investigated because of their high energy that suggests that they may be unstable. Finally, ($1\bar{1}01$)/basal interfaces were considered since they are local minima and may present properties similar to PB interfaces.

Interestingly, regardless of the specific interfaces chosen to construct the twinned subdomain (i.e. high energy or low energy), the direction of the strain (i.e. shear strain or strain normal to the twinning plane or in the direction of the plane), the amount of strain (up to 10%) and the geometry of the nucleus (i.e. from 2-layers to more than 15-layers thick and from a ratio length/thickness from about 2 to about 10), none of these nuclei grew. The fact that we did not succeed in making a {$\bar{1}101$} twin nucleus grow suggests that {$\bar{1}101$} twin transverse propagation might not be driven by a particular interface, in contrast with {$\bar{1}102$} twins.

3. Thickening As stated above, twin thickening is mediated by the propagation of disconnections on the twin interfaces. The rate of twin growth is then a direct consequence of (1) the geometrical characteristic of the disconnections (e.g. their Burgers vector and step height), (2) the process of disconnection generation and (3) the mobility of the disconnections. To appreciate the potential difference in growth rate of tensile and compressive twins, these three points are addressed in what follows.

The following potential thickening defects were considered: two-layer steps, four-layer steps and finite size interfaces. For the latter case, the PB interface was chosen for the {$\bar{1}102$} twin while the basal/($1\bar{1}01$) interface—naturally forming as a result of energy minimization—was selected for the {$\bar{1}101$} twin, with either 6- or 8-layer thickness.

To study the steps, we consider a bi-crystal geometry illustrated in Figure 4 with one half of the crystal twinned with respect to the other half. Figure 4 is given as an illustration, however the real simulation cells were

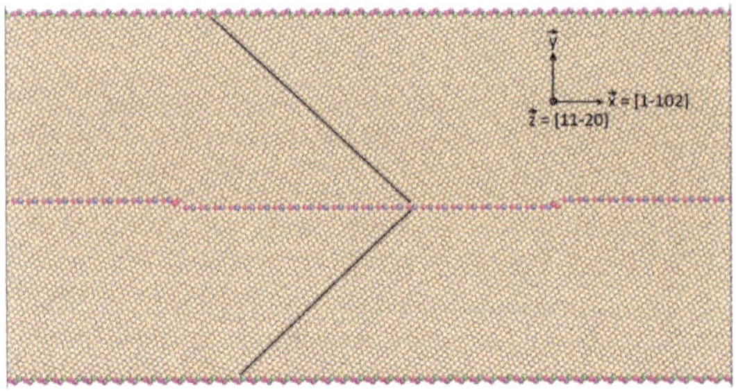

Figure 4. Simulation cell used to study the propagation of $\{\bar{1}102\}$ twins. The black lines indicate the orientations of the basal planes in the two parts of the crystal.

chosen large enough to avoid border effects and interaction between the steps. Due to the symmetry with respect to the twinning plane, periodic boundary conditions along the y-axis were removed. We introduced a pair of steps on the interface in order to maintain periodicity along the x-axis. The steps on the $\{\bar{1}101\}$ interfaces were created using the method described earlier in this letter. For $\{\bar{1}102\}$ twins, we repositioned the atoms as tracked by Wang et al. [37] during a nucleation event.

Following energy minimization, steps higher than two layers on $\{\bar{1}102\}$ twin interfaces systematically decompose into two-layer steps, even under an applied shear strain. This is particularly interesting as it shows that the stability of PB interfaces depends on the geometry of their extremities. Indeed, our calculations show that PB interfaces are not stable when they connect two parallel $(1\bar{1}02)$ twin interfaces. However, they were found to be stable when they are connected with other PB interfaces or with $(1\bar{1}02)$ conjugate planes, as seen in the nuclei of Figure 1. As a result, the PB interface cannot be regarded solely as a stack of two-layer disconnections. For the sake of the study on twin thickening though, in which case PB interfaces are unstable, the only defect pertaining to tensile twins considered in what follows is the two-layer step.

On $\{\bar{1}101\}$ twin interfaces, all steps reached stable configurations after relaxation. The four- and eight-layer steps were always stable, regardless of the amount and direction of the applied strain, while six-layer steps, depending on the initial configuration used before energy minimization—and in particular the initial position of the screw components of the disconnections—, may decompose into either a two-layer step and a four-layer step or into a two-layer step and an eight-layer step. Note that the later was found to be a stack of two four-layer steps.

The energies involved in disconnection generation were investigated using the nudged elastic band (NEB) method.[52] This method consists in minimizing the energy on intermediate configurations between an initial and a final state in order to identify the energy barrier to overcome when going from one state to the other. To study the nucleation of disconnections on both types of interfaces, we started from a perfectly flat twin interface and ended with two steps—to be consistent with the periodic conditions (Figure 4). The energies necessary to create these two steps are given in Figure 5(a). We found that the energy barrier is about three times higher for $\{\bar{1}101\}$ twins so that it is easier for steps to nucleate on $\{\bar{1}102\}$ twin interfaces. The energies necessary to create these two steps are given in Figure 5(a). We found that the energy barrier is about three times higher for $\{\bar{1}101\}$ twins so that it is easier for steps to nucleate on $\{\bar{1}102\}$ twin interfaces. As a consequence, we infer that two-layer steps on $\{\bar{1}102\}$ twin interfaces may nucleate by an intrinsic process not requiring, for example, interaction of the interface with glide dislocations. The functional dependence of the step nucleation energy on stress will require further study to decipher the relative roles of stress and temperature on twin thickening. Note that such thermal effect may not exist for the generation of the twin nucleus. On the contrary, steps on $\{\bar{1}101\}$ twin interfaces are probably mediated by extrinsic defects, possibly the interaction and absorption of screw dislocations.[46,53,54]

Finally, we focused on the mobility of these disconnections by calculating the energies necessary for their propagation using the NEB method. These calculations, shown in Figure 5(b) and 5(c), revealed that the energy necessary to propagate four-layer steps on $\{\bar{1}101\}$ twin interfaces is about ten times higher than the energy necessary to propagate two-layer steps (Figure 5(b)), confirming the stability of four-layer steps. The different nature of the two kinds of steps is also underlined by a metastable configuration for four-layer steps that does not exist for two-layer steps. This type of curve is consistent with the results of Wang et al. [47] for two-layer steps. For four-layer steps, the order of magnitude differs and the metastable configuration is more pronounced in our case. We checked the accuracy of our calculations by running the simulations again using the potential developed by Sun et al. [55] and found a similar behaviour. The structure of the eight-layer step was identified to be equivalent to two superimposed four-layer steps of either the stable or the metastable configuration. The intermediate configurations computed by the NEB method reveal that the propagation of an eight-layer step is equivalent to the simultaneous propagation of two four-layer steps. The differences in energies involved in step propagation suggest that $\{\bar{1}101\}$ twin growth, which is already difficult to activate as discussed previously, is hindered by the stability of high steps. As a consequence, these twins appear to be thin.

The critical resolved shear stress associated with the propagation of two-layer steps on $\{\bar{1}101\}$ twin interfaces, directly measured by applying a shear stress to the simulation cell, was found to be 310 MPa with a positive stress and −400 MPa with a negative stress. The difference between these two values can be explained by the non-symmetric character of the step with respect to

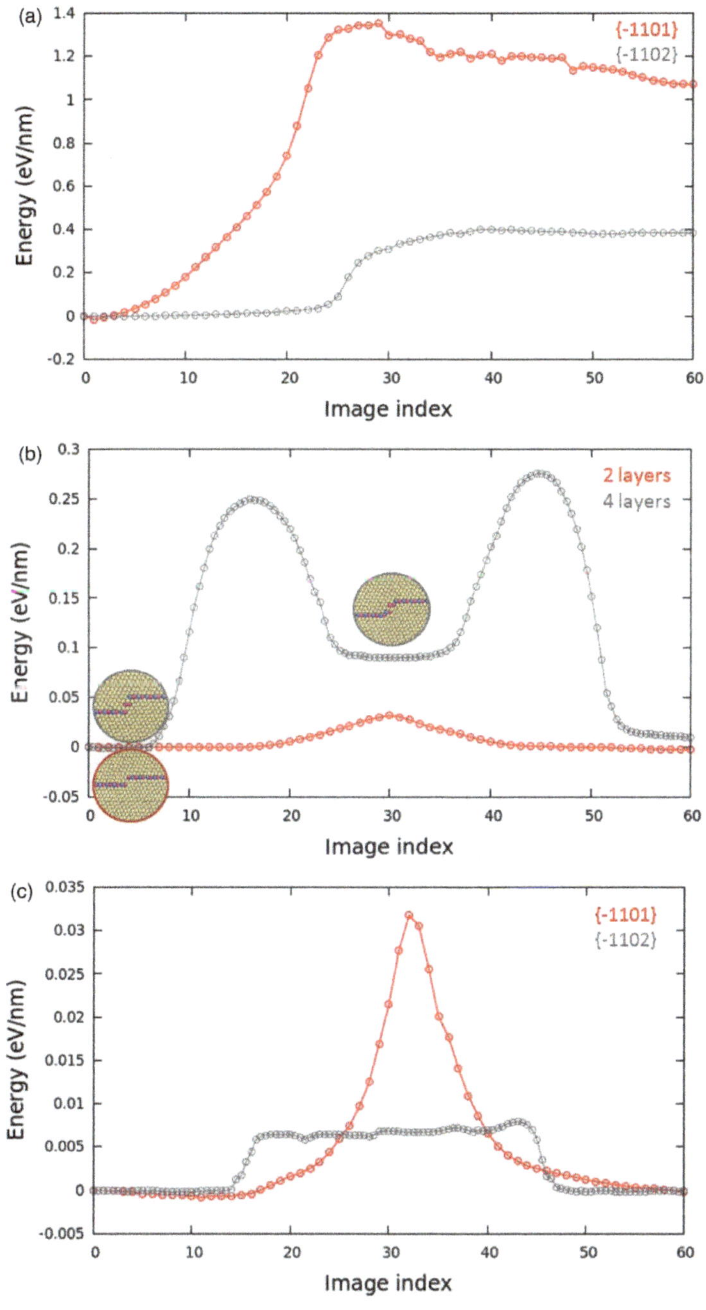

Figure 5. Energy necessary to nucleate two-layer steps on {$\bar{1}101$} and {$\bar{1}102$} twin interfaces obtained using the NEB method (a), propagation of two-layer and four-layer steps on {$\bar{1}101$} twin interfaces with atomic configurations (b) and energy necessary to propagate two-layer steps on {$\bar{1}101$} and {$\bar{1}102$} twin interfaces (c).

the *y*-axis. Incidentally, this asymmetry may be related to detwinning. However, it has to be kept in mind that our simulations are conducted at 0 K, such that we have no temperature effect and in particular no effect of thermal activation on step mobility. Also, our calculations were performed in two dimensions, forcing the steps to move as straight lines, while in three dimensions, a step may glide by a nucleation/propagation mechanism similar to the kink-pair mechanism of high Peierls stress dislocations.[56]

Regarding tensile twins, it was found that the energy barrier to be overcome to propagate two-layer steps on {$\bar{1}102$} twin interfaces is about 0.007 eV/nm (Figure 5(c)), which is about four times lower that for

$\{\bar{1}101\}$ twins, hence its easy propagation. The associated critical resolved shear stress was found to be 190 MPa when we apply a positive stress and −230 MPa when the stress is negative. These values stay below the CRSS of $\{\bar{1}101\}$ twins, which is consistent with the lower energy involved in the propagation of this type of twins.

These energy considerations provide two more arguments in favour of the more frequent appearance of $\{\bar{1}102\}$ twins since it is easier both to nucleate and to propagate two-layer steps on their interfaces. Moreover, while four-layer steps on $\{\bar{1}101\}$ twin interfaces significantly hinders the growth process, four-layer steps on $\{\bar{1}102\}$ twin interfaces decompose into mobile two-layer steps, which profits to twin growth.

4. Conclusion The different natures of $\{\bar{1}101\}$ and $\{\bar{1}102\}$ twins have been underlined, providing an explanation to their experimentally observed differences. The transverse propagation of $\{\bar{1}102\}$ twins was found in [40] to be driven by the propagation of PB interfaces. By way of contrast, the new results presented earlier in the present article show that $\{\bar{1}101\}$ twin transverse propagation may not be driven by such a process, which increases the singularity of $\{\bar{1}102\}$ twins. Focusing on twin thickening, we found that the glide of two-layer steps is the main mechanism controlling growth in both cases. Energy considerations on step propagation revealed that it is much easier to nucleate and propagate steps on $\{\bar{1}102\}$ twin interfaces, pointing towards an intrinsic mechanism involving thermal activation. On the contrary, the nucleation of $\{\bar{1}101\}$ twins has to be mediated, possibly by the interaction with lattice dislocations or other twins.

Acknowledgements The authors acknowledge the support of the French National Research Agency under the project MAGTWIN and the Georgia Institute of Technology Lorraine.

References

[1] Christian JW, Mahajan S. Deformation twinning. Prog Mater Sci. 1995;39(1–2):1–157.
[2] Hirth JP, Lothe J. Theory of dislocations. Melbourne, FL: Krieger; 1992.
[3] Zhang J, Joshi SP. Phenomenological crystal plasticity modeling and detailed micromechanical investigations of pure magnesium. J Mech Phys Sol. 2012;60(5):945–972.
[4] Cahn RW. Plastic deformation of alpha-uranium; twinning and slip. Acta Metall. 1953;1(1):49–70.
[5] Bell RL, Cahn RW. The nucleation problem in deformation twinning. Acta Metall. 1953;1(6):752–753.
[6] Bell RL, Cahn RW. The dynamics of twinning and the interrelation of slip and twinning in zinc crystals. Proc R Soc Lond A. 1957;239(1219):494–521.
[7] Cáceres CH, Sumitomo T, Veidt M. Pseudoelastic behaviour of cast magnesium AZ91 alloy under cyclic loading-unloading. Acta Mater. 2003;51(20):6211–6218.
[8] Mann GE, Sumitomo T, Cáceres CH, Griffiths JR. Reversible plastic strain during cyclic loading-unloading of Mg and Mg-Zn alloys. Mater Sci Eng A. 2007;456(1–2):138–146.
[9] Wang H, Wu PD, Wang J. Modeling inelastic behavior of magnesium alloys during cyclic loading-unloading. Int J Plast. 2013;47:49–64.
[10] Stanford N, Barnett MR. Solute strengthening of prismatic slip, basal slip and $\{10\bar{1}2\}$ twinning in Mg and Mg-Zn binary alloys. Int J Plast. 2013;47:165–181.
[11] Nave MD, Barnett MR. Microstructures and textures of pure magnesium deformed in plane-strain compression. Scr Mater. 2004;51(9):881–885.
[12] Syed B, Geng J, Mishra RK, Kumar KS. [0001] compression response at room temperature of single-crystal magnesium. Scr Mater. 2012;67(7–8):700–703.
[13] Barnett MR. Twinning and the ductility of magnesium alloys: part I: "tension" twins. Mater Sci Eng A. 2007;464(1–2):1–7.
[14] Barnett MR. Twinning and the ductility of magnesium alloys: part II: "contraction" twins. Mater Sci Eng A. 2007;464(1–2):8–16.
[15] Koike J, Sato Y, Ando D. Origin of the anomalous $\{\bar{1}102\}$ twinning during tensile deformation of Mg alloy sheet. Mater Trans. 2008;49(12):2792–2800.
[16] Barnett MR, Keshavarz Z, Beer AG, Ma X. Non-schmid behaviour during secondary twinning in a polycrystalline magnesium alloy. Acta Mater. 2008;56(1):5–15.
[17] Dudamell NV, Ulacia I, Gálvez F, Yi S, Bohlen J, Letzig D, Hurtado I, Pérez-Prado MT. Twinning and grain subdivision during dynamic deformation of a Mg AZ31 sheet alloy at room temperature. Acta Mater. 2011;59(18):6949–6962.
[18] Wu L, Jain A, Brown DW, Stoica GM, Agnew SR, Clausen B, Fielden DE, Liaw PK. Twinning-detwinning behavior during the strain-controlled low-cycle fatigue testing of a wrought magnesium alloy, ZK60A. Acta Mater. 2008;56(4):688–695.
[19] Kelley EW, Hosford WF. Plane-strain compression of magnesium and magnesium alloy crystals. Trans Metall Soc AIME. 1968;242(1–2):5–13.
[20] Roberts C. Magnesium and its alloys. New York: Wiley; 1960.
[21] Knezevic M, Levinson A, Harris R, Mishra RK, Doherty RD, Kalidindi SR. Deformation twinning in AZ31: influence on strain hardening and texture evolution. Acta Mater. 2010;58(19):6230–6242.
[22] Li J. Atomeye: an efficient atomistic configuration viewer. Modell Simul Mater Sci Eng. 2003;11(2):173–177.
[23] Kitahara T, Ando S, Tsushida M, Kitahara H, Tonda H. Deformation behavior of magnesium single crystals in c-axis compression. Key Eng Mater. 2007;345–346(1–2):129–132.
[24] Wonsiewicz BC. Plasticity of magnesium crystals [dissertation]. Massachusetts Institute of Technology; 1966.
[25] Yoshinaga H, Horiuchi R. Deformation mechanisms in magnesium single crystals compressed in the direction parallel to hexagonal axis. Trans Jpn Inst Met. 1963;4:1–8.
[26] Lilleodden E. Microcompression study of Mg (0001) single crystal. Scr Mater. 2010;62(8):532–535.
[27] Obara T, Yoshinga H, Morozumi S. $\{11\bar{2}2\} < \bar{1}\bar{1}23 >$ slip system in magnesium. Acta Metall. 1973;21(7):845–853.
[28] Byer CM, Li B, Cao B, Ramesh KT. Microcompression of single-crystal magnesium. Scr Mater. 2010;62(8): 536–539.
[29] Ando S, Harada N, Tsushida M, Kitahara H, Tonda H. Temperature dependence of deformation behavior in

magnesium and magnesium alloy single crystals. Key Eng Mater. 2007;345–346(1–2):101–104.

[30] Molnár P, Jäger A, Lejček P. Twin nucleation at grain boundaries in Mg-3 wt.% Al-1 wt.% Zn alloy processed by equal channel angular pressing. Scr Mater. 2012;67(5):467–470.

[31] Ghaderi A, Barnett MR. Sensitivity of deformation twinning to grain size in titanium and magnesium. Acta Mater. 2011;59(20):7824–7839.

[32] Thompson N, Millard DJ. Twin formation in cadmium. Philos Mag. 1952;43(339):422–440.

[33] Mendelson S. Zonal dislocations and twin lamellae in h.c.p. metals. Mater Sci Eng. 1969;4(4):231–242.

[34] Mendelson S. Dislocation dissociations in hcp metals. J Appl Phys. 1070;41(5):1893–1910.

[35] Ghazisaeidi M, Curtin WA. Analysis of dissociation of $< c >$ and $< c + a >$ dislocations to nucleate $\{10\bar{1}2\}$ twins in Mg. Modell Simul Mater Sci Eng. 2013;21: 1–13.

[36] Capolungo L, Beyerlein IJ. Nucleation and stability of twins in hcp metals. Phys Rev B. 2008;78:024117.

[37] Wang J, Yadav SK, Hirth JP, Tomé CN, Beyerlein IJ. Pure-shuffle nucleation of deformation twins in hexagonal-close-packed metals. Mater Res Lett. 2013;1:1–7.

[38] Wang J, Hoagland RG, Hirth JP, Capolungo L, Beyerlein IJ, Tomé CN. Nucleation of a twin in hexagonal close-packed crystals. Scr Mater. 2009;61(9):903–906.

[39] Luque A, Ghazisaeidi M, Curtin WA. Deformation modes in magnesium (0001) and $(01\bar{1}1)$ single crystals: simulations versus experiments. Modell Simul Mater Sci Eng. 2013;21(4):045010.

[40] Xu B, Capolungo L, Rodney D. On the importance of prismatic/basal interfaces in the growth of twins in hexagonal close packed crystals. Scr Mater. 2013;68(11): 901–904.

[41] Zhang XY, Li B, Wu XL, Zhu YT, Ma Q, Liu Q, Wang PT, Horstemeyer MF. Twin boundaries showing very large deviations from the twinning plane. Scr Mater. 2012;67(10):862–865.

[42] Westlake DG. On {10.1} twinning in the hcp structure. Acta Metall. 1966;14:443–445.

[43] Kronberg ML. Atom movements and dislocation structures for plastic slip in single crystals of α-uranium. J Nucl Mater. 1959;1(1):85–95.

[44] Serra A, Bacon DJ, Pond RC. The crystallography and core structure of twinning dislocations in h.c.p. metals. Acta Metall. 1988;36(12):3183–3203.

[45] Serra A, Pond RC, Bacon DJ. Computer simulation of the structure and mobility of twinning disclocations in h.c.p. metals. Acta Metall Mater. 1991;39(7):1469–1480.

[46] Serra A, Bacon DJ. Computer simulation of screw dislocation interactions with twin boundaries in H.C.P. metals. Acta Metall Mater. 1995;43(12):4465–4481.

[47] Wang J, Beyerlein IJ, Hirth JP, Tomé CN. Twinning dislocations on $\{\bar{1}101\}$ and $\{\bar{1}103\}$ planes in hexagonal close-packed crystals. Acta Mater. 2011;59(10):3990–4001.

[48] Wang J, Beyerlein IJ, Hirth JP. Nucleation of elementary $\{\bar{1}101\}$ and $\{\bar{1}103\}$ twinning dislocations at a twin boundary in hexagonal close-packed crystals. Modell Simul Mater Sci Eng. 2012;20:024001.

[49] Pond RC, Bacon DJ, Serra A, Sutton AP. The crystallography and atomic structure of line defects in twin boundaries in hexagonal-close-packed metals. Metall Trans A. 1991;22(6):1185–1196.

[50] Liu XY, Adams JB, Ercolessi F, Moriarty JA. EAM potential for magnesium from quantum mechanical forces. Modell Simul Mater Sci Eng. 1996;4:293–303.

[51] Eshelby JD. The determination of the elastic field of an ellipsoidal inclusion, and related problems. Proc R Soc Lond A. 1957;241:376–396.

[52] Henkelman G, Uberuaga BP, Jónsson H. A climbing image nudged elastic band method for finding saddle points and minimum energy paths. J Chem Phys. 2000;113:9901–9904.

[53] Yuasa M, Masunaga K, Yoshida T, Mabuchi M, Chino Y. Interactions of a screw dislocation with a $\{10\bar{1}1\} - \{10\bar{1}2\}$ double twin in Mg. Acta Mater. 2013;61(13):4714–4725.

[54] Wang J, Beyerlein IJ. Atomic structures of symmetric tilt grain boundaries in hexagonal close packed (hcp) crystals. Simul Mater Sci Eng. 2012;20(2):024002.

[55] Sun DY, Mendelev MI, Becker CA, Kudin K, Haxhimali T, Asta M, Hoyt JJ, Karma A, Srolovitz DJ. Crystal-melt interfacial free energies in hcp metals: a molecular dynamics study of Mg. Phys Rev B. 2006;73:024116.

[56] Proville L, Rodney D, Marinica MC. Quantum effect on thermally activated glide of dislocations. Nat Mater. 2012;11(10):845–849.

19

Effect of Microstructure on the Deformation Mechanism of Friction Stir-Processed $Al_{0.1}CoCrFeNi$ High Entropy Alloy

M. Komarasamy[a], N. Kumar[a], Z. Tang[b], R.S. Mishra[a]*, and P.K. Liaw[b]

[a]*Department of Materials Science and Engineering, Center for Friction Stir Processing, University of North Texas, Denton, TX 76203, USA;* [b]*Department of Materials Science and Engineering, The University of Tennessee, Knoxville, TN 37996, USA*

Grain refinement from several millimeters in as-received (AR) condition to the range of 0.35–15 μm was achieved by friction stir processing (FSP). Due to the sluggish nature of atomic diffusion in high entropy alloys (HEAs), the FSP region exhibited an immense variation in microstructure which was directly attributed to the accumulated plastic strain during FSP. In accordance with the Hall–Petch relationship, yield strength (YS) has increased by a factor of four after grain refinement while maintaining large uniform elongation (UE). The Kocks–Mecking plot indicated different deformation mechanisms operative in both FSP and AR conditions.

Keywords: Grain Refinement, Strain Gradient, Work Hardening, Twinning

High entropy alloys (HEAs) are new class of materials, which consist of equimolar or near-equimolar compositions of minimum five principal alloying elements. They tend to form random solid solutions with simple crystal structures [face-centered-cubic (FCC) and/or body-centered-cubic (BCC)], synthesized based on Boltzmann's hypothesis of maximizing the configurational entropy by increasing the number of elements added in the alloy system.[1–8]

So far, microstructure and mechanical properties have been evaluated for various HEA systems, but in-depth analysis on the deformation mechanism of HEA is still lacking.[9–12] Recently, Otto et al. [13] studied the underlying deformation mechanisms in CoCrFeMnNi HEA in both fine-grained (FG) and coarse-grained (CG) materials, and elucidated the role of twinning when deformed at 77 K. So far, cold rolling and subsequent annealing treatments have been carried out to obtain grain refinement in HEAs.[13–15] Due to severe lattice distortion and sluggish diffusion, HEAs exhibit rather a high recrystallization temperature ($\sim 0.62T_m$).[14] Moreover, grain coarsening in CoCrFeMnNi HEA was observed only when the recrystallized sample was annealed at 850°C.[15] Generally, these factors will play a crucial role in the microstructural evolution during any thermo-mechanical processing.

Friction stir processing (FSP) [16] has proven to be an effective tool for microstructural modification of Al,[17] Mg,[18] Ti [19] alloys and steels.[20] So far, no one has attempted to refine the grain size in HEA using an FSP technique. The Hall–Petch relationship predicts a dependence of yield strength (YS) on grain size. Hence, FSP-assisted grain refinement is expected to increase the YS of the alloy. Moreover, there is a lack of understanding about the work-hardening or deformation mechanism as a function of grain size (both CG and FG regimes) in HEAs. In this letter, we are reporting (i) grain refinement using FSP and a detailed study on the microstructural evolution at various locations in the processed region using scanning electron microscopy (SEM). In addition, (ii) the effect of grain refinement on the tensile properties and the underlying deformation mechanisms in both CG and FG materials have been addressed.

A vacuum-induction melted and hot-isostatically pressed ~4.0 mm thick $Al_{0.1}CoCrFeNi$ HEA was subjected to FSP. A cylindrical tungsten carbide tool with 1.5 mm pin height was used to process the material. The

*Corresponding author. Email: rajiv.mishra@unt.edu

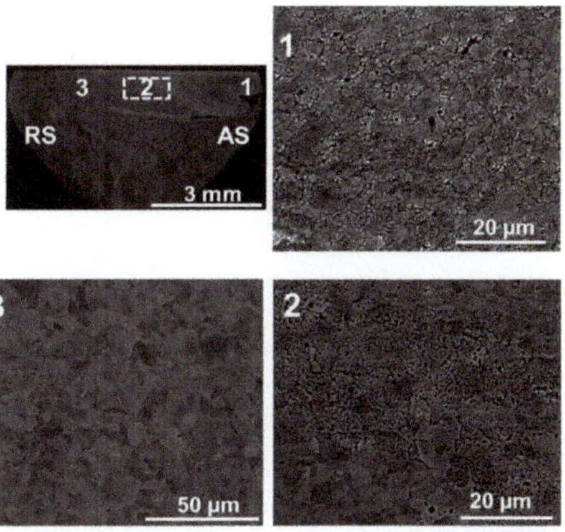

Figure 1. SEM micrographs taken at various locations in the friction stir-processed region. The locations are marked along with the location of cross-section of the mini-tensile sample gage on the macro view of the processed region.

following FSP process parameters were used: a tool rotation rate of 1000 revolutions per minute, tool traverse speed of 25.4 mm per minute, tool tilt angle of 0.5° and a plunge depth of 1.80 mm. The cross-section of the processed region was polished and analyzed using SEM. Mini-tensile testing was carried out in both CG and FG conditions at an initial strain rate of $1 \times 10^{-3}\ s^{-1}$ at room temperature. The mini-tensile sample location in FG condition is shown as a dotted white rectangle on the macro cross-section of the processed region, in Figure 1. The dotted rectangular region represents the gage cross-section of the mini-tensile sample, with the width and thickness being 1.25 and 0.6 mm, respectively. The gage length of the tested sample was 2 mm and it was oriented along the processed direction. Transmission electron microscopy (TEM) samples were extracted from the fractured tensile sample in the CG condition and prepared using a regular procedure. The final thinning was done using a precision ion polishing system. TEM analysis of the failed sample was carried out in Technai G2 F20 STEM. The surface reliefs on the failed sample in FG condition were analyzed using SEM. X-ray diffraction (XRD) was used for phase characterization.

Based on both the XRD and SEM analysis, the studied HEA is a single phase alloy with an FCC crystal structure. In as-received (AR) condition, the grains were on the order of few millimeter in size. In this microstructural condition, the alloy was subjected to FSP to refine the grain size. Figure 1 shows the SEM micrographs taken at various locations in the processing region. The micrographs are marked with numbers, and the corresponding locations are shown in the macro view of the processed region. Micrographs 1, 2 and 3 were taken at the advancing side (AS), the center and the retreating side (RS) of the processing region, respectively. The presence of filling defect at the AS can be seen in the macro view of the nugget. Grain size in the AS was in the range of 0.35–3 µm and in the RS, it was in the range of 2–13.5 µm. Several annealing twins were also observed in the RS, which were absent in the AS. Although the microstructural evolution was inhomogeneous in the processed region; overall, a significant grain refinement was achieved in HEA by FSP in a single step.

Figure 2(a) shows the hardness variation across the processed region. The hardness variation can be directly correlated to the variation in grain size in the processed region. There were three levels of hardness observed and marked as regions 1, 2 and 3. Region 1 exhibited higher hardness due to finer grains in the AS, followed by region 2, a transition region, which was due to the non-recrystallized zones observed in between recrystallized regions and then by region 3, which possessed lower hardness due to the presence of relatively coarse grains. The observed hardness variation in the processed region is unique to the HEA and was not observed in the friction stir-processed region of any other alloy systems.[21–23]

Hence, a better understanding of the microstructural evolution at various regions becomes necessary, since it directly impacts the mechanical properties. Generally, microstructural evolution, dynamic recrystallization and

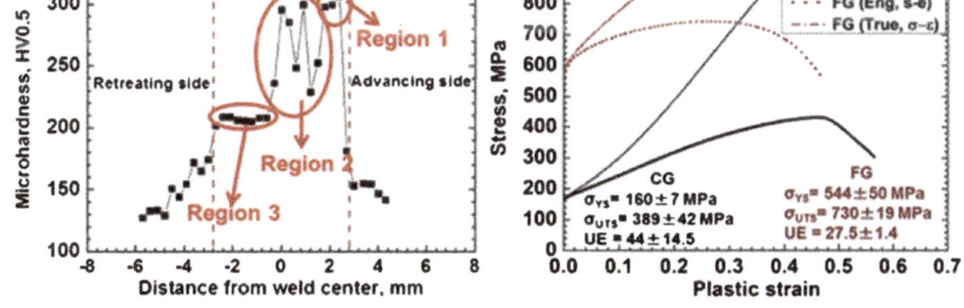

Figure 2. (a) The variation of hardness across the processed region on the transverse cross section and (b) the plastic stress–strain curves in both CG and FG conditions.

grain growth during FSP is controlled by strain, strain rate and temperature experienced by that local region. It is widely known that the dynamic recrystallization during thermo-mechanical processing of low stacking fault energy (SFE) materials is mainly via discontinuous dynamic recrystallization (DDRX).[24] Zaddach et al. [25] calculated the SFE of NiFeCoCr HEA using DFT calculations and XRD analysis, and found it to be as low as 17 mJ/m^2. Therefore, it is highly likely that DDRX was a dominant recrystallization mechanism during FSP of HEA. Moreover, the low SFE limits the role of dynamic recovery during thermo-mechanical processing, hence results in a higher fraction of stored dislocations. These dislocation structures and grain boundaries act as nucleation sites for the recrystallized nuclei during both dynamic and static recrystallization. In a given region, the number density of the nucleation sites will increase with the accumulated plastic strain. In FSP, AS experiences larger plastic strain as compared with the RS [20]; hence, results in a higher and lower fraction of recrystallized nuclei in the AS and RS, respectively.

Upon the completion of dynamic recrystallization, grain growth might occur during the cooling cycle, depending on both the thermodynamic and kinetic factors; such as temperature, time, total grain boundary energy and diffusivity of the atoms. In single phase materials with high self-diffusivity, significant grain growth after dynamic recrystallization is unavoidable in the processed region during the cooling cycle. On the contrary, sluggish diffusion of atoms in HEA systems will likely to have a major impact on the grain growth during the cooling cycle of the friction stir process. Therefore, in the absence of second-phase particles to pin the grain boundaries (the present alloy is a single-phase FCC material), it is the sluggish diffusion that controlled and retarded the grain growth during the cooling cycle, which is unique to materials like HEAs. Hence, the observed variation in grain size in the processed region is justified, where AS displayed very fine grain size and RS displayed comparatively coarser grains. Given the small size of the processed region, the differences in the thermal cycles of AS and RS locations are likely to be insignificant. As shown, such variations in microstructure locally also result in local variation in mechanical properties [Figure 2(a)].

Figure 2(b) shows the engineering and true stress–strain curves in both CG and FG conditions. YS, ultimate tensile strength (UTS) and uniform elongation (UE) values in both CG and FG conditions are shown in Figure 2(b). Since the FS processed region possessed inhomogeneous microstructure, great care was taken to extract a representative gage section which contains both ultrafine and fine grains. First point to be noted here is the effect of the grain size on YS, which was nearly fourfold higher after the grain refinement. Secondly, the extensive work-hardening ability observed in both CG and FG HEA materials. Thirdly, the difference in the shape of the stress–strain curves in both FG and CG conditions, which indicates the on-going deformation mechanisms. Detailed reasoning for the observed deformation mechanism will be discussed in the forthcoming section.

Work-hardening curves in both CG and FG conditions are shown in Figure 3. Though, there has been discussions on HEA possessing higher work-hardening ability, so far, no one has reported on the work-hardening behavior, $d\sigma/d\varepsilon$, and discussed the deformation mechanisms operating at various stages of the work-hardening curve. A simple hardening curve (a polynomial fit to the true stress–strain curve) was chosen rather than power-law-hardening curve (power law fit to the true stress–strain curve) to exaggerate the inflection point and the variation in slope. The effect of grain size on the work-hardening behavior can be clearly seen. In the CG material, three distinct stages of work hardening were observed, marked as stages A, B and C. The decrease in the work-hardening rate in stage A signifies the dislocation-controlled plasticity, and it resembles stage III of high SFE material, a recovery-dominated regime. In Figure 3, point 'a' marks the transition from stage A to B. Detailed TEM analysis has been carried out in the past to correlate this transition to the onset of twinning deformation.[26] Stage B is characterized by an increase in the work-hardening rate due to the introduction of twins in the microstructure. The presence of twins in the microstructure presents an effective barrier to dislocation motion and reduces the mean free path of the dislocation. In addition, the dislocations will interact with twin boundaries resulting in increased dislocation storage inside the grains at the twin–matrix interface. This appears to be the main reason for the exceptional work hardening observed in stage B of the CG material. By comparing the work-hardening curve of HEA in the CG condition with other low SFE materials, such as Cu alloys and TWIP steels,[27] the increase in work-hardening rate during stage B is very high and is unique to HEA. Point 'b' marks the transition from stage B to C, which is again a slip-dominated regime and can be seen from the decreasing work-hardening rate.

The effect of grain size on twinning has received due attention in the past in low SFE materials. It has been proved that with a reduction in grain size, twin nucleation is suppressed in TWIP steels and Cu alloys.[27] Though, the work-hardening behavior of the FG material is different from the CG material, it still exhibited three stages of hardening. The FG material exhibited the same work-hardening behavior as the CG material only in stage C. The slope, rate of recovery, in stage A is shallower in the FG material as compared with the CG material. The interesting part in the work-hardening

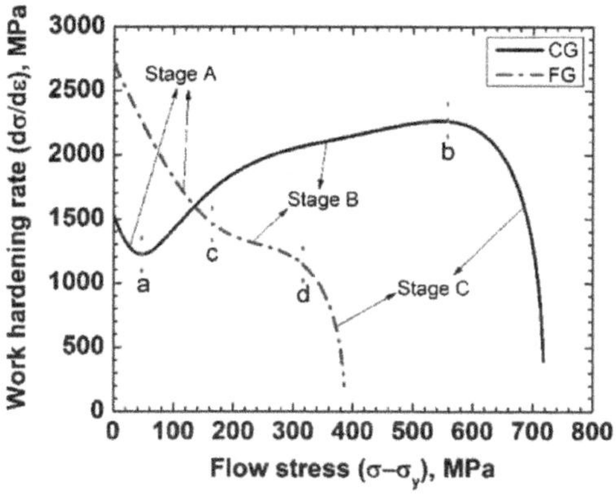

Figure 3. The variation of the work-hardening rate against the flow stress in both CG and FG conditions is shown. Note the difference in stage B of the two curves.

curve of FG material as compared with the CG material behavior is stage B. The work-hardening rate in stage B was continuously decreasing with stress, but at a slower rate as compared with the rate drop related to any recovery-related phenomena. From the literature observations of stage B in various low SFE alloys,[27] the work-hardening behavior observed in stage B of the FG material does indeed have contribution from twinning. The observed stage B behavior in the FG can be explained in two possible ways. First, the reduction in the twin density in the FG material reduced the barrier to dislocation motion. Second possibility is based on the interaction between the twins and dislocations. If the twin orientation was parallel rather than at 70.5° to the primary slip system, and if there was very limited activation of the secondary slip system, then the interaction between the dislocations and twins will be reduced. This will greatly impact the work-hardening rate in the twin-dominated regime.

It is well established that both yield stress (σ_y) and twinning stress (σ_T) greatly depend on the grain size and follows the Hall–Petch type relationship [28]:

$$\sigma_y \text{ or } \sigma_T = \sigma_0 + K_{S \text{ or } T} d^{-1/2}, \qquad (1)$$

where σ_0 is the friction stress, d is the grain size, and K_s and K_T are the Hall–Petch constants for slip and twinning, respectively. This model is based on the critical pile-up stress required at the grain boundary to nucleate a twin or to cause slip in the neighboring grain. For a given material, the stress to nucleate twins increases with the reduction in grain size. Based on the stress–strain curve, the twinning stress, initiation of stage B, in the FG material was 140 MPa (point c in Figure 3) and is quite large, as compared with the twinning stress in the CG material, which was 43 MPa (point a in Figure 3). The effect of grain size on the change in the deformation mechanism of the HEA can be clearly seen from this study.

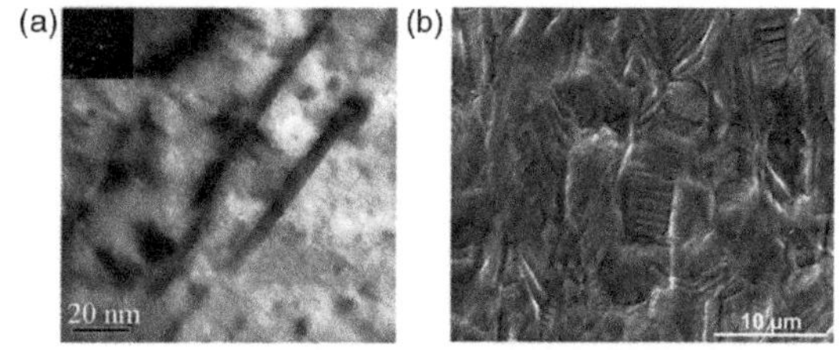

Figure 4. (a) TEM multi-beam bight field image of a deformed specimen showing the presence of twins in the CG material. (b) SEM surface micrograph showing the slip traces in the FG material after deformation to failure.

Figure 4(a) shows a TEM micrograph of the post deformation analysis carried out in the failed CG tensile sample. The presence of nanoscale deformation twins can be seen from the TEM micrograph, and the corresponding selected area diffraction pattern is shown in the inset. There were no micron-sized twins present in the deformed tensile sample, and only one set of twin orientation was observed. The observation of nanosized deformation twins explains the exceptional work-hardening rate observed in stage B, and its uniqueness as compared with other low SFE materials, which had micron-sized twins. Figure 4(b) shows the SEM micrograph of surface slip markings observed in the FG material after tensile deformation. From the analysis of the slip traces in various grains, there was no observation of slip lines oriented in multiple directions or intersecting with each other, which implies that only the primary slip system was active in most of the grains. This will greatly reduce the twin–dislocation interactions, which could probably be the reason for the reduced work-hardening rate in stage B of the FG material.

In summary, extensive grain size reduction was obtained by FSP. The processed region exhibited a unique variation in the grain size, as compared with other metallic alloys. Variation in microstructural evolution across the processing region was attributed to the difference in the strain accumulation and sluggish nature of diffusion. Extensive work-hardening ability in the CG material was due to the presence of nanoscale deformation twins. A significant increase in YS was observed after grain refinement. The change in the work-hardening mechanism with the reduction in grain size was explained in terms of the limited dislocation–twin interactions.

Acknowledgement We would like to thank Center for Advanced Research and Technology (CART) for the microscopy facilities.

References

[1] Yeh JW, Chen SK, Lin SJ, Gan JY, Chin TS, Shun TT, Tsau CH, Chang SY. Nanostructured high-entropy alloys

with multiple principal elements: novel alloy design concepts and outcomes. Adv Eng Mater. 2004;6:299–303.
[2] Hemphill MA, Yuan T, Wang GY, Yeh JW, Tsai CW, Chuang A, Liaw PK. Fatigue behavior of $Al_{0.5}$CoCrCuFeNi high entropy alloys. Acta Mater. 2012;60:5723–5734.
[3] Senkov ON, Wilks GB, Miracle DB, Chuang CP, Liaw PK. Refractory high-entropy alloys. Intermetallics. 2010;18:1758–1765.
[4] Tang Z, Gao MC, Diao H, Yang T, Liu J, Zuo T, Zhang Y, Lu Z, Cheng Y, Zhang Y, Dhamen KA, Liaw PK, Egami T. Aluminum alloying effects on lattice types, microstructures, and mechanical behavior of high-entropy alloys systems. J Mater Sci. 2013;65:1848–1858.
[5] Tang Z, Huang L, He W, Liaw PK. Alloying and processing effects on the aqueous corrosion behavior of high-entropy alloys. Entropy. 2014;16:895–911.
[6] Zhang Y, Zhou YJ, Lin JP, Chan GL, Liaw PK. Solid-solution phase formation rules for multi-component alloys. Adv Eng Mater. 2008;10:534–538.
[7] Zhang Y, Zuo T, Cheng Y, Liaw PK. High-entropy alloys with high saturation magnetization, electrical resistivity, and malleability. Sci Rep. 2013;3:1455–1461.
[8] Zhang Y, Zuo TT, Tang Z, Gao MC, Dahmen KA, Liaw PK, Lu ZP. Microstructures and properties of high-entropy alloys. Prog Mater Sci. 2014;61:1–93.
[9] Zhang KB, Fu ZY, Zhang JY, Wang WM, Wang H, Wang YC, Zhang QJ, Shi J. Microstructure and mechanical properties of CoCrFeNiTiAl_x high-entropy alloys. Mater Sci Eng A. 2009;508:214–219.
[10] Gali A, George EP. Tensile properties of high- and medium-entropy alloys. Intermetallics. 2013;39:74–78.
[11] Shun TT, Du YC. Microstructure and tensile behaviors of FCC $Al_{0.3}$CoCrFeNi high entropy alloy. J Alloys Compd. 2009;479:157–160.
[12] Wang YP, Li BS, Ren MX, Yang C, Fu HZ. Microstructure and compressive properties of AlCrFeCoNi high entropy alloy. Mater Sci Eng A. 2008;491:154–158.
[13] Otto F, Dlouhý A, Somsen Ch, Bei H, Eggeler G, George EP. The influences of temperature and microstructure on the tensile properties of a CoCrFeMnNi high-entropy alloy. Acta Mater. 2013;61:5743–5755.
[14] Bhattacharjee PP, Sathiaraj GD, Zaid M, Gatti JR, Lee C, Tsai CW, Yeh JW. Microstructure and texture evolution during annealing of equiatomic CoCrFeMnNi high-entropy alloy. J Alloys Compd. 2014;587:544–552.
[15] Liu WH, Wu Y, He JY, Nieh TG, Lu ZP. Grain growth and the Hall–Petch relationship in a high-entropy FeCrNiCoMn alloy. Scripta Mater. 2013;68:526–529.
[16] Mishra RS, Mahoney MW, McFadden SX, Mara NA, Mukherjee AK. High strain rate superplasticity in a friction stir processed 7075 Al alloy. Scripta Mater. 1999;42:163–168.
[17] Kumar N, Mishra RS, Huskamp CS, Sankaran KK. Microstructure and mechanical behavior of friction stir processed ultrafine grained Al–Mg–Sc alloy. Mater Sci Eng A. 2011;528:5883–5887.
[18] Chang CI, Du XH, Huang JC. Achieving ultrafine grain size in Mg–Al–Zn alloy by friction stir processing. Scripta Mater. 2007;57:209–212.
[19] Su J, Wang J, Mishra RS, Xu R, Baumann JA. Microstructure and mechanical properties of a friction stir processed Ti–6Al–4V alloy. Mater Sci Eng A. 2013;573:67–74.
[20] Sato YS, Nelson TW, Sterling CJ, Steel RJ, Pettersson CO. Microstructure and mechanical properties of friction stir welded SAF 2507 super duplex stainless steel. Mater Sci Eng A. 2005;397:376–384.
[21] Zhang Y, Sato YS, Kokawa H, Park SHC, Hirano S. Microstructural characteristics and mechanical properties of Ti–6Al–4V friction stir welds. Mater Sci Eng A. 2008;485:448–455.
[22] Reynolds AP, Tang W, Khandkar Z, Khan JA, Lindner K. Relationships between weld parameters, hardness distribution and temperature history in alloy 7050 friction stir welds. Sci Technol Weld Join. 2005;10:190–199.
[23] Saeid T, Abdollah-zadeh A, Assadi H, Ghaini FM. Effect of friction stir welding speed on the microstructure and mechanical properties of a duplex stainless steel. Mater Sci Eng A. 2008;496:262–268.
[24] Wang Y, Shao WZ, Zhen L, Zhang XM. Microstructure evolution during dynamic recrystallization of hot deformed superalloy 718. Mater Sci Eng A. 2008;486:321–332.
[25] Zaddach AJ, Niu C, Koch CC, Irving DL. Mechanical properties and stacking fault energies of NiFeCrCoMn high-entropy alloy. J Mater Sci. 2013;65:1780–1789.
[26] Asgari S, El-Danaf E, Kalidindi SR, Doherty RD. Strain hardening regimes and microstructural evolution during large strain compression of low stacking fault energy fcc alloys that form deformation twins. Metall Mater Trans A. 1997;28:1781–1795.
[27] El-Danaf E, Kalidindi SR, Doherty RD. Influence of grain size and stacking-fault energy on deformation twinning in fcc metals. Metall Mater Trans A. 1999;30:1223–1233.
[28] Meyers MA, Vöhringer O, Lubarda VA. The onset of twinning in metals: a constitutive description. Acta Mater. 2001;49:4025–4039.

In Situ Investigation of High Humidity Stress Corrosion Cracking of 7075 Aluminum Alloy by Three-Dimensional (3D) X-ray Synchrotron Tomography

S. S. Singh[a], J. J. Williams[a], M. F. Lin[a], X. Xiao[b], F. De Carlo[b] and N. Chawla[a]*

[a]*Materials Science and Engineering, Arizona State University, Tempe, AZ 85287-6106, USA;* [b]*Advanced Photon Source, Argonne National Laboratory, Argonne, IL, USA*

In situ X-ray synchrotron tomography was used to investigate the stress corrosion cracking behavior of under-aged Al–Zn–Mg–Cu alloy in moisture. The discontinuous surface cracks (crack jumps) mentioned in the literature are actually a single continuous and tortuous crack when observed in three dimension (3D). Contrary to 2D measurements made at the surface which suggest non-uniform crack growth rates, 3D measurements of the crack length led to a much more accurate measurement of crack growth rates.

Keywords: Stress Corrosion Cracking, X-ray Tomography, Moisture, Aluminum Alloys

Al 7075 alloys are used extensively due to their high strength-to-weight ratio.[1] However, they are susceptible to stress corrosion cracking (SCC) and show a significant increase in crack growth rate in the presence of moisture, especially in the peak-aged and under-aged condition.[2,3] Hydrogen embrittlement has been attributed as a possible mechanism for this increase in crack growth rate.[3,4]

Several studies have looked at the crack growth behavior of aluminum alloys. Discontinuous surface cracks (crack jumps) were observed during SCC of aluminum alloys.[5] These discontinuous surface cracks have also been observed in steel during SCC.[6,7] It was suggested that the cracking is discontinuous where surface cracks nucleate in front of the main crack, grow separately and then combine with the main crack.[5,7] Many studies have used surface measurements (using optical techniques, for example) to quantify SCC behavior.[8–10] This approach may be inaccurate since the crack growth behavior in the interior of the specimen and at the surface may be quite different due to differences in stress state.[11]

It is clear that a thorough understanding of the cracking behavior in SCC requires a quantification of the crack growth behavior in three dimensions (3D). Furthermore, the experiments must be carried out *in situ* to verify whether the 'crack jumps' are indeed jumps. X-ray tomography is an excellent technique to study the behavior of materials under mechanical loading in 3D and 4D (the fourth dimension here is time). In particular, *in situ* mechanical testing is more promising where visualization and quantification of the microstructure changes can be performed with time. *In situ* experiments under tension,[12] cyclic loading,[13,14] corrosion-fatigue,[15] and SCC [16–18] have been conducted using X-ray tomography.

In this study, we have used X-ray synchrotron tomography to study the crack growth behavior of under-aged AA7075 in moisture. We were able to unequivocally determine whether the SCC cracks were discontinuous or simply a single tortuous crack. The SCC crack growth rates were measured and compared using the surface measurement (2D) and bulk measurement (3D). By using the *in situ* tomography approach it was possible to quantify the evolution of crack growth under SCC.

X-ray synchrotron tomography was performed at the 2-BM beamline of the Advanced Photon Source (APS) at Argonne National Laboratory. The details of the tomography system at 2-BM have been described elsewhere.[19,20] A monochromatic beam with energy of 24.3 keV was focused on the specimen. A LuAG:Ce scintillator screen was used to convert the transmitted X-rays to visible light. This was coupled with a 5× objective lens and a CCD camera (PCO Tech edge camera, Romulus, Michigan) to achieve pixel sizes of 1.3 μm. 2D projections were collected at an angular increment

*Corresponding author: Email: nchawla@asu.edu

of 0.12° for a total of 1500 projections. These 2D projections were then reconstructed using the Gridrec fast Fourier transform-based algorithm. The grayscale images obtained after reconstructions were segmented using conventional thresholding using image analysis software (ImageJ, Bethesda, MD). 3D rendering of these segmented SCC cracks was performed in Avizo® Fire (VSG, Burlington, MA) for visualization purposes.

A commercially available AA7075-T651 was solution-heat-treated at 475 °C for 2 h followed by quenching in water and then aging for 80 min at 120 °C to obtain an under-aged condition (Vickers hardness of ~138 HV). Single-edge-notch specimens (20 mm × 2.8 mm × 0.8 mm) were machined by electro-discharge machining such that the loading direction was along the short-transverse (S-T) orientation. The longest dimension (20 mm) is along the loading direction and the lowest dimension (0.8 mm) is the thickness. A width of 2.8 mm was chosen due to limited field of view (~3 mm, necessary to achieve a 2 μm resolution at the APS). The thickness of 0.8 mm was chosen on the basis of two factors: The ability to easily nucleate a crack on the microforce testing system (MTS Tytron 250) having a maximum capacity of ~240 N and the ability to observe the long growth of the crack while maintaining the diagonal dimension of the sample to be less than 3 mm. Fatigue precracking was performed *ex situ* on a microforce testing system (MTS Tytron 250) in tension–tension fatigue (4 Hz, $R = 0.1$). The precrack length was about 1 mm. The SCC experiments were performed in the *in situ* loading stage as shown in Figure 1(a). The upper grip was attached to the stepper motor with a linear actuator, which imposes a displacement on the sample to achieve the desire load. The linear actuator was capable of 8 μm per step and a total stroke of 25 mm. The load cell had a capacity of 500 N. The load was transmitted from the top of the stage to the bottom through a polymer sleeve (Poly methyl methacrylate or PMMA). The PMMA material was chosen due to its transparency to X-rays. The actuator can be controlled automatically using feedback from the load cell.

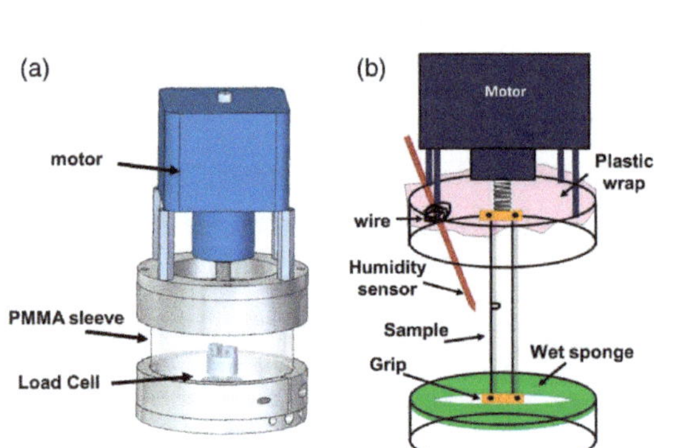

Figure 1. (a) *In situ* mechanical loading stage, and (b) schematic of *in situ* stage for SCC experiments in moisture. The PMMA sleeve is not shown in (b) to show the arrangements clearly.

A schematic of the *in situ* loading stage to perform the SCC experiments in moisture is shown in Figure 1(b). The PMMA sleeve has not been shown here for clear visualization of the arrangements inside the stage. The sample was loaded on both sides using grips attached to the actuator on one side and the load cell on the other. To introduce moisture into the cell, an annular (ring shape) wet sponge was placed around the load cell at the base of the loading stage. Since SCC experiments require long exposures to moisture, maintaining a constant humidity inside the chamber is a big challenge as fluctuations can lead to variability in results. In order to minimize the loss of moisture and to maintain constant relative humidity, the top part of the loading stage was covered with a plastic wrap as shown in Figure 1(b). A humidity sensor was placed in front of the notch to measure the relative humidity throughout the test. These arrangements led to the constant humidity of 95–96% inside the system throughout the test. Samples were placed inside the deionized water for about a week before starting the SCC test in moisture at Argonne National Laboratory.

Figure 2 shows the 3D rendering of the SCC crack in moisture, where each color represents time (cumulative) during which the stress corrosion cracks grew at a constant load of 110 N. Two independent cracks (not connected in 3D) were observed, where the bottom crack almost covers the entire thickness. It is interesting to note that there is a significant variation in the crack length through the thickness and therefore variation in crack growth rate.

The discontinuous cracks on both sides of the specimen can clearly be seen from the 2D slices obtained from X-ray tomography after about 25 h of SCC, as shown in Figure 3(a) and 3(b). These 2D images indicate that the second crack might have initiated in front of the first crack and then grew separately. However, 3D rendering of the crack in Figure 3(c) and 3(d) actually show that these discontinuous surface cracks were connected in three dimensions. Figure 3(e)–(i) shows the step-wise growth of the crack and the appearance of discontinuous surface cracks, observed in Figure 3(a). The time interval is shown in the legend. It can be seen that the main crack grew from inside toward the surface (Figure 3(e) and 3(f)) and finally appeared as a discontinuous crack on the surface (Figure 3(g)), and then continued to grow. The appearance of discontinuous surface cracks can clearly be seen in Figure 3(j), which was not in Figure 3(i). Both, Figure 3(i) and 3(j) are the top views of Figure 3(f) and 3(g).

The crack growth rates were measured using the average of the crack lengths at both sides (2D measurement) and also the average crack lengths though thickness (3D measurement), as shown in Figure 4. Using 2D slices of the 3D crack, the crack length increment (da) was measured. This increment in crack length was then divided

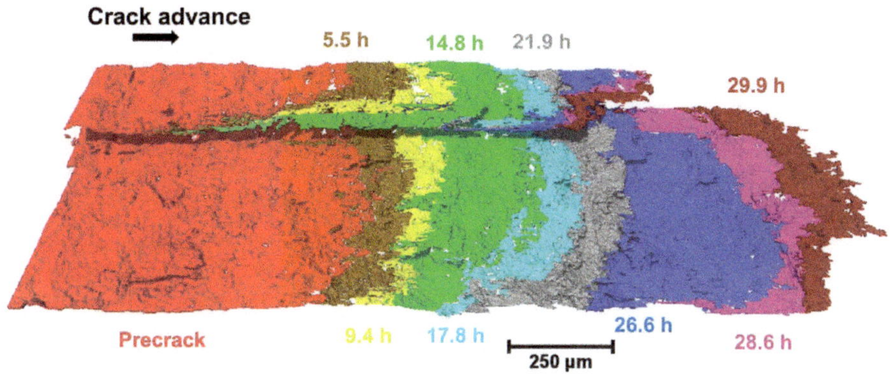

Figure 2. 3D SCC crack growth profile in moisture as a function of time. Significant variation of crack length through the thickness was observed.

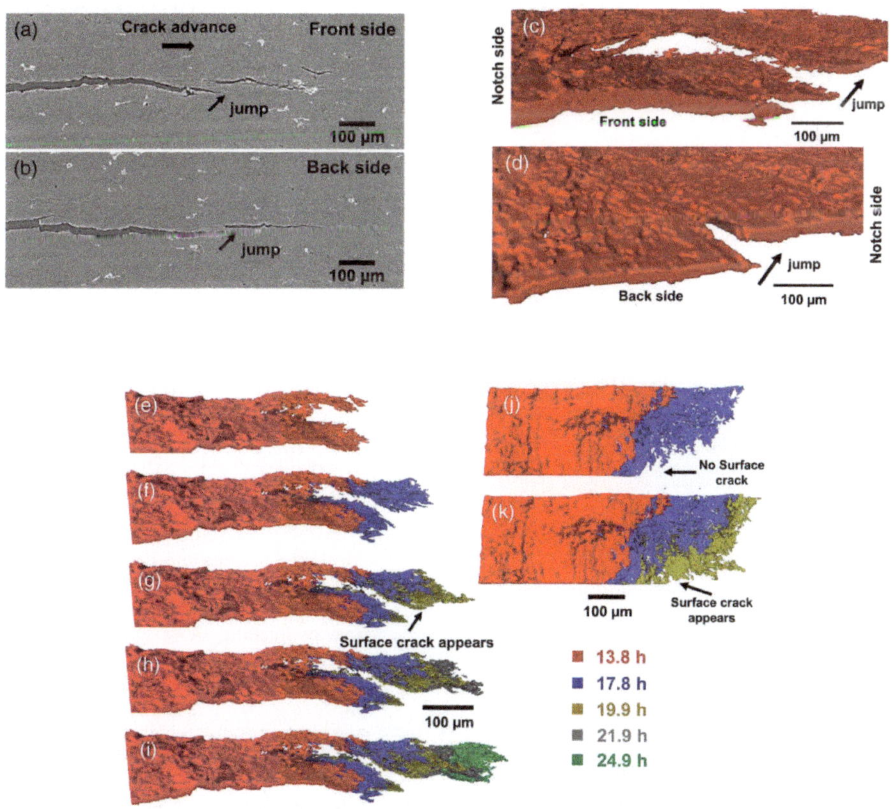

Figure 3. (a) and (b) are 2D slices of the sides of the specimen showing the discontinuous cracks (crack jump), (c) and (d) are the 3D reconstructions of part of the crack showing that the discontinuous crack at the surface are actually connected in 3D, (e)–(i) step wise growth of crack showing the appearance of the discontinuous crack (g) and then further growth, (j)–(k) top view of (f) and (g) showing that the surface crack does not appear in (j) but appears at (k).

by the time increment (dt) to obtain the crack growth rate. The same time interval (dt) was used for both types of measurement. It can be seen that the crack growth rate has a significant amount of variability when surface crack lengths were used for the measurement (2D) compared to the growth rate when the average crack length through thickness was used (3D). Moreover, a large deviation in the surface crack lengths resulted in significant differences in stress intensity factor, as shown as error bars in Figure 4. These can be attributed to the highly non-uniform growth of the SCC crack at the surface which includes formation of discontinuous cracks and independent growth of the two cracks. The stress intensity factor (K) was calculated using the equations derived by Ahmad et al. [21] for single-edge-notched specimens with fixed grip loading. Since the SCC experiment was performed at constant load, the stress intensity factor increases with an increase in the crack length. For a

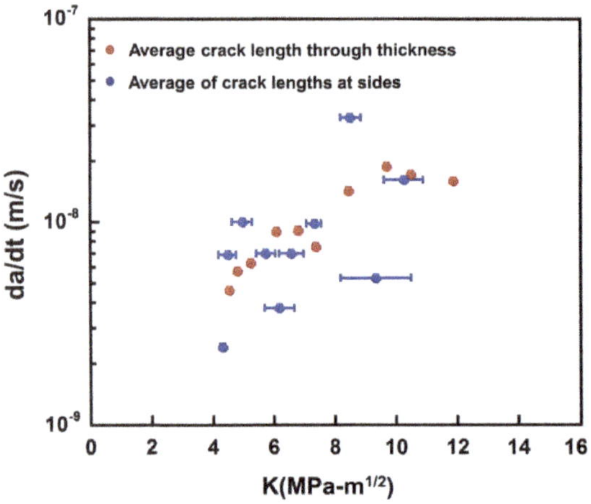

Figure 4. Crack growth rate measurement from average of the crack lengths (2D) and average of the crack length through thickness (3D). The 3D crack growth rate is quite uniform than that of the measurement based on 2D.

particular crack growth rate value (da/dt), the stress intensity factor value (in Figure 4) was taken as the average of the stress intensity factor values corresponding to the initial and final crack lengths for that particular time interval.

In summary, using the X-ray synchrotron tomography, we have shown that the discontinuous surface cracks formed during SCC of AA7075 in moisture were actually a single crack connected in three dimensions. These discontinuous surface cracks were formed when the internal main crack grew from inside the specimen and appeared on the surface. The crack growth rates as determined by measured crack lengths at the surface (2D) led to highly variable crack growth rates than when measured using average crack lengths through thickness (3D). The results obtained here show the importance and necessity of 3D techniques, such as X-ray synchrotron tomography, to study the crack growth behavior during SCC.

Acknowledgements The authors are grateful for the financial support from the Office of Naval Research (ONR) under Contract No. N00014-10-1-0350 (Dr A. K. Vasudevan, Program Manager).

References

[1] Strake EA, Staley JT. Application of modern aluminum alloys to aircraft. Progr Aerosp Sci. 1996;32:131–172.
[2] Speidel MO. Stress corrosion cracking of aluminum alloys. Metall Trans A. 1975;6:631–651.
[3] Holroyd NJH, Scamans GM. Crack propagation during sustained-load cracking of Al–Zn–Mg–Cu aluminum alloys exposed to moist air or distilled water. Metall Mater Trans A. 2011;42:3979–3998.
[4] Burleigh TD. The postulated mechanisms for stress corrosion cracking of aluminum alloys: a review of the literature 1980–1989. Corrosion. 1991;47:89–98.
[5] Zhang J, Kalnaus S, Behrooz M, Jiang Y. Effect of loading history on stress corrosion cracking of 7075-T651 aluminum alloy in saline aqueous environment. Metall Mater Trans A. 2011;42:448–460.
[6] Masuda H. SKFM observation of SCC on SUS304 stainless steel. Corros Sci. 2007;49:120–129.
[7] Qiao LJ, Gao KW, Volinsky AA, Li XY. Discontinuous surface cracks during stress corrosion cracking of stainless steel single crystal. Corros Sci. 2011;53:3509–3514.
[8] Dorward RC, Hasse KR. Incubation effects in precracked stress corrosion specimens from Al–Zn–Mg–Cu alloy 7075. Corros Sci. 1979;19:131–140.
[9] Connolly BJ, Deffenbaugh KL, Moran AL, Koul MG. Environmentally assisted crack growth rates of high-strength aluminum alloys. J Mater. 2003;55:49–52.
[10] Arnold EM, Schubbe JJ, Moran PJ, Bayles RA. Comparison of SCC thresholds and environmentally assisted cracking in 7050-T7451 aluminum plate. J Mater Eng Perform. 2012;21:2480–2486.
[11] Meyers M, Chawla KK. Mechanical behavior of materials. 2nd ed. Cambridge University Press; 2008.
[12] Williams JJ, Chapman NC, Jakkali V, Tanna VA, Chawla N, Xiao X, De Carlo F. Characterization of damage evolution in SiC particle reinforced Al alloy matrix composites by *in situ* X-ray synchrotron tomography. Metall Mater Trans A. 2011;42:2999–3005.
[13] Williams JJ, Yazzie KE, Padilla E, Chawla N, Xiao X, De Carlo F. Understanding fatigue crack growth in aluminum alloys by *in situ* X-ray synchrotron tomography. Int J Fatigue. 2013;57:79–85.
[14] Williams JJ, Yazzie KE, Phillips NC, Chawla N, Xiao X, De Carlo F, Iyyer N, Kittur M. On the correlation between fatigue striation spacing and crack growth rate: a three-dimensional (3-D) X-ray synchrotron tomography study. Metall Mater Trans A. 2011;42:3845–3847.
[15] Singh SS, Williams JJ, Xiao X, De Carlo F, Chawla N. *In situ* three dimensional (3D) X-ray synchrotron tomography of corrosion fatigue in Al 7075 alloy. In: Srivatsan TS, Imam AM, Srinivasan R, editors. Fatigue of materials II: advances and emergences in understanding. Pittsburgh: Materials Science and Technology; 2012. p. 17–25.
[16] Babout L, Marrow TJ, Engelberg D, Withers PJ. X-ray microtomographic observation of intergranular stress corrosion cracking in sensitised austenitic stainless steel. Mater Sci Tech. 2006;22:1068–1075.
[17] Marrow TJ, Babout L, Jivkov AP, Wood P, Engelberg D, Stevens N, Withers PJ, Newman RC. Three dimensional observations and modelling of intergranular stress corrosion cracking in austenitic stainless steel. J Nucl Mater. 2006;352:62–74.
[18] King A, Ludwig W, Engelberg D, Marrow TJ. Diffraction contrast tomography for the study of polycrystalline stainless steel microstructures and stress corrosion cracking. Rev Metall. 2011;108:47–50.
[19] Williams JJ, Flom Z, Amell AA, Chawla N, Xiao X, De Carlo F. Damage evolution in SiC particle reinforced Al alloy matrix composites by X-ray synchrotron tomography. Acta Mater. 2010;58:6194–6205.
[20] De Carlo F, Albee P, Chu YS, Mancini DC, Tieman B. High-throughput real-time X-ray microtomography at the advanced photon source. In: Bonse U, editor. Developments in X-Ray tomography III, Proc. SPIE, Vol. 4503. San Diego, California; 2002. p. 1–13.
[21] Ahmad J, Papaspyropoulos V, Hopper AT. Elastic-plastic analysis of edge-notched panels subjected to fixed grip loading. Eng Fract Mech. 1991;38:283–294.

Layer Stability and Material Properties of Friction-Stir Welded Cu–Nb Nanolamellar Composite Plates

Josef Cobb[a*], Shraddha Vachhani[b], Robert M. Dickerson[c], Patricia O. Dickerson[c], Weizhong Han[d,e], Nathan A. Mara[c,e], John S. Carpenter[c] and Judy Schneider[a]

[a]Mechanical Engineering Department, Mississippi State University, Starkville, MI 39762, USA; [b]Mechanical Engineering Department, Georgia Institute of Technology, Atlanta, GA 30332, USA; [c]Materials Science and Technology Division, Los Alamos National Laboratory, Los Alamos, NM 87545, USA; [d]Center for Advancing Materials Performance from the Nanoscale, State Key Laboratory for Mechanical Behavior of Materials, Xi'an Jiaotong University, Xi'an 710049, People's Republic of China; [e]Center for Integrated Nanotechnologies, Los Alamos National Laboratory, Los Alamos, NM 87545, USA

Initial efforts to friction-stir weld (FSW) Cu–Nb nanolamellar composite plates fabricated via accumulative roll bonding are reported in this study. Parent material layers within the composite were nominally 300 nm and exhibited a hardness of 2.5 GPa. After FSW, two types of microstructures were present: a refined layered structure, and an equiaxed nanocrystalline microstructure with grain diameters on the order of 7 nm. The type of microstructure was dependent on location within the FSW nugget and related to varying amounts of strain. Material hardness increased with refinement, with the equiaxed microstructure reaching a maximum hardness of 6.0 GPa.

Keywords: Nanograin, Nanolayers, Friction-Stir Welding, Copper Niobium

Nanolamellar Cu–Nb bimetallic composites have been extensively studied due to their significantly higher strength and radiation damage tolerance as compared with their bulk counterparts.[1–7] These improvements are reported to increase as the layer thickness (h) decreases to the nanometer (nm) scale with a maximum strength occurring around a thickness of 2 nm.[2] These properties are reported to emanate from a combination of heterophase interface character (ω) and density (ρ), with the strength largely dominated by ρ.[5,6] Cu–Nb nanolamellar bimetallic composites have been studied for materials produced via two fabrication routes: physical vapor deposition (PVD) [4] and accumulative roll bonding (ARB).[8–10] While the PVD method can currently only produce a small volume of laboratory scale material, the ARB process allows the production of much larger volume, engineering-scale quantities of material.[5,9] Previous studies indicate that ρ can be controlled and ω stabilized during ARB processing via manipulation of processing parameters.[10] Thus, ARB opens the door for the fabrication of nanocomposite materials with properties tailored for specific applications via industry scalable manufacturing techniques. To further realize this goal, the stability of ω and ρ under commercially available joining processes must be understood. Friction-stir weld (FSW), a solid-state joining process, is explored as a possible joining process for maintaining high ρ and hence the strength of the material within the weld region. Simulations show that the stability of ω, observed in previous studies,[7,11,12] is derived from the ability of specific interfacial structures to accommodate the plastic deformation of the ARB process. Since the FSW process will introduce additional severe plastic deformation in a manner that differs considerably from that used to synthesize the initial composite, the stability of the preexisting ω is uncertain.

In FSW, a weld tool, consisting of a pin and shoulder, is rotated and plunged into the panel thickness, and traversed the length of the work piece, and hereby joining the material.[13,14] The two sides of the weld joint are

*Corresponding author. Email: jbc251@msstate.edu

referred to as the advancing side (AS), the side where tool rotation aligns with the travel direction, and the retreating side (RS), the side where tool rotation is opposite the travel direction. While the literature reports the use of FSW to join dissimilar alloys in butt and lap welds, these studies address the joining of different homogenous bulk material on either side of the weld.[15–17] This study investigates FSW as a possible joining method for a multi-layered, nano-scaled bimetal composite where both phases are resistant to chemical intermixing due to their positive heats of mixing in the solid phase.

FSW were made in a nominal 0.6 mm thick sheet of ARB Cu–Nb. The ARB material began with a single 2 mm thick sheet of polycrystalline Nb (99.97% pure, ATI-Wah Chang, Albany, OR) and two 1 mm thick sheets of polycrystalline Cu (99.99% pure, Southern Copper and Supply). The Nb layer was sandwiched between the two Cu layers and a 60% reduction in overall thickness was performed in a single step utilizing a two-high rolling mill (Waterbury-Farrel, Brampton, Canada). Layer thickness was reduced and number of layers increased through a repetitive sequence of cutting, stacking, and roll bonding. Prior to each roll bonding step, a surface treatment consisting of a 5-min ultrasonic acetone bath was performed followed by wire brushing. Further details of the ARB process can be found in the literature.[9,10] This process was repeated until a panel consisting of 50 vol.% Cu/50 vol.% Nb of alternating layers with average $h \approx 300$ nm was fabricated. A single piece, smooth tapered FSW tool was made from a lanthanated tungsten W-La_2O_3 alloy with a flat 2 mm diameter shoulder, initial pin length of 0.5 mm, upper diameter of 1 mm, and a taper of 27°mm. All FSWs were made in a bead on plate configuration using displacement control at a travel rate of 25.4 mm/m and a tool rotation of 1500 rpm.

Microstructure images were obtained using an FEI (Hillsboro, OR) Inspect F scanning electron microscope (SEM) and a FEI TITAN 300 keV transmission electron microscope (TEM). Hardness data were obtained on a Nano-XP 30 Indenter (Agilent, Santa Clara, CA) with a nominal 150 nm radius Berkovich tip. Load vs. displacement data were collected in continuous stiffness monitoring mode for indentation depths between 200 and 1100 nm and these data were used for hardness calculations.[18] The FSW panels were cross-sectioned normal to the welding direction using a low-speed diamond saw. SEM/indentation samples were mounted in a low heat epoxy resin and metallurgically ground and polished. TEM foils were extracted and thinned utilizing a FEI Helios focused ion beam (FIB).

Figure 1 shows a representative SEM-based secondary electron image (SEI) of a transverse section of the FSW. Regions of interest include the parent material (PM), the thermal mechanical affected zone (TMAZ), and the weld nugget (WN). All FSW, regardless of process parameters, exhibited a weld defect on the AS crown surface. The images in Figure 2 document the change in layer direction and thickness within the TMAZ region. Details of the representative regions in Figure 1 are shown in the SEI in Figure 2(a)–(c). Superimposed on these images are the nanohardness data with values in GPa. To resolve the finer layers, TEM bright field imaging (BFI), on the RS, was used to obtain Figure 2(d). It shows layers with an h of 6 nm, and a variation in material flow resulting from the asymmetrical flow field of the FSW.[13]

FIB was used to extract an additional foil from the region labeled by vertical red line in Figure 2(a). Figure 3 shows the higher magnification BFI TEM and accompanying ring selected area (electron) diffraction pattern (SADP), which indicates a nanocrystalline structure. Changes in diffraction contrast in the BFI TEM indicate an equiaxed grain structure approximately 5–10 nm in diameter. The ring SADP in Figure 3 confirms a nanocrystalline structure, which exhibits distinct Cu and Nb phases. The mottled contrast suggests that the individual phases are randomly distributed. The equiaxed grains

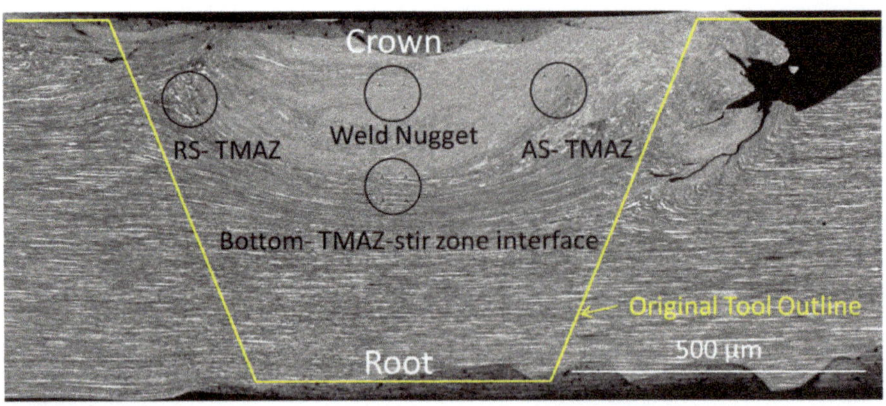

Figure 1. SEI of a transverse section of FSW with regions-of-interest marked. Note this sample was taken after 5 cm of travel with significant tool wear having occurred.

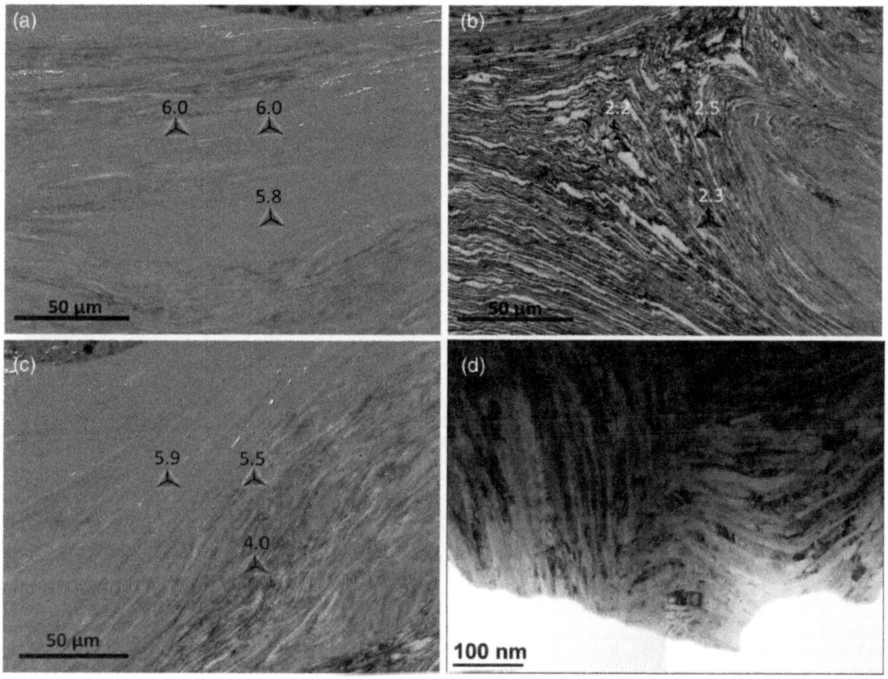

Figure 2. Details of FSW transverse seen in Figure 1. (a) FSW nugget, (b) RS TMAZ, and (c) AS TMAZ. Location of nanoindentations and hardness values in GPa are shown in (a)–(c) with (d) representing a BFI TEM of the RS WN in a layered region. The red line in (a) indicates the location where the TEM foil was extracted for Figure 3.

on the order of the refined h suggest a pinch-off mechanism may be responsible for the grain refinement.[19] In this mechanism, the torsional motion of the FSW process causes the sub-boundaries extending across the thickness of the elongated grains to pinch-off forming equiaxed grains.

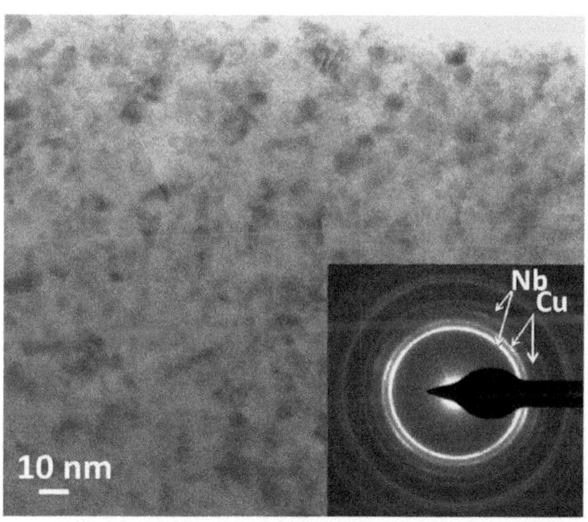

Figure 3. BFI TEM of WN exhibiting a nanocrystalline structure with equiaxed grains on the order of 5–10 nm in diameter. The inset accompanying SADP shows a ring structure, consistent with a polycrystalline grains, and the presence of both Cu and Nb phases.

The layered structure of the PM was maintained in parts of the FSW region as shown in Figure 2, especially on the RS and show changes in orientation resulting from the asymmetric material flow. Figure 2(d) shows a refined layered structure in the WN region, with the majority of the layers ranging from 6 to 12 nm. Equation (1) can be used to calculate the true strain (ε_t) input in the layered regions, and, assuming layer thicknesses in Figure 3(d) (which was close to the equiaxed nanograin region) of $\geq$6 nm to be stable, a minimum true strain required to form the equiaxed nanograin structure can also be calculated. For Equation (1); t_1 is the initial h of the PM and t_2 is the final h within the FSW nugget.

$$\varepsilon_t = \ln\left(\frac{t_1}{t_2}\right). \tag{1}$$

Thus for the observed reduction of h from 300 to 6 nm, the FSW process input a minimum strain of 3.9 in the equiaxed nanograin WN region. While this strain value is similar to strain ranges reported in other FSW studies of aluminum,[20–22] other studies indicate that higher strains are possible which vary as a function of location.[13]

A 2.5 GPa with a standard deviation of 0.13 GPa average hardness value was obtained from the 18 nanoindentations within the PM. Within the FSW region, the hardness values varied with respect to the refinement in microstructure. A steep increase in hardness was

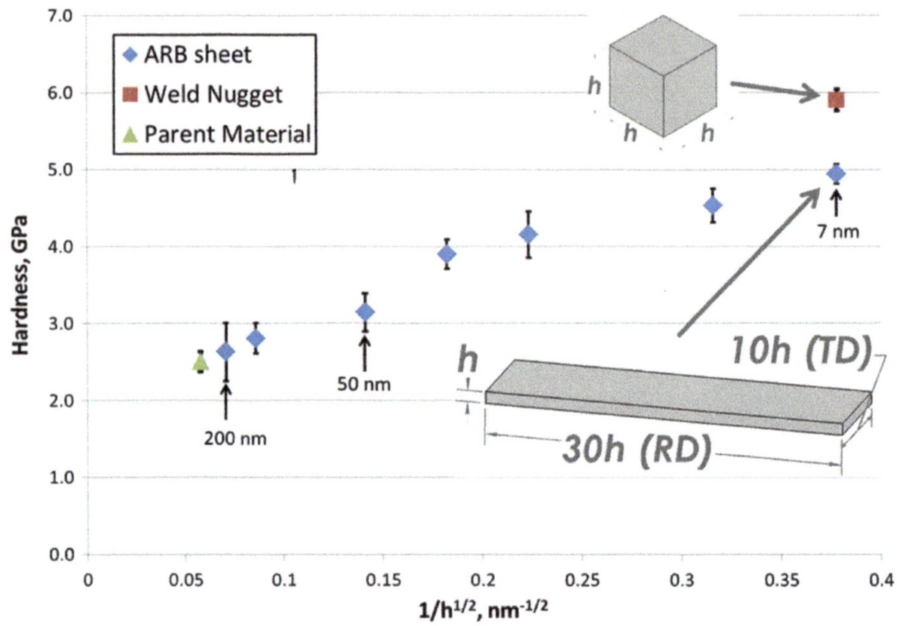

Figure 4. Hardness $v1/h^{1/2}$ plot for ARB Cu–Nb,[5] across length scales from the PM to the FSW WN. The insets illustrate the difference between the grains of an as rolled ARB panel (elongated grains) and the WN (equiaxed grains) at the finest length scale. Grain dimensions for the ARB panel are estimated from the literature.[10]

observed in the TMAZ region, due to the gradient of h. The average hardness of 5.9 GPa was measured in the equiaxed grain region of the WN.

The nanoindentation results performed on multiple FSW showed increases in hardness as ρ increases, consistent with observations from ARB Cu–Nb sheet material.[5,6,10] Figure 4 plots Cu–Nb ARB hardness vs. $h^{1/2}$ across the length scale from the PM to the FSW WN.[5] It also illustrates the difference in grain geometry between the WN and an as rolled ARB panel.[5]

Figure 4 shows a 20% increase in hardness in the equiaxed 7 nm grain size region of the WN compared to that of the an ARB panel with $h = 7$ nm. This increase in hardness likely arises due to different deformation mechanisms in the equiaxed grains of the WN.[23–28] Increased hardness suggests that friction-stir processing of Cu–Nb plates could be used to form a three-dimensional (3D) nanocrystalline structure to further increase the maximum obtainable strength. The significance in a 20% increase in hardness over an already extremely hard material (considering its bulk counterparts) could be of particular importance in high wear applications, and other areas where an even further increase in surface, or through-thickness, hardness is desired.

Although resolvable with TEM, the small size presented by the grains precluded an electron backscatter diffraction study of ω within the WN. Studies of ARB sheets had well-defined layer directions, which allowed the use of a normal direction inverse pole figure to act as a measure of the interface plane normal distribution. With no layer directions present in the WN, bulk texture measurements via neutrons or X-rays will not provide an understanding of ω. With grain diameters on the order of 7 nm, this also limits the applicability of local texture measurements such as precession electron diffraction within a TEM, a technique used in the past for investigations of Cu–Nb nanoscale multi-layers.[29,30]

With regards to the FSW process, there was a defect on the AS of the FSW region. This type of defect is typically seen in FSW when the insufficient down-force is applied.[31] Although the shoulder was in contact with the material during welding, a substantial influence on the material flow from the shoulder was not observed as evidenced by the minimal material movement on the RS seen in Figure 1. Weld material on the AS, beyond the defect and not shown in Figure 1, was also in intimate contact with the shoulder, but material-movement/shoulder-interaction was virtually nonexistent. It is possible that adding features to the shoulder would help increase the lateral material-movement and help eliminate the volumetric defect.[14]

Tool wear during welding was significant with pin length decreasing by 50% after 5 cm of FSW length. Different material selections for the tool should improve tool life and preserve features added to improve material flow. Although improvements in tool life and preventing defect formations are still areas of concern, FSW seems to be a viable option for joining this type of material while maintaining its enhanced strength.

FSW of the ARB panels lead to both refined layers, and a nanocrystalline structure. Where layers were present, directional changes were observed due to the non-symmetric flow field of the FSW.[32,33] The initial h of 300 nm was reduced to grains with 3D nanoscale diameters of 7 nm following the FSW process. Within the WN hardness values of the equiaxed grain region exceeded those of ARB Cu–Nb panels with h equal to the equiaxed grain diameter. Hardness values were noted to increase from 2.5 GPa in the starting PM and increased to a maximum of 6.0 GPa as the length scale decreased.

Acknowledgements This work is supported by the Los Alamos National Laboratory Directed Research and Development (LDRD) project 20130764ECR. This work was performed, in part, at the Center for Integrated Nanotechnologies, an Office of Science User Facility operated for the U.S. Department of Energy (DOE) Office of Science. Los Alamos National Laboratory, an affirmative action equal opportunity employer, is operated by Los Alamos National Security, LLC, for the National Nuclear Security Administration of the US Department of Energy under contract DE-AC52-06NA25396. Electron microscopy was performed at the Los Alamos Electron Microscopy Laboratory.

References

[1] Mara NA, Bhattacharyya D, Dickerson P, Hoagland RG, Misra A. Deformability of ultrahigh strength 5 nm Cu/Nb nanolayered composites. Appl Phys Lett. 2008;92(23):231901.

[2] Misra A, Hirth JP, Hoagland RG. Length-scale-dependent deformation mechanisms in incoherent metallic multi-layered composites. Acta Mater. 2005;53(18):4817–4824.

[3] Misra A, Demkowicz MJ, Zhang X, Hoagland RG. The radiation damage tolerance of ultra-high strength nanolayered composites. JOM. 2007;59(9):62–65.

[4] Han W, Demkowicz MJ, Mara NA, Fu E, Sinha S, Rollett AD, Wang Y, Carpenter JS, Beyerlein IJ, Misra A. Design of radiation tolerant materials via interface engineering. Adv Mater. 2013;25(48):6975–6979.

[5] Beyerlein IJ, Mara NA, Carpenter JS, Nizolek T, Mook WM, Wynn TA, McCabe RJ, Mayeur JR, Kang K, Zheng S, Wang J, Pollock TM. Interface-driven microstructure development and ultra high strength of bulk nanostructured Cu–Nb multilayers fabricated by severe plastic deformation. J Mater Res. 2013;28(13):1799–1812.

[6] Carpenter JS, Zheng SJ, Zhang RF, Vogel SC, Beyerlein IJ, Mara NA. Thermal stability of Cu–Nb nanolamellar composites fabricated via accumulative roll bonding. Philos Mag. 2013;93(7):718–735.

[7] Zheng S, Beyerlein IJ, Carpenter JS, Kang K, Wang J, Han W, Mara NA. High-strength and thermally stable bulk nanolayered composites due to twin-induced interfaces. Nat Commun. 2013;4, Article number 1696. doi: 10.1038/ncomms2651.

[8] Lee S-B, LeDonne JE, Lim SCV, Beyerlein IJ, Rollett AD. The heterophase interface character distribution of physical vapor-deposited and accumulative roll-bonded Cu–Nb multilayer composites. Acta Mater. 2012;60(4):1747–1761.

[9] Carpenter JS, Vogel SC, LeDonne JE, Hammon DL, Beyerlein IJ, Mara NA. Bulk texture evolution of Cu–Nb nanolamellar composites during accumulative roll bonding. Acta Mater. 2012;60(4):1576–1586.

[10] Carpenter JS, McCabe RJ, Vogel SC, Rollett AD, Mara NA, Beyerlein IJ. Process parameter influence on texture evolution in Cu–Nb multilayer composites fabricated via accumulative roll bonding. Met Mater Trans. A 2014. doi:10.1007/s11661-013-2162-4

[11] Bronkhorst CA, Mayeur JR, Beyerlein IJ, Mourad HM, Hansen BL, Mara NA, Carpenter JS, McCabe RJ, Sintay SD. Meso-scale modeling the orientation and interface stability of Cu/Nb-layered composites by rolling. JOM. 2013;65(3):431–442.

[12] Hansen BL, Carpenter JS, Sintay SD, Bronkhorst CA, McCabe RJ, Mayeur JR, Mourad HM, Beyerlein IJ, Mara NA, Chen SR, Gray III GT. Modeling the texture evolution of Cu/Nb layered composites during rolling. Int J Plast. 2013;49:71–84.

[13] Schneider JA, Nunes, AC, Jr. Characterization of plastic flow and resulting microtextures in a friction stir weld. Metall Mater Trans B. 2004;35(4):777–783.

[14] Mishra RS, Ma ZY. Friction stir welding and processing. Mater Sci Eng R Rep. 2005;50(1–2):1–78.

[15] Schneider C, Weinberger T, Inoue J, Koseki T, Enzinger N. Characterisation of interface of steel/magnesium FSW. Sci Technol Weld Join. 2011;16(1):100–107.

[16] Galvão I, Leal RM, Loureiro A, Rodrigues DM. Material flow in heterogeneous friction stir welding of aluminium and copper thin sheets. Sci Technol Weld Join. 2010;15(8):654–660.

[17] Sun YF, Fujii H, Takaki N, Okitsu Y. Microstructure and mechanical properties of dissimilar Al alloy/steel joints prepared by a flat spot friction stir welding technique. Mater Des. 2013;47:350–357.

[18] Oliver WC, Pharr GM. Measurement of hardness and elastic modulus by instrumented indentation: advances in understanding and refinements to methodology. J Mater Res. 2004;19(01):3–20.

[19] Doherty RD, Hughes DA, Humphreys FJ, Jonas JJ, Jensen DJ, Kassner ME, King WE, McNelley TR, McQueen HJ, Rollett AD. Current issues in recrystallization: a review. Mater Sci Eng A. 1997;238(2):219–274.

[20] Arora A, Zhang Z, De A, DebRoy T. Strains and strain rates during friction stir welding. Scripta Mater. 2009;61(9):863–866.

[21] Chen ZW, Cui S. Strain and strain rate during friction stir welding/processing of Al–7Si–0.3Mg alloy. IOP ConfSer Mater Sci Eng. 2009;4(1):012026.

[22] Buffa G, Hua J, Shivpuri R, Fratini L. A continuum based FEM model for friction stir welding—model development. Mater Sci Eng A. 2006;419(1–2):389–396.

[23] Yang W, Wang H. Mechanics modeling for deformation of nano-grained metals. J Mech Phys Solids. 2004;52(4):875–889.

[24] Van Swygenhoven H, Caro A, Farkas D. A molecular dynamics study of polycrystalline fcc metals at the nanoscale: grain boundary structure and its influence on plastic deformation. Mater Sci Eng A. 2001;309–310:440–444.

[25] Fu H-H, Benson DJ, Meyers MA. Analytical and computational description of effect of grain size on yield stress of metals. Acta Mater. 2001;49(13):2567–2582.

[26] Yamakov V, Wolf D, Phillpot SR, Mukherjee AK, Gleiter H. Dislocation processes in the deformation of nanocrystalline aluminium by molecular-dynamics simulation. Nat Mater. 2002;1(1):45–49.

[27] Meyers MA, Mishra A, Benson DJ. Mechanical properties of nanocrystalline materials. Prog Mater Sci. 2006;51(4):427–556.

[28] Raj R, Ashby MF. On grain boundary sliding and diffusional creep. Metall Trans. 1971;2(4):1113–1127.

[29] Carpenter JS, Liu X, Darbal A, Nuhfer NT, McCabe RJ, Vogel SC, LeDonne JE, Rollett AD, Barmak K, Beyerlein IJ, Mara NA. A comparison of texture results obtained using precession electron diffraction and neutron diffraction methods at diminishing length scales in ordered bimetallic nanolamellar composites. Scripta Mater. 2012;67(4):336–339.

[30] Liu X, Nuhfer NT, Rollett AD, Sinha S, Lee S-B, Carpenter JS, LeDonne JE, Darbal A, Barmak K. Interfacial orientation and misorientation relationships in nanolamellar Cu/Nb composites using transmission-electron-microscope-based orientation and phase mapping. Acta Mater. 2014;64:333–344.

[31] Kim YG, Fujii H, Tsumura T, Komazaki T, Nakata K. Three defect types in friction stir welding of aluminum die casting alloy. Mater Sci Eng A. 2006;415(1–2):250–254.

[32] Murr LE, Li Y, Flores RD, Trillo EA, McClure JC. Intercalation vortices and related microstructural features in the friction-stir welding of dissimilar metals. Mater Res Innov. 1998;2(3):150–163.

[33] Li Y, Murr LE, McClure JC. Solid-state flow visualization in the friction-stir welding of 2024 Al to 6061 Al. Scripta Mater. 1999;40(9):1041–1046.

22

Coherent Interfaces Increase Strain-Hardening Behavior in Tri-Component Nano-Scale Metallic Multilayer Thin Films

R.L. Schoeppner[a,b], J.M. Wheeler[c], J. Zechner[c], J. Michler[c], H.M. Zbib[a] and D.F. Bahr[b]*

[a]School of Mechanical and Materials Engineering, Washington State University, Pullman, WA 99164, USA; [b]School of Materials Engineering, Purdue University, West Lafayette, IN 47907, USA; [c]Empa, Swiss Federal Laboratories for Materials Science and Technology, Laboratory for Mechanics of Materials and Nanostructures, Thun CH-3602, Switzerland

Strain-hardening in tri-component nano-scale metallic multilayers was investigated using nanoindentation and micro-pillar compression. Cu/Ni/Nb films were made in tri-layer structures as well as bi-layers consisting of an alloy of Cu–Ni/Nb. Strain-hardening increases as the layer thickness decreases, with 5 nm layers exhibiting higher strengths and hardening coefficients than 30 nm layers. The experimental evidence is described in light of the confined layer slip model, and supports the hypothesis that coherent interfaces with a modulus mismatch in the tri-layer system are responsible for additional deformation mechanisms that can lead to hardening in excess of that found in bi-layer systems.

Keywords: Multilayers, Nanoindentation, Micro-Pillar Compression, Strain-Hardening

Computational simulations of tri-layer nano-scale metallic multilayer (NMM) systems having a combination of coherent and incoherent interfaces have suggested that these systems will possess significant strain-hardening ability above that of their bi-layer counterparts.[1–4] However, the experimental validation of the strain-hardening behavior [2] has only been carried out at a single tri-layer thickness period. With significant evidence in bi-layer NMMs that layer thickness plays a substantial role in controlling the deformation processes in the confined layer slip (CLS) regime,[5–7] this effect of layer thickness on tri-layer systems needs to be evaluated. Briefly, as described in [1], the presence of two adjacent FCC layers with different elastic moduli causes shear stresses to develop at the chemically distinct, but coherent, interface as a threading dislocation moves through the FCC layers. This shear stress can lead to cross-slip at higher stresses, assisted by the high strength imparted by the incoherent interface at the FCC/BCC layers. The cross-slip process, which in a bulk material would weaken the system through additional deformation at a fixed stress, leads to additional dislocation debris within the FCC layers blocking the motion of subsequent threading dislocations and leading to additional hardening. As the density of interfaces increases, the density of potential cross-slip sites also increases; therefore, it may be expected that reducing layer thickness in the CLS regime of a tri-layer system would increase both the strength- and strain-hardening behavior for these materials systems.

Multiple experimental techniques can be used to investigate a material's relative strain-hardening ability including tensile, compression, and nanoindentation testing. Indentation testing is a high-throughput method which leads to a large test population and provides valuable statistics as opposed to testing limited numbers of samples in thin film compression or tensile testing; however, the indentation results require a more complex and less determinant analysis. The most common indentation technique used to probe the strain-hardening relationship in a material compares the hardness measured using tips with different included angles that create different local strains in the material directly beneath the contact.[8] The Berkovich and cube-corner indenter geometries, corresponding to equivalent cone angles of 70.3° and 42.3°, result in effective strains of approximately 8% and 22%, respectively.[9] A material that exhibits strain-hardening between 8% and 22% strain will show larger

*Corresponding author. Email: dfbahr@purdue.edu

hardness values with the cube-corner tip over the hardness measured with a Berkovich tip, whereas an elastic-perfectly plastic material should have the same hardness values regardless of indenter geometry.

Micro-pillar compression testing provides a close to uniaxial stress state which allows for a more accurate determination of the strain-hardening exponent of these NMMs, but requires sample preparation techniques which both limit the number of experiments which can be performed as well as adding the concern that milling with Ga ions can alter the subsequent material properties. Additionally, micro-pillar compression tests can access the strain regime from initial yield to the 8% value probed by a Berkovich tip. A limitation of micro-pillar compression testing is that shear often is initiated at the edge of the punch/pillar contact, leading to the inability to generate uniaxial stress–strain conditions past 10% in many materials. Therefore, the combination of indentation and compression testing will allow a more complete study of strain-hardening in these novel multi-component multilayers.

Tri-component films consisting of Cu (lattice parameter = 3.61 Å), Ni (lattice parameter = 3.52 Å), and Nb (lattice parameter = 3.30 Å) are the basis of this study. Cu/Ni/Nb NMM films used in this study were deposited using magnetron sputtering on (100) oriented Si starting with the Nb layer and ending with the Ni layer to a total thickness of 3 μm with equal individual layer thicknesses of 5 and 30 nm. Preliminary transmission electron microscopy investigations of the 30 nm multilayers show a mixture of coherent FCC/FCC boundaries as well as some grains which show renucleation at the boundary. A second set of samples was fabricated to investigate the specific role of the FCC/FCC interfaces while maintaining the same overall chemistry of the tri-layer; these samples began with a 5 or 30 nm Nb layer, but then used co-sputtering to create a 10 or 60 nm thick FCC alloy layer. The alloy system was deposited to a total thickness of 1.6 μm, the repeating period of layers was constant between the alloy and tri-layer system. X-ray diffraction characterization of these films shows very strong textures with Nb predominantly (110) and Cu and Ni (and CuNi alloy) predominately (111) orientation. Since Cu and Ni are both FCC materials and the lattice constants are within 3% of each other, this combined effect can promote epitaxial growth across the interface, leading to the desired coherent interface. Previous studies have also shown that epitaxial growth is possible between Cu/Ni multilayers deposited using magnetron sputtering,[10–13] further supporting the assumption that these layers are deposited with coherent interfaces.

Micro-pillars with aspect ratios of 1:3 were fabricated using a focused Ga ion beam at an accelerating voltage of 30 keV in a Tescan Vela FIB instrument. Initially, currents of 4 nA were used to mill rough pillar shapes; exact dimensions were achieved after low current polishing to help minimize irradiation damage with currents ranging from 1 nA to 100 pA. The micro-pillars were imaged using a Hitachi S4800 high-resolution scanning electron microscope (HRSEM) to determine the taper due to FIB milling. These particular milling conditions resulted in a taper angle of 3.25°, the effects of which were subtracted from initial load–displacement curves using a first-order correction.[14]

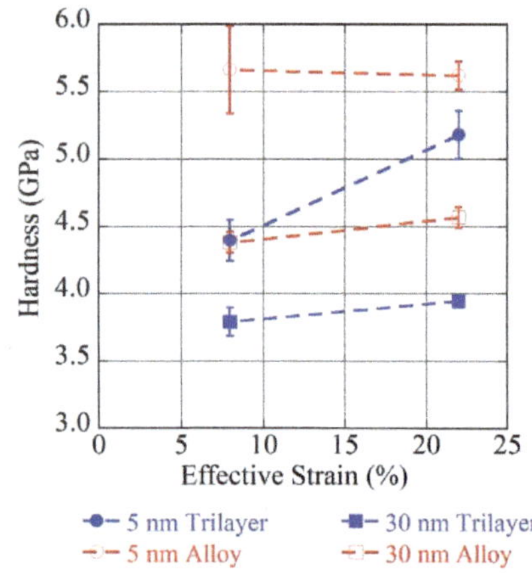

Figure 1. Hardness measurements from nanoindentation experiments conducted using a Berkovich tip (effective strain 8%) and cube-corner tip (effective strain 22%), approximating the hardening response of multilayer films with mixed interfaces (tri-layer) and incoherent interfaces (alloy).

Nanoindentation was performed using a Hysitron UBI indenter using the load-controlled partial unload technique to a maximum load of 10 mN. Hardness values were determined from similar contact areas, corresponding to depths of approximately 140 nm for the Berkovich and 400 nm for the cube-corner, to suppress differences due to indentation size effects. No less than 20 indents were obtained on each sample to ensure statistical reliability. Figure 1 shows the nanoindentation hardness results obtained from the two different tips on all four multilayers. All eight sets of experiments showed similar modulus values of approximately 140 GPa ensuring the variations in hardness are not due to tip calibration errors. The substantial increase in hardness for the tri-layer samples with a 5 nm layer thickness indicates a significant strain-hardening ability, whereas little to no strain-hardening is observed in the 5 nm bi-layer alloy. However, due to the large deviation in the Berkovich values for the 5 nm alloy film, this result is inconclusive. Both samples with a 30 nm layer thickness show a small increase in hardness for both the tri-layer and alloy film, which indicates some strain-hardening has occurred. Prior results by our group and collaborators [15] on similar films (but grown under different conditions) showed similar change in hardness of 0.6 GPa between 5 and 30 nm layer thickness tri-layers,

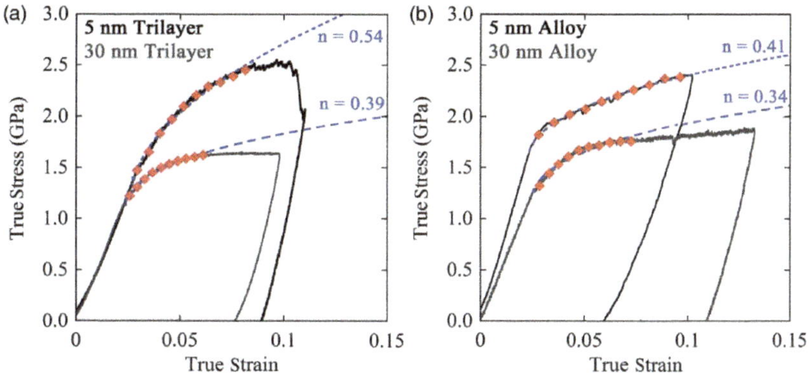

Figure 2. Representative stress–strain curves from micro-pillar compression testing of NMMs for (a) tri-layer (mixed interfaces) and (b) the bi-layer alloy (incoherent interface only) films.

though the measured hardness in the current study is lower than those prior published values.

To gain a more accurate measurement of the strain-hardening response at lower strains, micro-pillar compression tests were conducted in a Zeiss DSM 962 SEM with a modified Alemnis *in situ* indenter first developed by Rabe et al.,[16] modified and improved by Wheeler and Michler,[17] using displacement control at strain rates of approximately 1e – 3/s. The taper corrections previously described were applied to the load–depth curves and resulted in the true stress–true strain curves shown in Figure 2. While the first-order taper correction is more accurate than using no correction at all, there are still limitations to this simplified correction. First and foremost, this correction assumes elastic–perfectly plastic deformation which, in the case of samples which exhibit strain-hardening behavior, is a gross over correction. Additionally, since the deformation behavior of these pillars is different (5 nm tri-layers do not deform to a taper-free pillar before shearing while the alloy films barrel significantly), the relative taper correction of those conditions would only show a stronger trend than the one reported here, with tri-layers hardening more than alloy films and thinner layers (specifically for the tri-layer samples) hardening more than thicker layers.

Slight differences in the initial loading slope or rounding at the initial contact in Figure 2 are likely due to slight misalignment of the pillar with the tip,[18,19] different roughness or roundness of the top of the pillar which would alter the initial contact, or different amounts of residual stresses.[20] Due to these issues, the elastic modulus in micro-pillar testing is more often taken from an unloading slope, where the initial contact is not as strong of an effect.[21] In this case, the substrate stiffness was sufficiently high, so that the effect of pillar sink-in into the substrate compliance was found to be negligible.[22] To address any non-linear loading due to rounding of the top of the pillar from FIB milling, the curves are offset so that the extrapolated elastic portion crosses the origin.

Uniaxial strain-hardening behavior follows the well-known Ludwik relationship between stress and strain, as shown in other compression tests on nanolaminates [23]:

$$\sigma = \sigma_{ys} + K\varepsilon_p^n, \tag{1}$$

where σ_{ys} is the yield strength, K is the strength index, and n is the strain-hardening exponent. The portion of each curve used in curve fitting is emphasized in red diamonds and overlaid over the stress–strain curve with the curve fits shown as dashed lines. All curve fits have R-values of at least 0.98, suggesting these reasonably describe the strain-hardening relationship of the films during the initial strain-hardening (up to $\approx 8\%$). A summary of the mechanical properties extracted from micro-pillar compression tests is listed in Table 1.

The trend seen from the indentation study follows closely with that observed in the compression results, with thinner layers exhibiting stronger strain-hardening behavior than thicker layers. The stress–strain curves shown in Figure 2 indicate that by 8% strain, both 30 nm films have reached their maximum flow stress, while 5 nm films continue to harden until approximately 11% strain. This explains why the 30 nm films do not show significant increases in the hardness values when indented using Berkovich and cube-corner tip geometries (Figure 1). Since the 5 nm tri-layer film continues to harden up until the pillar shears at 11% strain, the increased hardening is reflected in the higher hardness values observed when using the sharper tip in the indentation study. The only discrepancy in the strain-hardening behavior between the two techniques is seen in the 5 nm alloy film, which shows obvious hardening in the compression results, but not in the indentation results. However, since the Berkovich indentation results (Figure 1) show a large spread (0.7 GPa) for the 5 nm alloy film, the hardening could likely be masked by the scatter in the data. Additionally, since the alloy films exhibit some barreling in the post mortem images (Figure 3), some of the additional strain-hardening observed in compression testing could be an artifact of the increasing cross-sectional area.

The maximum strength (σ_{max}) of the two different interface systems follows the expected strength trend with the 5 nm films showing higher strengths than the 30 nm films.[1] The yield strength of the films is more similar between samples, with the average strength of the 5 nm layers being only slightly higher than the 30 nm layers. Since the initial yield of the composite signifies the strength required to initiate dislocation motion in the softest layer, and thus is an inherent material property, it is not surprising that the yield strength is not as strongly affected by the layered structure. This was also seen by Abdolrahim et al.[24] The alloy films also have additional solid solution strengthening component, which is on the order of 60 MPa for the CuNi alloy.[25,26] This is on the order of the increase observed in the hardness and yield strength for the alloy system in Table 1. Differences in layer thickness in the FCC layers between the alloy (60 nm) and the tri-layer (30 + 30 nm) make an exact comparison difficult, but it appears that there is a measurable impact from solid solution strengthening in the alloy systems. While a few of the tri-layer compression strength values are larger than the alloy samples, this is an issue of statistics as opposed to a physical relationship. With only two pillars for each compression test, the observed variations could be caused by misalignment, errors in taper correction, or actual differences found in point to point positions in the films (i.e. the observed scatter in the nanoindentation data). Nonetheless, in all cases, the ranking of hardness and flow stress follow between the indentation and compression testing, with the acknowledgement that the limited statistics in the compression testing means that there may not be statistically significant results in the compression testing. For instance, the average maximum stress for the 30 nm tri-layer is less than that of the 30 nm alloy, but not to a level that would be significant; however, when taken in light of the indentation testing the different testing methods do appear to support each other.

In general, these results show that the strain-hardening exponents of 5 nm films are larger than the 30 nm films in both material systems. Additionally, the higher strain-hardening exponents observed in the tri-layer samples indicate that the presence of the

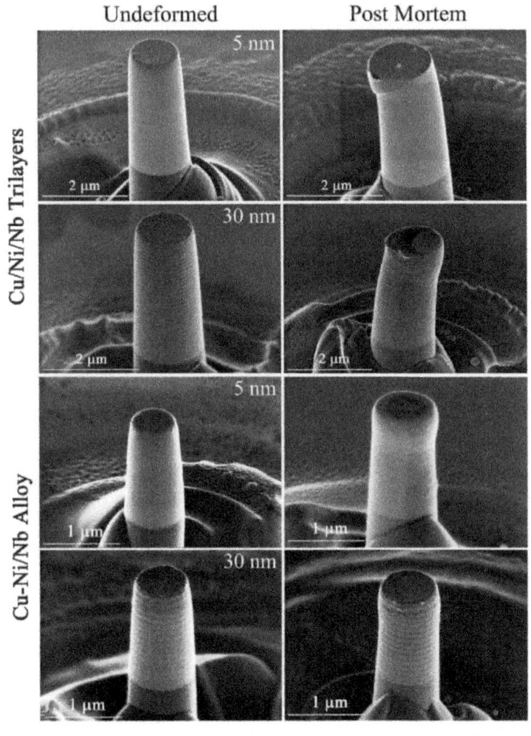

Figure 3. Undeformed (left) and post mortem (right) pillars of mixed interface tri-layer films and incoherent interface alloy films showing difference in deformation as a result of interface type and layer thickness.

Table 1. Strength- and work-hardening summary for tri-layer and alloy films tested using micro-pillar compression and nanoindentation.

		Micro-pillar compression				Nanoindentation
Sample		σ_{ys} (GPa)	σ_{max} (GPa)	n, Ludwik	Eng. stress at 8% (GPa)	Berkovich-8% (GPa), H/2.7 [8]
5 nm Tri-layer	P1	1.48	2.17	0.50	2.01	1.63 ± 0.06
	P2*	1.57	2.32	0.54	2.23	
30 nm Tri-layer	P1	1.44	1.85	0.35	1.83	1.4 ± 0.12
	P2*	1.18	1.53	0.39	1.51	
5 nm Alloy	P1	1.64	2.60	0.42	2.34	2.1 ± 0.04
	P2*	1.82	2.24	0.41	2.15	
30 nm Alloy	P1	1.32	1.75	0.31	1.68	1.62 ± 0.03
	P2*	1.36	1.68	0.34	1.65	

Stress–strain curves shown in Figure 3 are noted with *.

FCC/FCC interfaces adds an additional hardening mechanism, which does not occur in the bi-layer samples. While only a small difference is observed between the strain-hardening exponents in the 30 nm films, the 5 nm tri-layers display significantly more hardening than the alloy film with an increase in the hardening coefficient from 0.4 to 0.5.

The strain-hardening in tri-component multilayers has been previously [27] shown to be a result of increased dislocation content that is deposited along the interface as dislocations propagate through an individual layer. These deposited dislocations act as barriers to further deformation and also can act as nucleation sources for more dislocations, leading to an increase in the dislocation density of the films. Furthermore, a previous study conducted by Abdolrahim et al. [24] indicates decreasing layer thickness should result in an increase in the number of interfacial interactions and dislocation nucleation sites, which would increase strain-hardening in the system as dislocations interact with one another and hinder further deformation. This trend is verified in the current study for both interface structures; however, it is in direct contrast to recent investigations on incoherent and semi coherent interface multilayers. These studies have observed strain-hardening exponents of multilayer films decrease with layer thickness, while the strain-hardening rate increased up to a critical layer thickness before decreasing.[23,28] The decreasing hardening ability of multilayers with decreasing layer thicknesses was attributed to the dislocation storage capacity of the material, with higher strength materials (multilayers with smaller individual layer thicknesses) having a higher probability of dislocations being absorbed by the interface.[23] However, a direct comparison between these studies is difficult due to the significant variation in results possibly due to differences in sample production, mechanical testing apparatus, and data analysis procedures. Further research is needed to elucidate the cause of this disparity.

For this specific tri-layer system, it is believed that deformation is driven by 'superthreader' dislocations that have penetrated the coherent interface and propagate through the Cu and Ni layers.[15] Due to the different moduli of the two FCC materials, a coherency stress is created at the boundary that causes the threading dislocation to lag at this interface. Additionally, the portions of the dislocation in each of the two layers propagate at different speeds as a result of the different moduli, which causes instability in the dislocation at the coherent boundary. The addition of a BCC layer adds additional strength to the system as a whole, and provides enough internal shear stress to allow for cross-slip of the superthreading dislocation at the coherent interface. This in turn can act as dislocation pinning sites or a dislocation source, both of which lead to subsequent strain-hardening in this tri-layer NMM. The defects produced from cross-slip debris have a minimum stable size and therefore will act as a larger relative barrier to dislocation motion in thinner layers, since there is less room to allow for dislocation bowing around the defect. Additionally, as the layer thickness decreases the interface density also increases allowing this mechanism to occur more often in the same volume. Thereby, when the individual layer thickness is reduced there are more dislocation interactions and an increase in both strength- and strain-hardening ability.

In summary, complementary measurement techniques have verified that NMM tri-component films with smaller layer thicknesses exhibit a greater strain-hardening ability and a greater hardness than those with larger layer thicknesses in both mixed and incoherent interface structures. Nanoindentation using different indenter tips suggests that strain-hardening of the 5 nm layer thickness sample is significantly larger than that of the 30 nm thickness samples, specifically for the tri-layer sample. Micro-compression testing demonstrated that the strain-hardening exponents of samples with layer thicknesses of 5 nm are $\approx 50\%$ greater than that of the material with layer thicknesses of 30 nm layer. In both layer thicknesses used in this study, the addition of a modulus mismatched FCC/FCC interfaces provides additional strain-hardening over compared with an FCC/BCC bi-layer with an equivalent composition. This provides experimental support of the proposed hardening mechanism based on cross-slip at the FCC/FCC interfaces in these nanolaminate systems.

Acknowledgements The authors acknowledge access, through an approved user project, to the Center for Integrated Nanotechnologies (CINT), a DOE Office of Basic Energy Sciences user facility. The assistance of J. Kevin Baldwin at CINT in sample synthesis, and Nathan Mara from Los Alamos National Laboratory for his taper correction method is greatly appreciated.

Disclosure Statement No potential conflict of interest was reported by the authors.

Funding Statement This work was supported by the by the US Department of Energy, Office of Basic Energy Sciences, Division of Materials Sciences under Grant No. DE-FG02-07ER46435.

References

[1] Zbib HM, Overman CT, Akasheh F, Bahr D. Analysis of plastic deformation in nanoscale metallic multilayers with coherent and incoherent interfaces. Int J Plast. 2011;27:1618–1639.

[2] Bellou A, Overman CT, Zbib HM, Bahr DF, Misra A. Strength and strain hardening behavior of Cu-based bilayers and trilayers. Scr Mater. 2011;64:641–644.

[3] Mastorakos IN, Zbib HM, Bahr DF. Deformation mechanisms and strength in nanoscale multilayer metallic composites with coherent and incoherent interfaces. Appl Phys Lett. 2009;94:173114.
[4] Shao S, Zbib HM, Mastorakos IN, Bahr DF. Deformation mechanisms, size effects, and strain hardening in nanoscale metallic multilayers under nanoindentation. J Appl Phys. 2012;112:044307.
[5] Mckeown J, Misra A, Kung H, Hoagland RG, Nastasi M. Microstructures and strength of nanoscale Cu–Ag multilayers. Scr Mater. 2002;46:593–598.
[6] Verdier M, Huang H, Spaepen F, Embury JD, Kung H. Microstructure, indentation and work hardening of Cu/Ag multilayers. Philos Mag. 2006;86:5009–5016.
[7] Huang H, Spaepen F. Tensile testing of free-standing Cu, Ag and Al thin films and Ag/Cu multilayers. Acta Mater. 2000;48:3261–3269.
[8] Tabor D. The hardness of metals. Oxford: Oxford University Press; 1951.
[9] Bucaille JL, Stauss S, Felder E, Michler J. Determination of plastic properties of metals by instrumented indentation using different sharp indenters. Acta Mater. 2003;51:1663–1678.
[10] Liu Y, Bufford D, Wang H, Sun C, Zhang X. Mechanical properties of highly textured Cu/Ni multilayers. Acta Mater. 2011;59:1924–1933.
[11] Misra A, Verdier M, Lu YC, Kung H, Mitchell TE, Nastasi M, Embury JD. Structure and mechanical properties of Cu-X (X, Nb, Cr, Ni) nanolayered composites. Scr Mater. 1998;39:555–560.
[12] Carpenter JS, Misra A, Uchic MD, Anderson PM. Strain rate sensitivity and activation volume of Cu/Ni metallic multilayer thin films measured via micropillar compression. Appl Phys Lett. 2012;101:051901.
[13] Carpenter JS, Misra A, Anderson PM. Achieving maximum hardness in semi-coherent multilayer thin films with unequal layer thickness. Acta Mater. 2012;60: 2625–2636.
[14] Mara NA, Bhattacharyya D, Dickerson P, Hoagland RG, Misra A. Deformability of ultrahigh strength 5 nm Cu/Nb nanolayered composites. Appl Phys Lett. 2008;92:231901.
[15] Mastorakos IN, Bellou A, Bahr DF, Zbib HM. Size-dependent strength in nanolaminate metallic systems. J Mater Res. 2011;26:1179–1187.
[16] Rabe R, Breguet J-M, Schwaller P, Stauss S, Haug F-J, Patscheider J, Michler J. Observation of fracture and plastic deformation during indentation and scratching inside the scanning electron microscope. Thin Solid Films. 2004;469–470:206–213.
[17] Wheeler JM, Michler J. Elevated temperature, nano-mechanical testing in situ in the scanning electron microscope. Rev Sci Instrum. 2013;84:045103.
[18] Soler R, Molina-Aldareguia JM, Segurado J, LLorca J. Effect of misorientation on the compression of highly anisotropic single-crystal micropillars. Adv Eng Mater. 2012;14:1004–1008.
[19] Zhang H, Schuster BE, Wei Q, Ramesh KT. The design of accurate micro-compression experiments. Scr Mater. 2006;54:181–186.
[20] Verdier M, Niewczas M, Embury JD, Hawley M, Nastasi M, Kung H. Plastic behaviour of Cu/Ni multilayers. Mat Res Soc Symp Proc. 1998;522:77–82.
[21] Choi I, Gan Y, Kaufmann D, Kraft O, Schwaiger R. Measurement of Young's modulus of anisotropic materials using microcompression testing. J Mater Res. 2012;27:2752–2759.
[22] Sneddon IN. The relation between load and penetration in the axisymmetric Boussinesq problem for a punch of arbitrary profile. Int J Eng Sci. 1965;3:47–57.
[23] Lei S, Zhang JY, Niu JJ, Liu G, Zhang X, Sun J. Intrinsic size-controlled strain hardening behavior of nanolayered Cu/Zr micropillars. Scr Mater. 2012;66: 706–709.
[24] Abdolrahim N, Zbib HM, Bahr DF. Multiscale modeling and simulation of deformation in nanoscale metallic multilayer systems. Int J Plast. 2014;52:33–50.
[25] Roth HA, Davis CL, Thomson RC. Modeling solid solution strengthening in nickel alloys. Metall Mater Trans A. 1997;28:1329–1335.
[26] Dieter G. Mechanical metallurgy. New York: McGraw-Hill; 1988.
[27] Misra A, Hirth JP, Hoagland RG. Length-scale-dependent deformation mechanisms in incoherent metallic multilayered composites. Acta Mater. 2005;53: 4817–4824.
[28] Zhang JY, Li J, Liang XQ, Liu G, Sun J. Achieving optimum mechanical performance in metallic nanolayered Cu/X (X = Zr, Cr) micropillars. Sci Rep. 2014;4:4205.

23

Solid Argon Precipitation in a Metallic Glass: Does Free-Volume Help?

T. Miyauchi[a], H. Kato[b] and E. Abe[a]*

[a]*Department of Materials Science & Engineering, University of Tokyo, Tokyo 113-8656, Japan;*
[b]*Institute for Materials Research, Tohoku University, Sendai 980-8577, Japan*

We find that small particles (~1 nm) of solid argon are densely precipitated in a $Pd_{40}Ni_{20}P_{20}$ metallic glass matrix after argon ion-beam irradiation with an accelerating voltage of 4 keV. This is in sharp contrast to the case of noble-gas implantations into crystalline metals, for which high-energy beam (50–300 keV) has been essential to promote creations of precursor defects, such as voids or vacancy aggregates for initial confinements of noble-gas atoms. Therefore, the present low-energy implantation implies that the argon confinements have occurred at pre-existing low-density regions in the glass matrix, providing an important clue to phenomenological free-volume entities.

Keywords: Metallic Glass, Electron Microscopy, Microstructure

Noble-gas atoms exist as monoatomic molecule gas at ordinary temperature/pressure conditions, and generally cryogenic and/or high-pressure techniques are required to solidify these gases.[1] Interestingly, even at room temperature the noble gases form small solid precipitates when implanted into metals (e.g. aluminum) by ion-beam irradiations,[2–5] since the noble-gas elements are essentially inert and hence insoluble to the host matrix. These noble-gas precipitates are typically at nanometer dimensions and effectively stabilized by the relevant matrix cavity that acts as a high-pressure cell.[5] Namely, the host cavity provides a pseudo-high-pressure, P, yielded to balance with the surface tension γ of the matrix, described as $P = 4\gamma/d$ where d is a diameter of the spherical cavity (note that the pseudo-pressure originates from a shrinking force against the unit normal surface of the cavity, resulting in an effective internal pressure to the substances that are embedded into the cavity). This fact in turn indicates that local precursors to form such nanometer-scale cavity are essential for the noble-gas precipitations, and therefore high-energy ion-beam implantations with typical accelerating voltages 50–300 keV [2–5] have been commonly used to promote creations of voids or vacancy aggregates, which provide sufficient spaces for the initial condensation of noble-gas atoms. So far Ne, Ar, Xe and Kr were indeed precipitated as solid particles in the crystalline metal of mostly aluminum, and these nanometer-scale solids were successfully probed by electron diffraction,[2–5] electron energy loss spectroscopy (EELS) [2] and high-resolution electron microscopy.[4,6–8] Note in particular that phase-contrast lattice imaging clearly showed up distinct fringes of the Xe precipitates grown epitaxially on the host aluminum lattice,[4] providing direct evidences about their crystalline order and even structural dynamics (e.g. coalescence behaviors [7]).

In the present paper, we describe that Ar atoms precipitate as disordered/amorphous solids in a metallic glass $Pd_{40}Ni_{40}P_{20}$, the feature of which has clearly emerged by utilizing the state-of-the-art scanning transmission electron microscopy (STEM) combined with EELS at the sub-nanometer scale. The remarkable fact is that the present Ar implantations have been achieved, for the first time, with significantly low-energy Ar ion-beam irradiations during conventional thin-specimen preparations for STEM (known as *ion-milling*; the present Ar solids have been accidentally discovered rather than a well-designed experiment). Consequently, the present results have significant implications on the pre-existing cavity/void precursors in the glass structure, providing a direct clue to phenomenological free-volume entities

*Corresponding author. Email: abe@material.t.u-tokyo.ac.jp

[9–11] whose realistic configurations/distributions are not fully understood yet.

Master alloys of nominal composition $Pd_{40}Ni_{40}P_{20}$ were synthesized by arc melting pure metals and pre-sintered Pd-P alloy with B_2O_3 flux in a purified argon atmosphere. Ribbon specimens of metallic glass were produced by a single roller melt spinning method with the wheel rotating at a speed of approximately 40 m/s. Thin specimens for STEM observations were prepared by standard Ar ion-beam milling with an accelerating voltage of 4 keV, during which the Ar atom implantations were unexpectedly made to a total dose of approximately 2×10^{25} ions/m^2 (beam current 1.3×10^{-4} A, total milling time ~6 h). We used an aberration-corrected STEM (JEM-ARM200F) operated at an accelerating voltage of 200 kV, equipped with the EELS detector to carry out elemental mapping with a high-spatial resolution less than 0.2 nm.

Figure 1(a) shows the high-angle annular dark-field (HAADF)-STEM image, where a large number of dark particles with an average diameter of ~1 nm are formed in the $Pd_{40}Ni_{40}P_{20}$ glass matrix. Since the HAADF provides significantly enhanced atomic-number-dependent contrast [12] (Z-contrast), at first glance one may intuitively attribute the dark particles to be possibly phosphides if concerning only the constituent elements of Pd ($Z = 46$), Ni ($Z = 28$) and P ($Z = 15$). Note that these particles are hardly detectable by phase-contrast imaging (Figure 1(b)), and no distinct lattice fringes were observed at any places both in the phase-contrast and Z-contrast atomic-resolution images, as exemplified in Figure 1(b) and 1(c), respectively. Unambiguous identification of the particles comes uniquely from the high-spatial resolution EELS, which performs simultaneous acquisitions of spectrum at each probe position during ADF imaging with STEM, generating a data-cube composed of ADF intensity map (x, y) and energy-loss spectrum (ΔE) dimensions. By summing up the spectra obtained at the dark particle positions, a distinct peak showing the presence of Ar ($L_{2,3}$-edge: ~245 eV) clearly emerges in the reconstructed spectrum in Figure 2(a), while such an Ar peak becomes remarkably weak for the spectrum reconstructed using all the spectra in the entire scanned area. Details of the element distributions are confirmed in the reconstructed maps in Figure 2(b), where it is immediately noticed that the Ar enrichment indeed occurs at the dark particles in the corresponding ADF image. We here note that the other elements become reasonably deficient in the particles; the constituent elements of Pd, Ni and P (Figure 2(b)) and even the impurity oxygen (Supplementary Material, Figure S1) tend to be rejected from the particles, as confirmed by their positive cross-correlation with reference to the ADF intensity map, even though the correlations appear to be very weak (diffuse) for Ni and P owing to their insufficient signal-to-noise ratio (similar weak-correlation feature is also seen for Fe in Figure S1). On the basis of these facts, the dark particles in the HAADF image (Figure 1(a)) are evidently identified as pure Ar precipitates. It was confirmed that no particles were observed when the thin specimen is prepared not by Ar ion-milling but by electrochemical polishing, or when the very thin edge of whiskers in the as-solidified ribbon is directly used for STEM observations.

Stabilization of the present Ar precipitates in the metallic glass matrix can be understood by the pseudo-high-pressure scenario described earlier. Given the particle diameter d ~1.1 nm (Figure 1(a)) and assuming γ of the $Pd_{40}Ni_{40}P_{20}$ glass as being 1.0 N/m (this is

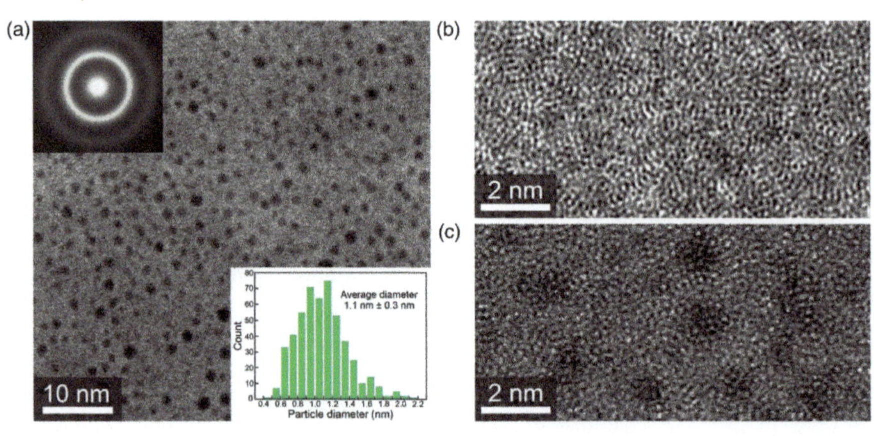

Figure 1. (a) HAADF-STEM image of a $Pd_{40}Ni_{40}P_{20}$ metallic glass irradiated by 4 keV Ar ion-beam, commonly used for a standard ion-milling for thin-specimen preparations. Inner angle of the annular detector was set to be 70 mrad, which is sufficiently large to achieve the chemical-sensitive Z-contrast condition. Histogram of dark-particle size is constructed from ~500 particles, evaluating an averaged diameter of 1.1 ± 0.3 nm. Extra peaks are hardly observed in the corresponding electron diffraction pattern, which merely shows halo rings of a $Pd_{40}Ni_{40}P_{20}$ amorphous structure. Atomic-resolution (b) phase-contrast and (c) Z-contrast images of the Ar-irradiated $Pd_{40}Ni_{40}P_{20}$ glass. Specimen thickness of the observed region was estimated to be approximately 18 ± 2 nm based on the EELS log-ratio method.

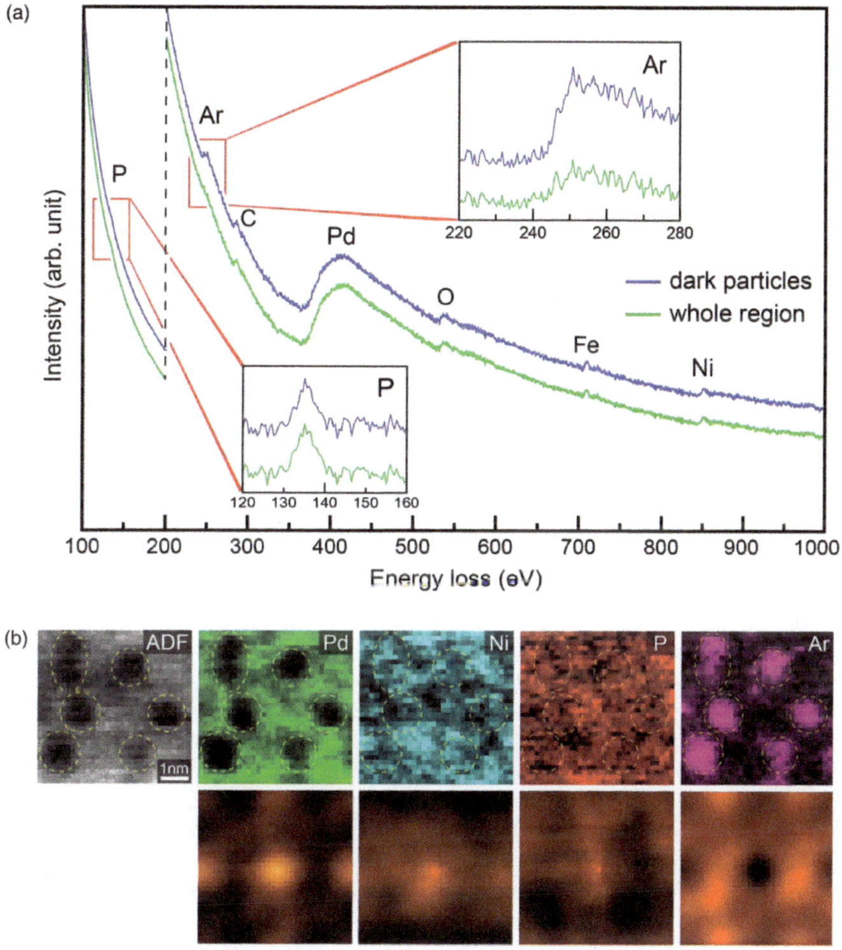

Figure 2. STEM-EELS analysis of the dark particles in Figure 1(a). Spectra and the corresponding ADF image (detector inner angle; ~20 mrad) were simultaneously obtained using the electron probe approximately 0.2 nm in diameter with a current of ~0.2 nA (~5.0 nA/nm^2), scanned over the 5.5 × 5.5 nm (32 × 32 pixel) area with a speed of 0.1 s/pixel. Multivariate statistical analysis using MSA software (HREM Research Inc.) was applied for the obtained spectra to reduce the noise effectively. (a) Reconstructed EEL spectra; shown in blue is obtained by summing 161 spectra taken at the dark-particle regions in the ADF image, while green is a summation of all the 1024 spectra in the whole scanned region (intensities are normalized for comparison). Enlarged peaks are inserted for P and Ar, after removing the background according to conventional power-law function extrapolations. Carbon originates from contaminations during scanning the electron beam, and oxygen and iron are involved as impurities in the present specimen. (b) Reconstructed element maps of Pd, Ni, P and Ar. Cross-correlation maps with reference to the ADF image are shown at the bottom, showing distinct positive/negative correlations between ADF intensity and Pd/Ar map. Maps of the impurity elements are shown in the Supplementary Material, Figure S1.

determined for the high-temperature liquid [13] and hence may differ somewhat from the precise γ value of the room-temperature glass), the internal pressure provided by the relevant matrix cavity is estimated to be approximately 3.6 GPa, which is sufficiently beyond the critical pressure for Ar solidification at room temperature (about 1.15 GPa,[1] which can be yielded by the cavity size up to d ~3.5 nm with γ ~ 1.0 N/m). In fact, in Figure 1(a) there are no particles larger than this maximum size. It should be noted that the present Ar solid precipitates are not distinctly crystalline but significantly disordered or amorphous, as in the following reasons. Suppose that the Ar particles form face-centered-cubic (*fcc*) structure under a high pressure,[1] the lattice parameter is expected to be about $a = 0.48$ nm with the pseudo-pressure of 3.6 GPa. Then, the relevant diffraction peaks and/or the occasional *fcc* lattice fringes should be observed; though, neither of these appears both in the diffraction pattern (Figure 1(a)) and the high-resolution images (Figure 1(b) and 1(c)). Besides, the Ar precipitates may well reveal faceted morphology if having distinct crystalline order;[14] this is evidently not consistent with the spherical morphology of the present precipitates (Figure 1(a)). These considerations lead to a conclusion

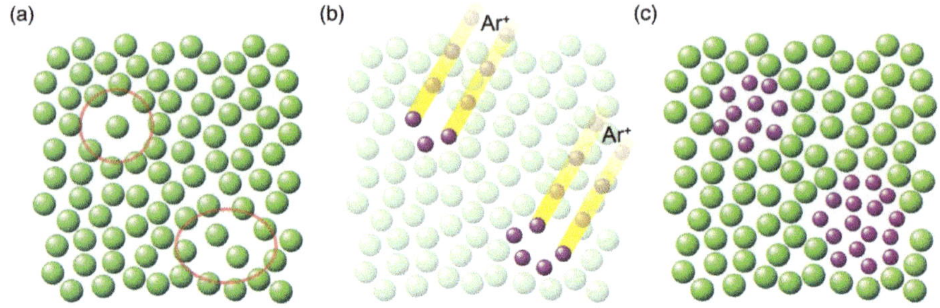

Figure 3. Schematic illustrations of an effective Ar trapping at 'free-volume' in the metallic glass matrix. The free-volume structure is represented as local low-density configurations (a), where the primary confinements of Ar atoms are achieved (b) and finally small Ar solids precipitate as being stabilized by nanometer-scale cavities (c).

that the structure of the present Ar precipitates is highly disordered or amorphous. Perhaps this may naturally be expected from the fact that the present particles are embedded in an amorphous matrix. Remember that there are at most 50–60 Ar atoms contained in the particles as small as 1 nm in diameter (under nearly dense-packing conditions), and more than half of the constituent atoms are exposed to the particle surface and directly attached to the host atoms of the amorphous matrix. In such a specific condition, the Ar atoms in the particles may take the form of amorphous as an energy-minimized configuration. The icosahedral clusters,[15] e.g. so-called as Mackay-type,[16] may also be favored so as to fit the spherical morphology. Unfortunately, further details of the Ar atom configurations within the particles are hardly determined by the present experiments. It is also noteworthy that the embedded Ar atoms could be a rather fluid state (i.e. less dense condition), and hence the noble-gas precipitates were frequently referred to as 'bubbles'.[4] In fact, Ne in the crystalline Al matrix was likely to be of fluid state since there were no distinct diffraction peaks,[2] and Xe precipitates in Al crystals behaved fluidly with partial crystalline order at the particle/matrix interfaces.[8]

Occurrence of the present Ar precipitations with the remarkably low-energy ion-beam implantations gives significant implications in terms of local structures of a metallic glass. As mentioned earlier, noble-gas precipitations in crystalline metals essentially require the high-energy ion-beam (50–300 keV) irradiations in order to generate precursory defects of voids or vacancy aggregates, to which the noble-gas atoms are primarily confined. The present accelerating voltage of 4 keV is far below by 1–2 orders of magnitude for promoting such internal voids in the matrix, suggesting that the Ar precipitation has occurred at pre-existing void-like low-density regions in the glass. Such local low-density zones are believed to be intrinsically involved in the glass structures, known as 'free-volume' that had been originally introduced as a phenomenological entity to understand thermodynamic [9,10] and mechanical [11] behaviors of the glass. Although realistic features of the free-volume are not fully understood, yet its structural concept can be schematically represented as in Figure 3(a). Suppose that most of the relevant zones have been used for Ar confinements, the random distributions of the particles (Figure 1(a)) may well trace an initial distribution of the free-volume in the specimen. Given that the penetration depth of the implanted Ar atoms is about 8 nm (calculated by Monte-Carlo simulation [17] with the present accelerating voltage 4 keV; details are described in Supplementary Material, Figure S2), the number density (ND) of the Ar particles is estimated to be about $\sim 1.0 \times 10^{-2}\,\mathrm{nm}^{-3}$. This might be comparable with a typical ND of the free-volume zone in the glass; i.e. ND is only reported for the polymer glass to be a order of $\sim 1.0\,\mathrm{nm}^{-3}$,[14] according to which ND of the metallic glass can be estimated as being a order of 10^{-1}–$10^{-2}\,\mathrm{nm}^{-3}$ if simply weighted by their relevant total free-volume ratio (~10% [14] and ~0.1% [18,19] for polymer and metallic glasses, respectively. Note that for these glasses, mean volume of one free-volume (v_f [9]) is reasonably assumed to be roughly in a same order of magnitude). As for the comparison of their total free-volumes, the volume fraction of the Ar particles (V_{Ar}) may overestimate the pre-existing total free-volume (V_{fv}), since V_{Ar} is a consequence of initial confinements and subsequent growth of the Ar precipitates; see Figure 3. The latter stage indeed involves expanding the original free-volume regions into cavities, which is perhaps accomplished by subsequent assembling of intrinsic vacancies. Although a precise value of V_{fv} cannot be deduced from V_{Ar}, it is nevertheless noteworthy that V_{Ar} is estimated to be about ~0.8% based on the present particle distributions (Figure 1(a)), the value of which is reasonably comparable with $V_{fv} \sim 0.3\%$ reported for the Pd-based bulky glass phases [18,19] (note that V_{fv} of the present rapidly cooled ribbon should be larger than that of the bulky glass prepared by slow-cooling; therefore, the true V_{fv} of the present specimen can even be closer to the present V_{Ar}). To test further the idea of an effective Ar trapping at free-volume zones in the $Pd_{40}Ni_{40}P_{20}$ glass, we check possible

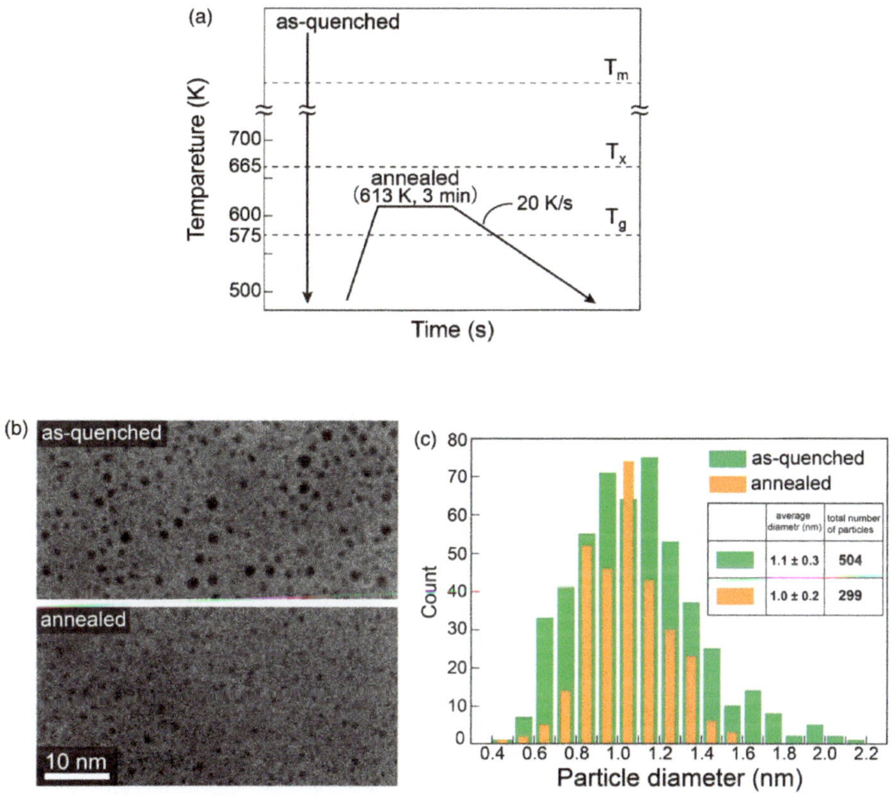

Figure 4. (a) Subsequent heating–cooling cycle applied to the as-quenched $Pd_{40}Ni_{40}P_{20}$ ribbon specimen. Specimen is heated above glass transition temperature (Tg ~575 K) up to 613 K (supercooled liquid state), holding there for 3 min, followed by cooling with 20 K/s. (b) HAADF-STEM images of as-quenched (part of Figure 1(a)) and subsequently annealed $Pd_{40}Ni_{40}P_{20}$ glass specimens. (c) Histogram of dark-particle size in the annealed specimen (yellow) is constructed from ~300 particles, evaluating an averaged diameter of 1.0 ± 0.2 nm. Histogram for the as-quenched specimen (green; same as that inserted in Figure 1(a)) is also shown for comparison.

changes of the Ar precipitation behaviors after the subsequent heating–cooling cycle across grass transition (Tg, Figure 4(a)), during which reductions of the free-volume take place due to a slower cooling (20 K/s) than the original fast-cooling (approximately $\sim 10^{4-5}$ K/s) during the melt spinning solidification. As confirmed in Figure 4(b) and 4(c), it is evident that the total number of the distinct Ar particles is reasonably decreased after the heating–cooling treatment across Tg; the ND and the volume fraction become about 5.8×10^{-3} nm^{-3} and 0.4%, respectively. This reasonably supports a validity of the idea that the Ar precipitations have occurred at the free-volume zones in the glass.

We find that the Ar precipitations by low-energy ion-beam irradiation do not occur for all the metallic glasses; e.g. preliminary observations show that there are no Ar precipitates in some of the Zr-based metallic glasses (e.g. $Zr_{70}Ni_{30}$ and $Zr_{70}Cu_{30}$) irradiated by the same Ar ion-beam as the present condition. This suggests that, although a further study is necessary, local variations of the free-volume structures are perhaps substantially different between the present Pd–Ni–P and the Zr-based glass phases. Regarding the issue, we note an implicative fact that they reveal remarkably different relaxation behaviors. A secondary relaxation with low activation energy (~1.0 eV), referred to as slow-β type,[20] is distinctly observed for the Pd–Ni–P glasses [21] ('fragile' type [22]), while the Zr-based glasses reveal predominantly the common α type relaxation (glass transition) driven by higher activation energy of 5–10 eV ('strong' type,[23] though accompanying weak β relaxation signs that are almost buried in the relevant α relaxation peak tails). Therefore, it may be presumed that the cavities necessary for the Ar confinements have grown at the local slow-β relaxation centers, where the atoms are reasonably mobile via vacancy-mediated diffusion jumps [24] that can be activated by ion bombardments (the activation energy of the slow-β relaxation for the present $Pd_{40}Ni_{40}P_{20}$ glass is determined to be about 1.47 eV, as described in Supplementary Material, Figure S3). For such low-energy relaxation events, the relevant free-volume zone is likely to possess sufficiently sparse atomic configurations. Existence of such local configurations is in fact suggested as 'holes' [25] in the

$Pd_{52}Ni_{32}P_{16}$ glass, and the related high atomic mobility regions (weakly- bonded soft regions) are also confirmed in terms of a microstructural heterogeneity [26] in the $Pd_{42.5}Ni_{7.5}Cu_{30}P_{20}$ glass. In case that the critical ion-beam energy for the noble-gas confinements is turned out to be very sensitive to slight local density differences (i.e. different pore structural types), energy-variable ion-beam irradiations may be possibly used as an effective probe for tracing the free-volume variations [25,27] responsible for the different relaxation events of the metallic glass. By combining with the positron annihilation technique, it provides an interesting testing ground for elucidating details of the realistic free-volume features in various metallic glasses [28].

Acknowledgements This work is partly supported by KAKENHI No. 22360263 from the Ministry of Education, Culture, Sports, Science and Technology of Japan (MEXT).

References

[1] Finger LW, Hazen RM, Zou G, Mao HK, Bell PM. Structure and compression of crystalline argon and neon at high pressure and room temperature. Appl Phys Lett. 1981;39:892–894.
[2] Vom Felde A. Fink J, Muller-Heinzerling TH, Pfluger J, Scheerer, B, Linker, H. Pressure of neon, argon, and xenon bubbles in aluminum. Phys Rev Lett. 1984;53:922–925.
[3] Templier C, Gaboriaud RJ, Garem H. Precipitation of implanted xenon in aluminum. Mater Sci Engg. 1985;69:63–66.
[4] Donnelly SE, Rossouw CJ. Lattice images of solid xenon precipitates in aluminum at room temperature. Science. 1985;230:1272–1273.
[5] Birtcher RC, Jäger W. Precipitation and phase stability of solid Kr in cavities after room-temperature implantation of Al. Ultramicroscopy. 1987;22:267–280.
[6] Mitsuishi K, Kawasaki M, Takeguchi M, Furuya K. Determination of atomic positions in a solid Xe precipitate embedded in an Al matrix. Phys Rev Lett. 1999;82: 3082–3084.
[7] Birtcher RC, Donnelly SE, Song M, Furuya K, Mitsuishi K, Allen, CW. Behavior of crystalline Xe nanoprecipitates during coalescence. Phys Rev Lett. 1999;83;1617–1620.
[8] Donnelly SE, Birtcher RC, Allen CW, Morrison I, Furuya K, Song M, Mitsuishi K, Dahmen U. Ordering in a fluid inert gas confined by flat surfaces. Science. 2002;296: 507–510.
[9] Cohen MH, Turnbull D. Molecular transport in liquids and glasses. J Phys Chem. 1959;31:1164–1169.
[10] Cohen MH, Grest GS. Liquid-glass transition, a free-volume approach. Phys Rev B. 1979;20:1077–1098.
[11] Ye JC, Lu J, Liu CT, Wang Q, Yang Y. Atomistic free-volume zones and inelastic deformation of metallic glasses. Nature Mater. 2010;9:619–623.
[12] Pennycook SJ, Boatner LA. Chemically sensitive structure-imaging with a scanning transmission electron microscope. Nature. 1988;336:565–567.
[13] Mukherjee S, Johnson WL, Rhim WK. Noncontact measurement of high-temperature surface tension and viscosity of bulk metallic glass-forming alloys using the drop oscillation technique. Appl Phys Lett. 2005;86:014104-1–014104-3.
[14] Bohlen J, Wolff J, Kirchheim, R. Determination of free-volume and hole number density in polycarbonates by positron lifetime spectroscopy. Macromolecules. 1999;32:3766–3773.
[15] Rodriguez A, Amiens C, Chaudret B. Synthesis and isolation of cuboctahedral and icosahedral platinum nanoparticles. Ligand-dependent structures. Chem Mater. 1996;8:1978–1986.
[16] Mackay AL. A dense non-crystallographic packing of equal spheres. Acta Cryst. 1962;15:916–918.
[17] Ziegler JF. Open software. Available from: http://www.srim.org/
[18] Haruyama O, Inoue A. Free volume kinetics during sub-Tg structural relaxation of a bulk $Pd_{40}Ni_{40}P_{20}$ metallic glass. Appl Phys Lett. 2006;88:131906-1–131906-2.
[19] Hu X, Mg SC, Feng YP, Li Y. Cooling-rate dependence of the density of $Pd_{40}Ni_{10}Cu_{30}P_{20}$ bulk metallic glass. Phys Rev B. 2001;64:172201-1–172201-4.
[20] Johari JP, Goldstein M. Viscous liquids and the glass transition. II. Secondary relaxations in glasses of rigid molecules. J Chem Phys. 1970;53:2372–2388.
[21] Zhao ZF, Wen P, Wang WH, Shek CH. Observation of secondary relaxation in a fragile $Pd_{40}Ni_{10}Cu_{30}P_{20}$ bulk metallic glass. Appl Phys Lett. 2006;89:071920-1–071920-3.
[22] Kato H, Wada T, Hasegawa M, Saida J, Inoue A, Chen HS. Fragility and thermal stability of Pt- and Pd-based bulk glass forming liquids and their correlation with deformability. Scripta Mater. 2006;54:2023–2027.
[23] Wang WH, Wen P, Liu, XF. The excess wing of bulk metallic glass forming liquids. J Non-Cryst Solids. 2006;352:5103–5109.
[24] Tang XP, Geyer U, Busch R, Johnson WL, Wu Y. Diffusion mechanisms in metallic supercooled liquids and glasses. Nature. 1999;402:160–162.
[25] Sietsma J, Thijsse BJ. Characterization of free volume in atomic models of metallic glasses. Phys Rev B. 1995;52:3248–3255.
[26] Ichitsubo T, Matsubara E, Yamamoto T, Chen HS, Nishiyama N, Saida J, Anazawa K. Microstructure of fragile metallic glasses inferred from ultrasound-accelerated crystallization in Pd-based metallic glasses. Phys Rev Lett. 2005;95:245501-1–245501-4.
[27] Flores KM, Sherer E, Bharathula A, Chen H, Jean YC. Sub-nanometer open volume regions in a bulk metallic glass investigated by positron annihilation. Acta Mater. 2007;55:3403–3411.
[28] Wang WH. Correlation between relaxations and plastic deformation, and elastic model of flow in metallic glasses and glass-forming liquids. J Appl Phys. 2011;110:053521-1–053521-15.

24

In situ Observation of Defect Annihilation in Kr Ion-Irradiated Bulk Fe/Amorphous-Fe_2Zr Nanocomposite Alloy

K.Y. Yu[a,b*†], Z. Fan[c†], Y. Chen[b], M. Song[b], Y. Liu[b], H. Wang[d], M.A. Kirk[e], M. Li[f] and X. Zhang[b,c*]

[a]*Department of Materials Science and Engineering, China University of Petroleum-Beijing, Beijing 102249, People's Republic of China;* [b]*Department of Materials Science and Engineering, Texas A&M University, College Station, TX 77843-3123, USA;* [c]*Department of Mechanical Engineering, Texas A&M University, College Station, TX 77843-3123, USA;* [d]*Department of Electrical and Computer Engineering, Texas A&M University, College Station, TX 77843-3128, USA;* [e]*Materials Science Division, Argonne National Laboratory, Argonne, IL 60439, USA;* [f]*Nuclear Engineering Division, Argonne National Laboratory, Argonne, IL 60439, USA*

Enhanced irradiation tolerance in crystalline multilayers has received significant attention lately. However, little is known on the irradiation response of crystal/amorphous nanolayers. We report on *in situ* Kr ion irradiation studies of a bulk $Fe_{96}Zr_4$ nanocomposite alloy. Irradiation resulted in amorphization of Fe_2Zr and formed crystal/amorphous nanolayers. α-Fe layers exhibited drastically lower defect density and size than those in large α-Fe grains. *In situ* video revealed that mobile dislocation loops in α-Fe layers were confined by the crystal/amorphous interfaces and kept migrating to annihilate other defects. This study provides new insights on the design of irradiation-tolerant crystal/amorphous nanocomposites.

Keywords: Defect Cluster, Heavy Ion Irradiation, Absorption, Annihilation

Next-generation nuclear energy technology requires advanced structural materials that must withstand extreme irradiation environments.[1–3] One of the emerging approaches is to introduce abundant interfaces by forming nanostructures in the materials.[4,5] Consisting of two or more nanostructured heterophases, metallic nanolaminates exhibit not only excellent mechanical strength [6–9] and fatigue resistance,[10] but also impressive tolerance to ion irradiation.[11–13] He ion irradiation studies have been extensively reported in immiscible nanolaminates such as Cu/V,[14–16] Cu/Nb,[11] Cu/W,[17] Cu/Mo,[18] Ag/Ni[19] and Ag/V [20] and miscible Al/Nb [21] and Fe/W.[22] Defect density and irradiation-induced hardening in these systems show a strong size dependence on individual layer thickness (h), that is the magnitude of irradiation hardening and defect density both decrease at smaller h in immiscible nanolaminates. Fu et al. first showed prominent simultaneous reduction of irradiation hardening and He bubble density in the He ion-irradiated Cu/V system when h is smaller [16]; similar phenomena were later observed in several other systems, such as Ag/Ni,[19] Ag/V [20] and Fe/W.[22] Molecular dynamics simulations [23] pointed out that layer interfaces can act as effective defect sinks for irradiation-induced point defects.[24] Thus, a smaller h gives rise to a greater interfacial area per unit volume, which provides more sinks to irradiation-induced point defects and clusters.

The nature of such sinks is not quite well understood to date as there are only a handful of studies on irradiation damage in nanolaminates with different types of interfaces. In general, these studies can be classified into systems with miscible and immiscible interfaces. Miscible interfaces are prone to irradiation-induced mixing and hence may be less effective to sustain irradiation damage at higher doses and temperatures.[21,22] In immiscible systems, several types of interfaces have

*Corresponding authors. Emails: kyyu@cup.edu.cn; zhangx@tamu.edu

† *These authors contributed equally.*

been investigated, such as fcc/bcc,[12,14] bcc/bcc [22] and fcc/fcc,[19,25] where fcc and bcc stand for face-centred cubic and bcc-body centred cubic, respectively. The efficiency of the immiscible interfaces to capture and annihilate point defects might be related to the density of misfit dislocations at the interfaces.[12] Yu et al. recently reported *in situ* evidence of fast migration of defect clusters towards Ag/Ni layer interfaces decorated with high-density misfit dislocations. And the interactive energy needed to annihilate two loops with opposite nature was estimated to be ~0.4 eV.[25]

In situ irradiations inside a transmission electron microscope provide important insights for investigating dynamics of defect–defect and defect–interface interactions, thanks to real-time observation of microstructural evolution in irradiated materials. Numerous phenomena have been observed using the *in situ* irradiation technique, such as anisotropic void formation in Mg,[26] irradiation-induced grain growth in Fe and Zr,[27] defect absorption by triple junctions in nanocrystalline Ni,[28] defect removal by twin boundaries (TBs) and the irradiation-induced TB migrations in nanotwinned metals,[29,30] and defect migration kinetics in nanoporous metals.[31–33] Meanwhile several *in situ* mechanical testing studies on irradiated materials are also available.[34,35]

Most prior studies on multilayers focused on thin films (several μm in thickness), whereas the application in nuclear reactors typically calls for bulk nanocomposites.[2,3,5] Irradiation studies on bulk nanolaminate alloys remain scarce.[36] What is more, there are no investigations on irradiation response of nanocomposites with crystal/amorphous interfaces. In this study, we report on *in situ* studies on the heavy ion irradiation response of a bulk FeZr nanocomposites consisting of Fe/Fe_2Zr nanolaminates and submicron α-Fe grains prepared by the suction casting technique. Low-dose irradiation leads to amorphization of Fe_2Zr layers and subsequent formation of Fe/amorphous Fe_2Zr nanolaminates. The crystal/amorphous layer interface has extraordinary stability against irradiation and the confined migration of dislocation loops within α-Fe layers reduces their defect density significantly.

The Fe–4 at% Zr alloy was prepared by melting high-purity iron (99.95%) and zirconium (99.8%) using an Edmund Buhler AM arc melting machine. The raw materials were remelted multiple times (for homogeneity) in water-cooled Cu crucibles under an ultra high-purity (UHP) argon (Ar) atmosphere and solidified rapidly by the suction casting technique. The chamber was pumped down to a vacuum of 3×10^{-3} Pa prior to melting and later backfilled with UHP Ar at a pressure of 90 kPa during fabrication. The cooling rate of suction casting was estimated to be 1,000–4, 000°C/s.

The as-casted alloy ingots were cylindrical rods with a diameter of 3 mm. Samples for transmission electron microscopy (TEM) study were transversely cut from the rod to avoid contamination from the crucible, followed by mechanical grinding, chemical etching and polishing. The TEM specimens were then irradiated using the intermediate voltage electron microscope (IVEM) facility at Argonne National Laboratory where a Hitachi H-9000NAR microscope was equipped with a TANDEM accelerator. TEM specimens were irradiated with 1 MeV Kr^{++} ions up to 2 dpa (corresponding to a dose of 4×10^{14} ions/cm^2) at room temperature. The beam size was ~1.5 mm (without rastering) and was able to cover the entire thin area of TEM specimens during irradiation. A Tecnai F20 transmission electron microscope was operated at an acceleration voltage of 200 keV to examine the microstructure before and after the irradiation experiments. Scanning transmission electron microscopy (STEM) and energy-dispersive X-ray spectroscopy (EDX) analyses were conducted by using high-angle annular dark-field and Oxford detectors attached to the aforementioned TEM microscope.

As-casted specimens contained three major components: cellular zone, nanolaminates and large α-Fe grains as shown in the TEM micrograph in Figure 1(a). The volume fraction ratio of large grains to the eutectic phase is ~2 : 1. The selected area diffraction (SAD) pattern of the nanolaminate in the box in Figure 1(a) shows two major phases: α-Fe (examined from $\langle 110 \rangle$ zone axis) and fcc-like Fe_2Zr (C15 type Laves phase with Mg_2Cu as prototype) along the $\langle 112 \rangle$ zone axis (Figure 1(b)), consistent with X-ray diffraction profile (Figure S1). $Fe_{23}Zr_6$ is a minor phase as indicated by weak diffraction spots (connected by orange lines) in Figure 1(b). Index of diffraction patterns suggests that α-Fe $\{200\}$//$Fe_{23}Zr_6$ $\{800\}$. The interplanar spacing of α-Fe $\{200\}$ nearly equals that of Fe_2Zr $\{400\}$, consistent with the lattice parameter ratio of the two phases ($a_{Fe} =$ 0.287 nm, $a_{Fe2Zr} = 0.701$ nm),[37] and the d-spacing of α-Fe $\{200\}$ (0.145 nm) is similar to that of $Fe_{23}Zr_6$ $\{800\}$ (0.15 nm, $a_{Fe23Zr6} = 1.172$ nm). The streaking lines in Figure 1(b) indicate the formation of high-density stacking faults that are usually observed in the Fe_2Zr phase.[37,38] The STEM image in Figure 1(c) confirms the eutectic nanolaminate structure, and its line-scan compositional profile depicted using the EDX spectrum (as shown in Figure 1(d)) corresponds well with Fe_2Zr and α-Fe. As $Fe_{23}Zr_6$ is a minority phase, we will focus on α-Fe and Fe_2Zr in this irradiation study. Figure 1(e) shows a magnified TEM micrograph of nanolaminates with high-density stacking faults in the Fe_2Zr layers. In addition, growth defects decorating the Fe/Fe_2Zr interfaces were identified. High-resolution TEM (HRTEM) micrograph of the Fe/Fe_2Zr interface in Figure 1(f) reveals a typical incoherent configuration of the interface, where α-Fe $\{200\}$ is nearly parallel to Fe_2Zr $\{200\}$, as indicated by the inset fast Fourier

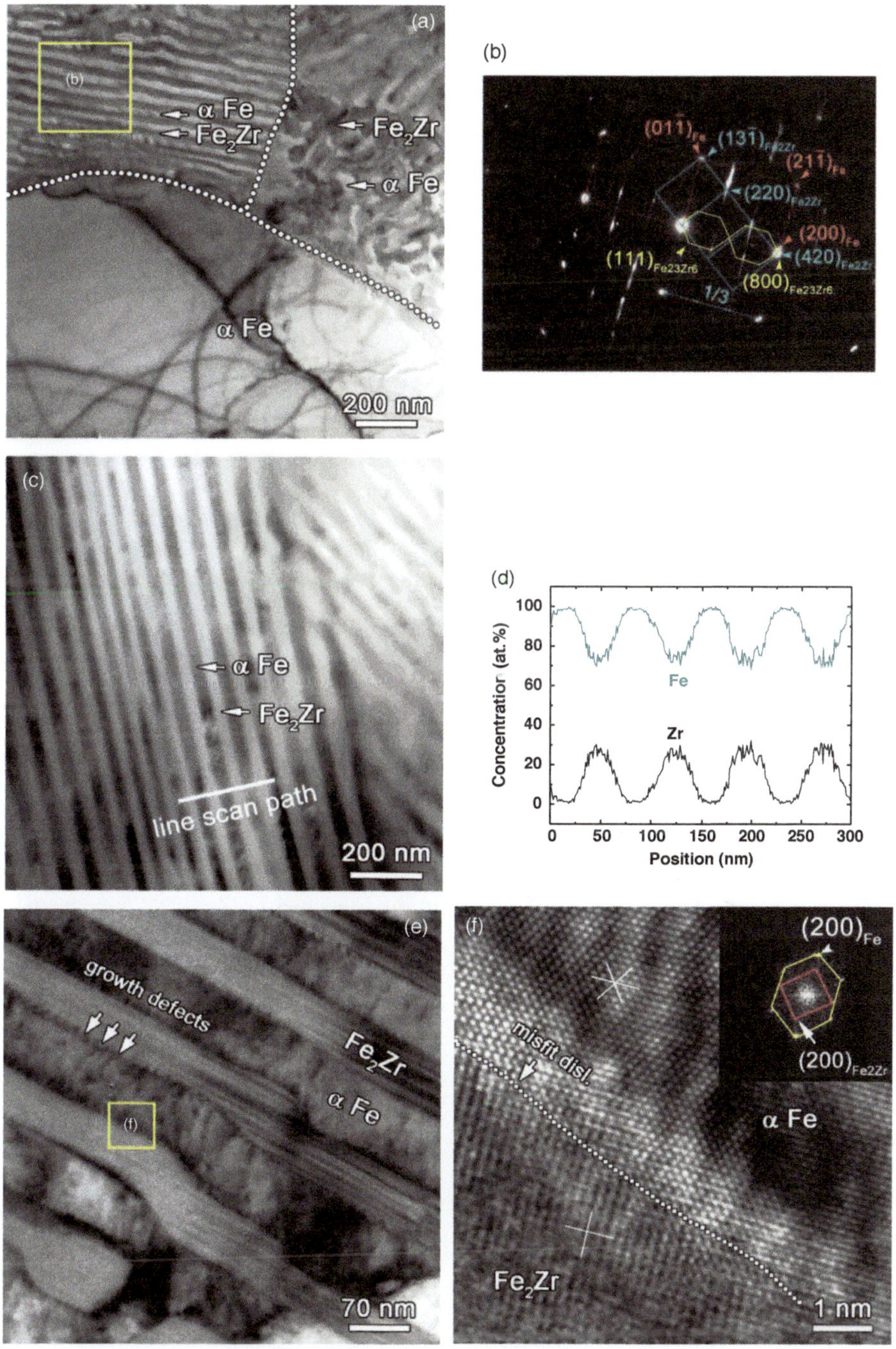

Figure 1. Microstructure and chemistry of unirradiated α-Fe/Fe_2Zr nanocomposites. (a) TEM micrograph showing three typical components: large α-Fe grains, cellular and nanolaminate phases consisting of α-Fe and Fe_2Zr layers. (b) The SAD pattern of nanolaminate in box b of Figure 1(a) shows three sets of diffraction dots, arising from α-Fe, Fe_2Zr and $Fe_{23}Zr_6$, respectively (see Supplementary Figure S1 and S2 for more detail), observed from $\langle 110 \rangle$ zone axes of bcc α-Fe and fcc-like $Fe_{23}Zr_6$, and $\langle 112 \rangle$ zone axis of fcc-like Fe_2Zr. The streaking lines indicate the formation of high-density stacking faults in the Fe_2Zr phase. (c) STEM image reveals the laminar α-Fe and Fe_2Zr with an average layer thickness of ~50 nm. (d) The alternating compositional line profile of nanolaminate determined by EDX spectrum corresponds to α-Fe and Fe_2Zr. (e) Magnified TEM micrograph shows stacking faults in Fe_2Zr layers and growth defects decorating the Fe/Fe_2Zr layer interfaces. (f) HRTEM micrograph reveals incoherent atomic configuration along the Fe/Fe_2Zr layer interface.

transformation of the image. Note that the interfacial relationship of Fe/Fe_2Zr varies frequently as shown in Figure S2.

In situ irradiation experiments were performed using 1 MeV Kr ions up to a dose of two displacements-per-atom (dpa). TEM micrograph of the Kr ion-irradiated specimen at 1 dpa (Figure 2(a)) shows abundant defect clusters in large Fe grains. In contrast, much fewer defects were generated in the cellular zones. Irradiated nanolaminates at 1 dpa (Figure 2(b)) have exceptionally low defect density, despite the sporadically visible dislocation loops bounded in the Fe layers. A magnified TEM image (Figure 2(c)) of nanolaminates shows defective Fe and clean Fe_2Zr layers. Moreover, the inserted SAD pattern shows a diffuse ring indicating the amorphization of Fe_2Zr. An HRTEM micrograph in Figure 2(d) shows significant amorphization of Fe_2Zr adjacent to interfaces in nanolaminates (see Supplementary Figure S3 for more details).

Statistical studies (in Figure 2(e)) show that defect density in large Fe grains increased rapidly during the initial stage of irradiation, and appeared to reach saturation by ~1.25 dpa, whereas defect density in Fe layers within nanolaminates was much lower and barely varied up to 2 dpa. Meanwhile in large Fe grains, defect size increased monotonically with dose to ~60 nm at 2 dpa (Figure 2(f)), in drastic contrast to nearly unchanged defect dimension of ~ 20 nm in Fe layers in nanolaminate (Figure S4). As defect clusters were absent in irradiated Fe_2Zr layers in nanolaminates, no statistical data were available for comparison for Fe_2Zr layers.

The saturated defect density in the Fe layers of Fe/Fe_2Zr nanolaminates is much lower than that in large Fe grains, implying that a large number of defects must have been annihilated during irradiation. In order to examine the evolution of microstructure of irradiated nanolaminates, we will show two important phenomena captured during *in situ* irradiation studies. The first one is low dose irradiation- (less than 0.1 dpa) induced amorphization of Fe_2Zr and subsequent formation of crystal/amorphous interfaces and the second case is a higher dose irradiation initiated dislocation loop migration in Fe layers.

Low-dose irradiation-induced amorphization of Fe_2Zr and the formation of crystal/amorphous interfaces in Fe/Fe_2 Zr nanolaminates. Figure 3(a) shows a series of snapshots of numerous events of dislocation absorption by the Fe/Fe_2Zr layer interface at a low dose (~0.01 dpa) (Supplementary Video 1). Before irradiation, crystalline Fe_2Zr layers contained high-density stacking faults. At 1 s, a dislocation formed in the Fe layer A and started to evolve and migrate towards the right layer interface, as indicated by a small arrow. At 1.1 s, the evolving dislocation became longer and well defined, and nearly reached layer interface. Meanwhile another dislocation line emerged in the Fe layer B. The two dislocations continued to migrate rightwards at 1.3 s and the dislocation in layer A was completely absorbed by layer interfaces at 1.4 s, followed by absorption of the dislocation in layer B. Furthermore, the crystalline Fe_2Zr layers became amorphous.

Schematics in Figure 3(b) illustrate the 'interface-assisted annihilation' of dislocations and subsequent amorphization of Fe_2Zr. First, irradiation-induced dislocation loops in Fe layers evolve into dislocation lines, which propagate towards the Fe/Fe_2Zr layer interfaces (Figure 3(b1)). The absorption of dislocations then leads to the distortion of the crystalline structure in Fe_2Zr layers. Due to the defect absorption and quenching of cascades, amorphization occurs in Fe_2Zr layers (Figure 3(b2)). Note that irradiation caused immediate destruction of stacking faults in the Fe_2Zr layers, followed by their amorphization at very low doses (<0.2 dpa). Meanwhile, the layer interface remained stable (chemically abrupt) during the entire irradiation experiments (at higher dose). Post-irradiation analysis also evidenced amorphization (Figure 2) and the absence of stacking faults throughout the irradiated specimen. Another example of low-dose irradiation-induced defect evolution is shown in Figure S5 and Supplementary Video 2. It is well known that irradiation breaks chemical bonds of compounds, such as Fe_2Zr in this study, and induces their amorphization. On the other hand, it is likely that the layer interface actively absorbed interstitials or interstitial clusters, creating rapid disordering in the irradiated Fe_2Zr layer. The layered Fe_2Zr might experience accelerated radiation-induced amorphization in comparison with the bulk counterpart. Thus, the reason for the amorphization of Fe_2Zr was a combination of defect kinetics and disordering of the chemical bonds induced by heavy ion bombardment.

High-dose irradiation-induced migration of dislocation loops in Fe layers in nanolaminates (up to 2 dpa). We observed different types of defect morphology in various zones of irradiated specimens. First, in α-Fe grains, discrete dislocation loops were frequently observed. Many individual loops also combine with one another to form dislocation segments at greater doses. Second, in the cellular zone, defect clusters were often bounded by layer interfaces (Figure 2(a)). Third, Fe_2Zr nanolayers in nanolaminates were frequently depleted with defect clusters, while defect clusters are typically sporadically distributed in α-Fe layers. And the amorphization of Fe_2Zr nanolayers is widespread in nanolaminates.

Numerous mechanisms could contribute to the drastic reduction of defect cluster density in Fe nanolayers. Misfit dislocations at layer interfaces (Figure 1(e)–(f)) may play a role in removing defects. Excess free volume at misfit dislocations can act as sinks to point defects. This mechanism has been proposed to

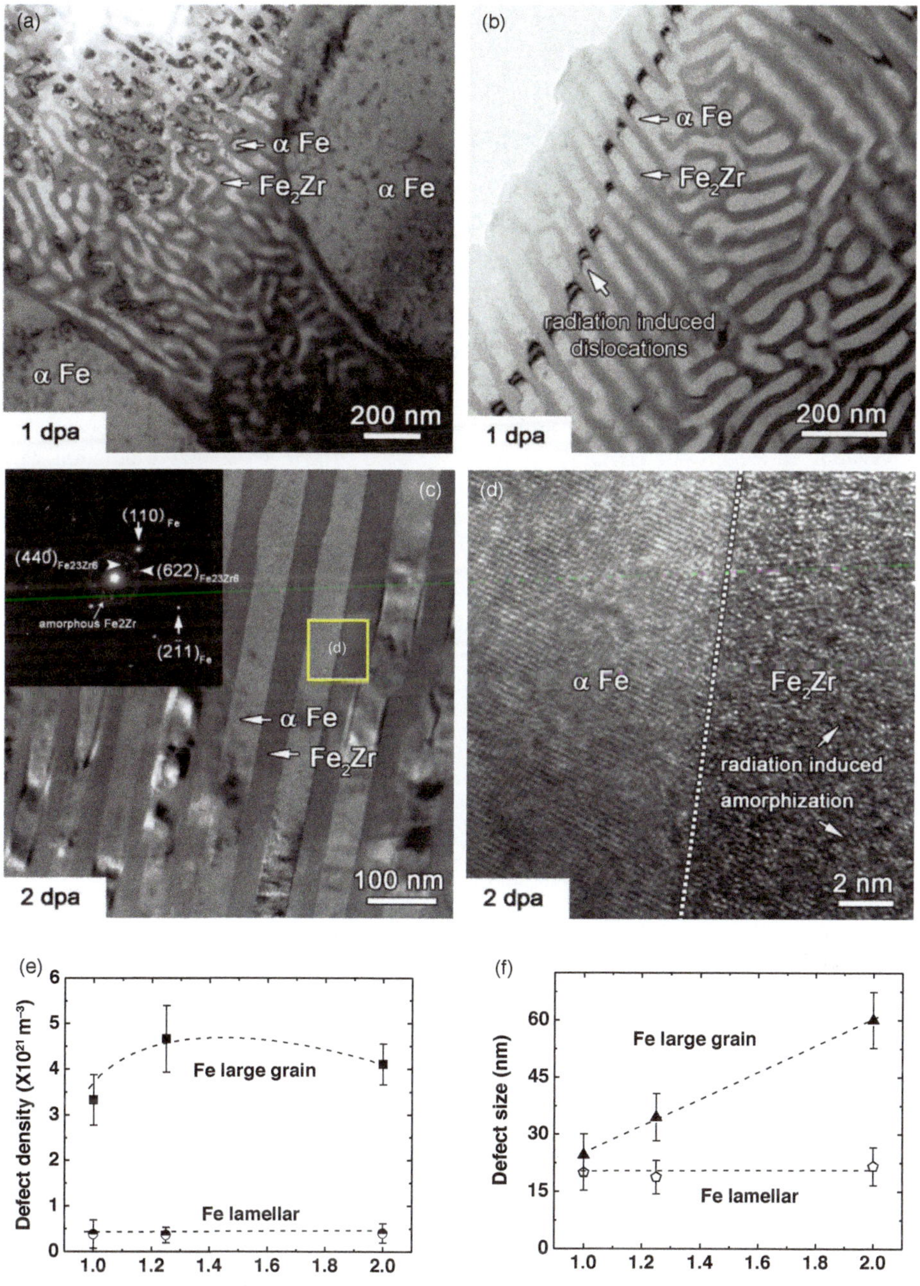

Figure 2. *Microstructure evolution of* $\alpha - Fe/Fe_2Zr$ *nanocomposites subjected to* in situ *Kr ion irradiation up to two displacements-per-atom (dpa)*. (a) At 1 dpa, TEM micrograph showing abundant defect clusters in the large α-Fe grains. In contrast, smaller defect clusters were observed in cellular areas, residing at the phase boundaries and remained stable during successive irradiation experiments. (b) α-Fe layers in irradiated nanolaminates were mostly clean, with sporadic dislocation loops. Little defect clusters were detected in the irradiated Fe_2Zr layers. (c) At 2 dpa, a TEM micrograph of irradiated nanolaminate shows the retention of sharp layer interfaces between α-Fe and Fe_2Zr. The halo ring in the inserted SAD pattern reveals the formation of amorphous Fe_2Zr. (d) An HRTEM micrograph along the interface showing amorphization of Fe_2Zr in irradiated nanolaminates. α-Fe remained a crystalline structure. *Comparison of dose-dependent evolution of defect clusters in* α*-Fe in large grains and nanolaminates*. (e) Defect density in α-Fe large grains increased rapidly during the initial irradiation, and appeared to reach saturation by ~1.25 dpa. In comparison, defect density in α-Fe in nanolaminate was nearly an order of magnitude lower and barely changed up to 2 dpa. (f) In large α-Fe grains, defect size increased monotonically with dose to ~60 nm up to 2 dpa, in drastic contrast to nearly unchanged defect dimension of ~20 nm in the α-Fe layers in nanolaminates.

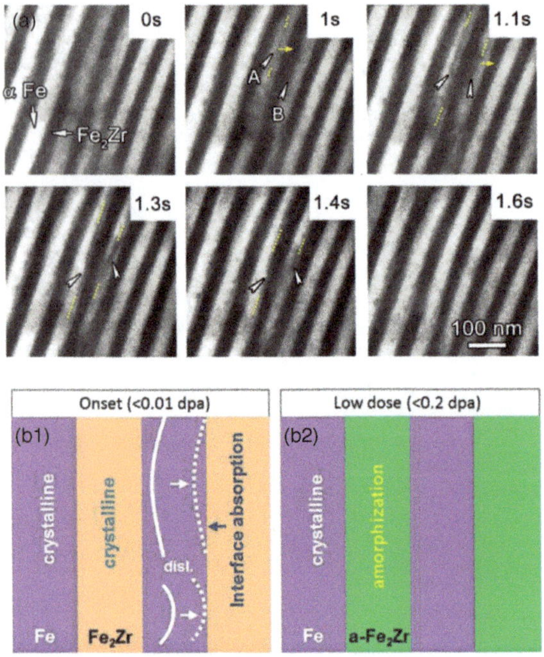

Figure 3. Defect annihilation by layer interface and amorphization of the Fe_2Zr phase in nanolaminates. (a)*In situ* snapshots of dislocation absorption by the Fe/Fe_2Zr layer interface at a dose of about 0.01 dpa (over less than 2 s). At 1 s, a dislocation formed in the α-Fe layer A and started to evolve and migrate towards the right layer interface, as indicated by the small arrow. The dotted lines mark the ends of the dislocation lines and the small arrows indicate the migration direction. At 1.1 s, the dislocation in layer A became longer and another dislocation line was formed in Fe layer B. The two dislocations continued to migrate rightwards at 1.2 and 1.3 s and were absorbed by layer interface at 1.4 s. By 0.2 dpa, Fe_2Zr became nearly amorphous. See Figure S5, Supplementary Videos 1 and 2 for more details. (b) Schematics of the 'interface-assisted annihilation' mechanism demonstrating the absorption of dislocations by layer interfaces and amorphization of Fe_2Zr at low irradiation dose.

explain size-dependent reduction of defect density in Kr ion-irradiated Ag/Ni,[25] and He-irradiated Cu/Nb,[12] Cu/V [16] and Fe/W [22] multilayers. However, unlike the previously observed rapid migration of defect clusters towards layer interface in the Ag/Ni system, we observed, through *in situ* studies, an interesting unexplored mechanism through which defect clusters were removed in α-Fe nanolayers (channels) in nanolaminates with crystal/amorphous layer interfaces.

In situ TEM snap shots in Figure 4(a) capture one such typical defect annihilation process in Fe nanolayers in nanolaminates. A dislocation loop A (outlined in circle in Figure 4(a)) nucleated at 0.755 dpa and started to migrate (Supplementary Video 3). Loop A migrated continuously during subsequent irradiation, while another dislocation loop B emerged at 0.849 dpa. Loop A migrated over a distance of 295 nm before it

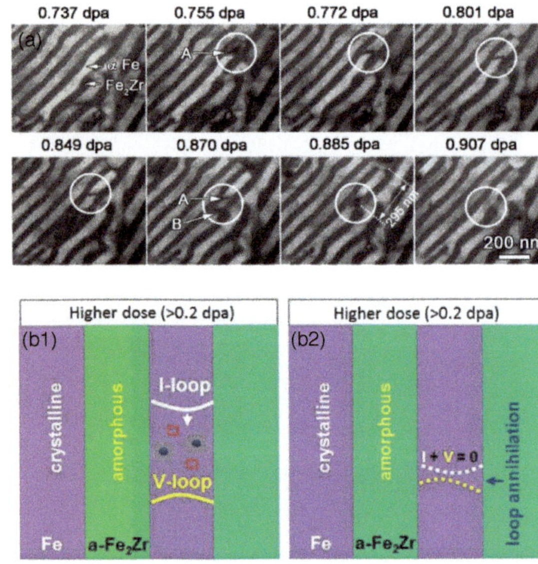

Figure 4. Annihilation of defect clusters via mobile dislocation loops in confined layers. (a) *In situ* TEM snapshots of dislocation loop migration over a dose range of 0.737–0.907 dpa (over 70 s) in confined nanolaminates. At 0.755 dpa, a dislocation loop A was generated and started to migrate downwards within the α-Fe layer, as outlined by the circle. Loop A migrated over a distance of 295 nm by 0.885 dpa before it encountered loop B. The two loops then interacted and combined with each other at 0.885 dpa and eventually disappeared at 0.907 dpa. More details are shown in Supplementary Videos 3 and 4. (b) Schematics of the defect removal mechanism illustrating the migration of dislocation loops in α-Fe channels and the recombination of dislocation loops with opposite nature. Notice that α-Fe layers are sandwiched between amorphous Fe_2Zr layers.

encountered loop B at 0.885 dpa. Subsequently, the two loops interacted and annihilated each other at 0.907 dpa.

The mechanism of such defect annihilation is shown schematically in Figure 4(b). First, defect clusters in Fe layers can evolve into a larger dislocation loop confined by crystal/amorphous layer interfaces. Mobile dislocation loops can migrate over hundreds of nanometres in long narrow α-Fe layers (channels), wherein point defects of the same (opposite) nature are collected (removed) (Figure 4(b1)). Meanwhile such mobile interstitial loops may eventually encounter and annihilate vacancy loops in the channel (Figure 4(b2)). Thus, these mobile interstitial loops could act as the 'sweeper' in the Fe layer to wipe out most other immobile defect clusters before they are annihilated by loops with the opposite Burgers vector. In contrast in large α-Fe grains, the dimension and density of defect clusters increase monotonically with dose as there is insufficient defect sinks to absorb these defects. In the cellular zones, the defects are bounded by limited layer interfaces with insufficient capability (or low probability) to nucleate mobile dislocation loops.

The role of amorphous Fe_2Zr nanolayers is critical. First, the amorphous layers have abundant excess free volume, which can accommodate a large number of irradiation-induced point defects in α-Fe. Second, such amorphous layers are not transparent to mobile dislocation loops. The mobile dislocation loops generated in confined α-Fe channels can thus propagate frequently to collect and remove other types of irradiation-induced defect clusters. Third, the crystal/amorphous interface is resistant to irradiation-induced intermixing up to at least 2 dpa, even though the adjacent layers include identical elements (Fe). Fourth, the amorphous Fe_2Zr interface curtails the growth of defect clusters in irradiated Fe layers.

In summary, rapidly solidified bulk $Fe_{96}Zr_4$ nanocomposite consists of Fe/Fe_2Zr nanolaminates, cellular zones and large Fe grains. Irradiated nanolaminates contained much lower defect density and smaller defect size than those in large Fe grains. Low-dose irradiation-induced amorphization of Fe_2Zr and the crystal/amorphous layer interfaces absorbed irradiation-induced defects and led to confinement of defect clusters. *In situ* video evidenced active defect annihilation due to the sweeping effect of mobile dislocation loops confined in Fe nanolayers. This study provides an encouraging outlook on the implementation of crystal/amorphous interfaces in the design of bulk irradiation-tolerant nanocomposites for extreme irradiation environments.

Acknowledgements We acknowledge financial support by NSF-DMR-Metallic Materials and Nanostructures Program under grant no. 1304101. ZF and YL are supported financially by NSF-CMMI 1161978. K.Y. Yu acknowledges financial support by Science Foundation of China University of Petroleum, Beijing (no. 2462014 YJRC019). XZ and HW also acknowledge Strategic research fund for new research initiatives provided by College of Engineering, Texas A&M University. We also thank Peter M. Baldo and Edward A. Ryan at Argonne National Laboratory for their help during *in situ* irradiation experiments. The IVEM facility at Argonne National Laboratory is supported by DOE-BES. Access to the DOE—Center for Integrated Nanotechnologies (CINT) at Los Alamos and Sandia National Laboratories and Microscopy and Imaging Center at Texas A&M University is also acknowledged.

References

[1] Grimes RW, Konings RJ, Edwards L. Greater tolerance for nuclear materials. Nat Mater. 2008;7:683–685.
[2] Zinkle SJ, Busby JT. Structural materials for fission and fusion energy. Mater Today. 2009;12:12–19.
[3] Zinkle SJ, Was GS. Materials challenges in nuclear energy. Acta Materialia. 2013;61:735–758.
[4] Sickafus KE, Grimes RW, Valdez JA, Cleave A, Tang M, Ishimaru M, Corish SM, Stanek CR, Uberuaga BP. Radiation-induced amorphization resistance and radiation tolerance in structurally related oxides. Nat Mater. 2007;6:217–223.
[5] Odette GR, Alinger MJ, Wirth BD. Recent Developments in Irradiation-Resistant Steels. Annu Rev Mater Res. 2008;38:471–503.
[6] Misra A, Verdier M, Lu YC, Kung H, Mitchell TE, Nastasi M, Embury JD. Structure and mechanical properties of Cu-X (X = Nb,Cr,Ni) nanolayered composites. Scripta Mater. 1998;39:555–560.
[7] Anderson P, Li C. Hall-Petch relations for multilayered materials. Nanostruct Mater. 1995;5:349–362.
[8] Hoagland R, Mitchell T, Hirth J, Kung H. On the strengthening effects of interfaces in multilayer fee metallic composites. Philos Mag A. 2002;82:643–664.
[9] Zhang X, Misra A, Wang H, Shen T, Nastasi M, Mitchell T, Hirth J, Hoagland R, Embury J. Enhanced hardening in Cu/330 stainless steel multilayers by nanoscale twinning. Acta Mater. 2004;52:995–1002.
[10] Zhang JY, Liu Y, Chen J, Chen Y, Liu G, Zhang X, Sun J. Mechanical properties of crystalline Cu/Zr and crystal–amorphous Cu/Cu–Zr multilayers. Mater Sci Eng A. 2012;552:392–398.
[11] Zhang X, Li N, Anderoglu O, Wang H, Swadener JG, Höchbauer T, Misra A, Hoagland RG. Nanostructured Cu/Nb multilayers subjected to helium ion-irradiation. Nucl. Instrum Methods Phys. Res. B, Beam Interact. Mater. At. 2007;261:1129–1132.
[12] Zhang X, Fu E, Li N, Misra A, Wang Y-Q, Shao L, Wang H. Design of Radiation Tolerant Nanostructured Metallic Multilayers. J Eng Mater Technol. 2012;134:041010.
[13] Misra A, Demkowicz M, Zhang X, Hoagland R. The radiation damage tolerance of ultra-high strength nanolayered composites. JOM J Miner Metal Mater Soc. 2007;59: 62–65.
[14] Fu EG, Misra A, Wang H, Shao L, Zhang X. Interface enabled defects reduction in helium ion irradiated Cu/V nanolayers. J Nucl Mater. 2010;407:178–188.
[15] Fu EG, Wang H, Carter J, Shao L, Wang YQ, Zhang X. Fluence-dependent radiation damage in helium (He) ion-irradiated Cu/V multilayers. Philos Mag. 2012;93: 883–898.
[16] Fu EG, Carter J, Swadener G, Misra A, Shao L, Wang H, Zhang X. Size dependent enhancement of helium ion irradiation tolerance in sputtered Cu/V nanolaminates. J Nucl Mater. 2009;385:629–632.
[17] Yuan W. Helium effect on the stability of the interface of Cu/W nanomultilayer. Acta Phys Sin. 2012;61: Article no. 176802. Available from http://wulixb.iphy.ac.cn/EN/abstract/abstract50594.shtml
[18] Li N, Carter JJ, Misra A, Shao L, Wang H, Zhang X. The influence of interfaces on the formation of bubbles in He-ion-irradiated Cu/Mo nanolayers. Philos Mag Lett. 2010;91:18–28.
[19] Yu KY, Liu Y, Fu EG, Wang YQ, Myers MT, Wang H, Shao L, Zhang X. Comparisons of radiation damage in He ion and proton irradiated immiscible Ag/Ni nanolayers. J Nucl Mater. 2013;440:310–318.
[20] Wei QM, Li N, Mara N, Nastasi M, Misra A. Suppression of irradiation hardening in nanoscale V/Ag multilayers. Acta Mater. 2011;59:6331–6340.
[21] Li N, Martin MS, Anderoglu O, Misra A, Shao L, Wang H, Zhang X. He ion irradiation damage in Al/Nb multilayers. J Appl Phys. 2009;105:123522–123528.
[22] Li N, Fu EG, Wang H, Carter JJ, Shao L, Maloy SA, Misra A, Zhang X. He ion irradiation damage in Fe/W nanolayer films. J Nucl Mater. 2009;389:233–238.

[23] Demkowicz MJ, Bellon P, Wirth BD. Atomic-scale design of radiation-tolerant nanocomposites. MRS Bulletin. 2010;35:992–998.
[24] Zhang X, Fu EG, Misra A, Demkowicz MJ. Interface-enabled defect reduction in He ion irradiated metallic multilayers. JOM. 2010;62:75–78.
[25] Yu KY, Sun C, Chen Y, Liu Y, Wang H, Kirk MA, Li M, Zhang X. Superior tolerance of Ag/Ni multilayers against Kr ion irradiation: an in situ study. Philos Mag. 2013;93:3547–3562.
[26] Xu W, Zhang Y, Cheng G, Jian W, Millett PC, Koch CC, Mathaudhu SN, Zhu Y. In-situ atomic-scale observation of irradiation-induced void formation. Nat Commun. 2013;4:2288.
[27] Kaoumi D, Motta AT, Birtcher RC. A thermal spike model of grain growth under irradiation. J Appl Phys. 2008;104:073525.
[28] Sun C, Song M, Yu K, Chen Y, Kirk M, Li M, Wang H, Zhang X. In situ Evidence of Defect Cluster Absorption by Grain Boundaries in Kr Ion Irradiated Nanocrystalline Ni. Metall Mater Trans A. 2013;43: 1966–1974.
[29] Yu KY, Bufford D, Sun C, Liu Y, Wang H, Kirk MA, Li M, Zhang X. Removal of stacking-fault tetrahedra by twin boundaries in nanotwinned metals. Nat Commun. 2013;4:1377.
[30] Yu KY, Bufford D, Khatkhatay F, Wang H, Kirk MA, Zhang X. In situ studies of irradiation-induced twin boundary migration in nanotwinned Ag. Scripta Materialia. 2013;69:385–388.
[31] Sun C, Bufford D, Chen Y, Kirk M, Wang Y, Li M, Wang H, Maloy S, Zhang X. In situ study of defect migration kinetics in nanoporous Ag with enhanced radiation tolerance. Sci Rep 3737. 2014;4:Article no. 3737.
[32] Fu EG, Caro M, Zepeda-Ruiz LA, Wang YQ, Baldwin K, Bringa E, Nastasi M, Caro A. Surface effects on the radiation response of nanoporous Au foams. Appl Phys Lett. 2012;101:191607.
[33] Bringa EM, Monk JD, Caro A, Misra A, Zepeda-Ruiz L, Duchaineau M, Abraham F, Nastasi M, Picraux ST, Wang YQ, Farkas D. Are Nanoporous Materials Radiation Resistant? Nano Lett. 2011;12:3351–3355.
[34] Kiener D, Hosemann P, Maloy S, Minor A. In situ nanocompression testing of irradiated copper. Nat mater. 2011;10:608–613.
[35] Landau P, Guo Q, Hattar K, Greer JR. The Effect of He Implantation on the Tensile Properties and Microstructure of Cu/Fe Nano-Bicrystals. Adv Graph Mag. 2013;23:1281–1288.
[36] Han W, Demkowicz MJ, Mara NA, Fu E, Sinha S, Rollett AD, Wang Y, Carpenter JS, Beyerlein IJ, Misra A. Design of Radiation Tolerant Materials Via Interface Engineering. Adva Mater. 2013;25:6975–6979.
[37] Stein F, Sauthoff G, Palm M. Experimental determination of intermetallic phases, phase equilibria, and invariant reaction temperatures in the Fe-Zr system. J Phase Equilib Diffus. 2002;23:480–494.
[38] Liu Y, Livingston JD, Allen SM. Room-temperature deformation and stress-induced phase. Metall Mater Trans A. 1992;23:3303–3308.

Strain-Induced Reverse Phase Transformation in Nanocrystalline Co-Ni Alloys

J.Y. Li[a,b], W. Li[b], M.J. Jin[b] and X.J. Jin[a,b]*

[a] *State Key Lab of Metal Matrix Composite, School of Materials Science and Engineering, Shanghai Jiao Tong University, Shanghai 200240, People's Republic of China;* [b] *Institute of Phase Transformation and Complex Microstructure, School of Materials Science and Engineering, Shanghai Jiao Tong University, Shanghai 200240, People's Republic of China*

A strain-induced reverse phase transformation from a hexagonal close-packed phase to a face-centered cubic phase was observed under tension at room temperature in electrodeposited nanocrystalline Co-Ni alloys with an average grain size of 15 nm. Such reverse phase transformation was previously observed only on heating. Investigation by transmission electron microscopy suggests a close relationship between the reverse phase transformation and grain growth during the tensile deformation.

Keywords: Nanocrystalline Materials, Strain-Induced Transformation, Grain Growth, Co-Ni Alloy

1. Introduction It is generally acknowledged that martensitic transformations can be triggered above the martensite start temperature (M_s) by plastic deformation.[1] The strain-induced martensitic transformation is one of the most important characteristics of shape memory alloys (SMAs), which has been widely investigated in Ni-Ti-based, Fe-based and other typical SMAs.[2] However, there are not many reports about strain-induced reverse transformation, which is an interesting topic.

Meanwhile, when the grain size is reduced to nano scale, nanocrystalline metals and alloys exhibit much different behaviors in comparison to their coarse-grained counterparts.[3–5] Phase transformations have been widely investigated in nanocrystalline Co processed by ball milling,[6,7] surface mechanical attrition treatment [8,9] and high-pressure torsion.[10,11]

Size dependence of the reverse transformation from hcp to fcc phase is still less understood. Binary Co-Ni alloys are selected as the experimental model material in this work for the reason that they exhibit a reversible fcc to hcp phase transformation when Ni content is less than 35at% Ni, as shown in the phase diagram (Figure 1).[12–15] The phase transformation temperature decreases with increasing Ni content.

Interaction between the reverse transformation and grain growth upon heating in nanocrystalline Co-Ni alloys was taken into consideration.[16,17] Since strain-induced grain growth has been extensively reported in nanocrystalline materials under indentation,[18,19] high-pressure torsion,[20,21] compression [22] and tensile test,[23,24] it is interesting to study the microstructure evolution accompanying a plastic deformation process.

In the present study, an interesting phenomenon of strain-induced phase transformation from hcp to fcc in nanocrystalline Co-31.0at%Ni alloy was observed. Pulse electrodeposition was employed to prepare nanocrystalline Co-Ni alloys with fewer defects and impurities. Phase change and grain growth were studied on Co-(20.6 ± 0.1at%)Ni, Co-(31.0 ± 0.1at%)Ni and Co-(48.0 ± 0.5at%)Ni under tension at room temperature. The close relationship between reverse transformation from hcp to fcc and grain growth during tensile tests was discussed and it suggested a possible mechanism for the strain-induced reverse transformation.

2. Experimental Nanocrystalline Co-Ni alloys were prepared by pulse electrodeposition. A saccharin-containing electrolyte with contents listed in Table 1

*Corresponding author. Email: jin@sjtu.edu.cn

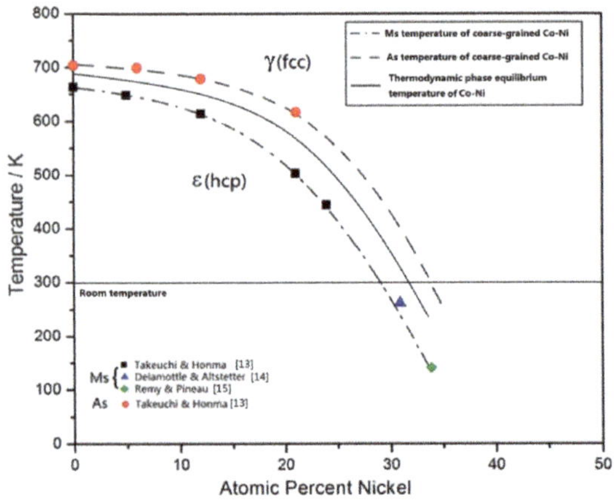

Figure 1. Phase diagram of Co-Ni.[12]

Table 1. Composition of electroplating baths used in the present study.

	$(mol \cdot l^{-1})$
$NiSO_4 \cdot 6H_2O$	0.87, 0.98, 1.09
$CoSO_4 \cdot 7H_2O$	0.30, 0.20, 0.10
Boric acid	0.50
Saccharin	$2.5(g \cdot l^{-1})$
Sodium dodecyl sulfate	$0.5(g \cdot l^{-1})$

Table 2. Pulse plating conditions leading to acceptable nanocrystalline Co-Ni deposits from saccharin-containing electrolyte.

Parameter	Value
On-time (ms)	30
Off-time (ms)	270
Peak current density $(mA \cdot cm^{-2})$	200
Temperature (°C)	40
pH	4.0

was used to produce electrodeposits of 20 μm thickness. Analytical grade chemicals and deionized water were used to prepare the solutions. The solution temperature was maintained at 40 ± 1°C by immersing the plating cell in a large volume thermostatted water bath. The cathode substrate was made of 304 stainless steel sheet with a size of 30 mm × 40 mm × 0.5 mm after a passivation treatment by immersion in nitric acid. A graphite sheet with surface area much larger than that of the cathode was used as the anode. Pulsed electrodeposition was carried out galvanostatically using cathodic square wave pulses with complete current cut-off during the interval between the pulses. The pulse plating parameters are given in Table 2.

Nickel contents in the Co-Ni samples were detected by energy dispersive X-ray analysis (EDX) on a scanning electron microscopy (SEM, FEI Sirion 2000 field emission microscope) and by using inductively coupled plasma optical emission spectrometer (ICP, Thermo iCAP6300). The structure of samples was determined by using X-ray diffraction (XRD, Rigaku D/max-2550/PC, Cu K_α) and the microstructures were investigated by transmission electron microscopy (TEM, JEOL-2100F). Tensile tests were performed using universal material experiment machine (Zwick/Roell Z020), where the strain was measured by using a knife-edge extensometer (makroXtens). Heat evolution was measured by differential scanning calorimetry (DSC, Perkin Elmer Diamond).

3. Results

3.1. Phases and Microstructure of as-deposited Co-Ni Alloys. Figure 2 shows the TEM images and XRD patterns of as-deposited samples with three typical compositions of Co-20.6at%Ni, Co-31.0at%Ni and Co-48.0at%Ni, respectively. It is clear from the bright-field and dark-field micrographs that these electrodeposits have a nanocrystalline structure with an average grain size of 15 nm and negligible porosity as confirmed by density measurements.[25] The selected-area electron diffraction (SAED) patterns indicate that the as-deposited Co-20.6at%Ni and Co-31.0at%Ni are mainly in the hcp phase with the diffraction rings from center to outwards corresponding to $\varepsilon(100)$, $\varepsilon(002)$, $\varepsilon(101)$, $\varepsilon(102)$, $\varepsilon(110)$ and $\varepsilon(103)$, respectively. On the contrary, Co-48.0at%Ni contains the fcc phase with the diffraction rings corresponding to $\gamma(111)$, $\gamma(200)$, $\gamma(220)$, $\gamma(311)$ and $\gamma(222)$. The results are also confirmed by the XRD analysis.

3.2. Tensile Tests at Room Temperature for Free-standing Films and Alloys with Substrates. Free-standing 20-μm-thick deposits can be mechanically stripped off the 304 stainless steel substrates and processed into tensile specimens as shown in Figure 3 by wire-electrode cutting. But mechanical testing of the free-standing films is difficult to perform due to the shear stress. The samples break quickly and the engineering strain is less than 0.4%. However, the soft substrates can stabilize the specimens and avoid strain localization during tensile tests.[26–28] As the samples were tested together with the 304 stainless steel substrates, ductility is not only a function of the film microstructure, but is also dependent on the overall adhesion of the substrate. Figure 3 shows the typical tensile engineering stress–strain curve for the free-standing sample (curve a) in comparison with that for the sample together with

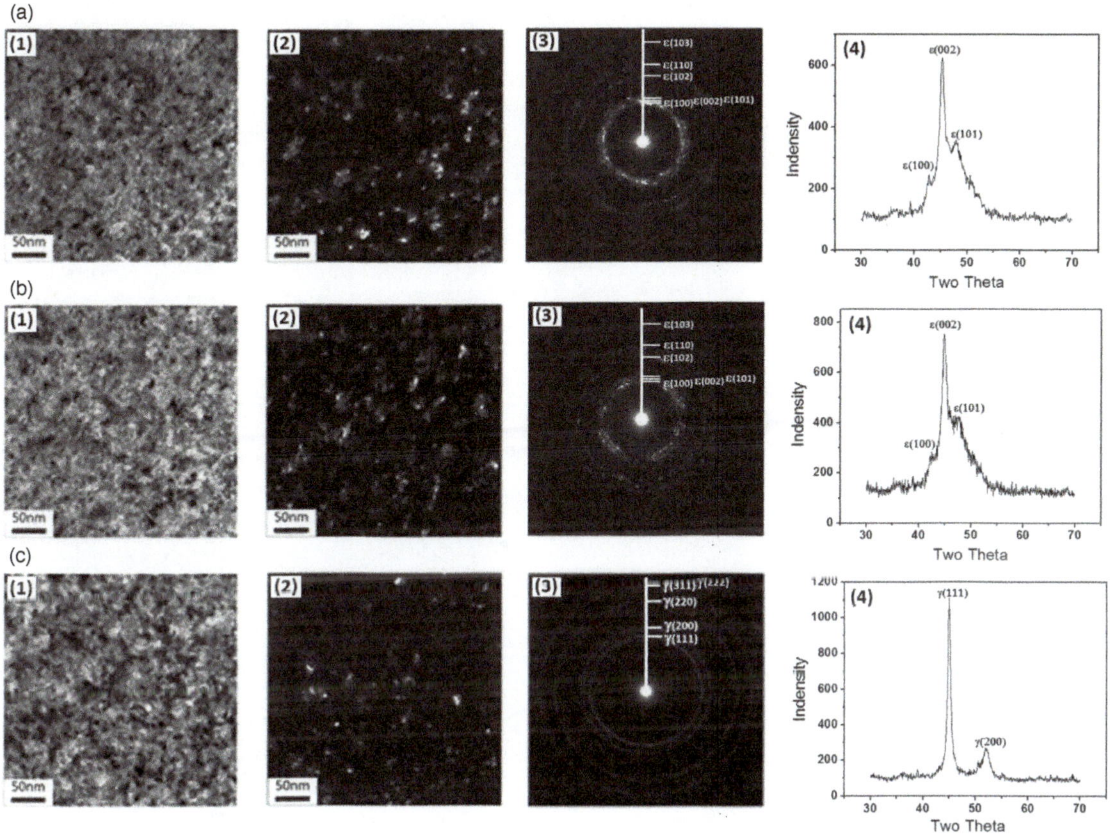

Figure 2. Initial structure (1.bright-field, 2.dark-field, 3.SAED, 4.XRD) of as-deposited samples with different nickel contents: (a) Co-20.6at%Ni; (b) Co-31.0at%Ni and (c) Co-48.0at%Ni.

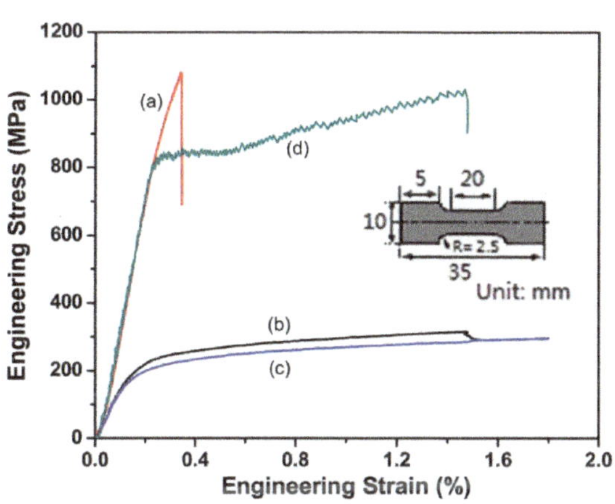

Figure 3. Typical tensile engineering stress–strain curves under a tensile rate of 0.5 mm/min: (a) free-standing sample; (b) sample together with substrate; (c) 304 stainless steel substrate and (d) curve of deposit layer estimated by ROM. The inset shows the geometry of tensile samples (R, radius of curvature).

the substrate (curve b) and the 304 stainless steel substrate (curve c) under a tensile rate of 0.5 mm/min. There is no synergetic and destructive interaction between the deposits and the substrate, so that the stress in the deposit layer (curve d) can be estimated by the rule of mixture (ROM) [29]:

$$\sigma_{\text{deposit}} = \frac{\sigma_{\text{total}} - \sigma_{\text{substrate}}(1 - V_{\text{deposit}})}{V_{\text{deposit}}}, \qquad (1)$$

where σ_{deposit} is the calculated stress of the deposit, σ_{total} is the measured stress of the deposit together with substrate, $\sigma_{\text{substrate}}$ is the measured stress of the substrate without deposit and V_{deposit} is the volume fraction of the deposit layer. As the 20-μm thick Co-Ni alloys deposited on a 0.5-mm thick 304 stainless steel substrate, the value of V_{deposit} is about 0.04. The elastic modulus of the deposit layer estimated by ROM is well in agreement with the measured value of the free-standing sample. Results show that the fracture strain of the sample with substrate is more than four times larger than that of the free-standing sample, while the fracture stress of the sample with substrate is smaller than that of the free-standing sample. Samples after tensile tests are collected for the next morphology observation. Disks of 3 mm diameter were punched for TEM observation.

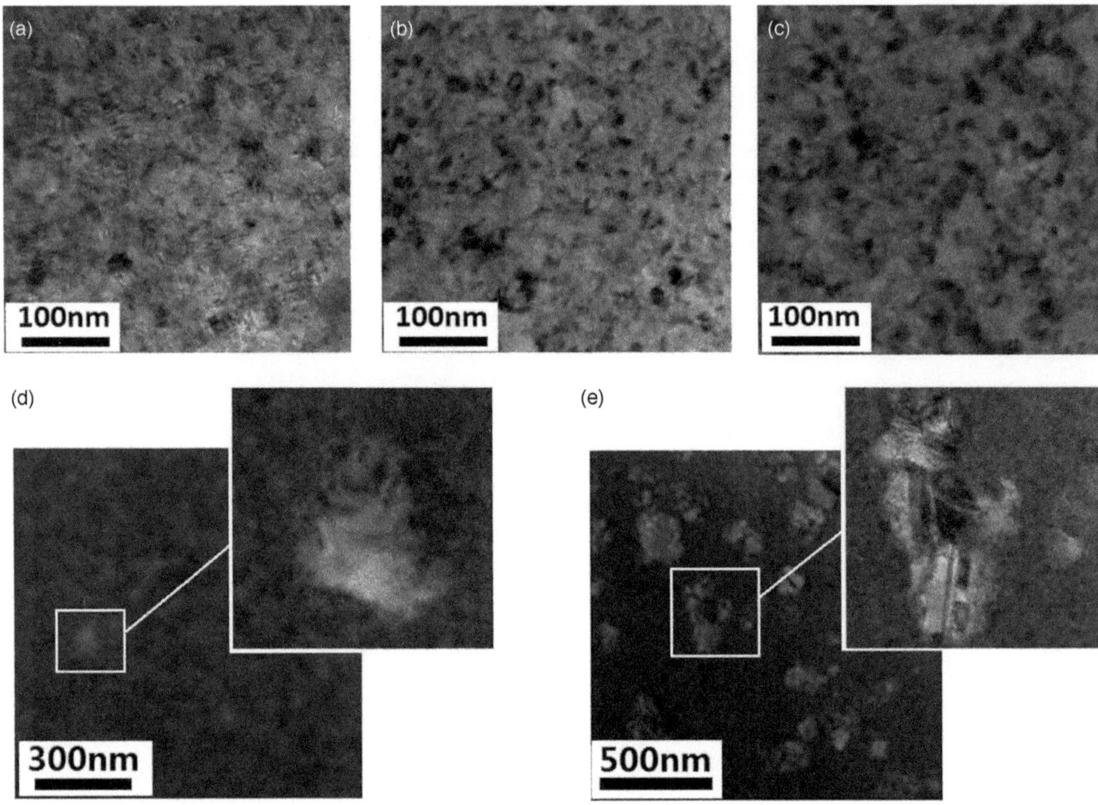

Figure 4. Bight-field TEM images of Co-31.0at%Ni after different tensile tests: (a) $\varepsilon = 0.35\%$ (free-standing sample); (b) $\varepsilon = 0.6\%$; (c) $\varepsilon = 1.0\%$; (d) $\varepsilon = 1.4\%$ and (e) $\varepsilon = 1.5\%$.

3.3. Microstructure of Co-Ni Alloys after Tensile Tests. Figure 4 shows the bright-field TEM images of Co-31.0at%Ni after different tensile tests. Five samples after tensile tests are collected for the morphology observation: (a) free-standing sample after tensile failure up to $\varepsilon = 0.35\%$; (b), (c) and (d) samples with substrates after tensile tests stop when $\varepsilon = 0.6\%$, $\varepsilon = 1.0\%$ and $\varepsilon = 1.4\%$; (e) sample with substrate after tensile failure up to $\varepsilon = 1.5\%$. The nanocrystalline structure with hcp phase remains unchanged in samples (a), (b) and (c). Just tiny amounts of grains began to grow when $\varepsilon = 1.4\%$ (Figure 4(d)), and more grains began to grow after tensile failure up to $\varepsilon = 1.5\%$ (Figure 4(e)). As shown in Figure 5, the background nano grains remain the hcp phase (Figure 5(c)), while an fcc nano-twinned structure is formed instead of the initial hcp phase in the grown grains (Figure 5(d)). Our previous studies [16] have reported that the nano-twinned fcc structure has an extraordinary stability within the temperature range from 4.2 to 1,000 K, and that the fcc phase will not transform back to the hcp phase. The formation of the fcc phase is also confirmed by XRD pattern (Figure 5(e)). The microstructures of all Co-20.6at%Ni and Co-48.0at%Ni samples remain unchanged after tensile tests, and are similar to the images shown in Figure 1, hence are not shown here.

4. Discussions It is well known that the reversible fcc to hcp transformation exhibits a negligible volume change.[30] The high degree of structural similarity between the two phases implies small entropy and enthalpy changes. Co-Ni alloys have low stacking fault energy with an approximate value of 31 and 18.5 kJ/m^2 for the hcp and fcc structures in pure Co, respectively.[31] Thus, the conversion between the two structures easily occurs via stacking fault formation.[32] These characteristics result in easier phase transformation in Co-Ni alloys [33] and provide the possibility of strain-induced phase transformation from hcp to fcc.

As shown in Figure 1, the reverse martensitic transformation from hcp to fcc in coarse-grained Co-31.0at%Ni alloy takes place at about 420 K. However, it happens at room temperature together with grain growth in electrodeposited nanocrystalline Co-31.0at%Ni sample under tension. The unusual strain-induced reverse transformation shows a close relationship with the grain growth.

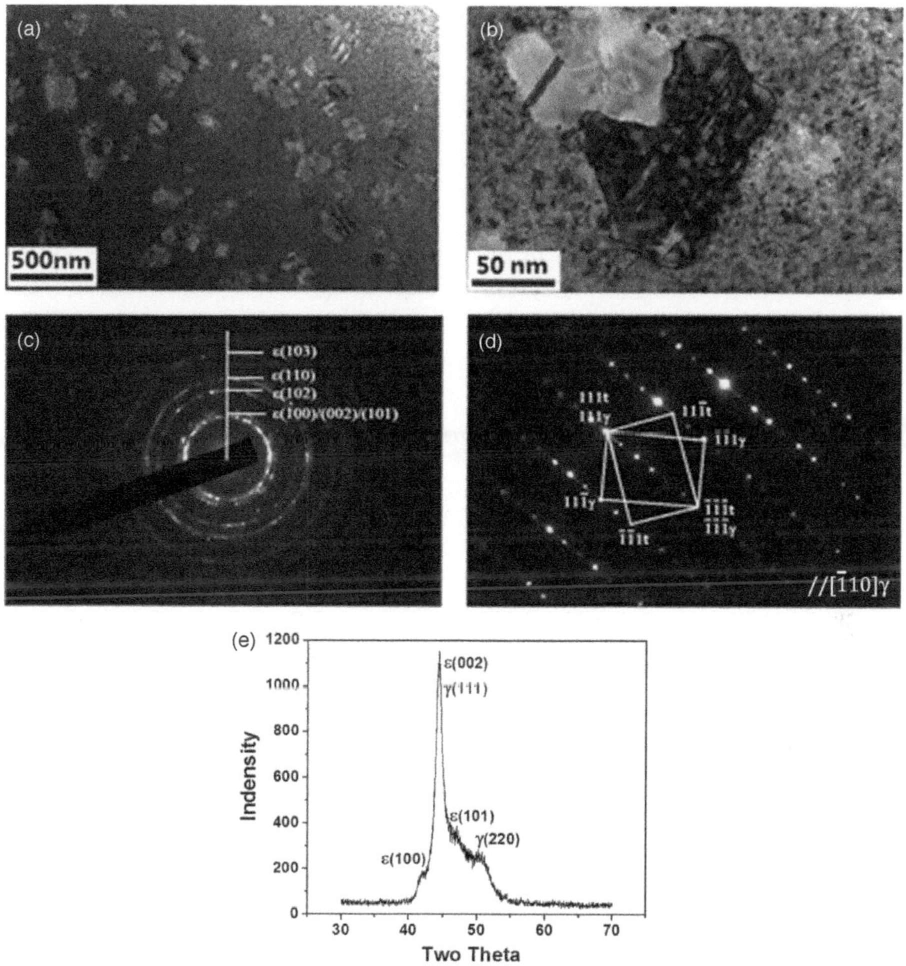

Figure 5. Structure of Co-31.0at%Ni samples after tensile failure up to $\varepsilon = 1.5\%$: (a) overall bright-field image; (b) bright-field image of the grown grains; (c) SAED pattern corresponding to the background; (d) SAED pattern corresponding to the grains grown up, which is measured by nano-beam diffraction (NBD) model and (e) X-ray diffraction pattern.

The high interfacial volume fraction in the nanocrystalline materials leads to significant driving forces for coarsening,[34] and there is a huge tendency for grain growth, especially at elevated temperatures. The strain energy provides a kind of driving force that the grain growth could occur during the deformation process. Strain-induced grain growth has been extensively studied in nanocrystalline materials and it is reported that there is a critical value of stress above which deformation-induced grain growth would occur.[21,22] However in the present study, grain growth during tensile test occurred only in the Co-31.0at%Ni with ϵ higher than 1.4%. Based on the engineering stress–strain curves shown in Figure 2, it is suggested that the grain growth was induced by sufficient strain rather than stress.

On the one hand, exothermic heat resulting from grain growth can promote reverse transformation. As shown in Figure 1, the thermodynamic phase equilibrium temperature (T_0) of the hcp and fcc phases for Co-31.0at%Ni is higher than room temperature, at which the free energies of the two phases are equal. It means that the phase transformation from hcp to fcc should not take place at room temperature unless there are other energies which can provide an additional driving force, that is, increasing the instant temperature for the sample. The resources of the extra energies may be provided by the thermal effect of the deformation process and grain growth.

We can estimate the temperature rise in the deformation process using the following equation [35]:

$$\Delta T = \frac{\beta}{\rho C_V} \int_{\varepsilon_1}^{\varepsilon_2} \sigma \, d\varepsilon, \qquad (2)$$

where $\beta = 0.9$ (assuming 90% of work of deformation was converted to heat), ρ is the density of the sample and C_V is the specific heat capacity. The specific heat capacity of Co-31at%Ni and 304 stainless steel at room temperature is about 0.44 J/(g·K) [36] and 0.50 J/(g·K),[37] respectively. As $\varepsilon = 1.4\%$, ΔT of the 304 stainless steel substrate is about 1 K and ΔT of Co-31at%Ni deposit layer is about 4 K. It means that the heat generated from the deformation process can be ignored.

Also, we can estimate the temperature rise in the grain growth process by measuring the heat evolution using DSC. The released heat can be estimated by calculating the area of exothermic peak. When the grains with an initial size of 15 nm grow up to about 300 nm,[16,17] the released heat is about 31 J/g. As the grain has grown up to about 150 nm after tension (Figure 5(b)), we can use 31 J/g as the value of released heat for estimation. Whereas the specific heat of Co-31at%Ni alloy is about 0.44 J/(g·K), the heat evolution may increase the temperature by about 70°, so that the temperature can reach T_0 and the reverse transformation can occur in Co-31.0at%Ni. This estimate is based on two assumptions. First, almost all the grains have grown up during heating, which is confirmed in our previous studies.[16,17] Second, the velocity of the phase transformation is quite rapid that the released heat of the localized grain growth has no time for dissipation, and then, when the extra heat diffuses to the whole specimen including substrate, there is no obvious temperature rise.

Meanwhile, it can be reasonably deduced that the grain boundary movement should occur more easily because of the high mobility of atoms during the phase transformation.[38] Omori et al. [39] reported an abnormal grain growth induced by cyclic heat treatment between a high-temperature single-phase region and a lower temperature two-phase region, and pointed out that the abnormal grain growth phenomenon results from the formation of a sub-grain structure introduced through phase transformation. As a consequence, we assume that the grain growth process in nanocrystalline Co-Ni alloys could be enhanced by phase transformation. Since the temperature for phase transformation from hcp to fcc in Co-20.6at%Ni is much higher than room temperature and the initial crystal structure of the as-deposited Co-48.0at%Ni is fcc, no phase transformation will happen in these samples. As a result, grain growth has not been observed in Co-20.6at%Ni and Co-48.0at%Ni after tensile tests. The compositional dependent behaviors imply that the grain growth and the phase transformation from hcp to fcc at an appropriate transformation temperature can take place in the samples during tensile tests with a value of engineering strain about 1.4%. A much larger strain may be necessary to induce grain growth in samples without a phase transformation.

From what has been discussed in this paper, it is a mutual promotion process that the phase transformation from hcp to fcc always takes place coupled with grain growth in nanocrystalline Co-Ni alloys. Our previous studies [16] have reported that the symbiotic transformation happens at about 600 K upon heating in electrodeposited nanocrystalline Co-Ni alloys, which is little relevant with the composition. However in the present study, we observe such symbiotic process during tensile tests performed at room temperature in electrodeposited nanocrystalline Co-31.0at%Ni. It is suggested that the symbiotic transformation can be triggered by a sufficiently high temperature or a sufficient applied strain. We can forecast that the strain-induced reverse transformation can be observed in nanocrystalline Co-Ni alloys with other components after tensile tests performed near the A_s temperature. More experiments of nanocrystalline Co-Ni alloys with different components at different temperatures would be carried out for verification.

5. Conclusion In summary, strain-induced reverse phase transformation from hcp to fcc in nanocrystalline Co-31.0at%Ni alloy was observed. Close relationship and conditions between the reverse phase transformation and grain growth during tensile test were confirmed based on the detail observations on Co-20.6at%Ni, Co-31.0at%Ni and Co-48.0at%Ni. It is suggested that the mutual promotion process of the reverse phase transformation coupled with grain growth can be triggered by a sufficiently high temperature or a sufficient applied strain.

Acknowledgement This work is supported by the National Natural Science Foundation of China (No. 51174251, No. 51201100 and No. 51201105) and 973 Program (No. 2011CB706604). Authors thank Dr. K. Zhang, University of Shanghai for Science and Technology, for assistance with the TEM observation.

Disclosure Statement No potential conflict of interest was reported by the authors.

References

[1] Pineau A, Pelloux R. Influence of strain-induced martensitic transformations on fatigue crack growth rates in stainless steels. Metall Trans. 1974;5(5):1103–1112.
[2] Otsuka K, Wayman CM. Shape memory materials. Cambridge: Cambridge University Press; 1998.
[3] Gleiter H. Nanocrystalline materials. Prog Mater Sci. 1989;33(4):223–315.
[4] Suryanarayana C. Nanocrystalline materials. Int Mater Rev. 1995;40(2):41–64.
[5] Lu K. Nanocrystalline metals crystallized from amorphous solids: nanocrystallization, structure, and properties. Mater Sci Eng R. 1996;16(4):161–221.
[6] Huang JY, Wu YK, Ye HQ, Lu K. Allotropic transformation of cobalt induced by ball milling. Nanostruct Mater. 1995;6(5):723–726.
[7] Sort J, Nogués J, Suriñach S, Baró MD. Microstructural aspects of the hcp-fcc allotropic phase transformation induced in cobalt by ball milling. Philos Mag. 2003;83(4):439–455.
[8] Wu X, Tao N, Hong Y, Liu G, Xu B, Lu J, Lu K. Strain-induced grain refinement of cobalt during surface mechanical attrition treatment. Acta Mater. 2005;53(3):681–691.

[9] Wu X, Tao N, Hong Y, Lu J, Lu K. $\gamma \rightarrow \varepsilon$ martensite transformation and twinning deformation in fcc cobalt during surface mechanical attrition treatment. Scr Mater. 2005;52(7):547–551.
[10] Edalati K, Toh S, Arita M, Watanabe M, Horita Z. High-pressure torsion of pure cobalt: hcp-fcc phase transformations and twinning during severe plastic deformation. Appl Phys Lett. 2013;102(18):181902-1–181902-4.
[11] Sort J, Zhilyaev A, Zielinska M, Nogués J, Surinach S, Thibault J, Baró MD. Microstructural effects and large microhardness in cobalt processed by high pressure torsion consolidation of ball milled powders. Acta Mater. 2003;51(20):6385–6393.
[12] Nishizawa T, Ishida K. The Co-Ni (Cobalt-Nickel) system. Bull Alloy Phase Diagr. 1983;4(4):390–395.
[13] Takeuchi S, Honma T. Studies on the $\beta \rightarrow \varepsilon$ transformation in Cobalt-Nickel alloys. I: Propagation process of the transformation. Sci Rep Res Inst Tohoku Univ. 1957;A9:492–507.
[14] Postnikov VS, Belko VN, Sharshakov IM. Temperature-dependence of internal friction in Cobalt-Nickel alloys. Fiz Met Metalloved. 1968;26(6):1051–1056.
[15] Remy L, Pineau A. Twinning and strain-induced fcc → hcp transformation on the mechanical properties of Co-Ni-Cr-Mo alloys. Mater Sci Eng. 1976;26(1):123–132.
[16] Li JY, Ni C, Liu JY, Jin MJ, Li W, Jin XJ. Extraordinary stability of nano-twinned structure formed during phase transformation coupled with grain growth in electrodeposited Co-Ni alloys. Mater Chem Phys. 2014;148(3):1202–1211.
[17] Li J, Liu J, Jin M, Jin X. Grain size dependent phase stability of pulse electrodeposited nano-grained Co–Ni films. J Alloys Compd. 2013;577:S151–S154.
[18] Zhang K, Weertman JR, Eastman JA. Rapid stress-driven grain coarsening in nanocrystalline Cu at ambient and cryogenic temperatures. Appl Phys Lett. 2005;87(6):061921-1–061921-3.
[19] Jin M, Minor AM, Stach EA, Morris Jr JW. Direct observation of deformation-induced grain growth during the nanoindentation of ultrafine-grained Al at room temperature. Acta Mater. 2004;52(18):5381–5387.
[20] Wen H, Zhao Y, Li Y, Ertorer O, Nesterov KM, Islamgaliev RK, Valievb RZ, Lavernia EJ. High-pressure torsion-induced grain growth and detwinning in cryomilled Cu powders. Philos Mag. 2010;90(34): 4541–4550.
[21] Liao XZ, Kilmametov AR, Valiev RZ, Gao H, Li X, Mukherjee AK, Bingert JF, Zhu YT. High-pressure torsion-induced grain growth in electrodeposited nanocrystalline Ni. Appl Phys Lett. 2006;88(2):021909-1–021909-3.
[22] Zhang Y, Sharon JA, Hu GL, Ramesh KT, Hemker KJ. Stress-driven grain growth in ultrafine grained Mg thin film. Scr Mater. 2013;68(6):424–427.
[23] Gianola DS, Van Petegem S, Legros M, Brandstetter S, Van Swygenhoven H, Hemker KJ. Stress-assisted discontinuous grain growth and its effect on the deformation behavior of nanocrystalline aluminum thin films. Acta Mater. 2006;54(8):2253–2263.
[24] Fan GJ, Fu LF, Qiao DC, Choo H, Liaw PK, Browning ND. Grain growth in a bulk nanocrystalline Co alloy during tensile plastic deformation. Scr Mater. 2006;54(12):2137–2141.
[25] El-Sherik AM, Erb U. Synthesis of bulk nanocrystalline nickel by pulsed electrodeposition. J Mater Sci. 1995;30(22):5743–5749.
[26] Kobler A, Lohmiller J, Schäfer J, Kerber M, Castrup A, Kashiwar A, Horst H, Kübel C. Deformation-induced grain growth and twinning in nanocrystalline palladium thin films. Beilstein J Nanotechnol. 2013;4(1):554–566.
[27] Lohmiller J, Woo NC, Spolenak R. Microstructure–property relationship in highly ductile Au–Cu thin films for flexible electronics. Mater Sci Eng A. 2010;527(29):7731–7740.
[28] Hommel M, Kraft O. Deformation behavior of thin copper films on deformable substrates. Acta Mater. 2001;49(19):3935–3947.
[29] Wu XL, Jiang P, Chen L, Zhang JF, Yuan FP, Zhu YT. Synergetic strengthening by gradient structure. Mater Res Letts. 2014;2(4):185–191.
[30] Taylor A. Latice parameters of binary nickel cobalt alloys. J Inst Met. 1950;77:585–594.
[31] Betteridge W. The properties of metallic cobalt. Prog Mater Sci. 1980;24:51–142.
[32] Hsu TY. Thermodynamic properties of $\beta(\gamma) \rightarrow \varepsilon$ martensitic transformation. Acta Metall Sin. 1980;16:430–434.
[33] Liu Y, Yang H, Tan G, Miyazaki S, Jiang B, Liu Y. Stress-induced FCC $\leftrightarrow$ HCP martensitic transformation in CoNi. J Alloys Compd. 2004;368(1):157.
[34] Wang N, Wang Z, Aust KT, Erb U. Isokinetic analysis of nanocrystalline nickel electrodeposits upon annealing. Acta Mater. 1997;45(4):1655–1669.
[35] Mishra A, Kad BK, Gregori F, Meyers MA. Microstructural evolution in copper subjected to severe plastic deformation: experiments and analysis. Acta Mater. 2007;55(1):13–28.
[36] Barin I. Thermochemical data of pure substances. Weinheim: VCH; 1995.
[37] Xavior MA, Adithan M. Determining the influence of cutting fluids on tool wear and surface roughness during turning of AISI 304 austenitic stainless steel. J Mater Process Tech. 2009;209(2):900–909.
[38] Ding X, Liu X. Correlation between anatase-to-rutile transformation and grain growth in nanocrystalline titania powders. J Mater Res. 1998;13(9):2556–2559.
[39] Omori T, Kusama T, Kawata S, Ohnuma I, Sutou Y, Araki Y, Ishida K, Kainuma R. Abnormal grain growth induced by cyclic heat treatment. Science. 2013;341(6153): 1500–1502.

26

Effects of Alloying Elements on Stacking Fault Energies and Electronic Structures of Binary Mg Alloys: A First-Principles Study

William Yi Wang[a,d]*, Shun Li Shang[a], Yi Wang[a], Zhi-Gang Mei[a], Kristopher A. Darling[c], Laszlo J. Kecskes[c], Suveen N. Mathaudhu[b], Xi Dong Hui[d] and Zi-Kui Liu[a]

[a]Department of Materials Science and Engineering, The Pennsylvania State University, University Park, PA 16802, USA; [b]Materials Science Division, U.S. Army Research Office, Research Triangle Park, NC 27709, USA; [c]U.S. Army Research Laboratory, Weapons and Materials Research Directorate, RDRL-WMM-B, Aberdeen Proving Ground, MD 21005, USA; [d]State Key Laboratory for Advanced Metals and Materials, University of Science and Technology Beijing, Beijing 100083, People's Republic of China

The growth, deformation, and extrinsic faults in binary Mg–X alloys are investigated via first-principles calculations. Here, the alloying elements X include Al, Ca, Cu, Fe, K, La, Li, Mn, Na, Nd, Pr, Si, Sn, Sr, Y, Zn, and Zr. In addition to stacking fault energies, the effect of the elements on the bond structure of Mg are studied in term of electron localization morphology. It is observed that rod-like directional bonds in non-fault planes transform into tetrahedral morphologies in fault planes and are strengthened by addition of Zn and Al, but weakened by Na.

Keywords: Magnesium Alloys, Alloying Elements, Stacking Faults, First-Principles Calculations

With a density that is two-thirds of aluminum or one-quarter of steel, Mg alloys are important structural materials for vehicles to improve fuel economy and reduce emissions.[1] The stacking fault energy, γ_{sf}, is an important property dictating deformation in Mg alloys.[2] It is closely related to the core structure and the mobility of the dislocations.[3–7] The reduction of γ_{sf} could enhance the twinnability of the material,[8,9] increase the ductility,[10–12] and reduce the steady-state creep rate.[13,14] The stacking sequence of an ideal HCP metal metal is ...ABABABAB... with A, B, (or C, see below) being different {0001} planes. There are three types of stacking faults commonly studied, i.e. growth fault (I1), deformation fault (I2), and extrinsic fault (EF). An I1 fault is formed by removing an A plane above a B plane, and then shearing the remaining planes above the B plane by $\frac{1}{3}[1\bar{1}00]$, resulting in ...ABA$\dot{B}$CBCB... with the fault layer marked with a dot on top of it (same below). An I2 fault is formed by a shearing of $\frac{1}{3}[1\bar{1}00]$, resulting in ...ABA$\dot{B}\dot{C}$ACA.... [4,15] An EF fault is generated by inserting an extra C plane into the ideal HCP structure, resulting in ...ABA$\dot{B}\dot{C}\dot{A}$BAB....

Recently, electronic structures have been used to gain insights into mechanical properties of materials.[16–19] For example, first-principles calculations based on density functional theory [20] indicate correlations between the elastic properties of Al, its bonding electrons, and the electrostatic potential distributions through the deformation electron density,[16] $\Delta\rho$, defined as the difference between the total electron density (ρ_{total}) and the electron density associated with unbounded atoms (ρ_{IAM}). In a previous work,[21] we found that the number of atomic layers affected by the presence of the stacking fault increases in the order of I1, I2, and EF in accordance with their stacking fault energies. Additionally, quantitative evaluations of charge density reveal that the redistribution range of deformation electron density along the [0001] direction is closely related to stacking fault energy.

Theoretical investigations of stacking fault energies of some binary Mg–X alloys have recently been reported in the literature [22–25] along with contour plots of electron density in 2D and electron density isosurface figures in 3D,[26–30] however, the detailed chemical bond structures of the fault and non-fault layers and the

*Corresponding author. Emails: yuw129@psu.edu, billyi.wang@gmail.com

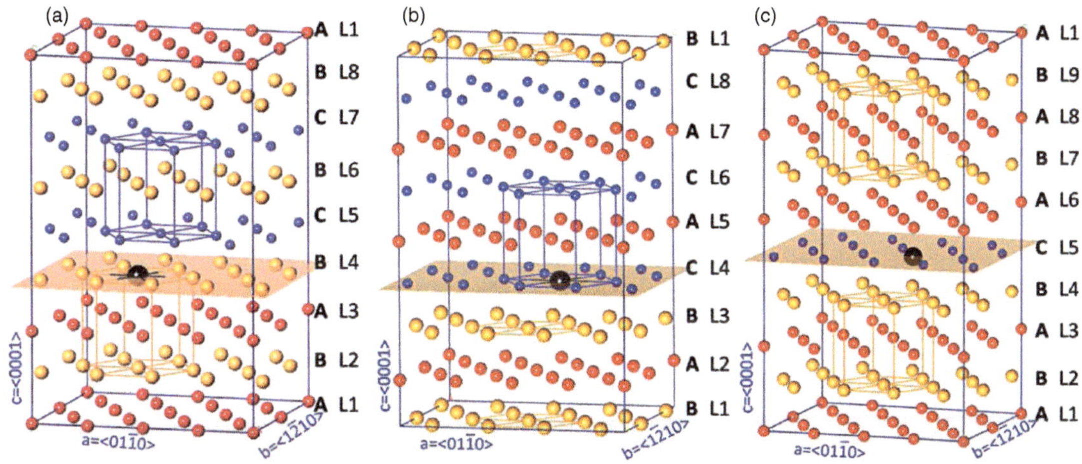

Figure 1. Schematic structure of the Mg–X alloys with different stacking faults (shown as (0001) miller plane), (a) growth fault—I1; (b) deformation fault—I2, and (c) EF. The atoms in different stacking layers are depicted with different color and size, with the alloying element being depicted as the largest size in the fault plane. The concentrations of the alloying element in fault planes are 6.25 at%.

contributions of alloying elements on the electronic structure of Mg with stacking faults have not been reported in the literature. In the present work, the effects of major alloying elements on the energies and bond structures of I1, I2, and EF of Mg–X alloys are systematically investigated through first-principles calculations in term of electron localization morphology. Based on their usage in advanced Mg alloys, the following alloying elements are included in the present study, i.e. Al, Ca, Cu, Fe, K, La, Li, Mn, Na, Nd, Pr, Si, Sn, Sr, Y, Zn, and Zr.

In the present first-principles calculations, the orthorhombic supercells of I1, I2, and EFs are designed as $4\alpha' \times 2\sqrt{3}\alpha' \times 4c'$, $4\alpha' \times 2\sqrt{3}\alpha' \times 4c'$, and $4\alpha' \times 2\sqrt{3}\alpha' \times 4.5c'$ with 128, 128, and 144 atoms, respectively, where a' and c' are the lattice parameters of primitive HCP Mg.[31] In Figure 1, the lattice vectors **a**, **b**, and **c** in the supercells are shown to be parallel to the directions of $\langle 01\bar{1}0\rangle$, $\langle 1\bar{2}10\rangle$, and $\langle 0001\rangle$ of HCP Mg, respectively.

Calculations of stacking fault energy at 0 K are conducted by employing the Vienna *ab initio* simulation package [32,33] with the generalized gradient approximation (GGA) [34] for the exchange-correction functional and the projector-augmented wave [35] for the electron–ion interaction. The wave functions are sampled on the Γ-centered 5 × 5 × 3 mesh. The plane wave cut-off energy is set as 1.4 times the default energy cut-off (ENCUT = 300 eV) and the energy convergence criterion of electronic self-consistency is 10^{-6} eV/atom. The structures are fully relaxed by the Methfessel–Paxton technique,[36] while the final total energy calculations are performed by the tetrahedron method incorporating Blöchl correction.[37] The total energies of I1, I2, and EF structures as a function of volume are fitted by a four parameters Birch–Murnaghan equation of state.[38]

From previously examined results looking at the effects of supercell size (on the [0001] direction),[39] it was seen that the difference between three and seven atomic layers separating the stacking faults (I1, I2 and EF) is less than 3%. The deformation electron density,[16] $\Delta\rho$, is calculated using the Harris–Foulkes functional,[40,41] also known as the reference state in the independent atom model. The isosurface and the contour plots of the charge density are generated using VESTA.[42]

The stacking fault energy (γ_{sf}) is defined as

$$\boldsymbol{\gamma_{sf}} = \frac{1}{\boldsymbol{A}}(\boldsymbol{E}_{\mathrm{sf}} - \boldsymbol{E}_{\mathrm{Bulk}}) \qquad (1)$$

where $\boldsymbol{E}_{\mathrm{sf}}$ and $\boldsymbol{E}_{\mathrm{Bulk}}$ are the total energies of supercells with and without stacking fault, respectively, and $\boldsymbol{A}$ the total area of the stacking fault planes. Due to the periodic boundary conditions used in our calculations, there are two identical interfaces in the I1 and I2 supercells and one in the EF supercell. Taking the stacking fault energy of pure Mg (γ_{sf}) as the reference, the segregation energy per area ($E_{seg}(X)$) is expressed as[43,44]

$$\begin{aligned} E_{\mathrm{seg}}(X) &= \frac{\lfloor (E_{\mathrm{sf}}(Mg_{n-m}X_m) - E_{\mathrm{sf}}(Mg_n)) - (E_{\mathrm{Bulk}}(Mg_{n-m}X_m) - E_{\mathrm{Bulk}}(Mg_n)) \rfloor / m}{A} \\ &= \gamma_{\mathrm{sf}}(X) - \gamma_{\mathrm{sf}} \end{aligned} \qquad (2)$$

where $\gamma_{\mathrm{sf}}(X)$ is the energy per area with alloying element in stacking faults, and m and n the number of solute atoms and the total number of atoms in the supercell.

Through the four parameters Birch–Murnaghan equation of state,[38] the total energies of I1, I2, and EF structures at equilibrium volume are obtained and

Table 1. First-principles calculated stacking fault energy (γ_{sf}, mJ/m^2), and segregation energy per unit area of the dopants ($E_{seg}(X)$, mJ/m^2) of Mg–X alloys with different structures at 0 K.

System	Growth fault (I1) Mg–0.78at%X		Deformation fault (I2) Mg–0.78at%X		Extrinsic Fault (EF) Mg–0.69 at%X	
	γ_{sf}	$E_{seg}(X)$	γ_{sf}	$E_{seg}(X)$	γ_{sf}	$E_{seg}(X)$
Mg	14.4 8.1–17.8[a], 11[b], 18[c,d], 16[e], 27[f], 22[f], 15[f], 17[g,h], 21[i]	–	48.2 38.3[a], 33.8[c], 23[b], 33[d], 34[e], 54[f], 44[f], 30[f], 36[j], 34[g,h]	–	99.3 80.6[g], 52.6[c], 36[b], 140[k], 59[e], 40[h]	–
Mg–Al	12.5 13[d], 19[i]	−1.8 −5[d], −3[i]	33.6 23[d], 21[j]	−14.6 −10[d], 16[j]	55.7 130[k]	−43.6 −10[k]
Mg–Ca	13.8 12[i]	−0.6 −9[i]	33.5	−14.7	49.6	−49.7
Mg–Cu	16.6 23[i]	2.3 2[i]	41.0 53[j]	−7.2 17[j]	66.4	−32.9
Mg–Fe	3.3 (17.48 FM) 37[i]	−11.1 (3.1) 16[i]	40.3 (14.8 FM) 52[j]	−7.9 (−33.4)	79.8 (119.0 FM)	−19.6 (19.7)
Mg–K	13.5 7[i]	−0.9 −14[i]	32.5	−15.7	47.5	−51.8
Mg–La	12.1 −3[i]	−2.3 −18[i]	24.7	−23.5	29.9	−69.5
Mg–Li	16.7 28[d], 23[i]	2.3 10[d], 2[i]	48.2 46[d], 47[j]	−0.0 13[d,j]	63.2	−36.1
Mg–Mn	13.5 33[i]	−0.9 12[i]	40.3 38[j]	−7.9 2[j]	81.4	−18.0
Mg–Na	15.7 18[i]	1.3 −3[i]	46.8	−1.4	66.7	−32.7
Mg–Nd	10.8 0[i], −2[h]	−3.6 −21[i], −19[h]	32.3 4[h]	−15.9 −29[h]	51.9 29[h]	−47.4 −10[h]
Mg–Pr	15.0 −1[i], −4[h]	0.6 −22[i], −21[h]	31.9 0.5[h],	−16.3 −34[h]	50.7 21[h]	−48.6 −19[h]
Mg–Si	13.8	−0.6	33.9	−14.3	50.5	−48.8
Mg–Sn	13.8 16[i]	−0.6 −5[i]	37.2 2[j]	−11.0 −32[j]	50.1 126[k]	−49.2 −14[k]
Mg–Sr	15.6 4[i]	1.3 −17[i]	33.2	−15.0	46.3	−53.0
Mg–Y	1.8 4[g], 5[i]	−12.5 −13[g], −16[i]	32.9 25[j], 16[g]	−15.3 −11[j], −12[g]	47.1	−52.3
Mg–Zn	8.5 18[g], 21[i]	−5.9 1[g], 0[i]	30.8 37[j], 35[g],	−17.5 1[j], 1[g]	58.8 18.9[l]	−40.5 −61.7[l]
Mg–Zr	9.8 13[i]	−4.5 −8[i]	35.3 26[j]	−12.9 −10[j]	67.5	−31.9

Note: Experimentally determined basal plane stacking fault energies of Mg are 78 mJ/m^2 [6] and ~50 mJ/m^2 [48]. The ferromagnetic (FM) moment of Fe atom is considered and listed in the brackets.

[a] Wang et al. first-principles calculation with GGA.[49]

[b] Chetty and Weinert first-principles calculation with Local density approximation (LDA).[4]

[c] Fan et al. first-principles calculation with GGA.[27]

[d] Han et al. first-principles calculation of Mg–1.67at%X alloys.[28]

[e] Wen et al. first-principles calculation with GGA.[15]

[f] Han et al. MD with different potentials.[28]

[g] Zhang et al. first-principles calculation of Mg–1at%X alloys with GGA and the concentration of alloying elements is ~11.1at% in the slip plane.[25]

[h] Zhang et al. first-principles calculation of Mg–2.08at%X alloys with GGA and the concentration of alloying elements is ~25at% in the slip plane.[24]

[i] Zhang et al. first-principles calculation of γ_{I1} with GGA-Perdew-Burke-Ernzerhof (PBE) and the concentration of alloying elements is 11at% in the fault plane.[22]

[j] Muzyk et al. first-principles calculation of Mg–2at%X alloys with GGA-PBE and the concentration of alloying elements is 25at% in the slip plane.[23]

[k] Wang et al. first-principles calculation of Mg–0.69at%X alloys with GGA and the stacking sequence is supposed to be ABACAB based on their cited literature by Datta et al.[50]

[l] Datta et al. first-principles calculation of Mg and Mg–2.08at%X alloys with LDA.[30]

summarized in Table 1 along with various computational data in the literature with the majority of data for I1 from Zhang et al. [22] and those of I2 from Zhang et al. [24] and Muzyk et al.[23] Correspondingly, the segregation behavior of the alloying element in stacking faults are obtained and shown in Figure 2. According to the crystal structures of each individual solute atom at room temperature, it can be seen that some alloying elements with HCP and FCC structures reduce the stacking fault energies the most. If a reduction of stacking fault energies results in a high density of stacking fault, the strength can be improved without significant losses in ductility.[45,46] Thus, by

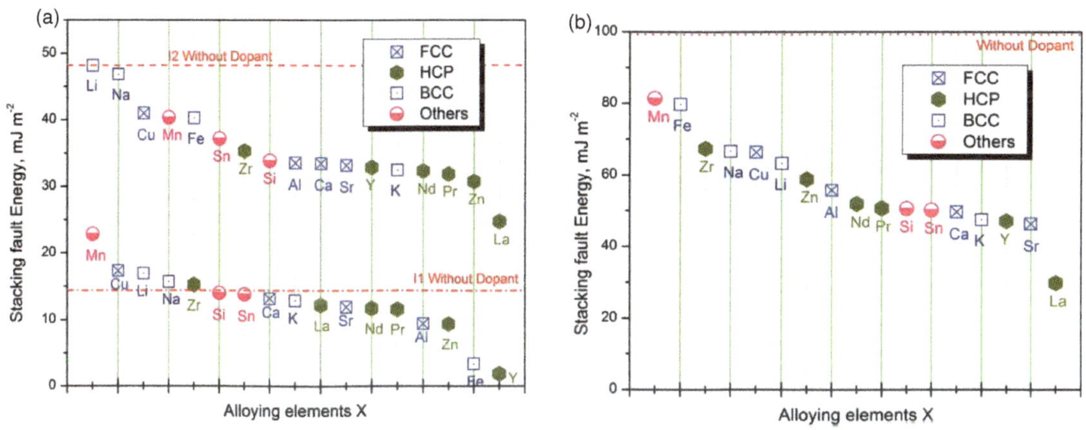

Figure 2. Contribution of alloying element on the formation energy of the basal plane stacking faults of binary Mg–X alloys, (a) Mg–0.78at%X alloys with growth (I1) and deformation (I2) faults; (b) Mg–0.69at%X alloys with EFs. Various symbols are used to identify the crystal structures of each individual solute atom at room temperature, such as FCC, BCC, HCP, and others.

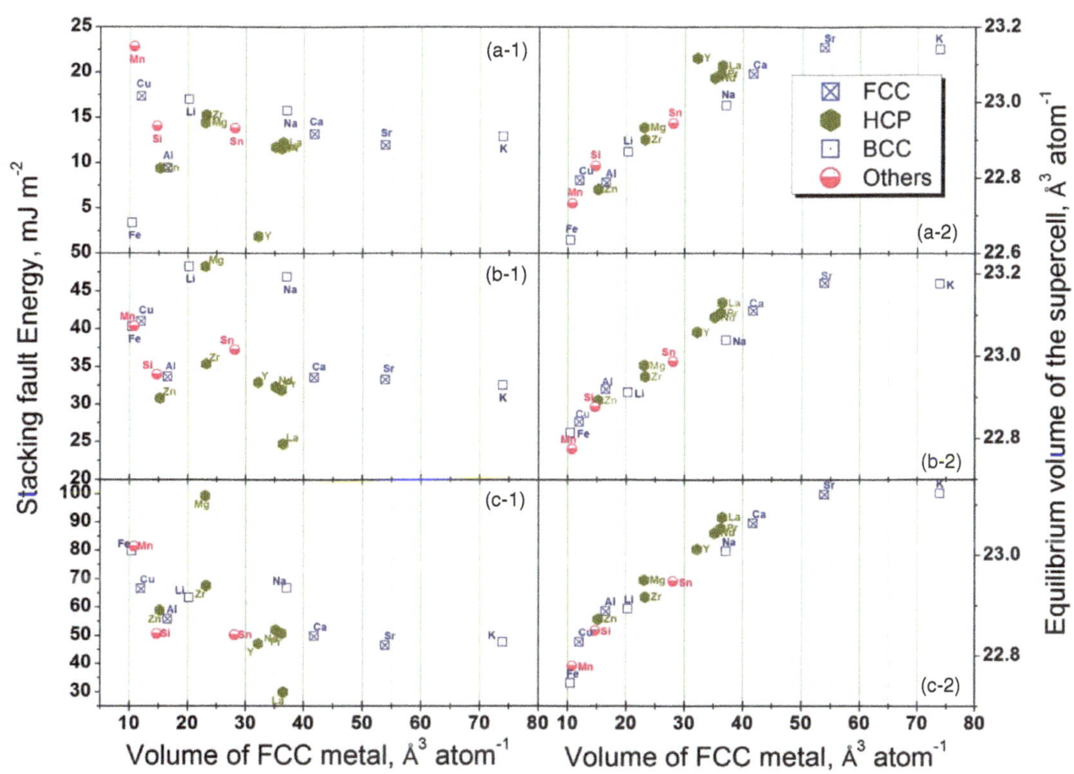

Figure 3. The calculated stacking fault energy and equilibrium volume of Mg–X alloys, referring to the alloying element X in FCC structure [31,38]; (a) growth fault, (b) deformation fault, and (c) EF. The numbers 1 and 2 are applied to characterize stacking fault energy and equilibrium volume in *y*-axis.

decreasing the width of Shockley partial dislocations in Mg,[6,47] the following alloying elements (La, Y, Zn, Al, Sr, Ca, Nd, and Pr) that are used in advanced Mg alloys are likely to be beneficial. It is noted that K reduces the stacking fault energies the most among all BCC alloying elements studied in the present work, probably due to its strong chemical reactivity.

It should be pointed out that the concentrations of alloying elements in a fault plane in the works by Muzyk et al. [23] and Zhang et al. [22,24] are 25 at% with all Mg atoms involved in Mg–X bonds and 11 at% with only one Mg–Mg bond, respectively, yielding strong interactions between alloying elements through X–Mg–X bonds in the fault plane. In the present work, the concentration

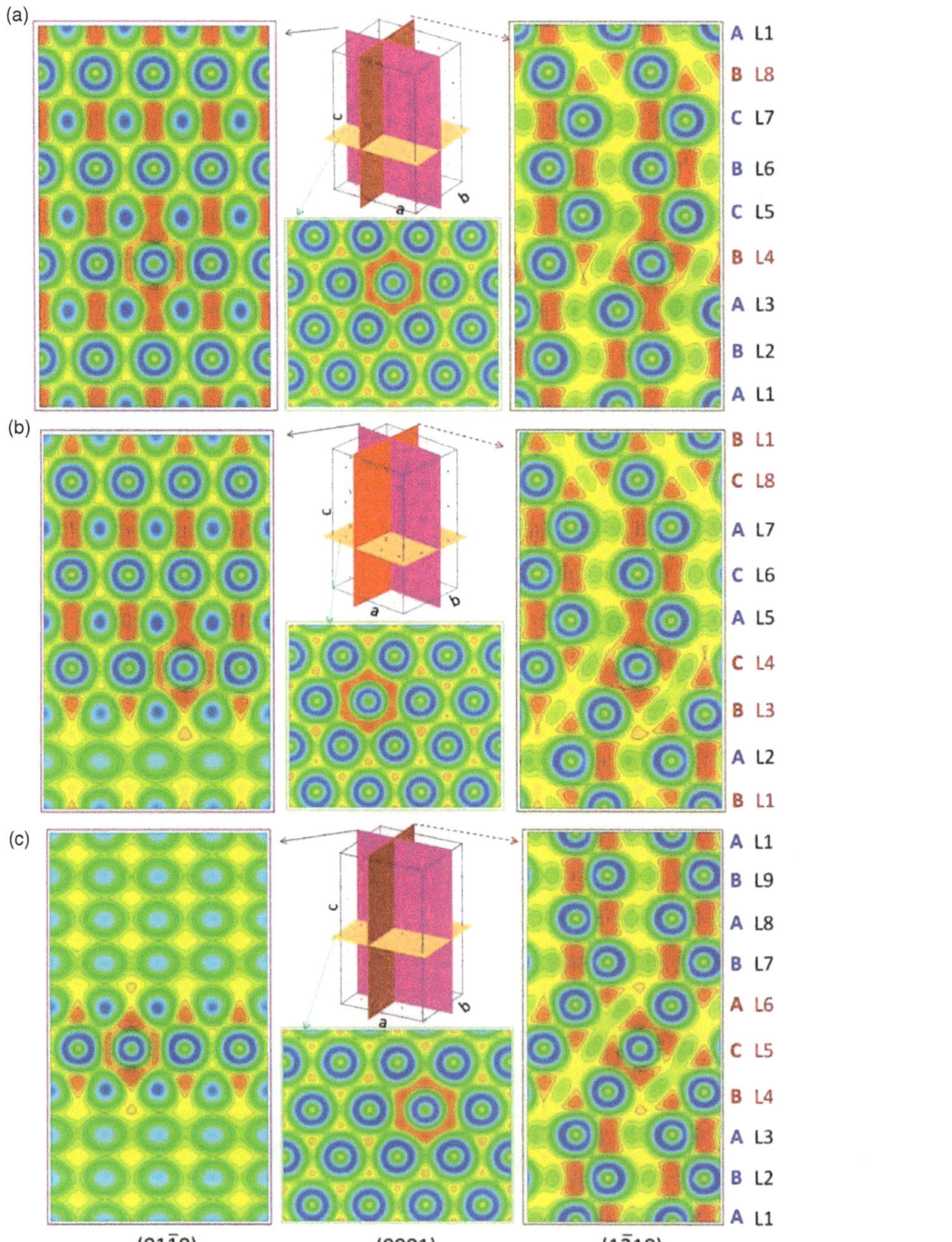

Figure 4. Contour plots of $\Delta\rho$ of Mg–Al in different views; (a) I1, (b) I2, and (c) EF with 0.0005 $e^-/Å^3$ intervals. Plots are generated using VESTA [42] with red for $\Delta\rho > 0$ and blue for $\Delta\rho < 0$. The coordinates of each plane are identified by the corresponding Miller plane in the supercell.

of the alloying elements in the fault plane is 6.25 at% without the X–Mg–X bonds in the fault plane. The distance between two alloying element is larger than 11 Å to avoid the interaction between the elements.[39,51] The present setting is also in line with the experimentally determined concentrations of alloying element in the fault layers of Mg alloys, typically lower than 10 at% such as 2 ± 1 at%Zn–4 ± 2 at%Y,[52] 7 at%Zn–6 at%Y,[53] 6 at%Zn–9 at%Y,[54] and 3 at%Zn–6at%Y [53] in the alloys of Mg–1 at%Zn–2 at%Y. The different concentration of alloying elements is likely the reason for the discrepancy in reported data from Muzyk et al. [23] and Zhang et al.[22,24]

Since the local chemical environments for alloying elements in I1 (ABC), I2 (ACB), and EF (ACB) are similar to that of FCC, the calculated stacking fault energies and the Mg–X supercell volumes are plotted in Figure 3 with respect to the volume of each individual alloying element X in the FCC structure.[31,38] It can be seen that the equilibrium volumes of Mg–X supercells increase almost linearly with the volume of alloying element X in the FCC structure. Consequently, it is plausible to select two alloying elements, both decreasing the stacking fault energy but changing the equilibrium volume in opposite way so that their co-segregation will be more energetically favorable due to the minimization of elastic lattice strain.[44]

In our previous study, isosurfaces of the deformation electron density (corresponding to $\Delta\rho_{max}$) of I1, I2, and EF in pure Mg were used to study the effects of stacking faults and alloying elements on the bond structure.[21] It was observed that the shape of the $\Delta\rho_{max}$ isosurface in fault planes is different from that in non-fault planes, and the number of atomic layers with altered electron density depends on the structure of stacking faults, i.e. 1, 2, and 3 for I1, I2, and EF, respectively. This indicates that the change of electron distribution is the largest for the EF and the smallest for I1 with that of I2 in between, in the same order as their stacking fault energies.

Figure 4 shows various contour plots of $\Delta\rho$ in I1, I2, and EF of the Mg–Al binary system. In the $(1\bar{2}10)$ plane

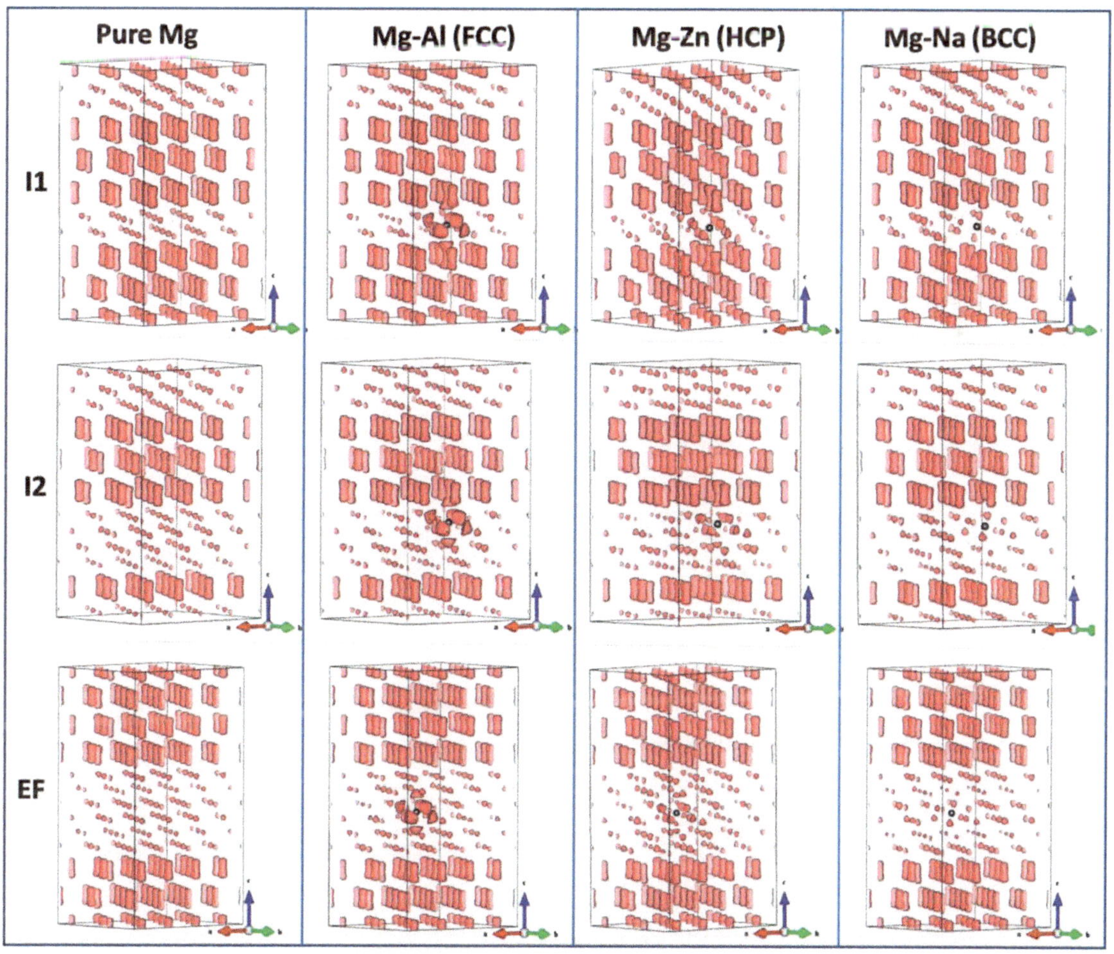

Figure 5. Deformation electron density isosurface ($0.75\Delta\rho_{max}$) of Mg, Mg–Al, Mg–Zn, and Mg–Na with I1, I2, and EF (from top to bottom). Only alloying elements in the supercells are displayed in black.

view, the rod-shaped directional bonds within the non-fault planes match well with the previous investigation by Blaha et al.,[55] while the triangle-shaped directional bonds are observed in the fault planes (L4 for I1, L3, and L4 for I2, and L4, L5, and L6 for EF). It can be seen that $\Delta\rho$ around the Al atomic basin has been increased compared with that without Al, i.e. Al and Mg atom in L5 of EF. In the (0001) plane view, $\Delta\rho$ around the Al atomic hexagonal basin is enhanced with respect to the case without Al, indicating the bond strength of basal plane is strengthened by Al. This is also consistent with the views of $(1\bar{2}10)$ and $(01\bar{1}0)$ planes. Furthermore, the number of atomic layers with altered electron density around the Al atom is one non-fault planes for I1, one fault plane and one non-fault for I2, and two fault planes for EF as shown in Figure 4.

Figure 5 shows the $0.75\Delta\rho_{max}$ (75% of the maximum value of $\Delta\rho$) isosurface of I1, I2, and EF with Al, Zn, and Na as alloying elements, respectively. It can be seen that the rod-like directional bonds have transformed into a tetrahedral shape with the addition of alloying elements in the fault plane. For alloying elements with the same crystallography structure, the similar electron localization morphology can be seen by plotting $\Delta\rho_{max}$ isosurface. For example, the $0.75\Delta\rho_{max}$ isosurface of I1, I2, and EF affected by FCC-Cu is similar to those of FCC-Al, shown as the second column of Figure 5. Moreover, the tetrahedral-shaped charge distributions in the fault planes are enhanced when alloyed with FCC-Al and HCP-Zn, see the second and third columns of Figure 5. On the contrary, the BCC elements show the reduced size of the tetrahedral-shaped charge distribution, for example, BCC-Na shown in the fourth column of Figure 5, agreeing with previous observations that Na reduces the bond strength in interfaces.[56,57]

In summary, the effect of 17 alloying elements on stacking fault energies of growth, deformation, and EFs in Mg is investigated through first-principles calculations. It is observed that all of them reduce the stacking fault energies of the deformation and EFs with La showing the least reduction for both. For the growth fault, five elements, i.e. Mn, Cu, Li, Na, and Zr, increase the stacking fault energy with the other elements decreasing the fault energy. As all stacking faults can be recognized as the local HCP–FCC transformation, the calculated results show that the bond strength is enhanced by alloying elements of FCC and HCP structures such as FCC-Al and HCP-Zn.

Acknowledgements This work was financially supported by National Science Foundation (Grant No. DMR-1006557), U.S. Army Research Lab (Project No. W911NF-08-2-0064) in United States, and National Natural Science Foundation of China (Grant No. 51071018). W.Y. Wang acknowledges the support from the Project Based Personnel Exchange Program with China Scholarship Council and American Academic Exchange Service ([2008]3072). First-principles calculations were carried out on the LION clusters at the Pennsylvania State University supported by the Materials Simulation Center and the Research Computing and Cyberinfrastructure unit at the Pennsylvania State University. Calculations were also carried out on the CyberStar cluster funded by NSF through grant OCI-0821527 and partially on resources of the National Supercomputer Center in Shenzhen.

References

[1] Shang S, Zhang H, Ganeshan S, Liu ZK. The development and application of a thermodynamic database for magnesium alloys. JOM. 2008;60:45–47.

[2] Shang SL, Wang WY, Wang Y, Du Y, Zhang JX, Patel AD, Liu ZK. Temperature-dependent ideal strength and stacking fault energy of fcc Ni: A first-principles study of shear deformation. J. Phys Condes Matter. 2012;24:155402-1–10.

[3] Hartford J, von Sydow B, Wahnstrom G, Lundqvist BI. Peierls barriers and stresses for edge dislocations in Pd and Al calculated from first principles. Phys Rev B. 1998;58:2487–2496.

[4] Chetty N, Weinert M. Stacking faults in magnesium. Phys Rev B. 1997;56:10844–10851.

[5] Argon AS, Moffatt WC. Climb of extended edge dislocations. Acta Metall Mater. 1981;29:293–299.

[6] Sastry DH, Prasad YVR, Vasu KI. On the stacking fault energies of some close-packed hexagonal metals. Scripta Matall Mater. 1969;3:927–930.

[7] Suzuki M, Kimura T, Koike J, Maruyama K. Effects of zinc on creep strength and deformation substructures in Mg-Y alloy. Mater Sci Eng A. 2004;387–389:706–709.

[8] Bernstein N, Tadmor EB. Tight-binding calculations of stacking energies and twinnability in fcc metals. Phys Rev B. 2004;69:094116-1–10.

[9] Van Swygenhoven H, Derlet PM, Froseth AG. Stacking fault energies and slip in nanocrystalline metals. Nat Mater. 2004;3:399–403.

[10] Lu K, Lu L, Suresh S. Strengthening materials by engineering coherent internal boundaries at the nanoscale. Science. 2009;324:349–352.

[11] Zhao YH, Zhu YT, Liao XZ, Horita Z, Langdon TG. Tailoring stacking fault energy for high ductility and high strength in ultrafine grained Cu and its alloy. Appl Phys Lett. 2006;89:121906-1–3.

[12] Zhu YT, Liao XZ, Wu XL. Deformation twinning in nanocrystalline materials. Prog Mater Sci. 2010;57:1–62.

[13] Li Y, Kong QP. On the relationship between creep rate and stacking-fault energy. Phys Status Solidi A-Appl Res. 1989;113:345–351.

[14] Guo Z, Miodownik AP, Saunders N, Schille JP. Influence of stacking-fault energy on high temperature creep of alpha titanium alloys. Scripta Mater. 2006;54:2175–2178.

[15] Wen L, Chen P, Tong ZF, Tang BY, Peng LM, Ding WJ. A systematic investigation of stacking faults in magnesium via first-principles calculation. Eur Phys J B. 2009;72:397–403.

[16] Nakashima PNH, Smith AE, Etheridge J, Muddle BC. The bonding electron density in aluminum. Science. 2011;331:1583–1586.

[17] Trinkle DR and Woodward C. The chemistry of deformation: how solutes soften pure metals. Science. 2005;310:1665–1667.

[18] Midgley PA. Electronic bonding revealed by electron diffraction. Science. 2011;331:1528–1529.

[19] Kioussis N, Herbranson M, Collins E, Eberhart ME. Topology of electronic charge density and energetics of planar faults in fcc metals. Phys Rev Lett. 2002;88:125501.
[20] Kohn W, Sham LJ. Self-consistent equations including exchange and correlation effects. Phys Rev. 1965;140:1133–1138.
[21] Wang WY, Shang SL, Wang Y, Darling KA, Mathaudhu SN, Hui XD, Liu ZK. Electron localization morphology of the stacking faults in Mg: a first-principles study. Chem Phys Lett. 2012;551:121–125.
[22] Zhang J, Dou Y, Liu G, Guo Z. First-principles study of stacking fault energies in Mg-based binary alloys. Comput Mater Sci. 2013;79:564–569.
[23] Muzyk M, Pakiela Z, Kurzydlowski KJ. Generalized stacking fault energy in magnesium alloys: density functional theory calculations. Scripta Mater. 2012;66: 219–222.
[24] Zhang Q, Fan TW, Fu L, Tang BY, Peng LM, Ding WJ. Ab-initio study of the effect of rare-earth elements on the stacking faults of Mg solid solutions. Intermetallics. 2012;29:21–26.
[25] Zhang Q, Fu L, Fan TW, Tang BY, Peng LM, Ding WJ. Ab initio study of the effect of solute atoms Zn and Y on stacking faults in Mg solid solution. Physica B. 2013;416:39–44.
[26] Li XQ, Schonecker S, Zhao JJ, Johansson B, Vitos L. Ideal strength of random alloys from first principles. Phys Rev B. 2013;87:214203-1-12.
[27] Fan TW, Tang BY, Peng LM, Ding WJ. First-principles study of long-period stacking ordered-like multi-stacking fault structures in pure magnesium. Scripta Mater. 2011;64:942–945.
[28] Han J, Su XM, Jin ZH, Zhu YT. Basal-plane stacking-fault energies of Mg: a first-principles study of Li- and Al-alloying effects. Scripta Mater. 2011;64:693–696.
[29] Tang PY, Tang BY, Peng LM, Ding WJ. Effect of Y and Zn substitution on elastic properties of 6H-type ABCBCB LPSO structure in Mg97Zn1Y2 alloy. Mater Chem Phys. 2012;131:634–641.
[30] Datta A, Waghmare UV, Ramamurty U. Structure and stacking faults in layered Mg-Zn-Y alloys: a first-principles study. Acta Mater. 2008;56:2531–2539.
[31] Wang Y, Curtarolo S, Jiang C, Arroyave R, Wang T, Ceder G, Chen LQ, Liu ZK. Ab initio lattice stability in comparison with CALPHAD lattice stability. CALPHAD. 2004;28:79–90.
[32] Kresse G, Furthmuller J. Efficient iterative schemes for ab initio total-energy calculations using a plane-wave basis set. Phys Rev B. 1996;54:11169–11186.
[33] Kresse G, Furthmuller J. Efficiency of ab-initio total energy calculations for metals and semiconductors using a plane-wave basis set. Comput Mater Sci. 1996;6:15–50.
[34] Kresse G, Joubert D. From ultrasoft pseudopotentials to the projector augmented-wave method. Phys Rev B. 1999;59:1758–1775.
[35] Wang Y, Perdew JP. Correlation hole of the spin-polarized electron gas, with exact small-wave-vector and high-density scaling. Phys Rev B. 1991;44:13298–13307.
[36] Methfessel M, Paxton AT. High-precision sampling for Brillouin-zone integration in metals. Phys Rev B. 1989;40:3616–3621.
[37] Blochl PE, Jepsen O, Andersen OK. Improved tetrahedron method for Brillouin-zone integrations. Phys Rev B. 1994;49:16223–16233.
[38] Shang SL, Saengdeejing A, Mei ZG, Kim DE, Zhang H, Ganeshan S, Wang Y, Liu ZK. First-principles calculations of pure elements: equations of state and elastic stiffness constants. Comput Mater Sci. 2010;48:813–826.
[39] Fan TW, Zhang Q, Tang BY, Peng LM, Ding WJ. Interaction between stacking faults in pure Mg. Eur phys J B. 2011;82:143–146.
[40] Harris J. Simplified method for calculating the energy of weakly interacting fragments. Phys Rev B. 1985;31: 1770–1779.
[41] Farid B, Heine V, Engel GE, Robertson IJ. Extremal properties of the Harris-Foulkes functional and an improved screening calculation for the electron-gas. Phys Rev B. 1993;48:11602–11621.
[42] Momma K, Izumi F. VESTA 3 for three-dimensional visualization of crystal, volumetric and morphology data. J Appl Crystallogr. 2011;44:1272–1276.
[43] Fabris S, Elsasser C. First-principles analysis of cation segregation at grain boundaries in alpha-Al2O3. Acta Mater. 2003;51:71–86.
[44] Nie JF, Zhu YM, Liu JZ, Fang XY. Periodic segregation of solute atoms in fully coherent twin boundaries. Science. 2013;340:957–960.
[45] Jian WW, Cheng GM, Xu WZ, Koch CC, Wang QD, Zhu YT, Mathaudhu SN. Physics and model of strengthening by parallel stacking faults. Appl Phys Lett. 2013;103:133108-1-4.
[46] Jian WW, Cheng GM, Xu WZ, Yuan H, Tsai MH, Wang QD, Koch CC, Zhu YT, Mathaudhu SN. Ultrastrong Mg alloy via nano-spaced stacking faults. Mater Res Lett. 2013;1:61–66.
[47] Zhu YM, Morton AJ, Weyland M, Nie JF. Characterization of planar features in Mg-Y-Zn alloys. Acta Mater. 2010;58:464–475.
[48] Couret A, Caillard D. An insitu study of prismatic glide in Magnesium. 2. Microscopic activation parameters. Acta Metall Mater. 1985;33:1455–1462.
[49] Wang Y, Chen LQ, Liu ZK, Mathaudhu SN. First-principles calculations of twin-boundary and stacking-fault energies in magnesium. Scripta Mater. 2010;62: 646–649.
[50] Wang H-Y, Zhang N, Wang C, Jiang Q-C. First-principles study of the generalized stacking fault energy in Mg-3Al-3Sn alloy. Scripta Mater. 2011;65:723–726.
[51] Yang S, Kiraly B, Wang WY, Shang S, Cao B, Zeng H, Zhao Y, Li W, Liu Z-K, Cai W, Huang TJ. Fabrication and characterization of beaded SiC quantum rings with anomalous red spectral shift. Adv Mater. 2012;24: 5598–5603.
[52] Abe E, Kawamura Y, Hayashi K, Inoue A. Long-period ordered structure in a high-strength nanocrystalline Mg-1 at% Zn-2 at% Y alloy studied by atomic-resolution Z-contrast STEM. Acta Mater. 2002;50:3845–3857.
[53] Itoi T, Seimiya T, Kawamura Y, Hirohashi M. Long period stacking structures observed in Mg 97Zn 1Y 2 alloy. Scripta Mater. 2004;51:107–111.
[54] Egusa D, Abe E. The structure of long period stacking/order Mg-Zn-RE phases with extended non-stoichiometry ranges. Acta Mater. 2012;60:166–178.
[55] Blaha P, Schwarz K, Dederichs PH. Electronic structure of hcp metals. Phys Rev B. 1988;38:9368–9374.
[56] Zhang S, Han Q, Liu ZK. Fundamental understanding of Na-induced high temperature embrittlement in Al-Mg alloys. Philos Mag. 2007;87:147–157.
[57] Lu G-H, Zhang Y, Deng S, Wang T, Kohyama M, Yamamoto R, Liu F, Horikawa K, Kanno M. Origin of intergranular embrittlement of Al alloys induced by Na and Ca segregation: grain boundary weakening. Phys Rev B. 2006;73:224115-1–5.

27

Coexistence of Ferromagnetic and a Re-entrant Cluster Glass State in the Layered Quaternary $(Cr_{1-x},Mn_x)_2GeC$

Q. Z. Tao[a], C. F. Hu[a]*, S. Lin[b], H. B. Zhang[c], F. Z. Li[a], D. Qu[a], M. L. Wu[a], Y. P. Sun[b], Y. Sakka[d] and M. W. Barsoum[e]

[a]*Ningbo Institute of Material Technology and Engineering, Chinese Academy of Sciences, Ningbo 315201, People's Republic of China;* [b]*Key Laboratory of Materials Physics, Institute of Solid State Physics, Hefei 230031, People's Republic of China;* [c]*Institute of Nuclear Physics and Chemistry, China Academy of Engineering Physics, Mianyang 621900, People's Republic of China;* [d]*Advanced Materials Processing Unit, National Institute for Materials Science, Tsukuba, Ibaraki 305-0047, Japan;* [e]*Department of Materials Science and Engineering, Drexel University, Philadelphia, PA 19104, USA*

Herein we present evidence for the coexistence of a Mn-rich ferromagnetic (FM) state and a Mn-poor reentrant cluster glass state in bulk, polycrystalline, layered $(Cr_{1-x},Mn_x)_2GeC$ samples, where x is varied between 0.01 and 0.1. The Mn-poor regions form a reentrant cluster glass state below $\sim$ 30 K. The Mn-rich regions become FM at Curie temperatures that increase with increasing Mn content. The interface coupling between these two regions gives rise to exchange anisotropy and a change in sign at 20 K resulting in, rarely observed, inverted hysteresis loops.

Keywords: MAX Phase, Cluster Glass, Inverted Loops

PACS numbers: 75.50.Lk, 75.30.Gw, 75.60.Ej, 75.70.Cn

The $M_{n+1}AX_n$ (MAX) phases—where M is an early transition metal, A is an A group element—mostly from groups 13 and 14—and X is carbon and/or nitrogen and $n = 1$ to 3—are a large (>60) family of thermodynamically stable nanolaminates, wherein $M_{n+1}X_n$ layers are interleaved with layers of pure A, with the X atoms occupying the octahedral sites between the M layers.[1] These phases, most of which were discovered by Nowotny et al.[2] in the 1960s, have attracted much interest recently since some of them exhibit a unique combination of properties, such as high thermal stability, excellent thermal and electrical conductivities, together with excellent machinability,[1,3] fully reversible dislocation-based deformation,[4] among other attributes.

Materials with inhomogeneities at the nanoscale are of fundamental interest as well as in applications. Disorder arises when the energy differences between the different ordered states is small.[5] Furthermore, magnetically, the interplay of different coexisting states can give rise to exotic magnetic states. This interplay is particularly important in multilayered structures that, for the most part, are artificially produced.[6]

With the substantial interest generated by magnetic multilayered structures and the importance of interfaces in these structures, the allure, and potential, of magnetic MAX phases, that are intrinsically nanolaminated, is obvious. The quest for magnetic MAX phases, led by Rosén and coworkers, is a few years old.[7–11] Starting with a theoretical paper predicting the existence of a stable magnetic $(Cr_{1-x},Mn_x)_2AlC$ compound, wherein nonmagnetic Cr layers, form an inherent structure of atomically thin, magnetic multilayers.[7] In that paper, it was also suggested that the degree of disorder between the Cr and Mn atoms in the compound could be used to tune the sign and magnitude of their coupling. In subsequent papers, it was theoretically shown that the most promising MAX phases are ones in the Cr–Mn–A–C system, were A is Al, Ga or Ge.[8–11] It was only recently, however, that Ingason et al., showed that $(Cr_{1-x},Mn_x)_2GeC$ thin films were ferromagnetic (FM) with Curie temperatures,

*Corresponding author. hucf@nimte.ac.cn

T_c that were higher than room temperature.[10] Based on DC magnetization measurement, it was concluded that the Mn atoms were disordered. Nevertheless, from the first principle calculation, the energy difference between the FM-ordered state and antiferromagnetic (AFM) disordered state was quite small, suggesting that they could co-exist.[7]

Coexistence of states with distinctly different magnetic behavior, especially at the nanoscale, often gives rise to intriguing phenomena due to complex interface exchange coupling. For example, the coexistence of ferrimagnetic (FI) and FM states results in zero field cooling, ZFC, exchange anisotropy,[12] and the interplay of FM and spin glassy states gives rise to large exchange anisotropies.[13] Another example is the coexistence of AFM and spin glass states in relaxor ferroelectrics.[5]

Even more recently, in what has to be considered a notable triumph of *ab initio* calculations, the same team not only predicted, but also synthesized Mn_2GaC and showed it to be FM at all temperatures < 230 K.[14] At 50 K, the saturation moment (at 5 T) was 0.29 μ_B per Mn atom and the remnant moment was 0.15 μ_B per Mn atom. Not surprisingly, and typical of thin FM films, the magnetic easy axis was shown to lie in the basal plane, where shape anisotropy is the dominant term in the anisotropy energy.[14]

This discovery is noteworthy because, to date, Mn was *not* a member of the transition metals known to form MAX phases. The addition of this element to the MAX phase family and the fact that it can endow them with magnetic properties is an exciting development. It also represents the second magnetic MAX phase.

In both papers by Ingason et al., the measurements were carried out on epitaxial thin films and no results were shown for temperatures below 50 K. The goal of this work was to try and synthesize, for the first time, a bulk MAX phase, i.e. one that was not in thin film form, and study its response to magnetic fields to temperatures as low as 2 K. Very recently, Nakamura et al. reported on Mn-induced itinerant-electron ferromagnetism in Cr_2GeC.[15] The Curie temperature was found to be proportional to the Mn concentration and varied from 43 to 270 K. They also applied spin fluctuation theory to describe it.

Like Nakamura et al., our starting point was the $(Cr_{1-x},Mn_x)_2GeC$ phase. To understand the physics behind the complex magnetic response observed herein, we focused on compositions with $x \leq 0.1$. The anticipated disordered structures are analogous to multilayers grown with different magnetic ordering such that interfaces between the various ordered states are established. Herein we present the evidence that the coexistence of Mn-rich and Mn-poor regions—that exhibit quite different magnetic states due to the aforementioned closeness in energies between the FM and AFM configurations—give rise to quite complex magnetic behavior.

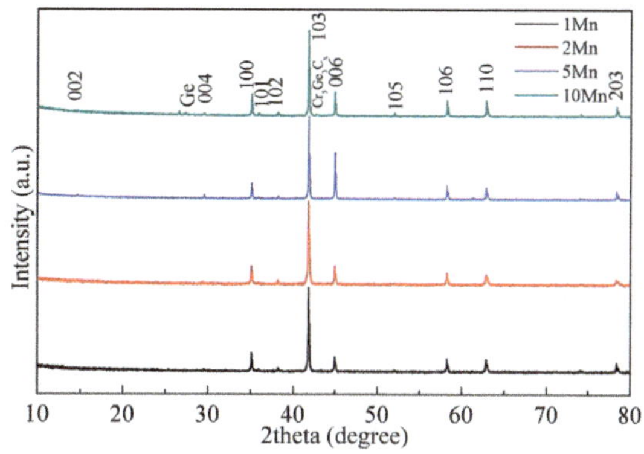

Figure 1. XRD patterns of $(Cr_{1-x}Mn_x)_2GeC$ samples with $x = 0.01, 0.02, 0.05$ and 0.1, denoted in this and all other figures as 1 Mn, 2 Mn, 5 Mn and 10 Mn, respectively. For all compositions, all major peaks belong to $(Cr_{1-x}Mn_x)_2GeC$. For the $x = 0.1$ sample, a few small peaks belonging to $Cr_5Ge_3C_x$ and possibly graphite were also observed. Interestingly, no distinct peak shifts were observed as a function of composition.

The powders used herein were graphite (99.9%, Cr (99.5%), Ge (99.99%) (Research Institute for Nonferrous Metals) and Mn (99%) Aladdin Industrial Inc.) To obtain the $(Cr_{1-x},Mn_x)_2GeC$ polycrystalline samples, with $x = 0.01, 0.02, 0.05$ and 0.1 stoichiometric amounts were mixed in an agate mortar and sealed in evacuated quartz tubes, heated to 1100°C and held at that temperature for 48 h. After furnace cooling, the samples were ground, pelletized, cold isostatically pressed to a pressure of 160 MPa, re-sealed in evacuated quartz tubes and annealed again at 1200°C for another 72 h.

The samples were characterized with X-ray powder diffraction, XRD. DC magnetization measurements were performed with a quantum design superconducting quantum interference device. AC magnetization measurements were carried out on a quantum design physical property measurement system.

All major peaks in the XRD patterns of the as prepared samples belonged to the MAX phase (Figure 1). For the $x = 0.1$ sample, a few small peaks belonging to $Cr_5Ge_3C_x$ were also observed (Figure 1). The *a*- and *c*-lattice parameters are calculated to be 2.953 and 12.086 Å, respectively. These values are quite similar to those of pure Cr_2GeC, with *a*- and *c*-lattice parameters of 2.954 and 12.08 Å, respectively.[16] In accordance with the result of Ingason et al., the incorporation of Mn into Cr_2GeC does not appear to affect the lattice parameters.[10] In Nakamura el al.'s recent work,[15] they found a nonlinear dependence of the lattice parameters on Mn content. For reasons that are unclear, however, their lattice constants in general, and those for pure Cr_2GeC ($a \approx 3.05$ Å and $c \approx 12.5$ Å) in particular are significantly higher than the reported values for Cr_2GeC ($a \sim 2.95$ Å, $c \sim 12.08$ Å),[16] and our results.

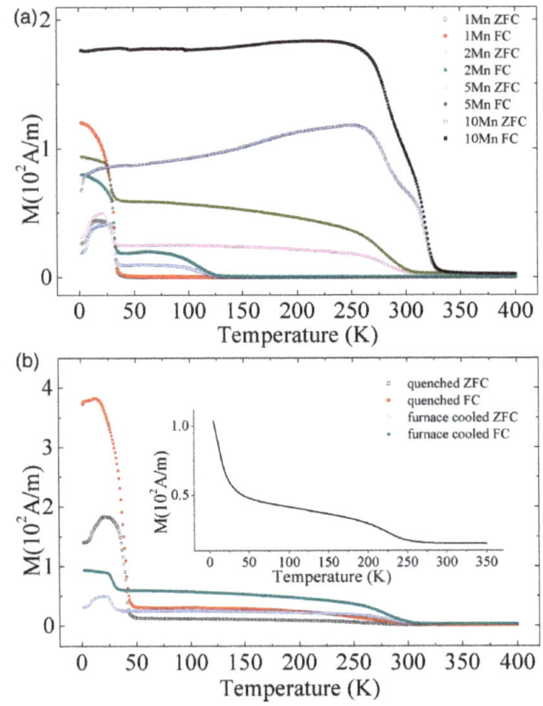

Figure 2. (a) Temperature dependence of zero field, ZFC, and field cooled, FC, DC magnetization curves for $(Cr_{1-x},Mn_x)_2GeC$ as a function of x. (b) Temperature dependence of zero field, ZFC, and field cooled, FC, DC magnetization curves for $(Cr_{0.95},Mn_{0.05})_2GeC$ as a function of cooling rate/quenching. The applied magnetic field was $5/4\pi\ 10^3\ Am^{-1}$ for (a) and (b). Inset in (b) shows the magnetization for $(Cr_{0.95},Mn_{0.05})_2GeC$ synthesized at 1000°C in an applied field of $5/4\pi\ 10^5\ Am^{-1}$.

The dependencies of the magnetization, M, on temperature and composition, under ZFC and field cooled (FC) modes in an applied field, of $5/4\pi\ 10^3\ Am^{-1}$ are shown in Figure 2(a). The $x = 0.01$ sample shows a reentrant cluster glass-like transition at 32.7 K; with increasing Mn content, an additional higher temperature transition appears. For the $x = 0.02$ and $x = 0.05$ samples, two transitions are evident: one at low temperature identical to the $x = 0.01$ sample, the others at $\sim$ 120 and at 286 K, respectively. For the $x = 0.1$ sample, the higher temperature transition is observed at 320 K, the lower temperature transition is not observed, but the cluster glass transition is still present. It is important to note that the fact that T_c is a function of Mn content rules out the possibility that the signal is coming from a magnetic impurity phase that in principle would possess a T_c that was independent of composition. In this case, the increases in T_c with Mn content most probably reflect a stronger interaction between the Mn atoms as the average distance between them shrinks (compare Figure 3(b) and 3(c)). In Figure 3, the Mn moments are assumed to lie in the basal plane because they do so in

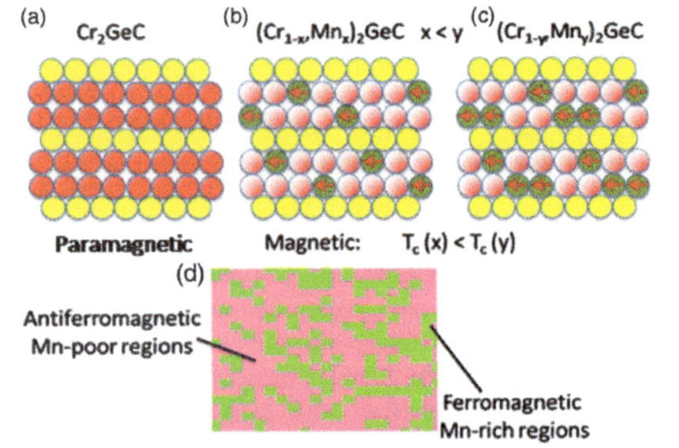

Figure 3. Schematic of magnetic structures inferred from the magnetic measurements and their temperature dependencies. (a) Paramagnetic Cr_2GeC, where the Cr atoms are depicted in red, Ge atoms in yellow and the C-atoms are not shown. (b) Substitution of a few Cr atoms with Mn atoms (green). Pink atoms represent a matrix that is Mn-poor and AFM; (c) same as B but at a higher Mn concentration. (d) Top view of basal planes showing Mn clustering (green) within an AFM matrix (pink). In this schematic, which is not to scale, for the sake of clarity, the magnetic moments of the Mn atoms are assumed to lie in the basal planes.

Mn_2GaC.[14] The actual situation is most probably significantly more complicated.[10] It is worth noting that although some theoretical calculation indicate that the undoped Cr_2GeC may have an AFM ground state,[17–19] Cr_2GeC, like other MAX phases [1] appears to be a typical Pauli paramagnetic solid.[15]

To further check the validity of this conclusion, a sample with $x = 0.05$, was quenched from 1100°C into water and its magnetization was measured. In this case, the high temperature T_c was significantly suppressed (Figure 2(b)), confirming the importance of having a critical Mn concentration for the high temperature or Mn-rich FM transition. Noteworthy, and consistent with this interpretation, is the large increase in the spin glass signal observed after quenching (Figure 2(b)).

What this result also strongly implies is a phase separation of the Mn atoms into Mn-rich and Mn-poor regions (Figures 2(a) and Figure 3(d)). The Mn-rich regions exhibit the higher ordering temperature that, as noted above, is a function of Mn content. The Mn-poor layer, on the other hand, orders at $\approx$ 30 K. Presumably, for the $x = 0.1$ sample, the Mn content is larger than the percolation threshold which in turn results in the formation of a more homogeneous material that exhibits a single transition (Figure 2(a)). At this time, the Mn segregation mechanism is unclear. First principles calculations indicate that the Mn–C–Mn configuration is energetically more favorable compared to the Mn–C–Cr configuration.[10] Furthermore, given the continuous and gradual changes in lattice parameters of the solid solutions, it is not unreasonable to assume that spinodal decomposition is occurring. This comment notwithstanding, more work is needed to better understand this crucial experimental variable.

The importance of this variable, however, is manifested in the actual T_c values In the Nakamura et al. paper,[15] the T_c values are lower than those presented herein. For example, their T_c at $x = 0.2$ is 270 K; herein, T_c at the lower Mn loading of 0.1 is 320 K. The only apparent difference is the synthesis temperature. We also note that the synthesis temperature is essential for a higher T_c. For example a $x = 0.05$ sample synthesized at 1000°C shows a transition at $\sim$ 224 K (inset of Figure 2(b)). This fact may relate to the formation of Mn-rich regions at elevated temperature. In our own study when $x \geq 0.2$, a substantial amount of impurity, most probably $(Mn_{1-x}Cr_x)_5Ge_3$, emerged and the diffraction pattern was identical to the impurity phase in Figure 1 of Ref. [15]. The $x = 0.2$ sample in our own work showed a sharp transition at 250 K. It is reasonable to conclude that the Mn containing intermetallic is responsible for the magnetism observed in the $x > 0.2$ sample.

For both the 0.02 and 0.05 compositions, below the 32 K transition, the ZFC magnetization curve shows a broad maximum at $\approx$ 22 K. Cooling below 10 K results in a drop in magnetization (Figure 2(a)), suggesting the existence of a spin cluster glass behavior at the lowest temperatures.[20]

To investigate the nature of this low temperature transition, AC-susceptibility measurements for the $x = 0.05$ sample were performed as a function of temperature from 3 to 350 K in an applied field of $5/4\pi\ 10^3\ \mathrm{Am^{-1}}$. The prominent frequency independent peak at 22.5 K (Figure 4(a)) agrees well with the maximum in DC magnetization observed at $\sim$ 22 K. The small bulge at $\sim$ 60 K might be due to AFM coupling within the Mn-rich layers. This feature is a probably a consequence of the complex competition between FM and AFM ordering mentioned above. The existence of AFM coupling is consistent with the small saturation moment (0.31 μ_B per Mn atom at 2 K).

The functional dependencies of the real, χ', (right y-axis) and imaginary, χ'', (left y-axis) components of

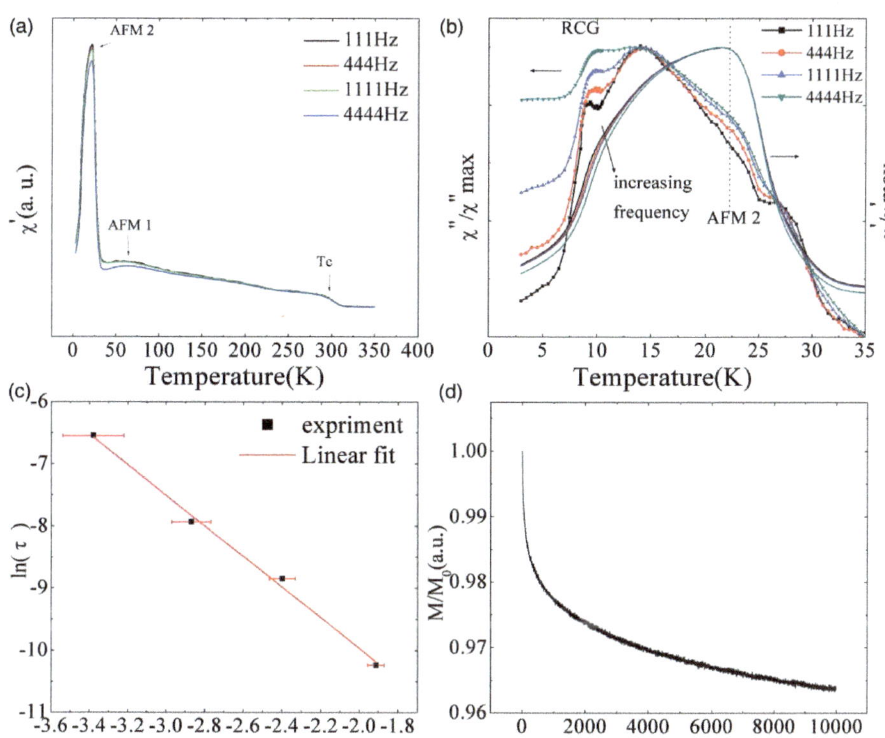

Figure 4. (a) Real components of ac magnetization as a function of temperature in an applied field of $5/4\pi\ 10^3\ \mathrm{Am^{-1}}$ as a function of frequencies shown in top right corner; (b) Real (right y-axis) and imaginary components (left y-axis) of AC magnetization as a function of temperature from 3 to 35 K in an applied field of $5/4\pi\ 10^3\ \mathrm{Am^{-1}}$ as a function frequencies shown in top right corner. (c) $\ln \tau$ vs. $\ln (T''_f - T_{RCG})/T_{RCG}$. (d) Time dependence of normalized magnetization at 2 K after cooling the sample in a field of $500/4\pi\ 10^3\ \mathrm{Am^{-1}}$, removing the field and measuring the magnetization vs. time when the magnetic field reach zero. All measurements are performed on $x = 0.05$ samples.

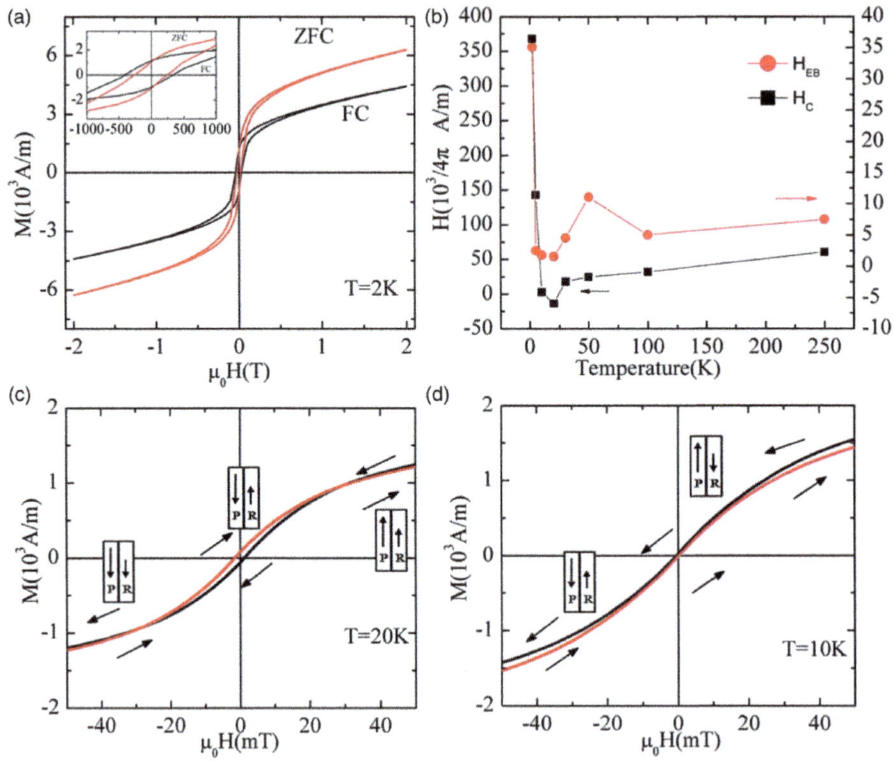

Figure 5. (a) ZFC and FC hysteresis loops at 2 K. Inset shows enlarged view of the low field part. Exchange anisotropy and coercivity enhancements are evident; (b) Coercive field, H_C, and exchange bias field, H_{EB}, obtained from the hysteresis loops. The uncertainty should mainly come from remanence of the superconducting magnet which is smaller than the symbols' size; (c) Reversed hysteresis loop obtained at 20 K. Arrows indicate the sweep direction. The arrows in the insets schematically indicate the magnetization reversal. (d) Normal hysteresis loop obtained at 10 K. All measurements were performed on $x = 0.05$ samples.

the AC susceptibility on frequency, in the 2–35 K temperature range, are shown in Figure 4(b). Both values are normalized by their maximum values.

The χ'' shows a frequency-dependent maximum at around 10 K. Note that at 0.1 K, the temperature step used, when the results obtained between 8.5 and 10.5 K, were taken, was quite small. Said otherwise, the heating rate was particularly low.

Figure 4(c) plots the relaxation times, τ, where τ equals to $1/2\pi f$, and f the frequency of the applied field, versus T_f, where T_f is the temperature at which χ'' is a maximum, assuming, the following relationship:[21]

$$\frac{\tau}{\tau_0} = \left(\frac{T''_f - T_{RCG}}{T_{RCG}}\right)^{-z\upsilon}$$

is operative. First, we determined $T_g = 8.8 \pm 0.05$ K, which gives the best linear fit and then, from the linear fit of the results shown in Figure 4(c), the dynamical critical exponent, $z\upsilon$, was found to be 2.46 ± 0.4, and the spin flipping time, τ_0 was $3.4 \pm 0.9 \times 10^{-7}$ s. This relatively large flipping time agrees well with other reentrant cluster glass system suggesting that the glassy nature is due to a cluster and *not* individual spins.[22] Figure 4(d) plots a typical relaxation curve at 2 K, of the magnetization, M (normalized by M_o, the magnetization at $t = 0$ when the magnetic field is removed) after FC in a magnetic field of $500/4\pi\, 10^3\ \mathrm{Am^{-1}}$. All these features are consistent with a cluster glass state at $T < 10$ K.

Based on these results, it is reasonable to assume that upon cooling, some of the Mn atoms, residing on the Cr sites, form Mn-rich regions in a Mn-poor matrix (Figure 3(d)). The former undergoes a paramagnetic to FM transition at a T_c that depends on the Mn concentration in that region. The Mn-poor matrix, on the other hand, undergoes a reentrant cluster glass transition at ~ 30 K.

For all temperatures studied, and regardless of whether the sample was ZFC or FC, M did not saturate (e.g. Figure 5(a)), indicating the presence of an AFM interaction, in accordance with our interpretation above.

To study the exchange coupling, the DC magnetization was measured at different temperatures after cooling from 350 K in either ZFC or FC mode. The corresponding hysteresis loops at 2 K are shown in Figure 5(a). The 2 K FC M-H loop exhibits a negative shift, whereas the ZFC M-H loop shows no shift (see inset in Figure 5(a)). The exchange bias field, H_{EB} and coercive fields, H_C obtained from the hysteresis loops, at various temperatures, are shown in Figure 5(b). Exchange anisotropy usually emerges as a result of coupling between FM–AFM, FM–FI or, as the case here, FM-spin glass states at the interface.[23] According to the above phase transition interpretation, it is reasonable to conclude that coupling between the Mn-rich, FM regions and the reentrant cluster glass Mn-poor matrix gives rise to the exchange anisotropy. As discussed below, from 10 to 30 K, the exchange bias field, H_{EB}, decreases greatly.

At 20 K, H_C is negative since the hysteresis loops are *inverted* (Figure 5(c)). Inverted hysteresis loops are rare in nature. Not coincidentally, they have been observed in exchange coupled multilayers.[24–26] Such inverted loops are a consequence of the interplay of magnetization reversal and magneto-crystalline anisotropy.[24] In other words, it is reasonable to assume that the interplay between Mn-rich regions and the reentrant cluster matrix is responsible for the inverted loops. The magnetization reversal mechanism is shown schematically in the insets shown in Figure 5(c); those for the normal hysteresis loop in the insets shown in Figure 5(d). For the 20 K hysteresis loop, the Mn-poor regions (P regions) reverse their magnetization first at positive fields, followed by the Mn-rich regions (R regions) at negative fields. For the 10 K, normal hysteresis loop, the reversal mechanism is different. It follows that the different magnetization reversal sequences of the Mn-rich regions (R regions) and the Mn-poor regions (P regions) or matrix, give rise to the inverted loops at 20 K and normal hysteresis loops at 10 K.

Since the degree of inhomogeneity is tunable—as shown by our quenching experiment, for example—the exchange anisotropy, in turn, can also be readily tuned. Thus as elucidated herein, $(Cr_{1-x}Mn_x)_2GeC$ can be synthesized to be a homogeneous [10] or an inhomogeneous complex system. Note that since the changes in the lattice parameters due to the incorporation of Mn in Cr_2GeC are negligible, the introduction of various interfaces can be carried out with little, to no, lattice mismatches.

To summarize, the plethora of magnetic phenomena observed in $(Cr_{1-x}Mn_x)_2GeC$ can be viewed as a consequence of the competition between intralayer FM and interlayer AFM order.[10] The intricate competition between these two states could in principle be tuned by altering the interlayer distances. Considering the rich chemistry of the MAX phases, there will indubitably be more magnetic phases. If some of them happen to be M_3AX_2 or M_4AX_3, then the effect of varying the interlayer distances can be explored since the latter have larger c-lattice parameters than the M_2AX phase studied herein. The c-lattice parameters of the M_2AX, M_3AX_2 and M_4AX_4 are, respectively, of the order of 1.3, 1.8 and 2.3 nm.[1] Accordingly, the aforementioned competition may result in new equilibrium states.

In addition to the discovery of Mn_2GaC, two additional new exciting recent developments bode well for our ability to fine tune the FM–AFM competition. The first is the possibility of exfoliating the MAX phases themselves into nanosheets comprising a few MAX layers, which has been demonstrated in the $Ti_3Si_{1-x}Al_xC_2$ system.[27] The second is the complete etching of the A-layers from in between the MX layers phases to form a totally new class of 2D materials labeled MXenes.[28,29] In the latter, the coupling between layers is quite weak. Based on the analysis outlined herein, it is reasonable to assume that both types of exfoliated layers would yield a multitude of quite interesting magnetic responses and tunabilities.

In conclusion, the introduction of nanoscale inhomogeneities in bulk polycrystalline $(Cr_{1-x}Mn_x)_2GeC$ samples, wherein Mn-rich layers are embedded within a Mn-poor matrix, gives rise to complex magnetic behavior. The Mn-rich layers are FM and the Mn-poor matrix undergoes a reentrant cluster glass transition at low temperatures. Due to the varying magnetic states of the Mn-poor region, the coupling between the two phases results in exchange anisotropy, and inverted hysteresis loops at 20 K. The tunable coupling of the layered magnetic $(Cr_{1-x},Mn_x)_2GeC$ and other magnetic MAX phases may be of great potential in spintronic applications.

Acknowledgements We thank for the support of the 'Chunlei' program in Ningbo Institute of Material Technology and Engineering in China, Ningbo Natural Science Foundation (2013A610128) and National Natural Science Foundation of China (U1232136). QZT also thanks H.L. Yang from NIMTE for helpful discussion. MWB was partially funded by National Science Foundation (DMR-1310245). SL was funded by The National Natural Science Foundation of China (51301165).

References

[1] Barsoum MW. MAX phases: properties of machinable carbides and nitrides. Weinheim: Wiley VCH GmbH & Co.; 2013.
[2] Jeitschko W, Nowotny H, Benesovsky F. Kohlenstoffhaltige ternäre Verbindungen (V-Ge-C, Nb-Ga-C, Ta-Ga-C, Ta-Ge-C, Cr-Ga-C und Cr-Ge-C). Monatshefte für Chemie und verwandte Teile anderer Wissenschaften. 1963;94(5):844–850.
[3] Eklund P, Beckers M, Jansson U, Högberg H, Hultman L The M_{n+1} AX_n phases: materials science and thin-film processing. Thin Solid Films. 2010;518(8):1851–1878.
[4] Barsoum M, Zhen T, Kalidindi S, Radovic M, Murugaiah A. Fully reversible, dislocation-based compressive deformation of Ti_3SiC_2 to 1 GPa. Nat Mater. 2003;2(2): 107–111.
[5] Kleemann W, Shvartsman VV, Borisov P, Kania A. Coexistence of antiferromagnetic and spin cluster glass order in the magnetoelectric relaxor multiferroic $PbFe_{0.5}Nb_{0.5}O_3$. Phys Rev Lett. 2010;105(25):257202.
[6] Grunberg P, Schreiber R, Pang Y, Walz U, Brodsky M, Sowers H. Layered magnetic structures: evidence for antiferromagnetic coupling of Fe layers across Cr interlayers. J Appl Phys. 1987;61(8):3750–3752.
[7] Dahlqvist M, Alling B, Abrikosov I, Rosén J. Magnetic nanoscale laminates with tunable exchange coupling from first principles. Phys Rev B. 2011;84(22):220403.

[8] Mockute A, Dahlqvist M, Emmerlich J, Hultman L, Schneider JM, Persson POÅ, Rosen J. Synthesis and ab initio calculations of nanolaminated $(Cr,Mn)_2AlC$ compounds. Phys Rev B. 2013;87(9):094113.

[9] Liu Z, Waki T, Tabata Y, Yuge K, Nakamura H, Watanabe I. Magnetic ground state of the $M_{n+1}AX_n$-phase nitride Cr_2GaN. Phys Rev B. 2013;88(13):134401.

[10] Ingason AS, Mockute A, Dahlqvist M, Magnus F, Olafsson S, Arnalds UB, Rosen J. Magnetic self-organized atomic laminate from first principles and thin film synthesis. Phys Rev Lett. 2013;110(19):195502.

[11] Jaouen M, Chartier P, Cabioc'h T, Mauchamp V, André G, Viret M. Invar like behavior of the Cr_2AlC MAX phase at low temperature. J Am Ceram Soc. 2013;96(12): 3872–3876.

[12] Nayak AK, Nicklas M, Chadov S, Shekhar C, Skourski Y, Winterlik J, Felser C. Large zero-field cooled exchange-bias in bulk Mn_2PtGa. Phys Rev Lett. 2013;110(12): 127204.

[13] Ma L, Wang WH, Lu JB, Li JQ, Zhen CM, Hou DL, Wu GH. Coexistence of reentrant-spin-glass and ferromagnetic martensitic phases in the $Mn_2Ni_{1.6}Sn_{0.4}$ Heusler alloy. Appl Phys Lett. 2011;99(18):182507(3). doi:10.1063/1.3651767

[14] Ingason A, Petruhins A, Dahlqvist M, Magnus F, Mockute A, Alling B, Hultman L, Abrikosov IA, Persson POA, Rosen J. A nanolaminated magnetic phase: Mn_2GaC. Mater Res Lett. 2014;2(2):89–93.

[15] Liu Z, Waki T, Tabata Y, Nakamura H. Mn-doping-induced itinerant-electron ferromagnetism in Cr_2GeC. Phys Rev B. 2014;89(5):054435.

[16] Amini S, Zhou A, Gupta S, DeVillier A, Finkel P, Barsoum MW. Synthesis and elastic and mechanical properties of Cr_2GeC. J Mater Res. 2008;23(8):2157–2165.

[17] Ramzan M, Lebègue S, Ahuja R. Electronic and mechanical properties of Cr_2GeC with hybrid functional and correlation effects. Solid State Comm. 2012;152(13): 1147–1149.

[18] Zhou W, Liu L, Wu P. First-principles study of structural, thermodynamic, elastic, and magnetic properties of Cr_2GeC under pressure and temperature. J Appl Phys. 2009;106(3):033501.

[19] Mattesini M, Magnuson M. Electronic correlation effects in the Cr_2GeC $M_{n+1}AX_n$ phase. J Phys Condens Matter. 2013;25(3):035601.

[20] Marcano N, Sal JG, Espeso J, Barquín LF, Paulsen C. Cluster-glass percolative scenario in $CeNi_{1-x}$ Cu_x studied by very low-temperature ac susceptibility and dc magnetization. Phys Rev B. 2007;76(22): 224419.

[21] Hohenberg PC, Halperin BI. Theory of dynamic critical phenomena. Rev Mod Phys. 1977;49(3):435–479.

[22] Hanasaki N, Watanabe K, Ohtsuka T, Kézsmárki I, Iguchi S, Miyasaka S, Tokura Y. Nature of the transition between a ferromagnetic metal and a spin-glass insulator in pyrochlore molybdates. Phys Rev Lett. 2007;99(8):086401.

[23] Ali M, Adie P, Marrows CH, Greig D, Hickey BJ, Stamps RL. Exchange bias using a spin glass. Nat Mater. 2006;6(1):70–75.

[24] Ziese M, Vrejoiu I, Hesse D. Inverted hysteresis and giant exchange bias in $La_{0.7}Sr_{0.3}MnO_3$ $SrRuO_3$ superlattices. Appl Phys Lett. 2010;97(5):052504(3).

[25] Dos Santos MC, Geshev J, Schmidt J, Teixeira S, Pereira L. Origin of the magnetization reversal of an Fe thin film on Si (111). Phys Rev B. 2000;61(2):1311.

[26] Valvidares S, Alvarez-Prado L, Martin J, Alameda J. Inverted hysteresis loops in magnetically coupled bilayers with uniaxial competing anisotropies: theory and experiments. Phys Rev B. 2001;64(13): 134423.

[27] Zhang X, Xu J, Wang H, Zhang J, Yan H, Pan B, Zhou J, Xie Y. Ultrathin nanosheets of MAX phases with enhanced thermal and mechanical properties in polymeric compositions: $Ti_3Si_{0.75}Al_{0.25}C_2$. Angew Chem Int Ed. 2013;52(16):4361–4365.

[28] Naguib M, Kurtoglu M, Presser V, Lu J, Niu J, Heon M, Hultman L, Gogotsi Y, Barsoum MW. Two-dimensional nanocrystals produced by exfoliation of Ti_3AlC_2. Adv Mater. 2011;23(37):4248–4253.

[29] Naguib M, Mashtalir O, Carle J, Presser V, Lu J, Hultman L, Gogotsi Y, Barsoum MW. Two-dimensional transition metal carbides. ACS Nano. 2012;6:1322–1331.

28

Swift Heavy Ion Shape Transformation of Au Nanocrystals Mediated by Molten Material Flow and Recrystallization

A. A. Leino[a]*, O. H. Pakarinen[a], F. Djurabekova[a], K. Nordlund[a], P. Kluth[b] and M. C. Ridgway[b]

[a]*Helsinki Institute of Physics and Department of Physics, University of Helsinki, PO Box 43, Helsinki, FI-00014, Finland;* [b]*Department of Electronic Materials Engineering, Research School of Physics and Engineering, Australian National University, Canberra, ACT 0200, Australia*

Swift heavy ion (SHI) irradiation of amorphous SiO_2 that contains metal nanocrystals can be used to transform the shape of the particles into peculiar asymmetric ones not easily achievable by other means. Using a molecular dynamics simulation framework augmented to include the electronic excitations of the SHIs, we predict that the reshaping of spherical particles into nanorods occurs continuously during consecutive ion impacts by a dynamic crystal–liquid–crystal phase transition of metal particle with the flow of liquid phase into an underdense track core in silica. The simulated nanocrystals are shown to have a saturation width that agrees with experiments.

Keywords: Nanoparticles, Nanocrystals, Elongation, Swift Heavy Ions, Silica

When silica composites that contain spherical metal nanocrystals are subjected to swift ($E_{\text{kin}} \geq 1$ MeV/amu) heavy ion (SHI) irradiation, the particles undergo a shape transformation to nanorods or prolate spheroids, so that they elongate in the direction parallel to the ion beam.[1–14] This effect is analogous to the well-established possibility to use multi-wall carbon nanotubes inside an electron microscope as pressure vessels to modify and study the properties of metal nanocrystals.[15,16] The swift ion processing has, however, the major advantage that the end result is stable over macroscopic timescales when taken out of the experimental processing chamber. Therefore, in addition to being a directly observable nanoscale test system for understanding the fundamentals of ion–solid interactions, the phenomenon has application potential. By controlling the dimensions of the crystal by SHI irradiation, control over surface plasmon resonance is gained,[17,18] which is desirable in the fabrication of plasmonic devices.[19] The method has the advantage of producing large arrays of equally aligned nanocrystals, which is difficult to achieve otherwise.

In this letter, we examine the nanocluster shape transformation mechanisms using classical molecular dynamics (MD) simulations, extended with the inelastic thermal spike model (i-TS),[20] and experiments. The experiments show a major elongation effect similar to that reported previously by other groups.[1–14] The MD simulations with the i-TS energy deposition reproduce the experimentally observed progressive elongation. The minor axis of the elongated particle is shown to saturate at a width correlated with the width of the ion track in silica,[4] in excellent agreement with experiments. The results show that the elongation is caused by the thermal expansion associated with heating and melting of *crystalline* metal, which causes it to flow within the track core in silica.

Although technologically promising, the elongation mechanism of nanocrystals by SHIs has not yet been identified. In the energy range of SHI's, the primary interactions between the ion and the irradiated material are dominated by collisions with electrons, and the kinetic energy of the ion is first consumed in the excitation of the electronic subsystem.[21] D'Orléans et al. [1] calculated that there is a large overpressure in the crystal after impact, which is released by the flow of molten matter to the track in silica. However, it was not clear why the nanocrystals resist deformation when embedded in a crystalline matrix.[22] Another proposed reason is the so called 'ion-hammering' effect,[23] which is characteristic for an amorphous matrix and leads to

*Corresponding author. Email: aleksi.leino@helsinki.fi

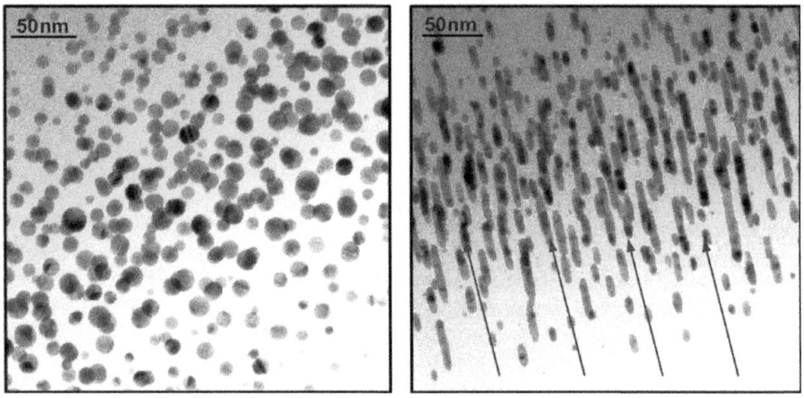

Figure 1. Modification of the shapes of Au nanocrystals embedded in silica into strongly elongated non-equilibrium shapes. Left: Cross-sectional transmission electron microscopy (XTEM) image of the initial shape of spherical nanocrystals. Right: XTEM image of nanocrystals with the same initial radius after SHI irradiation (direction indicated by arrows) with 54 MeV Au ions with a fluence of 2×10^{14} ions/cm^2. Thanks to the surrounding inert silica matrix, the final shape is very stable at ambient conditions.

an in-plane stress, which might deform the metal particle once melted. Klaumünzer [2] considered both ion hammering and melting, concluding that ion hammering could not be the cause of deformation, whereas a rough estimate showed that, at least within an order of magnitude, the high pressure in molten metal could explain the elongation. Dawi et al. [8] found the effect to be sensitive, under certain conditions, to the initial concentration of the nanoparticles. However, such behaviour could be associated with Ostwald ripening, which was avoided in [7] by forcing a large distance between individual Au nanoparticles. Results in this study confirmed that elongation can proceed through individual processes. Amekura et al. also observed that a small change in the shape of the nanocrystals was induced by single impacts.[12]

Au nanocrystals were formed in thermally grown SiO_2 by ion implantation and annealing. Samples were then irradiated with 185 MeV Au^{+13} ions at normal incidence and room temperature. Transmission electron microscope in a cross-section geometry was utilized to determine the dimensions of the elongated nanocrystals (Figure 1), measuring 300 crystals for reasonable statistics. Additional details on the experimental procedure is given in [9,11].

The computer simulations were performed by using a multiscale model implemented in the classical MD code PARCAS [24] (The main principles of the MD algorithms are presented in [25,26]. The adaptive time step is the same as in [27]) to simulate the elongation. Classical MD [28] can be used to study the transport of the atoms from given initial conditions, but it does not explicitly describe the electronic subsystem or its coupling to the atomic subsystem (see Supplementary Material for the advantages and limitations). The initial nanocrystal was generated following a procedure described by Djurabekova,[29] i.e cutting a sphere out of fcc Au, compressing it by 2% to avoid unrealistically low interatomic separations in the initial configuration, and inserting this sphere into a bigger void than the uncompressed sphere in silica (see Figure 2, case (0)). The resulting cell was then pressure and temperature relaxed for 50 ps to cancel the compression of the cluster (see above) and to obtain a stable, well-relaxed metal nanocrystal embedded in silica.

For silica interactions (Si-O, Si-Si, O-O), the Watanabe-Samela interatomic potential was used,[30,31] and for gold (Au-Au), the EAM potential.[32] The gold potential reproduces the parameters relevant to thermal expansion well (see Supplementary Material for details). Gold–silica interactions were implemented using pair potentials as explained in the Supplementary Material and previous work.[33]

The effect of the SHI was implemented by assigning randomly distributed velocities obtained from the i-TS to the atoms at the beginning of the simulation. The i-TS model is a phenomenological model which utilizes heat equations to describe the temperatures of the atomic and electronic subsystems as a function of time, and has yielded a good agreement with experiments over a wide range of irradiation conditions and materials.[20] The same combined simulation approach was used to study the fine structure of ion tracks in pure silica, revealing a low density core—high density shell structure in agreement with experiments.[4]

For silica, the distribution of kinetic energy vs. radial distance from the ion path was extracted from the i-TS model for a 164 MeV Au ion at 100 fs from impact by an equilibrium temperature to kinetic energy conversion (some details of the i-TS calculation are given in the Supplementary Material). Even though this approach does not include further heat transfer between atomic and electronic subsystems, it is justified within the i-TS model since most of the heat transfer from the electrons to the lattice occurs within this time. Since the Watanabe–Samela potential overestimated the glass transition temperature

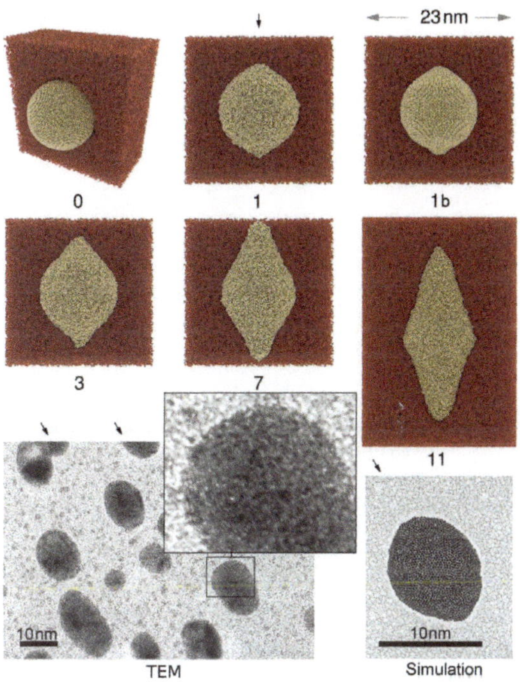

Figure 2. Elongation of Au nanocrystals in silica in MD simulations (coloured images). Each image shows a cross section of the simulation cell 50 ps after the ion impacts. The top row shows the initial cluster (0) and the shape after the recrystallization procedure (1b). Shown below each image is the total number of impacts. The energy deposited to the cell corresponds to an ion intersecting through the middle of the crystal, as indicated by the black arrow. The lower left image shows an experimental TEM image at an intermediate dose, and the lower right image shows the simulated shape after three non-overlapping impacts. The inset in the lower left image shows a zoom of one of the crystals, showing the crystalline lattice planes.

approximately a factor of 2 (we calculated 3500 ± 500 K, whereas the experimental [34] is 1995 K), the energy deposition from the i-TS model was scaled up by the same factor to match the homologous temperatures. This approach yielded track features in pure silica that matched those of experimentally measured tracks.[4] Motivated by the very rapid electronic heat conduction in Au, a constant energy of 0.5 eV/atom was deposited to Au nanocluster atoms at the same time as the deposition to silica atoms. Tests of other energy deposition schemes showed that the elongation mechanism is not sensitive to the exact energy deposition to Au, as long as sufficient energy is provided to melt the cluster (see Supplementary Material for details).

To mimic heat conduction to the bulk and to dampen pressure waves that travel through the periodic boundary, the Berendsen thermostat [35] was applied at the borders of the cubical simulation cell. The width of the boundary cooling region is about 10% of the total cell width, which was 23 nm. We extended the cell size to 33 nm in the ion beam direction to accommodate nanocrystals whose major axis had grown close to the original cell height. After the SHI energy deposition was performed, the evolution of the system was followed for 100 ps. During the last 50 ps, no significant changes in the cluster shape were found to occur.

In a single ion simulation, we observe that the gold nanocrystal first melts by the SHI energy deposition during the first few ps after the ion impact. During the same time scale, an underdense silica track core forms as described previously.[9] The crucial stage for elongation is about 5–20 ps after the impact when the molten, pressurized nanocluster expands (see Figures 3 and 4 and movie in the Supplementary Material) by a longitudinal flow of Au atoms into the underdense core of the silica track. After about 20 ps, silica cools down below the glass transition temperature and the evolution of the nanocluster shape stops.

The atomic structure in the cluster remained amorphous within the computed simulation time (100 ps). Close inspection of the experimental images of elongated clusters (see the experimental image inset in Figure 2)

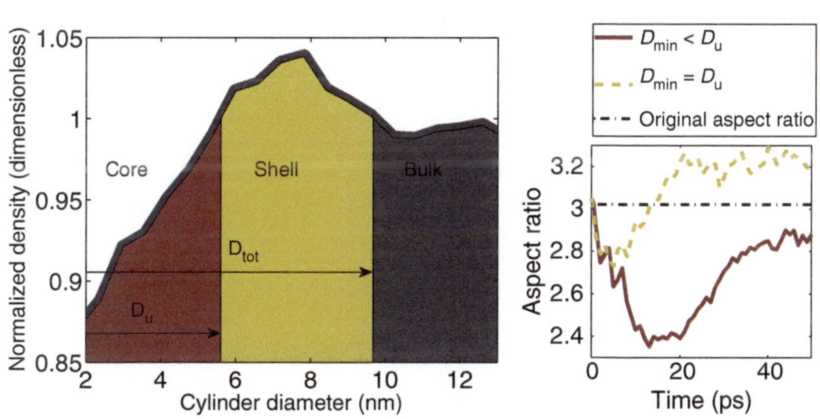

Figure 3. The final core-shell structure in the density profile of the silica track for the simulated 164 MeV Au ion and the evolution of the aspect ratio of nanocrystals whose diameter (D_{minor}) is smaller than the diameter of the underdense core (D_u) or equal to it.

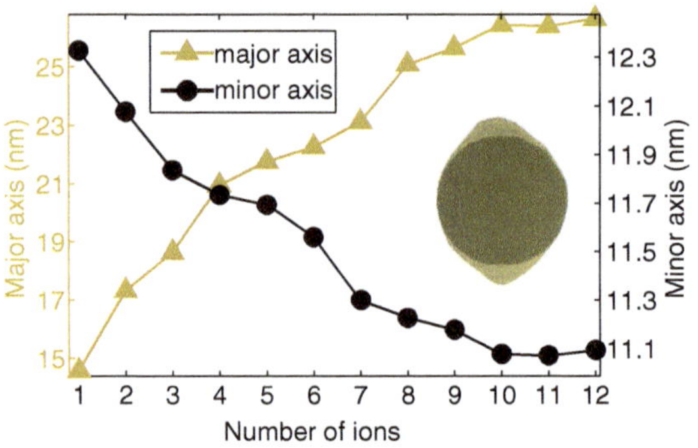

Figure 4. Evolution of the dimensions of $r = 6$ nm cluster as a function of ion impacts. Shown also the shape of the cluster before and after first impact. The overlap of these regions appear dark. Note that contrary to our previous study, the current study reproduces the loss of width of the cluster (see Figures 3 and 4 in [33]). This demonstrates the importance of the recrystallization.

shows that they are crystalline or polycrystalline, in agreement with previous work.[36] Test simulations on longer (ns) timescales showed that the cluster evolved towards a polycrystalline state, in agreement with the experiments, but simulation of full recrystallization was not practical due to the limited time scale of MD simulations.[28] From the experimental ion fluxes and known nanocrystal diameters [9,11], it is clear that the typical time scale between ion impacts on an individual nanocrystal is of the order of seconds. Therefore, it is straightforward to assume that the simulated nanoclusters would naturally crystallize before the next ion impact. Hence, we introduce a simple recrystallization scheme in which we construct a new simulation cell as the initial cell, but instead of using a crystalline sphere as the nanocrystal, the shape of the cluster 50 ps after every impact is used (Figure 2). The atom count of the particle is preserved within ±15 atoms.

On subsequent impacts, using the recrystallization procedure, the elongation proceeds gradually from a sphere to a rod-like shape (see Figures 2 and 4 and the movie in the Supplementary Material.) Without the inclusion of the recrystallization, the shape transformation does not continue after the first impact.[33] MD calculations using the EAM potential showed that an amorphous cluster has about 4% higher volume than a crystalline cluster with equal atom count (30,000) at 300 K. The gain in volume lowers the thermal expansion pressure of the cluster as the crystal melts during the next ion impact. We found this pressure to be essential for the shape transformation. Therefore, if the cluster is left amorphous, the elongation observed in the experiment can never be achieved. On the contrary, with the recrystallization step we obtain a continued elongation (Figure 4) which agrees quantitatively with experiments (Figure 5).

The simulation cells are initially stress-free (except for the transient thermal stresses introduced by the ion

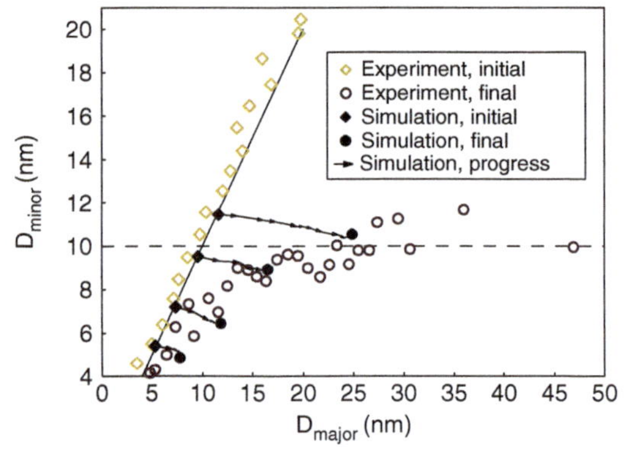

Figure 5. Plot of the dimensions of the nanocrystals before and after irradiation. Blank diamonds and circles indicate the dimensions of the initial experimental nanocrystal population and after a fluence of 2×10^{14} ions per cm^2, respectively. Filled diamonds and circles show the evolution of the simulated nanocrystals. Each arrow represents one impact. Dashed line shows the approximate saturation width of the largest nanocrystals in the experiments.

itself). Hence the current simulations show that the first stages of the elongation can be explained by thermal expansion and flow of matter from the nanocluster into the track core, and is not a consequence of an ion-hammering effect or any diffusion-like processes.

Some difference in the final elongated shapes obtained experimentally and in the simulations is evident. In the current simulations, all the SHI impacts were simulated through the centre of the nanocrystal to keep the focus on the main underlying mechanism for its elongation. At the same time, the experimental impacts cannot be expected to act likewise. Analysis of the experimental TEM image of the elongated nanocrystals (see TEM image in Figure 2) reveals shape asymmetry for some nanocrystals, which was also observed in our earlier simulations of random-position SHI impacts on Au nanoclusters.[33] Also, clusters in these simulations appeared more round at the bottom and at the top, as seen in the simulated image next to the TEM image in Figure 2. In some experimental works, the shapes of elongated nanocrystals similar to the currently simulated 'lemon-like' ones were also observed (e.g. in [14]).

We next turn to the question of why the elongation experiments show a critical width of the particle, below which no or little elongation occurs.[9,17] Plotting the experimental nanocrystal population on a graph according to their minor and major diameters (D_{minor}, D_{major}) before and after irradiation shows that the largest crystals have a width comparable to the track size in silica and that the small nanocrystals are not elongated as efficiently (Figure 5). After a dose of 2×10^{14} ions per cm^2, further irradiation induces no significant changes in the curve. This enables a comparison between the simulations and experiment. We prepared four simulation cells with spherical nanocrystals using different diameters and performed ion impact simulations as described above until no significant changes were observed in the dimensions of the nanocrystals. We then placed the simulated evolution of the maximal width and height of the crystals on the (D_{minor}, D_{major}) graph. The simulated final distribution is similar to the experimental one, as shown by the circles in Figure 5.

Since the simulations of multiple impacts on the same nanocrystal consume very large amounts of CPU time, we also constructed a test system designed to study specifically the saturation, i.e. a rod-like nanocrystal with $D_{minor} = 6\,nm$, $D_{major} = 12\,nm$. Two different energy deposition profiles were then used in an impact simulation. The first produces a track where the underdense core is larger than the minor axis of the cluster, and the second produces one which is equal to the minor axis.

These simulations enable us to identify the basic *reason behind saturation width dependence on the track size in silica* [10]: when the underdense track is larger than the minor axis of the nanocluster, it expands in all three directions. In the opposite case, the nanocluster is unable to expand in the directions perpendicular to the ion as efficiently as in the parallel direction. Therefore, the cluster is growing in aspect ratio. Due to the high temperatures in silica in the vicinity of the ion trail, the pressure exerted from silica to the cluster prevents the growth for a short time before the track core cools down, as seen from the evolution of the aspect ratios in Figure 3.

The saturation width of the largest nanocrystals matches quite well with the track radius, which includes both the underdense core and the overdense shell of the ion track (Figures 3 and 5), both in the simulations and experiments. Similar behaviour was experimentally observed by Ridgway et al. [11] for several other metal species, but some materials appear to saturate at a lower width (e.g. platinum).

In single impact simulations, the rod-like crystals whose widths are smaller than the underdense track core (Figure 3) are not growing in aspect ratio, whereas bigger ones are. It might be, therefore, expected that the saturation width is equal to the width of the underdense track. We note that the increment in the major axis dimension has to be sufficiently large for the nanocrystal to subsequently elongate, since some of the increment is lost during the recrystallization to minimize surface energy. When the minor axis of the crystal is equal to the underdense core, growth in aspect ratio occurs initially, but most of it is lost during the recrystallization. The major axis growth is sufficient for elongation only when the minor axis is clearly larger than underdense core. Hence the saturation width that is obtained in Figure 5 is larger than the diameter of the underdense core.

In conclusion, we have performed MD simulations in conjunction with the i-TS model that reproduce the experimentally observed elongation of Au nanocrystals under SHI irradiation. In the simulations, the elongation is caused by an anisotropic thermal expansion of the molten nanocluster within the underdense track core in silica. Nanocrystals whose minor axis is smaller than the diameter of the underdense track expand more isotropically and are not elongated, which leads to a saturation of the width of the crystals. The saturation width was also observed experimentally. The results imply that both formation of an underdense track in silica and shape-conserving crystallization of the molten cluster between ion impacts are necessary prerequisites for elongation of metallic nanocrystals by SHI impacts.

Acknowledgements This project was funded by the Academy of Finland and the National Doctoral Programme in Nanoscience (NGS) of Finland. The authors thank Marcel Toulemonde for providing the energy deposition profile for SiO_2 and CSC—IT Center for Science Ltd (Finland) for grants of computation time. P.K. and M.C.R. thank the Australian Research Council for support.

References

[1] D'Orléans C, Stoquert JP, Estournes C, Cerruti C, Grob JJ, Guille JL, Haas F, Muller D, Richard-Plouet M. Anisotropy of Co nanoparticles induced by swift heavy ions. Phys Rev B. 2003;67:220101.

[2] Klaumünzer S. Modification of nanostructures by high-energy ion beams. Nucl Instrum Methods Phys Res B. 2006;244(1):1–7.

[3] Rizza G, Dunlop A, Dezellus A. Behavior of metallic nanoparticles in Al matrix under high electronic energy deposition. Nucl Instrum Methods Phys Res B. 2007;256(1):219–223

[4] Kluth P, Schnohr CS, Pakarinen OH, Djurabekova F, Sprouster DJ, Giulian R, Ridgway MC, Byrne AP, Trautmann C, Cookson DJ, Nordlund K, Toulemonde M. Fine structure in swift heavy ion tracks in amorphous SiO_2. Phys Rev Lett. 2008;101:175503.

[5] Reyes-Esqueda JA, Rodriguez-Iglesias V, Silva-Pereyra HG, Torres-Torres C, Santiago-Ramirez AL, Cheang-Wong JC, Crespo-Sosa A, Rodriguez-Fernandez L, Lopez-Suarez A, Oliver A. Anisotropic linear and nonlinear optical properties from anisotropy-controlled metallic nanocomposites. Opt Express. 2009;17(5): 12849–12868.

[6] Awazu K, Wang X, Fujimaki M, Tominaga J, Aiba H, Ohki Y, Komatsubara T. Elongation of gold nanoparticles in silica glass by irradiation with swift heavy ions. Phys Rev B. 2008;78:054102.

[7] Awazu K, Wang X, Fujimaki M, Komatsubara T, Watanabe J, Matsumoto Y, Warisawa S, Ishihara S. The fabrication of aligned pairs of gold nanorods in SiO_2 films by ion irradiation. Nanotechnology. 2009;20:325303.

[8] Dawi EA, Rizza G, Mink MP, Vredenberg AM, Habraken FHPM. Ion beam shaping of Au nanoparticles in silica: particle size and concentration dependence. J Appl Phys. 2009;105:074305.

[9] Kluth P, Giulian R, Sprouster DJ, Schnohr CS, Byrne AP, Cookson DJ, Ridgway MC. Energy dependent saturation width of swift heavy ion shaped embedded Au nanoparticles. Appl Phys Lett. 2009;94(11):113107.

[10] Avasthi DK, Mishra YK, Singh F, Stoquert JP. Ion tracks in silica for engineering the embedded nanoparticles. Nucl Instrum Methods Phys Res B. 2010;268:3027–3034.

[11] Ridgway MC, Giulian R, Sprouster DJ, Kluth P, Araujo LL, Llewellyn DJ, Byrne AP, Kremer F, Fichtner PFP, Rizza G, Amekura H, Toulemonde M. Role of thermodynamics in the shape transformation of embedded metal nanoparticles induced by swift heavy-ion irradiation. Phys Rev Lett. 2011;106(9):095505.

[12] Amekura H, Ishikawa N, Okubo N, Ridgway MC, Giulian R, Mitsuishi K, Nakayama Y, Buchal C, Mantl S, Kishimoto N. Zn nanoparticles irradiated with swift heavy ions at low fluences: optically-detected shape elongation induced by nonoverlapping ion tracks. Phys Rev B. 2011;83(20):205401.

[13] Dufour C, Khomenkov V, Rizza G, Toulemonde M. Ion-matter interaction: the three-dimensional version of the thermal spike model. Application to nanoparticle irradiation with swift heavy ions. J Phys D Appl Phys. 2012;45:065302. doi:10.1088/0022-3727/45/6/065302

[14] Rizza G, Coulon PE, Khomenkov V, Dufour C, Monnet I, Toulemonde M, Perruchas S, Gacoin T, Mailly D, Lafosse X, Ulysse C, Dawi EA. Rational description of the ion-beam shaping mechanism. Phys Rev B. 2012;86:035450.

[15] Sun L, Banhart F, Krasheninnikov AV, Rodriguez-Manzo JA, Terrones M, Ajayan PM. Carbon nanotubes as high-pressure cylinders and nanoextruders. Science. 2006;312:1199.

[16] Sun L, Krasheninnikov AV, Ahlgren T, Nordlund K, Banhart F. Plastic deformation of single nanometer-sized crystals. Phys Rev Lett. 2008;101:156101.

[17] Mishra YK, Singh F, Avasti DK. Synthesis of elongated Au nanoparticles in silica matrix by ion irradiation. Appl Phys Lett. 2007;91:063103.

[18] Link S, El-Sayed MA. Spectral properties and relaxation dynamics of surface plasmon electronic oscillations in gold and silver nanodots and nanorods. J Phys Chem B. 1999;103(40):8410–8426.

[19] Barnes WL, Dereux A, Ebbesen TW. Surface plasmon subwavelength optics. Nature. 2003;424;824–830.

[20] Toulemonde M, Assmann W, Dufour C, Meftah A, Studer F, Trautmann C. Experimental phenomena and thermal spike model description of ion tracks in amorphisable inorganic insulators. Mat Fys Medd Kong Dan Vid Selsk. 2006;52:263.

[21] Ziegler JF, Biersack JP, Littmark U. The stopping and range of ions in matter. New York: Pergamon; 1985.

[22] Kerboua CH, Chicoine M, Roorda S. Gold nanoparticles resist deformation by swift heavy ion irradiation when embedded in a crystalline matrix. Nucl Instrum Methods Phys Res B. 2011;269(18):2006–2010.

[23] Benyagoub A, Löffler S, Rammensee M, Klaumünzer S, Saemann-Ischenko G. Plastic deformation in SiO_2 induced by heavy-ion irradiation. Nucl Instrum Methods Phys Res B. 1992;65(1–4):228–231.

[24] Nordlund K. PARCAS computer code. 2006.

[25] Nordlund K, Ghaly M, Averback RS, Caturla M, Diaz de la Rubia T, Tarus J. Defect production in collision cascades in elemental semiconductors and FCC metals. Phys Rev B. 1998;57(13):7556–7570.

[26] Ghaly M, Nordlund K, Averback RS. Molecular dynamics investigations of surface damage produced by keV self-bombardment of solids. Philos Mag A. 1999;79(4):795.

[27] Nordlund K. Molecular dynamics simulation of ion ranges in the 1–100 keV energy range. Comput Mater Sci. 1995;3:448.

[28] Allen MP, Tildesley DJ. Computer simulation of liquids. Oxford: Oxford University Press; 1989.

[29] Djurabekova F, Nordlund K. Atomistic simulation of the interface structure of Si nanocrystals embedded into amorphous silica. Phys Rev B. 2008;77:115325.

[30] Watanabe T, Yamasaki D, Tatsumura K, Ohdomari I. X. Improved interatomic potential for stressed Si, O mixed systems. Appl Surf Sci. 2004;234:207–213.

[31] Samela J, Nordlund K, Popok VN, Campbell EEB. Origin of complex impact craters on native oxide coated silicon surfaces. Phys Rev B. 2008;77:075309.

[32] Foiles SM, Baskes MI, Daw MS. Embedded-atom-method functions for the fcc metals Cu, Ag, Au, Ni, Pd, Pt, and their alloys. Phys Rev B. 1986;33(12):7983.

[33] Leino AA, Pakarinen OH, Djurabekova F, Nordlund K. A study on the elongation of embedded Au nanoclusters in SiO_2 by swift heavy ion irradiation using MD simulations. Nucl Instrum Methods Phys Res B. 2012;282:76–80.

[34] Haynes WM. CRC Handbook of chemistry and physics. 93rd ed. (Internet Version 2013). Boca Raton, FL: CRC Press/Taylor and Francis; 2012–2013.

[35] Berendsen HJC, Postma JPM, Gunsteren WF, DiNola A, Haak JR. Molecular dynamics with coupling to external bath. J Chem Phys. 1984;81(8):3684.

[36] Silva-Pereyra H, Arenas-Alatorre J, Rodriguez-Fernandez L, Crespo-Sosa A, Cheang-Wong J, Reyes-Esqueda J, Oliver A. High stability of the crystalline configuration of Au nanoparticles embedded in silica under ion and electron irradiation. J Nanoparticle Res. 2010;12:1787–1795.

Exceptional Friction Mitigation via Subsurface Plastic Shear in Defective Nanocrystalline Ceramics

H. Mohseni, B. A. Mensah, N. Gupta, S. G. Srinivasan and T. W. Scharf*

Department of Materials Science and Engineering, University of North Texas, Denton, TX 76207, USA

Single-crystal and microcrystalline ceramics classically fail via brittle fracture under loading. Here, we report that atomic layer deposition {0002}-textured ZnO films, with nanocolumnar grains and defective sub-stoichiometric structure, exhibit nanocrystalline plasticity and exceptionally low sliding friction and wear. These films were tailored to contain low-energy basal stacking fault planes parallel to the film surface. Cross-sectional high-resolution TEM studies inside friction-induced surfaces coupled with density functional theory calculations revealed the fundamental mechanism of this unexpected behavior. We show that sliding on {0002}-textured nanocrystalline ZnO films increases partial dislocation density through intrafilm shear, via glide of partial dislocations, on {0002}-stacking fault planes.

Keywords: Nanocrystalline Ceramics, Friction, Dislocations, Plastic Deformation, Density Functional Theory

Friction and the attendant wear of materials is a millennia old and, yet, a persistent problem, causing reduced efficiencies of engines, turbines, industrial machinery, and other technologies.[1] Friction arises whenever one material slides over another and is commonly measured macroscopically by the friction coefficient μ. Third bodies, such as liquid lubricants, atomically thin surfaces, and/or chemically and structurally altered wear surfaces, influence friction.[1–3] While entrapped in the moving interface, they often go unnoticed because they are buried in the contact. Nonetheless, they influence how materials accommodate the relative sliding motion, known as velocity accommodation modes (VAM), with interfacial sliding between the third body and the wearing surface being the most common.[3] The friction coefficient is controlled by a velocity accommodation parameter, often referred to as the interfacial shear strength, τ_o, where $\tau_o = \mu P$ and P is the contact pressure. This shear strength is related to, but not necessarily the same as, the shear strength of the material or third body. Godet [2] postulated the existence of other VAM, each with its own friction process. However, many of these have not been experimentally observed due to the lack of appropriate in situ triboscopy techniques or their occurrence in the subsurface.

Traditional liquid lubricants fail under extreme temperature, oxidative, and stress environments due to diminished intrinsic interplanar mechanical sliding. Therefore, the recent focus is to develop a new class of solid lubricants for use in extreme environments.[4] Nanocrystalline (NC) ZnO with a hexagonal wurtzite crystal structure is a promising candidate due to its low-friction deformation behavior.[4] This expectation is supported by theoretical [5] and computational [6–8] studies that show a change in deformation mechanisms from traditional dislocation-mediated slip to grain boundary sliding and rotation in randomly oriented NC ceramics with grains below 30 nm.

In this letter, we couple density functional theory (DFT) calculations, friction experiments, site-selective focused ion beam (FIB) microscopy, and cross-sectional high-resolution transmission electron microscopy (HRTEM) to probe friction-induced, buried subsurfaces in textured NC ZnO. We have identified a novel subsurface intrafilm shear VAM mechanism that significantly reduces friction and increases ductility. This mechanism is not found in either single-crystal or microcrystalline ZnO. We contend that the subsurface intrafilm shear mechanism can operate in other textured NC ceramics and could dramatically expand the

*Corresponding author. Email: scharf@unt.edu

set of lubricious materials available for use in extreme environments.

Single-crystal and microcrystalline ceramic oxides are inherently brittle, which results in cracking and subsequently increased abrasive frictional losses. NC ceramic oxides, in contrast, exhibit improved mechanical and tribological properties.[4,6,9–11] For instance, the friction coefficient of NC ZnO ($\mu = 0.1$–0.2) is significantly less than hot-pressed randomly orientated microcrystalline ZnO ($\mu = 0.7$). This is achieved by controlling the ZnO (0002)-textured grain structure and oxygen sub-stoichiometry.[4] This low friction was proposed to be due to the defective (0002)-nanocolumnar grain structure with subgrains that can accommodate plastic deformation.[4] Those authors suggested that site-selective TEM studies inside the worn subsurface are needed to conclusively determine the friction dissipation mechanisms. In this letter, we report the first results of such a characterization.

ZnO [12–14] and other hexagonal wurtzite crystals [15,16] are known to develop a high density of stacking faults (SFs) in combination with threading and partial dislocations (PDs) during film deposition. We hypothesize that these growth defects force dislocations to glide parallel to the interfacial sliding direction to accommodate shear. Interrogation of NC ZnO defect structure before and after sliding by TEM coupled with DFT calculations of the formation energies of these SFs is used to elucidate the fundamental plasticity mechanisms.

Atomic layer deposition (ALD) nanolaminate tri-layer stacks of ZnO (~100 nm thick)/Al_2O_3 (~20 nm)/ZrO_2 (~100 nm) were grown on Si (100) substrates as described elsewhere.[17] The amorphous Al_2O_3 below the ZnO layer helps the low surface energy (0002) plane of ZnO to grow with the *c*-axis preferred orientation by suppressing other competing higher surface energy planes, e.g. $(1\,0\,\bar{1}\,0)$ and $(1\,0\,\bar{1}\,1)$.[11] The underlying tetragonal ZrO_2 acts as a load-bearing layer (at moderate contact pressures of ~1 GPa) to minimize subsurface frictional shear stresses. The deposited nanolaminate was first annealed in vacuum at 400°C for 2 h to determine changes to defect structures (SFs and PDs) and friction. Previous X-ray diffraction (XRD) and cross-sectional HRTEM analyses confirmed that the as-deposited nanolaminates possess nanocolumnar ZnO grains of 28 nm average width and with random rotational orientation.[11,18] Annealing promoted limited grain growth to 34 nm as seen by XRD peak narrowing. The absence of XRD peak shifts in both the as-deposited and annealed nanolaminates indicates negligible lattice strain. Furthermore, the ZnO (0002)-orientated grains have a texture coefficient (TC) of 2.9 determined using the Harris method, where a TC > 1 indicates preferential orientation. X-ray photoelectron spectroscopy (XPS) revealed that the as-deposited $Zn_{0.51}O_{0.49}$ layer became more sub-stoichiometric ($Zn_{0.54}O_{0.46}$) with annealing. This has been linked to condensation of oxygen vacancies introducing a missing or extra ZnO (0002)-lattice plane and incoherent boundaries between adjacent nanocolumnar grains.[19] These are sources of ZnO growth basal stacking faults (BSF) and their crucial role in mitigating friction is discussed later.

Unidirectional sliding friction tests were conducted at room temperature in open air (~40% RH) using a stationary 1.59 mm diameter Si_3N_4 ball as a counterface material. Normal loads (L) of 0.245, 0.49, 0.98, and 2.45 N were applied, which correspond to initial mean Hertzian contact pressures of 0.46, 0.58, 0.72, and 0.98 GPa, respectively. The sliding velocity was 2.1 cm/s and at least five tests were performed for each condition. A transducer in the load arm measured the tangential load (F) over a sliding distance of 200 m. The friction coefficient μ is F/L and is in steady state when it changes by less than 5% from cycle to cycle. Wear rates (in $mm^3/N\,m$) were also determined as the wear surface volume divided by the normal load and by the total distance traveled by the Si_3N_4 counterface. It was computed by multiplying the averaged cross-sectional area of the wear track, measured by surface profilometry traces, by the circumference of the wear track (assuming uniform wear) and dividing the result by the normal load and by the total distance. In addition to the nanolaminates, monolithic (0001) ZnO single crystal (SC) (oxygen surface terminated, MTI Corp, Richmond, CA, USA) and ZnO with randomly orientated NC surface grains (Cerac Inc, Mayfield Heights, OH, USA) were tested under 0.72 GPa contact pressure.

The steady-state friction coefficients μ_{ss} of textured NC ZnO shown in Figure 1(a) were fitted to a power law equation of the normal load. The fits of $\mu_{ss} = L^{-0.32}$ and $\mu_{ss} = L^{-0.28}$ for as-deposited and annealed ZnO, respectively, agree well with the Hertzian contact model,

$$\mu = \tau_o \pi \left(\frac{3R}{4E^*}\right)^{2/3} L^{-1/3} + \alpha, \qquad (1)$$

where τ_o is the interfacial shear strength between the sliding surfaces, R is the ball radius, and E^* is the composite elastic modulus of the materials. The term α represents the pressure dependence of shear strength—it is the lowest attainable friction coefficient for the testing couples. Figure 1(a) shows that μ_{ss} values for both as-deposited and annealed NC ZnO layers decrease with increasing L (or decreasing inverse mean Hertzian contact pressure, P_H). This friction behavior contrasts Amonton's first law of friction wherein μ is independent of L. Similar non-Amontonian friction behavior was seen in molecular dynamic friction simulations of NC SiC sliding on diamond.[20] The linear relationship between μ and $L^{-1/3}$ in Equation (1) has been experimentally verified for a number of low-friction solid lubricant coatings.[21]

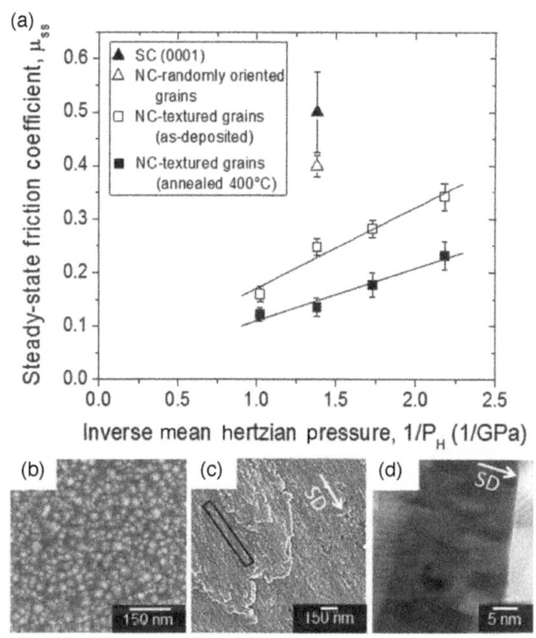

Figure 1. Linear regression fit of steady-state friction coefficients as a function of inverse mean Hertzian contact pressure for as-deposited and annealed ZnO NC-textured nanolaminate. Comparative data are also shown for SC-(0001) and randomly oriented NC ZnO. HRSEM surface images of (b) unworn annealed NC-textured ZnO, and (c) worn nanolaminate with friction coefficient denoted by the filled square symbol at $1.4\,GPa^{-1}$ in (a). SD is the sliding direction and the rectangle shows location of cross-sectional FIB-cut. (d) Corresponding HRTEM subsurface image acquired along the FIB lift-out in (c) shows a plastically deformed ZnO nanocolumnar grain with intrinsic SFs running parallel to the SD.

The steady-state friction coefficients μ_{ss} of SC ZnO and NC-randomly oriented ZnO grains plotted in Figure 1(a), in contrast, are significantly higher than the NC-textured ZnO grains. This suggests that texturing of (0002) grains plays an important role in reducing interfacial shear strength as opposed to just having NC-sized grains. Also, a favorable SC (0001)-basal plane orientation in ZnO by itself does not yield low friction, alluding to the importance of a high density of grain boundaries. These results also substantiate the sensitivity of the macroscopic friction measurements in differentiating between NC-textured and randomly orientated ZnO grains. Another important observation is that the annealed ZnO layer exhibited lower friction than the as-deposited layer at each corresponding contact pressure. Therefore, mechanistic studies are needed to explain these experimental results. Specifically, the exact crystal defect structure dependence of deformation mechanisms needs to be identified using advanced microscopy inside the deformed surfaces and subsurfaces, and this has been done below.

Ramping of the normal load during sliding also allows us to explore the pressure dependence of μ_{ss} on the interfacial shear strength via the equation

$$\mu_{ss} = \frac{F}{L} = \frac{A_R \tau}{L} = \frac{\tau}{P_H} = \frac{\tau_o}{P_H} + \alpha, \qquad (2)$$

where A_R is the real contact area according to Bowden and Tabor.[22] Fitting the μ_{ss} values in Figure 1(a) as a function of P_H yields mean τ_o (and α) values of 151 (0.021) and 98 (0.013) MPa for the as-deposited and annealed NC ZnO nanolaminates, respectively. These values are comparable to many low-friction solid lubricant films.[21] Note that the normal load dependency on the friction coefficient (Equation (1)) and the pressure-induced shear stress (Equation (2)) analyses for NC ZnO are valid because our cross-sectional HRTEM analysis shows that underlying Al_2O_3 and ZrO_2 nanolaminates and the Si substrate do not plastically deform.

In addition to friction coefficients, the resulting sliding wear rates are another important property. The wear rates at a normal load of 0.98 N (0.72 GPa contact pressure) are $2 \pm 0.8 \times 10^{-7}\,mm^3/N\,m$ and $8 \pm 2 \times 10^{-8}\,mm^3/N\,m$ for as-deposited and annealed NC-textured ZnO grains, respectively. The slightly lower wear rate with annealing is in good agreement with the trend in friction coefficients in Figure 1. In contrast, both the NC-randomly oriented ZnO grains and SC (0001) ZnO wear rates are higher in the range of $10^{-6}\,mm^3/N\,m$. Wear rates in the 10^{-7} to $10^{-8}\,mm^3/N\,m$ range are considered low wear regime materials, often exhibited by solid lubricant coatings, such as diamond-like carbon and nanocomposite-alloyed MoS_2.[21] Therefore, the NC-textured ZnO grain nanolaminate film exhibits a very good wear behavior at both room and elevated temperatures.

A HRSEM image of annealed nanolaminate surface prior to sliding shown in Figure 1(b) reveals surface grain sizes of ~30 nm, in agreement with our prior XRD results.[11,17] The surface morphology of these nanoscopic grains is relatively smooth with atomic force microscopy determining the average root mean square roughness to be 3.45 ± 0.07 nm. Therefore, surface roughness effects would likely have an insignificant contribution to sliding friction and wear under larger real contact areas. After sliding at 0.72 GPa contact pressure, the deformed ZnO surface in Figure 1(c) exhibits ductile layering/smearing but not brittle cracking. This implies that the individual ZnO nanocolumns plastically shear. The deformed surface morphology indicates an additive shear across neighboring grains accumulating to the microscale. Figure 1(d) is a cross-sectional TEM image acquired along the FIB lift-out region shown in Figure 1(c). It shows a diffracting, plastically deformed 20–25 nm wide ZnO nanocolumnar grain with BSFs parallel to the sliding direction. It should be noted that a careful FIB lift-out procedure was followed to circumvent Ga ion-beam damage during milling that could induce defect formation in the film. Therefore, the first step was to deposit Pt by using the electron-beam and then the ion-beam, which protects the TEM sample against beam damage. The second step includes milling and final polishing of the lifted-out sample using the Ga ion-beam at a conservatively low voltage of 5 keV and current at 70 pA to prevent Ga ion implantation.

Our DFT calculations of intrinsic basal and prismatic SF energies in wurtzite ZnO used the Vienna Ab Initio Simulation Package employing the projected augmented plane wave method.[23–25] Electron exchange and correlation are described within the generalized gradient approximation using the Perdew–Burke–Ernzerhof parameterization.[26] For a single unit cell ($1 \times 1 \times 12$ with dimensions of $3.29 \times 2.85 \times 63.66$ Å), the energy cut-off was 532 eV and the k-point spacing used was <0.185/Å. Basal and prismatic SFs were introduced into SC ZnO crystal via a suitable displacement of two halves of the crystal. The faulted structures are relaxed until the forces on the atoms were less than 0.02 eV/Å. The stacking fault excess energy, also called the stacking fault energy (SFE), is calculated by subtracting total energies of the perfect ZnO crystal from the energy of the relaxed faulted system, and dividing this excess energy by the area of the stacking fault.

We now explain the microstructural defect evolution during film growth and after sliding. A missing or extra ZnO (0002) plane in growth BSFs may originate from incoherent boundaries between adjacent nanocolumns and/or condensation of vacancies or interstitials in NC (0002)-texture ZnO since the annealed ZnO layer is oxygen deficient. This could likely weaken the Zn–O bond strength and influence the energetics of dislocation nucleation and glide. The TEM image in Figure 1(d), and also weak beam dark field cross-sectional TEM, shows BSFs originating at low angle (~15°) tilt grain boundaries. A cross-sectional HRTEM image acquired from the annealed, unworn ZnO layer shown in Figure 2(a) reveals a growth, intrinsic type I single BSF bordered by a PD. Type I intrinsic BSFs have only one violation in the stacking sequence.[16] Our DFT calculations show ······ *AaBbAaBbCcBbCc* ······ stacking sequence for type I BSFs as shown in Figure 2(b), where the capital and smaller letters denote a (0002) double layer occupied by Zn or O ions, respectively. The zinc blende structure forms locally with *AaBbCc* arrangement. This low-energy type I BSF is bordered by sessile Frank PDs with rigid body displacement (Burgers) vector equal to $\frac{1}{6}\langle 02\bar{2}\bar{3}\rangle$, shown in Figure 2(c). The intrinsic basal SFE was calculated to be 24 J/m^2, which is much lower and favorable than the calculated $\{10\bar{1}0\}$-prismatic SFE of 427 mJ/m^2. The average PD density bordering the BSF was based on counting the PDs in 10 different micrographs like the one in Figure 2(a). It is

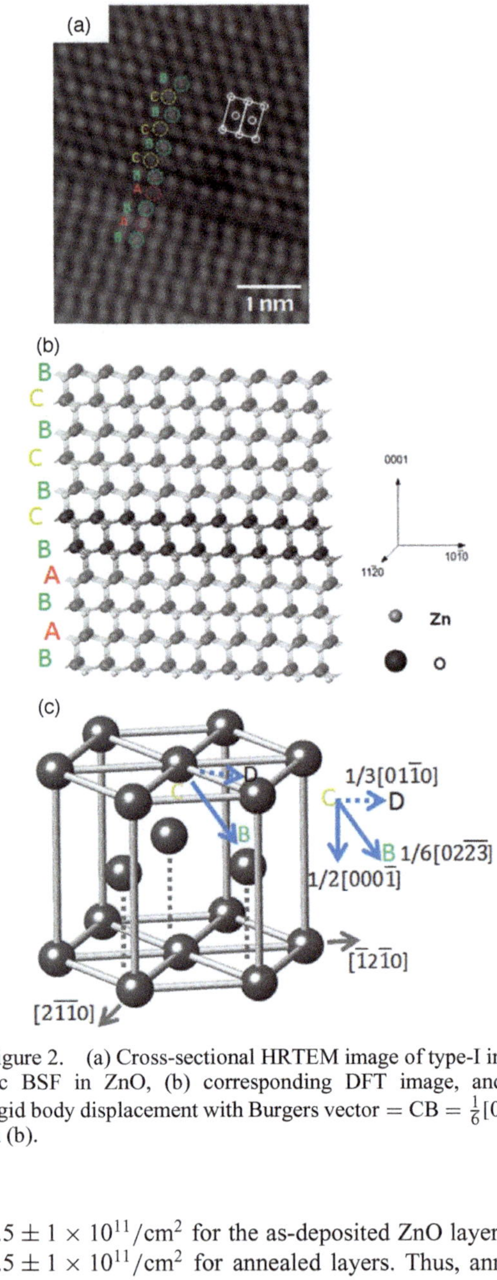

Figure 2. (a) Cross-sectional HRTEM image of type-I intrinsic BSF in ZnO, (b) corresponding DFT image, and (c) rigid body displacement with Burgers vector = CB = $\frac{1}{6}[02\bar{2}\bar{3}]$ in (b).

$1.5 \pm 1 \times 10^{11}$/cm^2 for the as-deposited ZnO layer and $6.5 \pm 1 \times 10^{11}$/cm^2 for annealed layers. Thus, annealing increases growth BSF likely due to an increase in oxygen vacancies determined by XPS, and correlates reduced friction to the increase in BSFs and PDs. In contrast, the higher friction observed for SC ZnO and NC-randomly orientated ZnO emphasizes the important role of grain boundaries as sources of BSFs and PDs in accommodating interfacial shear. The orientation of the PDs with respect to the sliding direction also has an effect as discussed below.

To determine if these mechanisms operate after frictional contact, cross-sectional HRTEM studies were done inside the ZnO subsurfaces of annealed nanolaminate with different NC grain orientations. Figure 3(a) shows

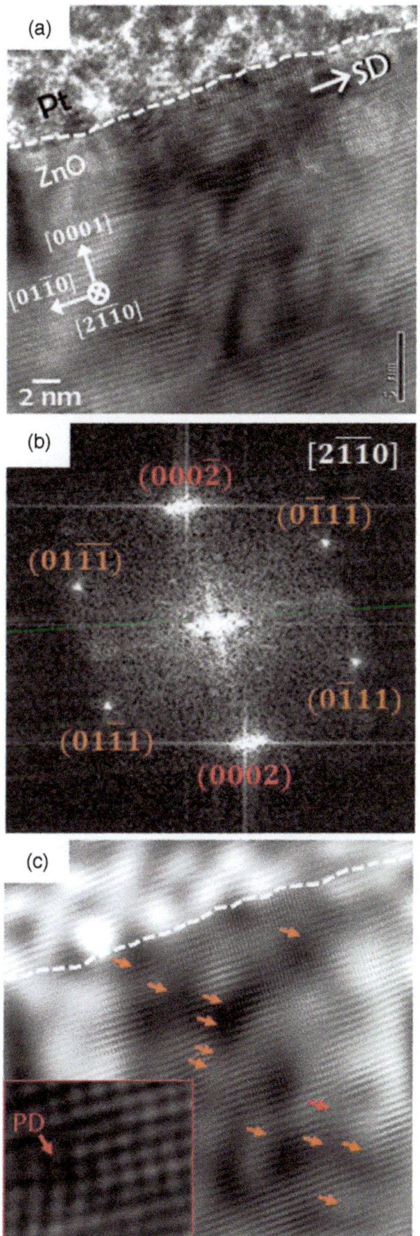

Figure 3. View of {0002} basal planes in a nanocolumnar ZnO grain. (a) Cross-sectional HRTEM image using $g = (01\bar{1}0)$ shows subsurface plastic deformation after frictional contact. SD is the sliding direction. (b) Corresponding FFT diffraction pattern shows that the basal planes are observed to be streaked along the $[2\bar{1}\bar{1}0]$ beam direction, which is consistent with the presence of stacking faults running normal to the $[2\bar{1}\bar{1}0]$ direction. (c) The identification of these dislocations is confirmed in the corresponding Fourier-filtered image using the {0002} diffraction spots, which demonstrate that each PD (shown by arrows) corresponds to one additional (0002) plane. The inset image shows a higher magnification view of the PD with the red arrow. The PD density increases with sliding (multiple slip of PDs by dislocation glide). The dashed lines represent the interface between the protective Pt coating and the ALD ZnO nanolaminate.

an image of near surface {0002}-orientated basal planes in a nanocolumnar ZnO grain with discernible small-scale contrast due to strain variation indicating a high density of dislocations. The strain is induced by PDs that terminate BSFs. It is evident that there is near surface ZnO grain deformation through subsurface-induced plastic shear with no evidence of brittle cracking. The corresponding fast Fourier transform (FFT) diffraction pattern in Figure 3(b) shows that the basal planes are observed to be streaked along the $[2\bar{1}\bar{1}0]$ beam direction. This is consistent with the presence of type I BSFs, since $\mathbf{g} \cdot \mathbf{b} \neq 0$ where $\mathbf{g}$ is $(01\bar{1}0)$ and $\mathbf{b}$ is $\frac{1}{6}\langle 02\bar{2}\bar{3}\rangle$, from Figure 2(c). The identification of these PDs is confirmed in the corresponding Figure 3(c) Fourier-filtered image using the {0002} diffraction spots, which demonstrate that each PD (shown by arrows) corresponds to one additional (0002) plane. The inset image in Figure 3(c) shows a higher magnification view of the PD denoted with the red arrow that terminates a BSF. The PD density increases by an order of magnitude to $1.2 \pm 1 \times 10^{12}/\mathrm{cm}^2$ as a result of frictional contact due to their nucleation at the nanocolumnar boundaries, which in turn increases the available pathways for dislocation glide. This averaged PD density was calculated by counting the number of PD from 12 similar TEM micrographs, like the one shown in Figure 3(c). The higher densities of growth BSFs also confine the slip to within each NC grain as inferred from non-observable gross-slip across multiple nanocolumnar grain boundaries and the absence of slip transmission across grain boundaries. Therefore, low friction and wear may be attributed to localized NC plasticity during sliding via a novel intrafilm shear velocity accommodation mode. This localized NC plasticity is single-crystal additive and accumulates to the microscale since the deformed surfaces, like the one shown in Figure 1(c), exhibit ductile layering/smearing.

Figure 4 shows another NC ZnO grain, but with $\{01\bar{1}0\}$ prismatic orientation. In contrast to the basal planes, the cross-sectional HRTEM image in Figure 4(a) shows no subsurface plastic deformation in the nanocolumnar grain after frictional contact. The corresponding FFT diffraction pattern shows that the prismatic planes are not streaked along the same $[2\bar{1}\bar{1}0]$ beam direction, which is consistent with no stacking faults running normal to the $[2\bar{1}\bar{1}0]$ direction. The absence of dislocations is confirmed in the corresponding Fourier-filtered image using the $\{01\bar{1}0\}$ diffraction spots, which means there are no prismatic stacking faults in the worn subsurface. This is consistent with our DFT calculated high SFE of $\sim$427 mJ/m^2 for the ZnO prismatic planes compared to $\sim$24 mJ/m^2 for the {0002}-basal SFE. Furthermore, these results imply that only basal slip is energetically favored, and that non-interacting slip systems translate to low interfacial shear and friction. A similar mechanism was postulated in HCP cadmium when the (0001) glide plane is parallel to surface and glide of the

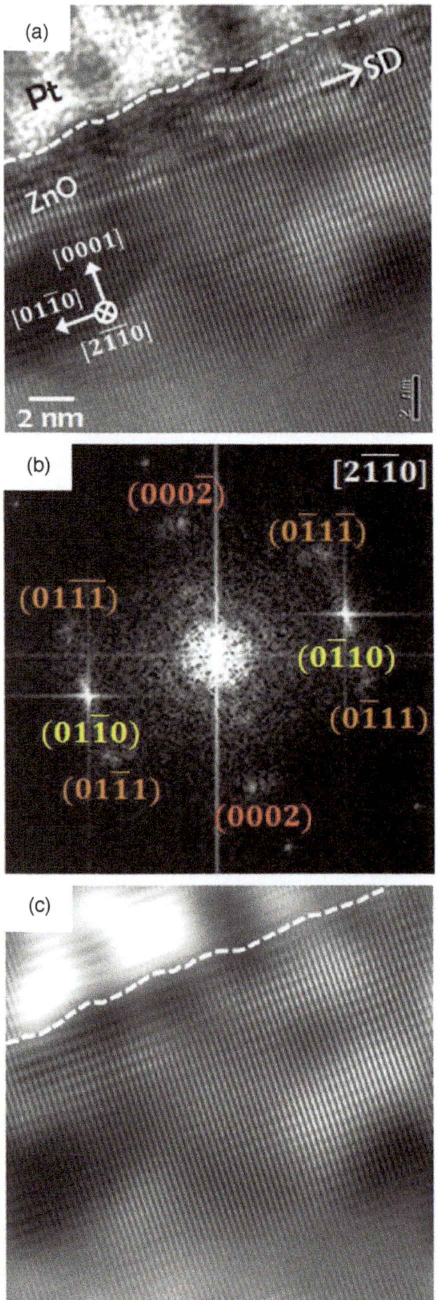

Figure 4. View of $\{01\bar{1}0\}$ prismatic planes in a nanocolumnar ZnO grain. (a) Cross-sectional HRTEM image using $g = (01\bar{1}0)$ showing no subsurface plastic deformation in a nanocolumnar grain after frictional contact. SD is the sliding direction. (b) Corresponding FFT diffraction pattern shows that the prismatic planes are not streaked along the $[2\bar{1}\bar{1}0]$ beam direction, which is consistent with no stacking faults running normal to the $[2\bar{1}\bar{1}0]$ direction. (c) The absence of dislocations is confirmed in the corresponding Fourier-filtered image using the $\{01\bar{1}0\}$ diffraction spots, which translates to no prismatic stacking faults in the worn subsurface (due to its high SFE). The dashed lines represent the interface between the protective Pt coating and ALD ZnO nanolaminate.

subsurface edge dislocations produce low shear.[27] In a low SFE system like cadmium, dissociation of PDs into widely spaced partials reduces the critical resolved shear stress required for their glide on the fault plane. However, in the case of SC ZnO, cross-sectional HRTEM revealed cleavage cracks along (0002) planes, cause brittle fracture and larger wear particle generation leading to a higher friction.[18] Since there are no low angle columnar grain boundaries for nucleation of BSFs and PDs, like in NC ZnO, we cannot achieve low interfacial shear. Higher friction and wear also occur in NC-randomly orientated ZnO grains, where many interacting slip systems would operate. Frictional contact in these randomly oriented grains likely caused the interaction and interlocking of slip systems, and ultimately localized hardening and brittle failure. This behavior is unlike other randomly oriented NC ceramics where grain boundary sliding and grain rotations contributed to enhanced plasticity.

It can be argued that the interfacial shear strength (τ_o) obtained from Equations (1) and (2) during sliding is that between the Si_3N_4 counterface and ZnO layer. Because there was no detectable ZnO third body transferred to the Si_3N_4 counterface, we can assert that τ_o is not the value for Si_3N_4 counterface and ZnO layer. Our TEM analysis also supports this assertion by revealing the subsurface intrafilm shear along the ZnO $\{0002\}$-basal planes parallel to the sliding direction. Therefore, while there is relative sliding between the Si_3N_4 counterface and ZnO layer, it is the defective ZnO–ZnO basal plane interface that determines the friction coefficients and interfacial shear strengths. Thus, it is likely this intrafilm VAM is independent of the counterface material as long as it is not plastically deforming and forming a third body transfer film.

In summary, we have shown that it is feasible to achieve low friction and wear ceramic oxides by controlling the nanometer-scale structure. NC ZnO with low SFE $\{0002\}$-textured grains plastically deform via shearing of subsurface BSFs parallel to the sliding direction. The resultant novel intrafilm shear VAM yields exceptionally low interfacial shear strengths and friction and prevents brittle fracture commonly observed in conventional ZnO, and has broad implications to other defective NC ceramics. The comparisons with SC ZnO of identical texture and randomly oriented NC ZnO further support this VAM. Therefore, we have uncovered an unexpected trend wherein controlling the texture, grain size, and defect structure can potentially transform a conventional brittle oxide into a ductile one during sliding.

Acknowledgements The authors would like to acknowledge support from the US National Science Foundation (Grant numbers CMMI-0700828, 0846444, and 1100648). We also acknowledge the UNT Center for Advanced Research and Technology (CART) and UNT's Talon cluster for computations.

References

[1] Singer IL, Pollock HM. Fundamental of friction: macroscopic and microscopic processes. Dordrecht: Kluwer Academic Publisher; 1992.

[2] Godet M. The third-body approach: a mechanical view of wear. Wear. 1984;100:437–452.

[3] Singer IL, Dvorak SD, Wahl KJ, Scharf TW. Role of third bodies in friction and wear of protective coatings. J Vac Sci Technol A. 2003;21:S232–S240.

[4] Zabinski JS, Sanders JH, Nainaparampil J, Prasad SV. Lubrication using a microstructurally engineered oxide: performance and mechanisms. Tribol Lett. 2000;8:103–116.

[5] Bobylev SV, Morozov NF, Ovid'ko IA. Cooperative grain boundary sliding and migration process in nanocrystalline solids. Phys Rev Lett. 2010;105:055504-1.

[6] Szlufarska I, Nakano A, Vashishta P. A crossover in the mechanical response of nanocrystalline ceramics. Science. 2005;309:911–914.

[7] Mo Y, Stone D, Szlufarska I. Strength of ultrananocrystalline diamond controlled by friction of buried interfaces. J Phys D: Appl Phys. 2011;44:405401.

[8] Koch CC, Ovid'ko IA, Seal S, Veprek S. Structural nanocrystalline materials fundamentals and applications. Cambridge: Cambridge University Press; 2007.

[9] Karch J, Birringer R, Gleiter H. Ceramics ductile at low temperature. Nature. 1987;330:556–558.

[10] Mayo MJ. High and low temperature superplasticity in nanocrystalline materials. Nanostruct Mater. 1997;9:717–726.

[11] Mohseni H, Collins PC, Scharf TW. Nanocrystalline orientation and phase mapping of textured coatings revealed by precession electron diffraction. Nanomater Energy. 2012;1:318–323.

[12] Vigué F, Vennéguès P, Vézian S, Laügt M, Faurie JP. Defect characterization in ZnO layers grown by plasma-enhanced molecular-beam epitaxy on (0001) sapphire substrates. Appl Phys Lett. 2001;79:194–196.

[13] Gerthsen D, Litvinov D, Gruber Th, Kirchner C, Waag A. Origin and consequences of a high stacking fault density in epitaxial ZnO layers. Appl Phys Lett. 2002;81:3972–3974.

[14] Yan Y, Dalpian GM, Al-Jassim MM, Wei SH. Energetics and electronic structure of stacking faults in ZnO. Phys Rev B. 2004;70:193206-1.

[15] Suzuki K, Takeuchi S. Electron microscopy of dislocations introduced into GaN by plastic deformation. Phil Mag Letts. 1999;79:423–428.

[16] Meng FY, Rao M, Newman N, Carpenter R, Mahajan S. Stacking faults in quaternary $In_xAl_yGa_{1-x-y}N$ layers. Acta Mater. 2008;56:4036–4045.

[17] Mohseni H, Scharf TW. Atomic layer deposition of $ZnO/Al_2O_3/ZrO_2$ nanolaminates for improved thermal and wear resistance in carbon–carbon composites. J Vac Sci Technol A. 2012;30:01A149-1.

[18] Mohseni H, Mensah BA, Gupta N, Srinivasan, SG, Scharf TW. On tailoring the nanocrystalline structure of ZnO to achieve low friction. Tribol Lubr Technol. 2012;1: 17–19.

[19] Kuo FL, Lin MT, Mensah BA, Scharf TW, Shepherd ND. A comparative study of the photoluminescence and conduction mechanisms of low temperature pulsed laser deposited and atomic layer deposited zinc oxide thin films. Phys Status Solidi A. 2010;207:2487–2491.

[20] Ivashchenko VI, Turchi PEA, Gonis A, Shevchenko VI, Ivashchenko LA. Tribology of amorphous, nanocrystalline, and crystalline slabs of Si, C, and SiC. Phys Rev B. 2005;72:115202-1.

[21] Scharf TW, Prasad SV. Solid lubricants: a review. J Mater Sci. 2013;48:511–531.

[22] Bowden FP, Tabor D. The friction and lubrication of solids. Oxford: Clarendon Press; 1986.

[23] Kresse G, Hafner J. *Ab initio* molecular dynamics for liquid metals. Phys Rev B. 1993;47:558–561.

[24] Kresse G, Furthmüller J. Efficient iterative schemes for ab initio total-energy calculations using a plane-wave basis set. Phys Rev B. 1996;54:11169–11186.

[25] Blöchl PE. Projector augmented-wave method. Phys Rev B. 1994;50:17953–17979.

[26] Perdew JP, Burke K, Ernzerhof M. Generalized gradient approximation made simple. Phys Rev Lett. 1996;77:3865–3868.

[27] Hirth JP, Rigney DA. Crystal plasticity and the delamination theory of wear. Wear. 1976;39:133–141.

30

Compressive Properties of ⟨110⟩ Cu Micro-Pillars after High-Dose Self-Ion Irradiation

J.A. Sharon[a], K. Hattar[a*], B.L. Boyce[a] and L.N. Brewer[a,b]

[a]*Sandia National Laboratories, PO Box 5800, Albuquerque, NM 87185, USA;* [b]*Naval Postgraduate School, 700 Dyer Rd., Monterey, CA 93943, USA*

Single-crystal Cu micro-pillars were self-ion irradiated up to 190 displacements per atom, a level commensurate with damage expected after long exposure in a reactor environment. Compression experiments performed along the ⟨110⟩ to 10% strain were compared against un-irradiated Cu. Two specimen configurations were explored: large 10 μm tall and small 4 μm tall pillars. Compared to un-irradiated Cu, the small irradiated pillars exhibited a flow stress increase of more than 500 MPa and were able to attain peak stresses approaching 1 GPa. These results are discussed in the context of an end of range effect, a damage gradient effect, and size effects.

Keywords: Micro-Pillar Compression, Copper, Radiation Damage, Mechanical Properties

'In service' radiation damage experienced by space and nuclear power components can be safely and feasibly simulated in a laboratory environment through proton, neutron, or ion irradiation. These accelerated aging experiments are critical for predicting material performance as it can take decades to reach high levels of radiation damage under operational neutron fluences. For ion irradiation, the damage zone typically extends only a few microns into the material; however, miniaturized mechanical testing methods can probe small volumes enabling site-specific property measurement to elucidate the radiation damage effects.[1]

In terms of damage, the energetic species impinging upon the target material cause displacement damage by knocking atoms out from their regular lattice position. Depending on the size of the resulting cascade, the stacking fault energy of the system, and the temperature of the irradiation event, the resulting damage often relaxes into a multitude of defect structures, including stacking fault tetrahedra (SFT) and dislocations loops [2–6] that serve as obstacles to dislocation motion. Hardening from irradiation, as well as decreases in ductility are well known in metals.[7] Cu in particular is interesting as it has been characterized on a variety of length scales in the un-irradiated state,[8–12] as well as at damage levels under ~50 dpa.[13–16] This paper reports micro-compression experiments of ⟨110⟩ oriented single-crystal Cu that has been self-ion irradiated with damage up to 190 dpa. Zinkle and Busby [17] and Zinkle and Was [18] outline that materials employed in future Gen IV energy systems will need to tolerate damage upwards of 200 dpa at high temperatures. This study's purpose is to extend the prior work on ion irradiation of Cu by providing new data in the regime of relatively high radiation damage commensurate with that expected in Gen IV reactors.

This study was performed on a ⟨110⟩ oriented single-crystal boule of Cu. The crystal was self-ion irradiated parallel to a ⟨110⟩ direction with Cu^{+5} at 30 MeV with an ~200 nA broad beam to a dose of 3.86×10^{16} ions/cm^2 at an average rate of 8.32×10^{12} ions/cm^2 s. The irradiation was performed at room temperature with no active heating or cooling of the Cu. The temperature was not monitored during the irradiation. Using the detailed calculations with full damage cascades in Stopping Range of Ions in Matter (SRIM) simulations,[19] the radiation damage profile was estimated. Damage increased in a nonlinear fashion from about 10 dpa at the surface to a peak damage of approximately 190 dpa at a depth of 4 μm and then subsided with significantly less radiation effects estimated beyond 5 μm. An electron transparent foil of the irradiated Cu was prepared with the focused ion beam (FIB) and inspected with a Phillips CM30 trans-

*Corresponding author. Email: khattar@sandia.gov

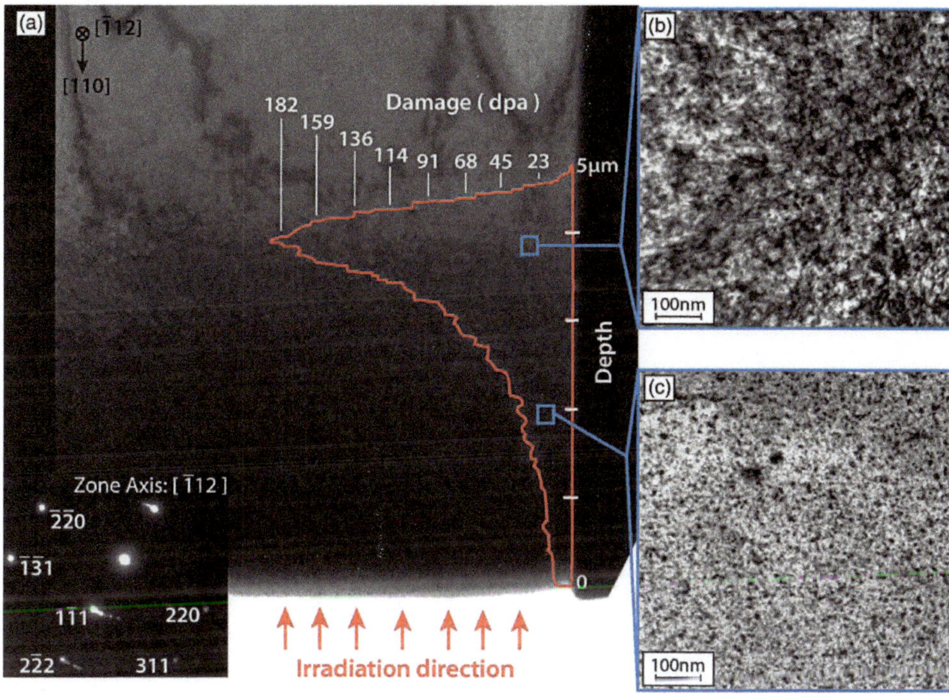

Figure 1. (a) Bright field TEM image of irradiated single-crystal Cu with overlay of the SRIM damage profile. The diffraction pattern was taken 90° from the [110] irradiation direction giving a [−112] zone axis. (b) Irradiated Cu near the peak damage site and (c) at ~2 μm in from the free surface.

mission electron microscope (TEM) operating at 300 kV in standard bright field mode. Figure 1(a) shows the irradiated Cu with an overlay of the damage profile. As predicted, radiation defects, evident by the dark contrast, are highly clustered at the peak damage site (Figure 1(b)) and become less dense towards the surface (Figure 1(c)). As Cu has a low-stacking fault energy, the majority of the defects are expected to be SFTs with a small amount of dislocation loops present as well.[13]

Following the approach of Uchic and Dimiduk [20] un-tapered micro-compression pillars with a nominal 1:2 diameter to height aspect ratio were machined using the FIB. All milling was performed with an accelerating voltage of 30 kV for the Ga^+ ion beam. While 30 keV Ga^+ ions have been shown to penetrate 50 nm into Cu,[21] this damage is considered insignificant: rudimentary SRIM estimates suggest that 30 keV Ga^+ will create ~788 vacancies per ion strike and have an end of range of 10–15 nm, whereas 30 MeV Cu^{+5} is estimated to generate ~2,505 vacancies per ion strike and have an end of range greater than 4 μm. Un-irradiated pillars were also manufactured on the same Cu boule, millimeters away from the well-identified local irradiation zone for direct side-by-side comparison of the irradiated and un-irradiated states. Two pillar geometries were machined for this study; a 'large' configuration with pillars 5 μm in diameter and 10 μm tall and a 'small' configuration having a 2 μm diameter and 4 μm height. Using a Hysitron Performech TriboIndenter operated in feedback displacement control, the pillars were compressed 10% with a 25 μm diameter flat ended cone indenter along the irradiation direction (Figure 2(a)) at a strain rate of 0.025 s^{-1}. The compressive engineering stress–strain curves for both pillar sizes are shown in Figure 2(b) and 2(c). Stress and strain were calculated from the raw load and displacement data using initial diameter and height measurements from electron images of the as-machined pillars. All data were adjusted for elastic displacement of the base using the Sneddon [22] correction.

Both irradiated and un-irradiated pillars had intermittent flow with load drops which, under displacement feedback control, have been linked to bursts of dislocation activity [23] and the appearance of glide steps.[9] Experiments, discrete dislocation simulations, and statistical analysis [24–27] suggest that a combination of factors to include specimen size, initial mobile dislocation density, and the distribution of pinning points govern the stochastics of source operation responsible for intermittent flow. As irradiation has been reported to frustrate dislocation motion,[28] it is reasoned that the subtle load drops for the small irradiated pillars stem from radiation defects obstructing dislocation glide thus preventing 'bursting' and easy escape at free surfaces which would presumably cause more substantial load drops.

There was only a slight distinction in the behavior for the large 10 μm tall, 5 μm diameter irradiated pillars compared to the same size un-irradiated pillars (Figure 2(a)). Consider the flow stress at 2% strain. For

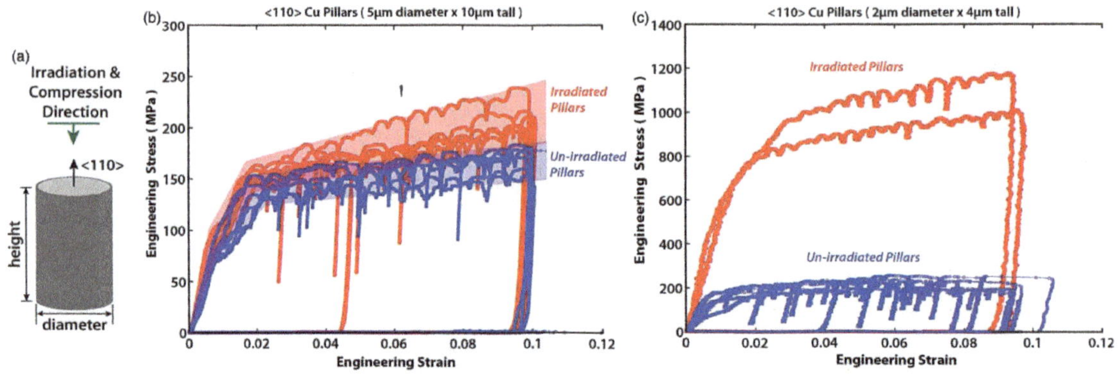

Figure 2. Compression stress-strain curves comparing irradiated and un-irradiated ⟨110⟩ Cu pillars. (a) Pillar schematic highlighting the crystallographic orientation of the Cu with respect to the irradiation and compression direction. Large pillar data presented in (b) with small pillar data presented in (c). Note the different scales of the *y*-axis in these two plots.

un-irradiated pillars, the flow stress at 2% strain was 122–153 MPa and the irradiated Cu had a slightly higher flow stress of 144–163 MPa. A small difference was also found to exist for the strain hardening rate, taken as the slope between the true stress—true strain data at 2% and 9% engineering strain. The hardening rate of the irradiated Cu was estimated to be 0.5 GPa, while the un-irradiated Cu was slightly lower at 0.3 GPa.

Unlike the large specimens, the small pillars that were only 4 μm tall with a 2 μm diameter exhibited a more substantial effect of radiation damage as evident in Figure 2(b). The flow stress at 2% strain for the un-irradiated small pillars was 159–211 MPa, while the irradiated pillars had flow levels more than three and a half times higher at 785–793 MPa. In terms of peak strength, the small irradiated pillars approach 1 GPa which is comparable to strengths observed for practically flaw free Cu micro-whiskers tested by Brenner [10,11] but weaker than the ⟨110⟩ Cu nano-whiskers investigated by Richter et al. [12] that reached theoretical strength levels.

In terms of strain hardening rate, the small 2 μm diameter un-irradiated Cu was similar to its larger counterpart at ~0.3 GPa but the estimate for the small irradiated Cu was much higher at 2.8 GPa. The hardening rate of un-irradiated single-crystal pillars is known to increase as sample diameter decreases from microns in size down to hundreds of nanometers.[29,30] The similar hardening behavior between the 4 and 2 μm diameter un-irradiated Cu may stem from the relatively small change in specimen size. As for the small irradiated pillars, the elevated hardening rate would appear to suggest that radiation damage not only suppresses initial dislocation source operation, but also renders source operation more difficult with increasing strain. However, it remains unclear if this hardening rate increase only manifests itself due to the combination of length scale effects and radiation damage. A study on much larger millimeter-scale irradiated polycrystalline tensile bars [15] found yield point increase but little change in hardening behavior compared

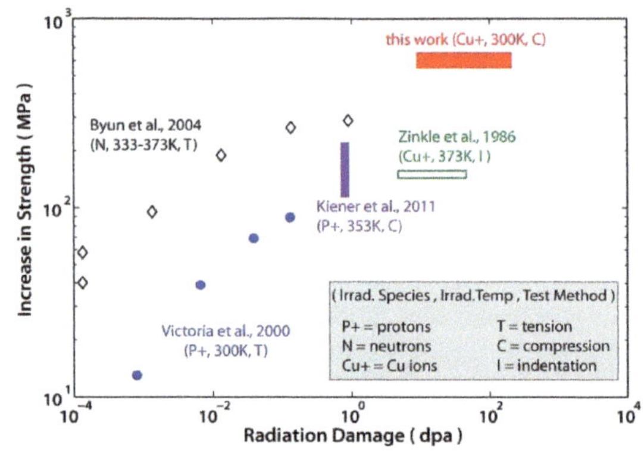

Figure 3. A compiled log-log plot comparing strength increase against various levels of radiation damage for Cu. Note that open symbols denote polycrystalline Cu, while solid markers represent single-crystal Cu.

to un-irradiated material. Additional studies are therefore needed to elucidate the individual impact size effects and radiation damage have on strain hardening.

Several studies have been published on the properties of single and polycrystalline Cu irradiated to damage levels under ~50 dpa at temperatures less than 200°C. Through tension,[13,15] compression,[14] and indentation,[16] these investigations all report strength increases for the irradiated material, however none exceed 300 MPa. The small 2 μm diameter irradiated pillars of this study had a larger strength enhancement with the radiation defects providing a 574–634 MPa increase in flow stress at 2% strain. Figure 3 is a log–log plot that assembles estimates of strength enhancement at yield or the 2% flow stress extracted from [13–16]. Comparisons from Figure 3 should not be taken as absolute as each of the aforementioned studies has differences in irradiation conditions (temperature, irradiation species, etc.). Nevertheless, it appears that higher levels of radiation damage generally corresponded to higher amounts of strengthening. This observation was slightly unexpected

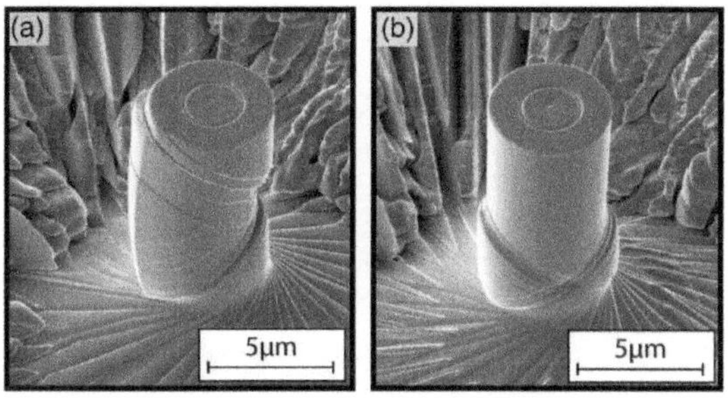

Figure 4. Comparison of large (5 μm diameter × 10 μm tall) un-irradiated (a) and irradiated (b) Cu pillars compressed by 10%. Images were captured with the specimen at a 45° tilt.

as several have reported that the density of radiation defects saturates in Cu at ~1 dpa [13,31,32]. It is then speculated that irradiation to higher damage levels perhaps evolves the saturated defects into configurations that supply additional strengthening. There is limited experimental and simulation results [33,34] that suggest active dislocations can interact with radiation defects such as full and truncated SFTs in such a manner that a dynamic mechanical response is predicted. Follow up *in situ* TEM irradiation analysis is needed to characterize the radiation defect–defect interactions that form from extensive irradiation and understand how they may afford substantial strengthening.

There are three key issues that must be given consideration when interpreting these experimental results: (1) an end of range effect, (2) a damage gradient effect, and (3) a pillar size effect. The end of range effect manifests itself in the large pillar data. Since the irradiation damage only extended about 5 μm into the pillar, a 10 μm pillar height for the large pillars means that the top half of the pillar is defect hardened, while the bottom half is nominally undamaged. Figure 4 compares deformed images of the large pillars. Note that the irradiated pillar (Figure 4(b)) had no slip traces in the upper region. It appears then that the experiments on the large irradiated pillars effectively test only the short, nominally un-irradiated bottom half. This is corroborated by the fact that the large irradiated pillars exhibited substantial load drops (Figure 2(b)) suggesting that dislocation glide was not heavily obstructed by radiation defects. Nevertheless some strengthening in Figure 2(b) is noticeable and may stem from the elastic base and irradiation hardened top that confine deformation to a smaller volume and also act as barriers to dislocation motion. This may also explain the slightly higher hardening rate for the irradiated large pillars at 0.5 GPa, while 0.3 GPa was measured for the un-irradiated Cu. In terms of a radiation effect, the only observation that can be drawn is that the flow stress at 2% strain for the irradiated Cu exceeds ~122 MPa. This end of range effect reiterates the message of Hosemann et al. [35] and Kiener et al. [36] that it is important to configure experiments to probe a well-known radiation affected zone.

The damage gradient effect refers to the fact that the irradiated Cu pillars have a gradient of defect agglomerations resulting from the radiation damage. For the small 4 μm tall, 2 μm diameter pillars, the experiments are not probing a structure with uniform damage but sampling over a range of damage levels from ~10 dpa at the surface to about 190 dpa at the pillar base. As roughly 50% of the pillar is at a damage level of less than 35 dpa, the inhomogeneity is not negligible. It is notable that the flow stress and hardening rate for the small pillars increased substantially raising the prospect that the strength might well exceed 1 GPa for a Cu single crystal with a mean damage level of 190 dpa. To avoid the damage gradient effect, some have suggested fabricating specimens normal to the irradiation direction.[36] A challenge with such an approach is that the region of relatively uniform damage is limited and may require a reduction in pillar size which could introduce a size effect.

Length scales such as physical specimen size and defect spacing can influence the deformation mechanisms governing strength.[37–39] Issues can arise if the test specimen dimensions are so small that the volume probed is not representative of the irradiated material. This issue is highlighted in the nano-pillar compression experiments by Kiener et al.[14] For pillars with a diameter less than ~400 nm, Kiener et al. [14] found no strength difference for un-irradiated and 0.8 dpa ⟨100⟩ Cu. It was suggested that the pillar size caused the deformation in both the irradiated and un-irradiated Cu to be limited by same factor; the ability to activate a dislocation source in a limited volume. For pillar diameters above 400 nm, the strength of the 0.8 dpa pillars remained constant as pillar diameter increased. The un-irradiated pillars on the other hand lost strength as size increased. It was suggested that this trend occurred because the deformation had become dominated by the interaction of dislocations with radiation defects. The small irradiated and un-irradiated pillars of this study do not have comparable strengths suggesting the 2 μm diameter pillars are a representative irradiated volume with the strength enhancement coming from the radiation-induced defects.

In summary, Cu micro-pillars, self-ion irradiated with damage spanning from 10 to 190 dpa, were compressed 10% and compared against un-irradiated Cu. Both large 5 μm diameter 10 μm tall and small 2 μm diameter 4 μm tall pillars were investigated. Since radiation damage primarily penetrated to a depth of 5 μm, only the small 2 μm diameter 4 μm tall pillars provided data purely characteristic of a radiation damaged microstructure. This work extends previous observations on radiation damage to higher dpa levels and appears to suggest that strengthening continues to scale with irradiation damage, even at levels commensurate with very long-term reactor damage.

Acknowledgements The authors thank M. Rye, G. Bryant, and B. Mckenzie for assistance with specimen fabrication and imaging as well as D. Buller and S. Van Deusen for assistance with the irradiation. LNB was funded by the Laboratory Directed Research and Development program, whereas BLB, JAS, and KH were supported primarily by the DOE Office of Basic Energy Sciences. Material charactzaiton was, in part, supproted by the Center for Integrated Nanotechnologies, an Office of Science User Facility operated for the US Department of Energy (DOE) Office of Science by Los Alamos National Laboratory (Contract DE-AC52-06NA25396) and Sandia National Laboratories (Contract DE-AC04-94AL85000). Sandia is a multiprogram laboratory owned and operated by Sandia Corporation, a Lockheed Martin Company, for the United States Department of Energy's National Nuclear Security Administration under contract DE-AC04-94AL85000.

References

[1] Hosemann P, Swadener JG, Kiener D, Was GS, Maloy SA, Li N. An exploratory study to determine applicability of nano-hardness and micro-compression measurements for yield stress estimation. J Nucl Mat. 2008;375(1): 135–143.

[2] Zinkle SJ, Matsukawa Y. Observation and analysis of defect cluster production and interactions with dislocations. J Nucl Mater. 2004;329–333(Part A(0)):88–96.

[3] Wirth BD. How does radiation damage materials? Science. 2007;318(5852):923–924.

[4] Wirth BD, Bulatov V, Diaz de la Rubia T. Atomistic simulation of stacking fault tetrahedra formation in Cu. J Nucl Mater. 2000;283–287(Part 2(0)):773–777.

[5] Osetsky YN, Bacon DJ. Defect cluster formation in displacement cascades in copper. Nucl Instrum Methods Phys Res B: Beam Interact Mater Atoms. 2001;180(1–4): 85–90.

[6] Osetsky YN, Bacon DJ. Atomic-scale modelling of primary damage and properties of radiation defects in metals. Nucl Instrum Methods Phys Res Sect B: Beam Interact Mater Atoms. 2003;202(0):31–43.

[7] Azevedo CRF. A review on neutron-irradiation-induced hardening of metallic components. Eng Failure Anal. 2011;18(8):1921–1942.

[8] Kiener D, Kaufmann P, Minor AM. Strength, hardening, and failure observed by in situ tem tensile testing. Adv Eng Mater. 2012;14(11):960–967.

[9] Kiener D, Grosinger W, Dehm G, Pippan R. A further step towards an understanding of size-dependent crystal plasticity: In situ tension experiments of miniaturized single-crystal copper samples. Acta Mater. 2008;56(3): 580–592.

[10] Brenner SS. Tensile strength of whiskers. J Appl Phys. 1956;27(12):1484–1491.

[11] Brenner SS. Plastic deformation of copper and silver whiskers. J Appl Phys. 1957;28(9):1023–1026.

[12] Richter G, Hillerich K, Gianola DS, Mönig R, Kraft O, Volkert CA. Ultrahigh strength single crystalline nanowhiskers grown by physical vapor deposition. Nano Lett. 2009;9(8):3048–3052.

[13] Victoria M, Baluc N, Bailat C, Dai Y, Luppo MI, Schäublin R, Singh BN. The microstructure and associated tensile properties of irradiated Fcc and Bcc metals. J Nucl Mater. 2000;276(1–3):114–122.

[14] Kiener D, Hosemann P, Maloy SA, Minor AM. In situ nanocompression testing of irradiated copper. Nat Mater. 2011;10(8):608–613.

[15] Byun TS, Farrell K. Plastic instability in polycrystalline metals after low temperature irradiation. Acta Mater. 2004;52(6):1597–608.

[16] Zinkle SJ, Oliver WC. Mechanical property measurements on ion-irradiated copper and Cu-Zr. J Nucl Mater. 1986;141–143(Part 1(0)):548–552.

[17] Zinkle SJ, Busby JT. Structural materials for fission & fusion energy. Mater Today. 2009;12(11):12–19.

[18] Zinkle SJ, Was GS. Materials challenges in nuclear energy. Acta Mater. 2013;61(3):735–758.

[19] Ziegler JF, Biersack JP, Littmark U. Stopping and ranges of ions in solids. New York: Pergamon Press; 1985.

[20] Uchic MD, Dimiduk DM. A methodology to investigate size scale effects in crystalline plasticity using uniaxial compression testing. Mater Sci Eng: A. 2005;400–401(0):268–278.

[21] Kiener D, Motz C, Rester M, Jenko M, Dehm G. Fib damage of Cu and possible consequences for miniaturized mechanical tests. Mater Sci Eng: A. 2007;459(1–2):262–272.

[22] Sneddon IN. The relation between load and penetration in the axisymmetric boussinesq problem for a punch of arbitrary profile. Int J Eng Sci. 1965;3(1):47–57.

[23] Minor AM, Asif SAS, Shan Z, Stach EA, Cyrankowski E, Wyrobek TJ, Warren OL. A new view of the onset of plasticity during the nanoindentation of aluminium. Nat Mater. 2006;5(9):697–702.

[24] Parthasarathy TA, Rao SI, Dimiduk DM, Uchic MD, Trinkle DR. Contribution to size effect of yield strength from the stochastics of dislocation source lengths in finite samples. Scr Mater. 2007;56(4):313–316.

[25] Rao SI, Dimiduk DM, Parthasarathy TA, Uchic MD, Tang M, Woodward C. Athermal mechanisms of size-dependent crystal flow gleaned from three-dimensional discrete dislocation simulations. Acta Mater. 2008;56(13):3245–3259.

[26] Ng KS, Ngan AHW. Stochastic nature of plasticity of aluminum micro-pillars. Acta Mater. 2008;56(8):1712–1720.

[27] Csikor FF, Motz C, Weygand D, Zaiser M, Zapperi S. Dislocation avalanches, strain bursts, and the problem of plastic forming at the micrometer scale. Science. 2007;318(5848):251–254.

[28] Briceño M, Fenske J, Dadfarnia M, Sofronis P, Robertson IM. Effect of Ion irradiation-produced defects on the mobility of dislocations in 304 stainless steel. J Nucl Mater. 2011;409(1):18–26.

[29] Frick CP, Clark BG, Orso S, Schneider AS, Arzt E. Size effect on strength and strain hardening of small-scale 111 nickel compression pillars. Mater Sci. Eng A—Struct Mater Prop Microstruct Process. 2008;489(1–2):319–329.

[30] Kiener D, Guruprasad PJ, Keralavarma SM, Dehm G, Benzerga AA. Work hardening in micropillar compression: In situ experiments and modeling. Acta Mater. 2011;59(10):3825–3840.

[31] Singh BN, Zinkle SJ. Defect accumulation in pure Fcc metals in the transient regime: a review. J Nucl Mater. 1993;206(2–3):212–229.

[32] Li N, Hattar K, Misra A. In situ probing of the evolution of irradiation-induced defects in copper. J Nucl Mater. 2013;439(1–3):185–191.

[33] Wirth BD, Bulatov VV, de la Rubia TD. Dislocation-stacking fault tetrahedron interactions in Cu. J Eng Mater Technol. 2002;124(3):329–334.

[34] Robach JS, Robertson IM, Lee HJ, Wirth BD. Dynamic observations and atomistic simulations of dislocation–defect interactions in rapidly quenched copper and gold. Acta Mater. 2006;54(6):1679–1690.

[35] Hosemann P, Kiener D, Wang Y, Maloy SA. Issues to consider using nano indentation on shallow ion beam irradiated materials. J Nucl Mater. 2012;425(1–3):136–139.
[36] Kiener D, Minor AM, Anderoglu O, Wang Y, Maloy SA, Hosemann P. Application of small-scale testing for investigation of ion-beam-irradiated materials. J Mater Res. 2012;27(21):2724–2736.
[37] Bei H, Shim S, Pharr GM, George EP. Effects of pre-strain on the compressive stress–strain response of Mo-Alloy single-crystal micropillars. Acta Mater. 2008;56(17):4762–4770.
[38] Girault B, Schneider AS, Frick CP, Arzt E. Strength effects in micropillars of a dispersion strengthened superalloy. Adv Eng Mater. 2010;12(5):385–388.
[39] Uchic MD, Shade PA, Dimiduk DM. Plasticity of micrometer-scale single crystals in compression. Annu Rev Mater Res. 2009;39(1):361–386.

31

Magnetic Anisotropy in the $(Cr_{0.5}Mn_{0.5})_2GaC$ MAX Phase

Ruslan Salikhov[a]*, Anna S. Semisalova[a,b,c], Andrejs Petruhins[d], Arni Sigurdur Ingason[d], Johanna Rosen[d], Ulf Wiedwald[a] and Michael Farle[a]

[a]*Fakultät für Physik and Center for Nanointegration (CENIDE), Universität Duisburg-Essen, Duisburg 47057, Germany;* [b]*Faculty of Physics, Lomonosov Moscow State University, Moscow 119991, Russia;* [c]*Institute of Ion Beam Physics and Materials Research, Helmholtz-Zentrum Dresden-Rossendorf, Dresden D-01328, Germany;* [d]*Thin Film Physics Division, Department of Physics, Chemistry and Biology (IFM), Linköping University, Linköping SE-581 83, Sweden*

Magnetic MAX phase $(Cr_{0.5}Mn_{0.5})_2GaC$ thin films grown epitaxially on MgO(111) substrates were studied by ferromagnetic resonance at temperatures between 110 and 300 K. The spectroscopic splitting factor $g = 2.00 \pm 0.01$ measured at all temperatures indicates pure spin magnetism in the sample. At all temperatures we find the magnetocrystalline anisotropy energy to be negligible which is in agreement with the identified pure spin magnetism.

Keywords: Magnetic MAX Phase, Ferromagnetic Resonance, *g*-Factor, Magnetic Anisotropy

MAX phase materials based on nanolaminated ternary compounds have been found to show unique and intriguing properties ranging from spectacular mechanical to self-healing and superconducting properties.[1–3] MAX phases ($M_{n+1}AX_n$, where $n = 1$, 2, or 3) are layered hexagonal structures, where M is an early transition metal (Sc, Ti, V, Cr, Zr, Nb, Mo, Hf, or Ta), A is an A-group element (e.g. Al, Ga, or Ge), and X is either C or N. Recently, Mn- and Cr-based MAX phases have attracted considerable interest due to their magnetic properties. Density functional theory calculations have predicted antiferromagnetic (AFM) long-range order in Cr_2AC (where A = Al or Ge) structures,[4,5] while ferromagnetic (FM) ordered Cr moments in an external magnetic field below 80 K [6,7] have been suggested from experiments. On the other hand, more recent work has shown that the Invar-like behavior discussed in Jaouen et al. [7] is quite common in similar structures, and probably cannot be attributed to the magnetic behavior.[8] Another experimental work suggested that the Cr_2GeC MAX phase can be classified as a correlated-electron paramagnetic metal in the vicinity of an FM quantum critical point.[9] The possibility to achieve a stable FM order at room temperature in nanolaminated MAX phases in bulk or thin film form makes these systems extremely interesting for both fundamental materials science and applications.[10] Recently, it has been demonstrated that magnetic order can be changed and stabilized up to room temperature by substituting Cr by Mn.[9,11–16] Stable $(Cr_{1-x}Mn_x)_2AC$ (A = Al, Ge or Ga) type MAX phase compounds can be considered as a new family of magnetic MAX phases, which exhibit FM properties. Furthermore, the Mn_2GaC magnetic compound has been synthesized as a heteroepitaxial thin film containing Mn as the exclusive M element.[10,17,18]

The magnetic MAX phases have very large saturation fields (larger than 5 T) and a low ratio of remanent-to-saturation magnetization.[15–18] The magnetization of these systems increases rapidly with decreasing temperature and usually does not show the characteristics of a conventional ferromagnet.[15–20] These findings have been interpreted to be the result of competing FM and AFM correlations,[10] which can lead to non-collinear or ferrimagnetic spin structures. The crystal structure of MAX phases is hexagonal with a repeated M-A-M-X-M-A-M-X atomic layer stacking in the *c* direction. First-principle calculations have shown that Cr-C networks (M-X-M) favor FM spin alignment in the Cr_2GeC

*Corresponding author. Email: ruslan.salikhov@uni-due.de

system, whereas Ge (in Cr-Ge-Cr bonds) supports AFM super-exchange like coupling between Cr atoms due to the mixing of the Cr 3d and Ge 4p states.[4] However, experiments show that the electron density of pure Cr_2GeC is not spin-polarized and only Mn doping leads to FM long-range order.[9] The spin structure of new Mn-doped magnetic MAX phases is currently under investigation. It is worth noting that the sign of exchange coupling and thus the ground state (AFM or FM) can be tuned by changing the interlayer distances, which makes these systems extremely interesting for magnetocaloric and spintronic applications.[10]

Due to the hexagonal crystal structure and the reduced dimensional (quasi 2D) layer structure, the spin–orbit coupling of Mn (Cr) 3d electrons can strongly influence the magnetic properties and spin structure. Magnetocrystalline anisotropy energy (MAE) is a qualitative measure of spin–orbit interaction and can be used to estimate anisotropic orbital magnetic moments.[21] A powerful method to quantify the magnetic anisotropy energy density is ferromagnetic resonance (FMR).[22] Here, we present the results of FMR measurements on 34 nm thick $(Cr_{0.5}Mn_{0.5})_2GaC$ films epitaxially grown on MgO(111) substrates. The selected MAX phase has to date the largest saturation magnetization among the magnetic MAX phase series and shows a FM response near room temperature.[16] The synthesis and structural properties of these films have been described elsewhere.[16] X-ray diffraction and cross-section transmission electron microscopy of these samples reveal a high crystal quality with a composition of $(Cr_{0.5}Mn_{0.5})_2GaC$. The films are phase-pure with a lateral grain size larger than 150 nm. The *c*-axis aligns parallel to the film normal.

Temperature- and angular-dependent FMR measurements were performed using a conventional FMR X-band setup at a microwave frequency $f = 9.23$ GHz. FMR spectra were recorded for different orientations of the static magnetic field with respect to the $(Cr_{0.5}Mn_{0.5})_2GaC$ film plane (out-of-plane geometry) and within the film basal plane (called in-plane geometry from this point forward) at temperatures between 110 and 300 K. Additionally, magnetic hysteresis loops at various temperatures were measured in both geometries using SQUID magnetometry.

FMR spectra of the $(Cr_{0.5}Mn_{0.5})_2GaC$ film measured at different temperatures (T) are presented in Figure 1 for in-plane (a) and out-of-plane (b) geometries. From the spectra the resonance field (H_r) was determined using the first derivative of the Lorentz function fitted to the experimental data. At $T = 105$ K the resonance field (H_r) is 282 ± 2 mT for in-plane and 436 ± 2 mT for out-of-plane geometry. This indicates that the easy axis of magnetization (M) is in the film plane.[22] With increasing T and decreasing magnetization, the resonance fields for in- and out-of-plane geometry shifts toward the same $\mu_0 H_r = 330$ mT (at $T \approx 250$ K). One can also see that with increasing temperature, the integral intensity (I) of the resonance line for both configurations decreases (Figure 1(c), black squares for in-plane

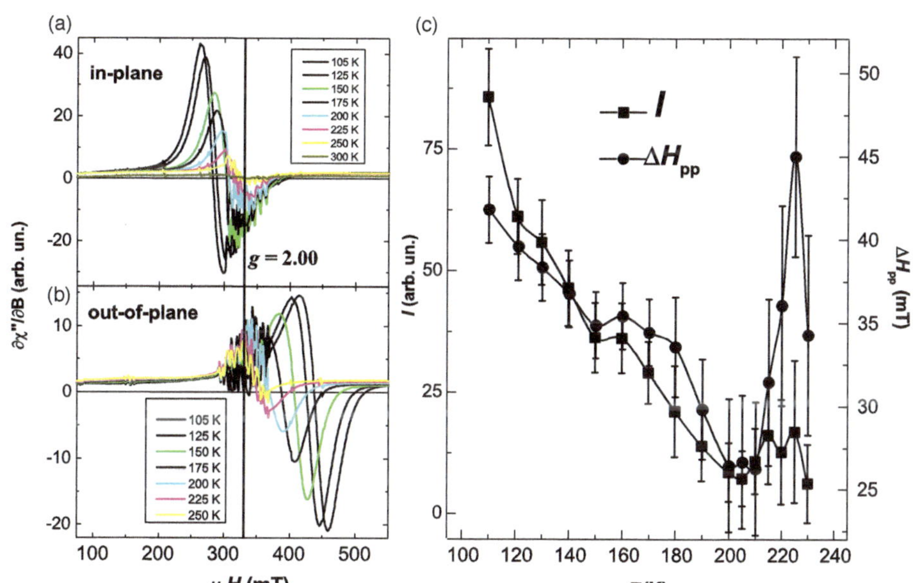

Figure 1. FMR spectra for different temperatures measured at $f = 9.2265(1)$ GHz with the magnetic field applied parallel (a) or perpendicular (b) to the $(Cr_{0.5}Mn_{0.5})_2GaC$ film plane. The 'noisy' features seen at 300–350 mT are paramagnetic resonance (EPR) signals from impurities in the MgO substrate. (c) Temperature dependence of the FMR integral intensity I (black squares and left scale) and peak-to-peak resonance linewidth ΔH_{pp} (blue circles and right scale), solid lines are guide to the eye.

geometry) up to 205 K representing the decrease in the magnetization.[23] The FMR peak-to-peak linewidth (ΔH_{pp}) at different temperatures (blue circles in Figure 1(c)) decreases with increasing temperature similar to the FMR integral intensity, indicating that magnetic relaxation processes in the FM $(Cr_{0.5}Mn_{0.5})_2GaC$ film are slowing down with the magnitude of M. At $T > 205$ K the intensity remains unchanged within the experimental error, but ΔH_{pp} rapidly increases to peak at 225 K. This is attributed to the crossover of the magnetic relaxation rates from critical speeding up just above the Curie temperature ($T > T_C$) to thermodynamical slowing down close to T_C, as has been previously reported for the metallic thin films near an FM to paramagnetic phase transition (see, for example, Li et al. [24] and Platow et al. [25] and references therein). A residual microwave absorption signal is seen up to 300 K. We also found that at temperatures above 240 K, the resonance fields for in- and out-of-plane geometry are identical within the error bar of 3 mT. This indicates that the magnetic anisotropy above $T = 240$ K becomes very small (or zero). The above findings allow the calculation of the spectroscopic g-factor using the equation $H_r = \omega/\gamma$, where $\omega = 2\pi f$, and $\gamma = g\mu_B/\hbar$ is the electron gyromagnetic ratio.[22] We arrive at $g = 2.00 \pm 0.01$, which suggests pure spin magnetism in the $(Cr_{0.5}Mn_{0.5})_2GaC$ film. The absence of an orbital contribution to the magnetic moment usually results in a small MAE. Indeed, the in-plane resonance field is independent of the azimuthal angle (not shown) at 110, 150 and 250 K within the error bar of 0.1 mT. Since the samples are epitaxial and textured with a lateral grain size larger than 150 nm,[16] we would expect that in-plane magnetic anisotropies would be reflected in the angular dependence of the resonance field.

In the following we discuss the *out-of-plane* angular dependence of the FMR field using the Smit-Beljers approach.[22,26,27] Considering that there is no significant magnetocrystalline anisotropy in the film plane as indicated by the lack of an in-plane FMR angular dependence in the epitaxial film, we calculate the resonance frequency using the effective uniaxial anisotropy field given by the shape anisotropy $(1/2)\mu_0(N_\perp - N_\parallel)M_{\text{eff}}^2$ for thin films, where M_{eff} is the effective magnetization. Note that M_{eff} (from FMR) and magnetization, M, measured by SQUID magnetometry can differ and such variations can be attributed to an out-of-plane anisotropy along the c axis. We use the demagnetization factors $N_\perp = 1$, $N_\parallel = 0$ of an infinitely thin film, which is appropriate for the respective thickness of our film.[28] The corresponding free energy (F) including Zeeman energy reads

$$F = \tfrac{1}{2}\mu_0 M_{\text{eff}}^2 \cos^2\theta - \mu_0 MH(\cos\theta\cos\theta_H + \cos(\varphi - \varphi_H)\sin\theta\sin\theta_H) \qquad (1)$$

with $(\theta,\ \varphi)$ and (θ_H, φ_H) being the polar (accounted from the film plane normal) and azimuthal (in-plane) angles of the magnetization M and the applied magnetic field H, respectively. The resonance frequency can then be calculated analytically using the double derivatives of the magnetic free energy F with respect to polar and azimuthal angles [22,26,27]:

$$\left(\frac{\omega}{\gamma}\right)^2 = \frac{1}{(M\sin\theta)^2}(F_{\theta\theta}F_{\varphi\varphi} - F_{\theta\varphi}^2). \qquad (2)$$

For FM films without significant in-plane MAE azimuthal angles in Equation (1) can be set to $\varphi = \varphi_H = 0$. The angles θ correspond to the equilibrium angle of the magnetization and can be determined from the condition $\mathrm{d}F/\mathrm{d}\theta = 0$ for each field angle θ_H at the corresponding resonance field $H = H_r$. Thus, for the out-of-plane angular dependence of resonance fields, the g-factor ($\gamma = g\mu_B/\hbar$) and M_{eff} are fitting parameters. In the simple case assuming that $\theta = \theta_H$ one finds from Equation (2) the frequency-field dispersions, known as Kittel equations:

$$\left(\frac{\omega}{\gamma}\right)^2 = \mu_0^2 H_{r\parallel}(H_{r\parallel} + M_{\text{eff}}), \qquad (3)$$

when the external magnetic field is applied parallel to the film plane ($\theta = \theta_H = 90^0$), and

$$\left(\frac{\omega}{\gamma}\right) = \mu_0(H_{r\perp} - M_{\text{eff}}) \qquad (4)$$

for the out-of-plane configuration ($\theta = \theta_H = 0^0$). $H_{r\parallel}$ and $H_{r\perp}$ are resonance fields detected at $\theta_H = 90^0$ and 0^0, respectively. The values of $\mu_0 M_{\text{eff}}$ calculated using Equations (3) and (4) for two configurations (with $g = 2$) are plotted in Figure 2. One can see that at temperatures between 125 and 200 K, $\mu_0 M_{\text{eff}}$ for the in-plane configuration is smaller than for the out-of-plane. This means that the effective magnetic anisotropy at these temperatures for the out-of-plane geometry is larger than for the in-plane geometry.

To quantify the effective magnetic anisotropy more precisely, we performed angular-dependent out-of-plane measurements at 110 K from in-plane ($\theta_H = 90°$) to out-of-plane ($\theta_H = 0°$) orientation of the external field with respect to the film plane (Figure 3). The Smit-Beljers approach described above was used as a fit to the experimental data (black circles). From the fit (solid line in Figure 3) we found $\mu_0 M_{\text{eff}} = 103 \pm 3$ mT in good agreement with the values in Figure 2 determined using Equations (3) and (4). The spectroscopic splitting $g = 1.99 \pm 0.01$ is within the error bar identical to the one determined at ambient temperature. The $\mu_0 M$ ($4\pi M$ in cgs units) at $\mu_0 H = 290$ mT measured by SQUID magnetometry in in-plane geometry at 110 K is

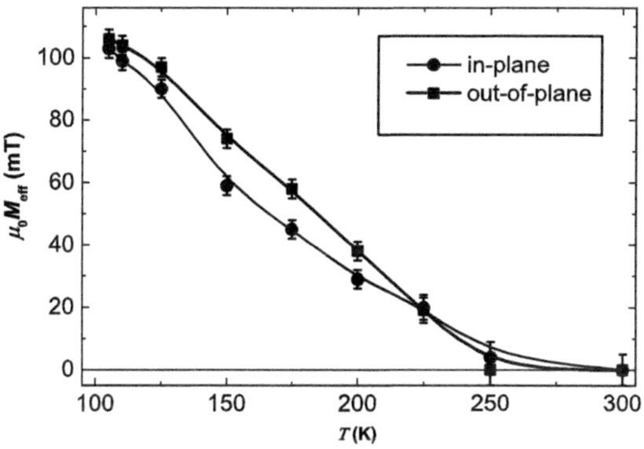

Figure 2. Effective magnetic anisotropy fields in the $(Cr_{0.5}Mn_{0.5})_2GaC$ film calculated using Equations (3) and (4) for in- and out-of-plane configurations at different temperatures. Solid lines are guide to the eye.

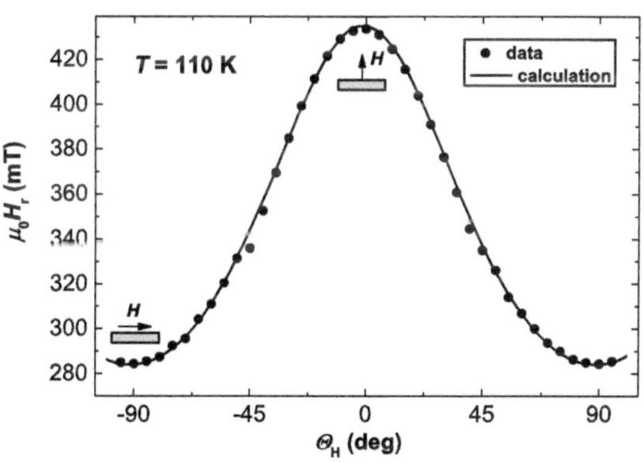

Figure 3. FMR field measured at 110 K (black circles, experimental error is approx. the dot size) as a function of angle between the applied external magnetic field and the axis normal to the film plane. The red solid line represents the result of fit using the Smit-Beljers approach with fit parameters: $\mu_0 M_{eff} = 103 \pm 3$ mT, and $g = 1.99 \pm 0.01$.

about 100 mT. This value is close to $\mu_0 M_{eff}$ at the same temperature, which leads us to following conclusions:

(1) the magnetocrystalline anisotropy along the c axis (sample normal) is negligibly small in the $(Cr_{0.5}Mn_{0.5})_2GaC$ MAX phase system and
(2) the Mn (Cr) spin moments are FM coupled and aligned parallel in a basal plane causing the shape anisotropy.

The possible AFM alignment of some magnetic moments in the basal plane would significantly reduce the FM long-range order, which in turn would reduce the magnetic shape anisotropy of an infinitely thin film ($N_\perp = 1$, $N_\| = 0$) down to the shape anisotropy of confined areas of locally FM coupled spins and for that case $\mu_0 M_{eff}$ is expected to be significantly smaller than $\mu_0 M$. Considering that the magnetocrystalline anisotropy is negligible one can conclude that the difference between $\mu_0 M_{eff}$ for in-plane and out-of-plane configurations (Figure 2) at the temperatures between 125 and 200 K is purely related to shape anisotropy.

According to Petruhins et al.,[16] epitaxial $(Cr_{0.5}Mn_{0.5})_2GaC$ films show a magnetic moment of 0.19 μ_B per M atom in a field of 5 T at 300 K and a remanent moment of 0.004 μ_B per M atom. This field-induced magnetization is also seen in microwave absorption as the signal above 205 K, which stays constant up to 300 K. The temperature dependencies of I, ΔH_{pp} and $\mu_0 M_{eff}$ in Figures 1(c) and 2 point toward a ferro- to paramagnetic phase transition at $T = 205$–230 K. The peak in $\Delta H_{pp}(T)$ at 225 K (above the ordering temperature) confirms the ferro- to paramagnetic phase transition, since is characteristic for critical fluctuations of the magnetization, which have also been observed at the ferro- to paramagnetic phase transition of Ni [24], for example. All aforementioned facts support the existence of FM order in the present samples. Based on the microwave data, however, it cannot be concluded unambiguously that itinerant ferromagnetism is encountered as has been concluded before for $(Cr_{1-x}Mn_x)_2GeC$ MAX phases.[9] The difference in shape anisotropy $\mu_0 M_{eff}$ between in- and out-of-plane configurations at temperatures between 125 and 200 K in Figure 2 indicates the anisotropy in FM order of magnetic moments. This anisotropy possibly can be explained considering the nanolaminated $(Cr_{0.5}Mn_{0.5})_2GaC$ system as a quasi 2D ferromagnet, where interlayer magnetic coupling supports 3D long-range ordering at finite temperatures.[29] In the out-of-plane configuration a larger magnetization appears due to enhanced interlayer coupling when changing the orientation of magnetic moments from parallel in the basal plane to perpendicular. The magnetometry data (not shown) support our present findings; however, the origin of the anisotropy of interlayer coupling remains open.

In conclusion, temperature- and angular-dependent FMR measurements of the new magnetic $(Cr_{0.5}Mn_{0.5})_2GaC$ MAX phase reveal that the g-factor is 2.00 ± 0.01 at ambient (300 K) and low temperatures (110 K). This suggests a small (or zero) orbital contribution to the magnetic moment. The in-plane and out-of-plane magnetocrystalline anisotropies are negligible. Furthermore, we provide evidence that magnetic moments in the basal plane of nanolaminated $(Cr_{0.5}Mn_{0.5})_2GaC$ are FM coupled and this magnetic material can most likely be classified as a quasi 2D laminar ferromagnet, where interlayer coupling depends on the magnetic field direction.

Disclosure statement No potential conflict of interest was reported by the authors.

Funding This work was supported by the EC contract REFreepermag no. 280670. A. S. Semisalova acknowledges financial support by UDE. J. R. acknowledges funding from ERC Grant agreement [number 258509], the Swedish Research Council (VR) [grant number 642-2013-8020] and from the SSF synergy grant FUNCASE.

References

[1] Barsoum MW. The M(N + 1)AX(N) phases: a new class of solids; thermodynamically stable nanolaminates. Prog Solid State Chem. 2000;28:201–281.

[2] Farle A-S, Kwakernaak C, van der Zwaag S, Sloof WG. A conceptual study into the potential of $M_{n+1}AX_n$-phase ceramics for self-healing for crack damage. J Europ Ceram Soc. 2015;35:37–45.

[3] Kuchida S, Muranaka T, Kawashima K, Inoue K, Yoshikawa M, Akimutsu J. Superconductivity in Lu $_2$SnC. Physica C: Supercond. 2013;494:77–79.

[4] Mattesini M, Magnuson M. Electronic correlation effects in the Cr_2GeC $M_{n+1}AX_n$ phase. J Phys: Cond Matt. 2013;25:035601(1–8).

[5] Dahlqvist M, Alling B, Rosen J. Correlation between magnetic state and bulk modulus of Cr_2AlC. J Appl Phys. 2013;113:216103(1–3).

[6] Jaouen M, Bugnet M, Jaouen N, Ohresser P, Mauchamp V, Cabioc'h T, Rogalev A. Experimental evidence of Cr magnetic moments at low temperature in Cr_2A(A = Al, Ge)C. J Phys Cond Matt. 2014;26:176002(1–6).

[7] Jaouen M, Chartier P, Cabioc'h T, Mauchamp V, Andre G, Viret M. Invar like behavior of the Cr_2AlC MAX phase at low temperature. J Am Ceram Soc. 2013;96:3872–3876.

[8] Caspi EN, Chartier P, Porcher F, Damay F, Cabioc'h T. Ordering of (Cr,V) layers in nanolamellar $(Cr_{0.5}V_{0.5})_{n+1}AlC_n$ compounds. Mater Res Lett. 2014; 3(2):100–106. doi:10.1080/21663831.2014.975294

[9] Liu Z, Waki T, Tabata Y, Nakamura H. Mn-doping-induced itinerant-electron ferromagnetism in Cr_2GeC. Phys Rev B. 2014;89:054435(1–6).

[10] Dahlqvist M, Ingason AS, Alling B, Magnus F, Thore A, Petruhins A, Mockute A, Arnalds UB, Sahlberg M, Hjörvarsson B, Abrikosov IA, Rosen J. Magnetically driven anisotropic structural changes in the atomic laminate Mn_2GaC (submitted).

[11] Mockute A, Dahlqvist M, Emmerlich J, Hultman L, Schneider JM, Persson PO, Rosen J. Synthesis and *ab initio* calculations of nanolaminated $(Cr, Mn)_2AlC$ compounds. Phys Rev B. 2013;87:094113(1–4).

[12] Mockute A, Lu J, Moon EJ, Yan M, Anasori B, May BS, Barsoum MW, Rosen J. Solid solubility and magnetism upon Mn incorporation in the bulk ternary carbides Cr_2AlC and Cr_2GaC. Mater Res Lett. 2015;3: 16–22.

[13] Mockute A, Persson POÅ, Lu J, Ingason AS, Magnus F, Olafsson S, Hultman L, Rosen J. Structural and magnetic properties of $(Cr_{1-x}Mn_x)$(5)Al-8 solid solution and structural relation to hexagonal nanolaminates. J Mater Sci. 2014;49:7099–7104.

[14] Mockute A, Persson POÅ, Magnus F, Ingason AS, Olafsson S, Hultman L, Rosen J. Synthesis and characterization of arc deposited magnetic $(Cr,Mn)_2AlC$ MAX phase films. Phys Status Solidi RRL. 2014;8(5): 420–423.

[15] Ingason AS, Mockute A, Dahlqvist M, Magnus F, Olafsson S, Arnalds U, Alling B, Abrikosov IA, Hjorvarsson B, Persson POÅ, Rosen J. Magnetic self-organized atomic laminate from first principles and thin film synthesis. Phys Rev Lett. 2013;110:195502(1–5).

[16] Petruhins A, Ingason AS, Lu J, Magnus F, Olafsson S, Rosen J. Synthesis and characterization of magnetic $(Cr_{0.5}Mn_{0.5})_2GaC$. J Mater Sci. 2015. doi:10.1007/s10853-015-8999-8

[17] Ingason AS, Petruhins A, Dahlqvist M, Magnus F, Mockute A, Alling B, Hultman L, Abrikosov IA, Persson POÅ, Rosen J. A nanolaminated magnetic phase: Mn_2GaC. Mater Res Lett. 2014;2:89–93.

[18] Thore A, Dahlqvist M, Alling B, Rosén J. First-principles calculations of the electronic, vibrational, and elastic properties of the magnetic laminate Mn_2GaC. J Appl Phys. 2014;116:103511(1–7).

[19] Tao QZ, Hu CF, Lin S, Zhang HB, Li FZ, Qu D, Wu ML, Sun YP, Sakka Y, Barsoum MW. Coexistence of ferromagnetic and a Re-entrant cluster glass state in the layered quaternary $(Cr_{1-x},Mn_x)_2GeC$. Mater Res Lett. 2014;2:192–198.

[20] Lin S, Tong P, Wang BS, Huang YN, Lu WJ, Shao DF, Zhao BC, Song WH, Sun YP. Magnetic and electrical/thermal transport properties of Mn-doped $M_{n+1}AX_n$ phase compounds $Cr_{2-x}Mn_xGaC$ ($0 \leq x \leq 1$). J Appl Phys. 2013;113:053502(1–6).

[21] Anisimov AN, Farle M, Poulopoulos P, Platow W, Baberschke K, Isberg P, Wäppling R, Niklasson AMN, Eriksson O. Orbital magnetism and magnetic anisotropy probed with ferromagnetic resonance. Phys Rev Lett. 1999;82:2390–2393.

[22] Farle M. Ferromagnetic resonance of ultrathin metallic layers. Rep Prog Phys. 1998;61:755–826.

[23] Platow W, Farle M, Baberschke K. What happens at the temperature-dependent magnetic order-disorder transition in Fe/Cu (001)? Europhys. Lett. 1998;43:713–718.

[24] Li Yi, Farle M, Baberschke K. Critical spin fluctuations and curie temperatures of ultrathin Ni(111)/W(110): a magnetic-resonance study in ultrahigh vacuum. Phys Rev B. 1990;41:9596–9599.

[25] Platow W, Anisimov AN, Dunifer GL, Farle M, Baberschke K. Correlations between ferromagnetic-resonance linewidths and sample quality in the study of metallic ultrathin films. Phys Rev B. 1998;58:5611–5621.

[26] Smit J, Beljers HG. Ferromagnetic resonance absorption in $BaFe_{12}O_{19}$, a highly anisotropic crystal. Philips Res Rep. 1955;10:113–130.

[27] Barsukov I, Mankovsky S, Rubacheva A, Meckenstock R, Spoddig D, Lindner J, Melnichak N, Krumme B, Makarov SI, Wende H, Ebert H, Farle M. Magnetocrystalline anisotropy and gilbert damping in iron-rich $Fe_{1-x}Si_x$ thin films. Phys Rev B. 2011;84:180405(R) (1–4).

[28] Farle M, Berghaus A, Baberschke K. Magnetic aniso tropy of Gd(0001)/W(110) monolayers. Phys Rev B. 1989;39:4838–4841.

[29] Ohta H, Noguchi D, Nabetani K, Katori HA. Itinerant electronic ferromagnetism in Sr_2ScO_3CoAs with largely spaced CoAs conduction layers. Phys Rev B. 2013;88:094441(1–6).

32

Duplex Precipitates and Their Effects on the Room-temperature Fracture Behaviour of a NiAl-Strengthened Ferritic Alloy

Zhiqian Sun[a], Gian Song[a], Jan Ilavsky[b] and Peter K. Liaw[a]*

[a] *Department of Materials Science and Engineering, The University of Tennessee, 1508 Middle Drive, Knoxville, TN 37996, USA;* [b] *X-ray Science Division, Advanced Photon Source, Argonne National Laboratory, Lemont, IL 60439, USA*

Duplex precipitates are presented in a NiAl-strengthened ferritic alloy. They were characterized by the ultra-small angle X-ray scattering and transmission electron microscope. Fine cooling precipitates with the size of several to tens of nanometres harden the matrix considerably at room temperature. Cracks are likely to initiate from precipitates, and coalesce and propagate quickly through the matrix due to the excessive hardening effect of cooling precipitates, which lead to the premature fracture of NiAl-strengthened ferritic alloys.

Keywords: NiAl-strengthened Ferritic Steels, Small-angle X-ray Scattering (SAXS), Precipitation, Fracture, Order Strengthening

NiAl-strengthened ferritic alloys have received considerable attention due to their good high-temperature creep resistance for applications in fossil-energy power plants.[1,2] The small lattice mismatch between the α-iron matrix and NiAl-type precipitates effectively delays the precipitate-coarsening kinetics at elevated temperatures ($\sim$700°C).[3,4] Studies have been performed to elucidate the creep mechanisms of NiAl-strengthened ferritic alloys.[5–9] However, a lack of the room-temperature ductility limits applications of NiAl-strengthened ferritic alloys.[6,10–13] Factors such as the precipitate-distribution parameters [6,10,11,13] (e.g. the precipitate volume fraction, f, average size, $\bar{r}$, and inter-particle distance, L) and solute atoms like aluminium and nickel in the matrix [11] were discussed. The room-temperature fracture mechanism of NiAl-strengthened ferritic alloys is not yet understood. Previous studies on NiAl-strengthened ferritic alloys revealed the presence of duplex precipitates (aged and fine cooling ones).[6,14] However, effects of fine cooling precipitates have not been investigated. In this study, we characterize duplex precipitates by the ultra-small angle X-ray scattering (USAXS) [15–17] and transmission electron microscope (TEM), and discuss their effects on the room-temperature fracture behaviour of NiAl-strengthened ferritic alloys.

A model NiAl-strengthened ferritic alloy, designated as FBB8, was investigated. An ingot with the size of $\sim 12.7 \times 25.4 \times 1.9$ cm was prepared by the Sophisticated Alloys, Inc., using the vacuum-induction melting. The nominal and measured compositions of the ingot are given in Table 1. The hot iso-thermal pressing (1,200°C/103 MPa/4 h) was conducted on the ingot to minimize casting porosity. Encapsulated specimens (in a quartz tube with the vacuum environment) were first solution treated at 1,200°C for 1 h, followed by air cooling (AC), and then, aged in the temperature range from 700°C to 950°C. The water quenching (WQ) and AC were used, following the age treatment. After the solution treatment (ST), precipitates with the size of 36 ± 5 nm were observed.[4] The nucleation and growth of NiAl-type precipitates are too fast to be suppressed by WQ, following ST.[4] The final grain size is ~ 100 µm. A small amount of segregation enriched in Zr is found along grain boundaries and within grains.

Readers are referred to Ref. [4,16] for the experimental details of USAXS. The size of cooling precipitates is assumed to follow the Gaussian distribution. The

*Corresponding author. Email: pliaw@utk.edu

Table 1. The compositions of the FBB8 ingot, weight per cent.

	Al	Ni	Cr	Mo	Zr	B	Fe
Nominal composition	6.5	10	10	3.4	0.25	0.005	Bal.
Measured composition	6.66	10.29	10.30	3.57	0.25	0.005	Bal.

form factor for spherical particles is employed in data modelling. Note that the contrast between the matrix and precipitates in USAXS originates from the composition difference (especially the aluminium content). The TEM images were taken, using a Zeiss Libra 200 model. Discs with the diameter of 3 mm were polished to the thickness of ~ 70 μm, and were further thinned in a Fischione twin-jet electropolisher, followed by ion milling with the energy of 3.5 keV for about 10 minutes. The electrolyte of the 5 volume-per cent HCl in ethanol was used at room temperature with the current of 6–8 μA.

Room-temperature tension/compression tests were conducted at the strain rate of 10^{-4} s^{-1}, using the hydraulic Materials Testing System machines. The 0.2% yield stress, σ_Y, was derived from three compression tests. The Vickers hardness of specimens was measured. At least six measurements were performed for each specimen. The fractured specimens were studied by scanning electron microscopy (SEM). Some deformed specimens were examined by TEM to reveal the deformation mechanisms.

Figure 1(a) and 1(b) shows the USAXS scattering curves, documenting the bimodal precipitate distribution in specimens aged for 100 h at 800°C, together with the corresponding TEM images. In this paper, the subscript of '1' stands for aged precipitates and '2' for cooling precipitates. Since the scattering vector, q, is inversely related to the real dimension of precipitates,[15] small precipitates show as features in USAXS curves at large q values. For example, in Figure 1(a), aged precipitates ($\bar{r}_1 \sim 166$ nm [4]) are represented by features (the Guinier knee and subsequent power-law slope) in the q range from 0.0003 to 0.02 $Å^{-1}$, while features from cooling precipitates ($\bar{r}_2 \sim 3.1 \pm 0.5$ nm) dominate the q regime from 0.02 to 0.15 $Å^{-1}$. Beyond 0.15 $Å^{-1}$, the scattering intensity from the background overwhelms that from precipitates. By comparing Figure 1(a) and 1(b), the size of cooling precipitates is strongly related to the cooling rate, following the ageing treatment. $\bar{r}_2$ $\sim 3.1 \pm 0.5$ and 7.6 ± 1.8 nm for WQ and AC, respectively, following the ageing treatment at 800°C. It is reported that the solubility of aluminium and nickel in α iron decreases, as the temperature decreases in the Fe–Ni–Al system.[18] Consequently, NiAl-type precipitates are likely to nucleate and grow in the cooling process due to supersaturated aluminium and nickel atoms in the matrix, and it is confirmed by the size dependence of fine precipitates on the cooling rate following the ageing treatment. The corresponding TEM images demonstrate that both aged and cooling precipitates have a B2-ordered structure (Figure 1(b)). The cooling precipitates-free zone around aged precipitates is observed. Specimens aged for a longer time at 800°C show a similar size distribution of cooling precipitates. $\bar{r}_2$ also presents the dependence on the ageing temperatures, as can be seen in Figure 1(c). Features of cooling precipitates for specimens aged at 700°C cannot be directly captured by USAXS due to the background scattering in the q regime greater than 0.1 $Å^{-1}$. When these specimens were aged for 50 h at 500°C, fine precipitates show their features in the relatively small q area (0.05–0.2 $Å^{-1}$), as shown in Figure 1(d). The double-ageing treatment probably does not change the volume fraction of fine precipitates because of the high precipitation rate of NiAl-type precipitates during cooling. Teng et al. [14] observed cooling precipitates ($\bar{r}_2$ ~ 1.5 nm) through the atom probe tomography for the FBB8 specimen aged for 100 h at 700°C + AC. The composition of these cooling precipitates was determined to be $Ni_{26.3}Al_{41.6}Fe_{26.9}Cr_{3.3}Mo_{1.7}$, which has an aluminium content similar to that of aged precipitates ($Ni_{41.2}Al_{43.6}Fe_{12.7}Cr_{0.8}Mo_{1.4}$). Assuming that aged and cooling precipitates have a similar contrast, relative to the matrix at the ageing temperatures of 700°C and 800°C, the volume fractions of cooling precipitates at both ageing temperatures can be estimated, based on the USAXS results, and the values are listed in Table 2. Note that quantitative TEM work could be helpful by comparing with the USAXS results.

Figure 2(a) presents the hardness of specimens subjected to different heat treatments. The hardness remains almost constant at ~ 457 HV for specimens aged at 800°C, while it slightly decreases at the beginning and stays at ~ 350 HV for specimens aged at 700°C. The hardness of specimens aged at 800°C is ~ 90 HV higher than that at 700°C. The fact agrees with the observed yield stresses at room temperature (1,015 $\pm$ 62 MPa for the specimen aged for 100 h at 700°C + AC (Condition 1); 1,166 $\pm$ 46 MPa for the specimen aged for 100 h at 800°C + WQ (Condition 2)). The solution-treated specimen has the highest hardness (474 HV). Note that the hardness is associated with the cooling rate, following the ageing treatment at 800°C. The air-cooled specimen is ~ 27 HV softer than the water-quenched one. On the contrary, specimens aged at 700°C do not show such strong dependence.

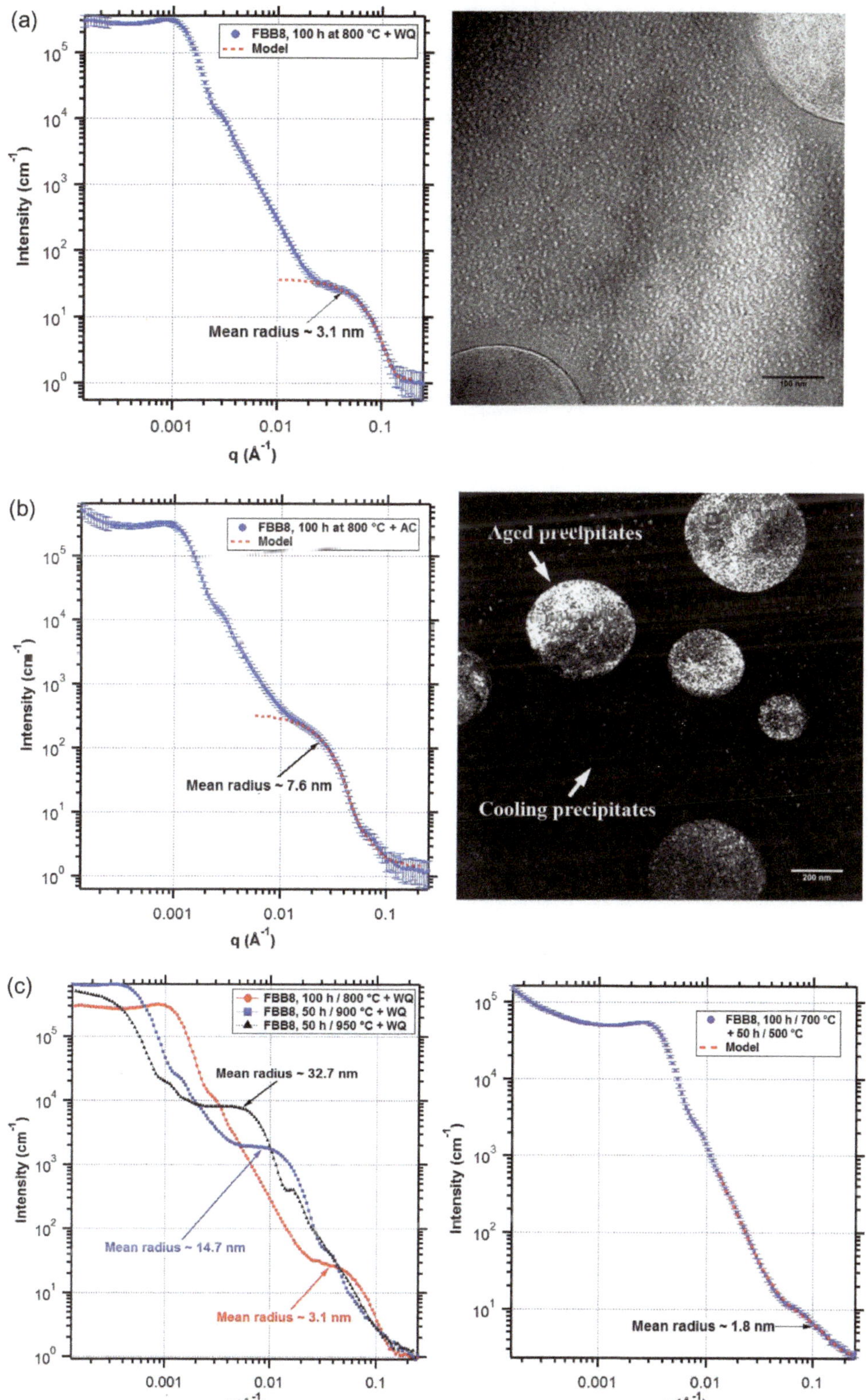

Figure 1. Cooling precipitates in FBB8 aged at 800°C for 100 h followed by WQ (a) and AC (b) characterized by USAXS and TEM. (c) The comparison of cooling precipitates for specimens aged at 800°C, 900°C, and 950°C. (d) Fine precipitates for the specimen aged for 100 h at 700°C + 50 h at 500°C. The dark-field image was taken using (100) reflection from the B2 ordered structure. For clarity, the experimental error bar is not included in (c).

Table 2. Aged and cooling precipitates for Conditions 1 and 2, and their hardening effects.

		Aged precipitates			Cooling precipitates		
Specimens	σ_Y (MPa)	f_1 (%)	$\bar{r}_1$ (nm)	$\Delta\sigma_{Or}$ (MPa)	f_2 (%)	$\bar{r}_2$ (nm)	$\Delta\sigma_{order}$ (MPa)
Condition 1	1,015 ± 62	15.5 ± 0.5 [14]	54 [4]	491 ± 14	~3.5	1.5 [14]	~322 (Equation 3)
Condition 2	1,166 ± 46	14.6 ± 0.9 [4]	166 [4]	181 ± 10	~5.0	3.1 ± 0.5	~639 (Equation 4)

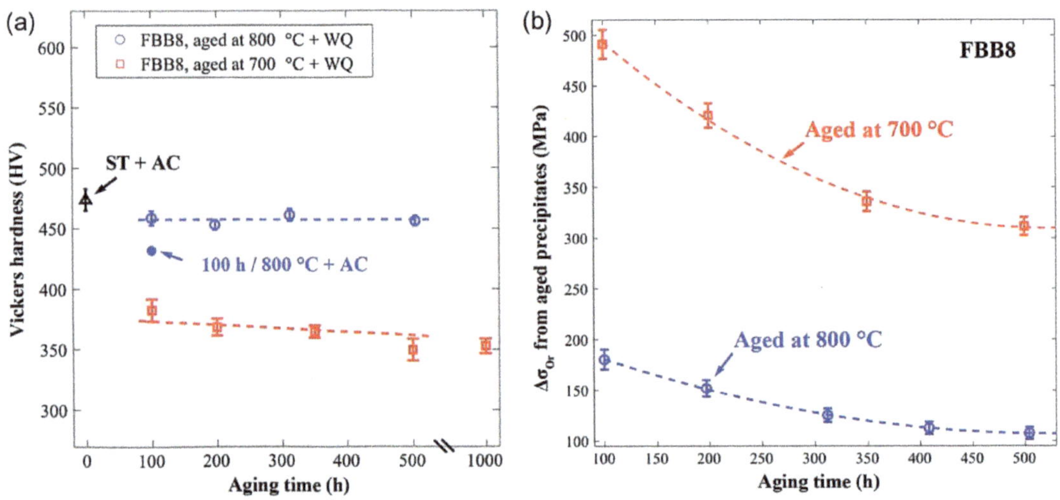

Figure 2. (a) The hardness of specimens subjected to different heat treatments. (b) The strengthening effect of aged precipitates as a function of ageing time.

The TEM micrographs of deformed specimens (Conditions 1 and 2) are shown in Figure 3. In Figure 3(a), dislocation loops (labelled as A) around aged precipitates are observed, indicating that dislocations bypass aged precipitates through the mechanism of Orowan bowing. Meanwhile, dislocations tend to form pairs (labelled as B) in the deformation process. It suggests that dislocations probably cut through fine ordered cooling precipitates.[19] The interaction between dislocations and cooling precipitates in Condition 2 is stronger than that in Condition 1, as shown in Figure 3.

By considering the mean planar inter-particle distance, dislocation dipole effect, and line tension of dislocations, Martin [20] proposed the strengthening effect by Orowan bowing, as described:

$$\Delta\sigma_{Or} = 0.81M\frac{Gb}{2\pi(1-\nu)^{1/2}}\frac{\ln(2r_s/r_0)}{(\lambda_s - 2r_s)}, \quad (1)$$

where G is the shear modulus, b is the length of Burgers vector, ν is the Poisson's ratio, r_s is the mean precipitate radius of intersections, r_0 is the inner cutting-off radius of a dislocation, λ_s is the average planar centre-to-centre spacing between a precipitate and its nearest neighbour, M is the Taylor factor (~3), and the factor of 0.81 is added by considering the effective precipitate spacing for a random array of obstacles.[20,21] Geometrically, we have the following relationships:

$$\lambda_s = \bar{r}_1\left(\frac{2\pi}{3f_1}\right)^{1/2} \text{ and } r_s = \sqrt{2/3}\,\bar{r}_1. \quad (2)$$

In this study, we take $G = 80$ GPa, $\nu = 0.3$, $b = 0.24982$ nm, and $r_0 = b$. The precipitate volume fractions are 0.155 ± 0.5 [14] and 0.146 ± 0.9 [4] for specimens aged at 700°C and 800°C, respectively. The coarsening behaviour of aged precipitates as a function of the ageing time at 700°C and 800°C was reported by Sun et al. [4]. Based on all these parameters, the hardening contribution from aged precipitates, $\Delta\sigma_{Or}$, is estimated, and the values are plotted against the ageing time at 700°C and 800°C (Figure 2(b)). As expected, the strengthening effect gradually decreases, as the ageing time increases at both temperatures. For example, $\Delta\sigma_{Or}$ goes down from 491 MPa for 100 h to 312 MPa for 500 h at 700°C. Note that $\Delta\sigma_{Or}$ is much high (~311–204 MPa) for specimens aged at 700°C than those aged at 800°C, given the same ageing time.

Dislocations probably cut through fine cooling precipitates, as suggested by Figure 3. Because of the small lattice and modulus mismatch between the matrix and precipitates, the order hardening could dominate the overall strengthening effect from cooling precipitates in this case.[22] The order-strengthening effect is given

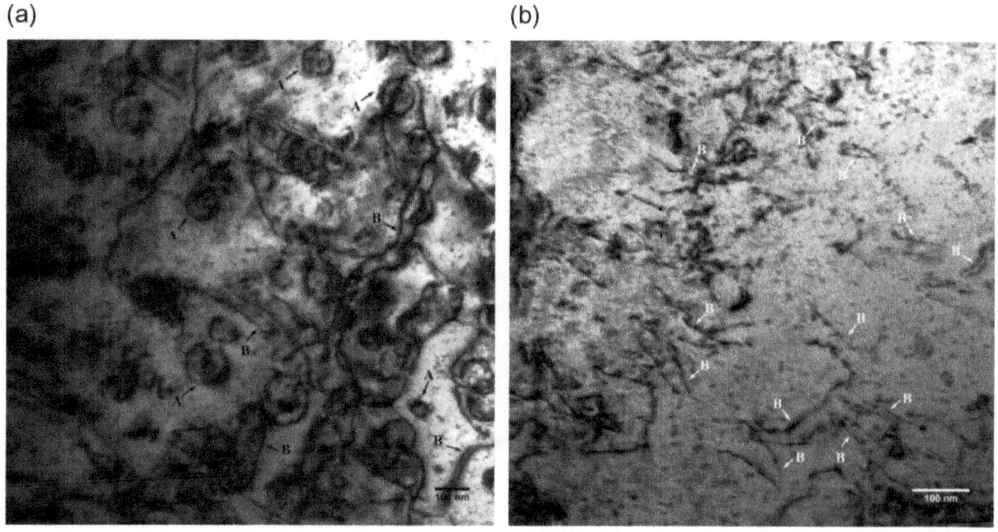

Figure 3. Bright-field images of dislocation structures (A: dislocation loops; B: dislocation pairs) for specimens (a) Condition 1, $\varepsilon \sim -1.7\%$; (b) Condition 2, $\varepsilon \sim -0.55\%$.

by [22]

$$\Delta\sigma_{\text{order}} = M\frac{\gamma_{\text{apb}}}{2b}\left(\frac{3\pi^2\gamma_{\text{apb}}f_2^{*}\bar{r}_2}{32\Gamma}\right)^{1/2}, \qquad (3)$$

where γ_{apb} ($\sim$0.5 J/m^2 [7,23]) is the anti-phase boundary energy on the (110) plane of NiAl-type precipitates, and Γ is the dislocation line tension ($\sim$ 1/2 Gb^2). The effective volume fraction of f_2^*, defined as $f_2^* = f_2/(1-f_1)$, is used instead of f_2, as suggested by Reppich et al.[24] When precipitates are strong enough (the breaking angle of dislocations $< \sim 120$°C [22]), the order-strengthening effect becomes [22]

$$\Delta\sigma_{\text{order}} = 0.81M\frac{\gamma_{\text{apb}}}{2b}\left(\frac{3\pi f_2^{*}}{8}\right)^{1/2}. \qquad (4)$$

Based on the TEM micrograph (Figure 3(b)), Equation (4) is probably more suitable to calculate the strengthening effect of cooling precipitates in Condition 2. The estimation of $\Delta\sigma_{\text{order}}$ of cooling precipitates in Conditions 1 and 2 is listed in Table 2, based on the microstructural information. $\Delta\sigma_{\text{order}}$ is $\sim$322 MPa for Condition 1, while it is approximately 2.5 times larger than $\Delta\sigma_{\text{Or}}$ in the case of Condition 2. We conclude that cooling precipitates harden the matrix considerably at room temperature in both conditions, besides the strengthening effect provided by aged precipitates.

The fractured specimens after tensile tests were examined by SEM. The specimen shows the river pattern of the cleavage transgranular fracture and cracked NiAl-type precipitates (labelled as C) just below the fracture surface are observed (Figure 4). NiAl is brittle at room temperature in both single-crystalline and polycrystalline forms.[25,26] It is very likely that cracks initiate from NiAl-type precipitates themselves. Cooling precipitates could have two impacts on the room-temperature fracture. First, they excessively harden the matrix, and, hence, the yield stress may exceed the fracture stress of NiAl-type precipitates. It is worth noticing that the stress is probably inhomogeneous on the microscopic scale during deformation.[20] Due to the excessive hardening of cooling precipitates, the plastic zone size ($\propto 1/\sigma_{\text{Y}}^2$) near the crack tip is expected to be small, which results in a small amount of energy absorption by plastic deformation.[27] Hence, once a crack forms, it could propagate quickly through the matrix. The coalescence and growth of cracks (labelled as P) are also observed in Figure 4. Teng et al. [11] studied the effect of the ageing time (up to 1,000 h at 700°C) on the room-temperature ductility of FBB8. They concluded that the ductility is independent of the ageing time, or equally the precipitate size in the studied range. $\bar{r}_1$ becomes 122 nm for the FBB8 specimen aged for 1,000 h at 700°C,[4] and the corresponding $\Delta\sigma_{\text{Or}}$ decreases to $\sim$250 MPa. However, the hardness of the specimen stays invariantly at $\sim$350 HV (Figure 2(a)) due to the strengthening effect of cooling precipitates. It is possibly why the room-temperature ductility of FBB8 does not improve even after a long-time ageing treatment in their study.[11] New composition designs by controlling the amounts of aluminium and nickel and their ratio in order to decrease the solubility of aluminium or nickel in the matrix at the ageing temperature are perhaps ways to improve the ductility of NiAl-strengthened ferritic alloys at room temperature. The refinement of grain size by the hot deformation and following heat treatment may also help.

In conclusion, duplex precipitates in FBB8 were observed by USAXS and TEM. Fine precipitates

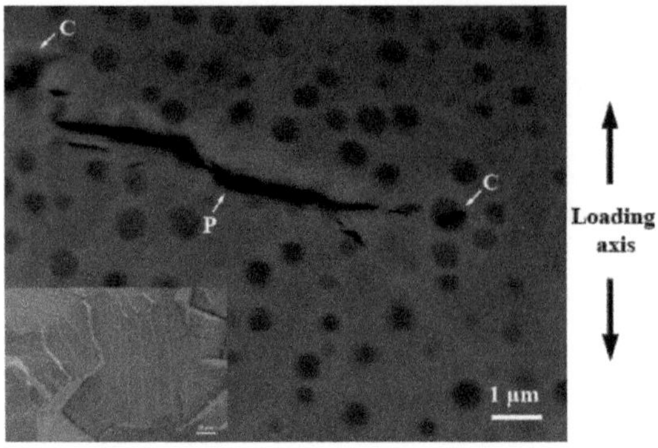

Figure 4. The cracked precipitates just below the fracture surface (C: cracked precipitates; P: coalescence and growth of cracks). The specimen was aged for 207 h at 800°C. Typical fractography is also shown in the inset.

nucleate and grow in the cooling process, following the ageing treatment. The average radius of these cooling precipitates ranges from several to tens nanometres, depending on the cooling rate and ageing temperature. At room temperature, dislocations bypass aged precipitates through the mechanism of Orowan bowing, while fine cooling precipitates are possibly cut through by dislocations. Cooling precipitates harden the matrix considerably at room temperatures, and the order-strengthening effects from cooling precipitates are estimated to be 322 and 639 MPa for Conditions 1 and 2, respectively. For the room-temperature fracture of FBB8, cracks probably initiate from NiAl-type precipitates, and they could propagate quickly through the matrix due to the excessive hardening contribution from cooling precipitates.

Disclosure statement No potential conflict of interest was reported by the authors.

Funding The research was supported by the Department of Energy, Office of Fossil Energy, under Grants of DE-09NT0008089 and DE-FE0005868 with Richard Dunst, Vito Cedro, and Patricia Rawls as the program managers. The ChemMatCARS Sector 15, located at the Advanced Photon Source of the Argonne National Laboratory, is principally supported by the National Science Foundation under the grant number of NSF/CHE-1346572. The use of the Advanced Photon Source was supported by the U.S. Department of Energy, Office of Science, Office of Basic Energy Sciences, under Contract No. DE-AC02-06CH11357.

References

[1] Bhadeshia HKDH. Design of ferritic creep-resistant steel. ISIJ International. 2001;41:626–640.
[2] Viswanathan R, Henry JF, Tanzosh J, Stanko G, Shingledecker J, Vitalis B, Purgert R. U.S. program on materials technology for ultra-supercritical coal power plants. J Mater Eng Perform. 2005;14(3):281–292.
[3] Calderon H, Fine ME. Coarsening kinetics of coherent NiAl-type precipitates in Fe-Ni-Al and Fe-Ni-Al-Mo alloys. Mater Sci Eng. 1984;63:197–208.
[4] Sun Z, Song G, Ilavsky J, Ghosh G, Liaw PK. Coarsening behavior of NiAl-type precipitates in a ferritic alloy. Manuscript submitted to Acta Mater. 2014.
[5] Huang S, Brown DW, Clausen B, Teng Z, Gao Y, Liaw PK. In situ neutron-diffraction studies on the creep behavior of a ferritic superalloy. Metall Trans A. 2011;43(5):1497–1508.
[6] Stallybrass C, Schneider A, Sauthoff G. The strengthening effect of (Ni,Fe)Al precipitates on the mechanical properties at high temperatures of ferritic Fe–Al–Ni–Cr alloys. Intermetallics. 2005;13(12):1263–1268.
[7] Vo NQ, Liebscher CH, Rawlings MJS, Asta M, Dunand DC. Creep properties and microstructure of a precipitation-strengthened ferritic Fe–Al–Ni–Cr alloy. Acta Mater. 2014;71:89–99.
[8] Sun Z, Liebscher CH, Huang S, Teng Z, Song G, Wang G, Asta M, Rawlings M, Fine ME, Liaw PK. New design aspects of creep-resistant NiAl-strengthened ferritic alloys. Scripta Mater. 2013;68(6):384–388.
[9] Huang S, Gao Y, An K, Zheng L, Wu W, Teng Z, Liaw PK. Deformation mechanisms in a precipitation-strengthened ferritic superalloy revealed by in situ neutron diffraction studies at elevated temperatures. Acta Mater. 2015;83:137–148.
[10] Teng ZK, Liu CT, Ghosh G, Liaw PK, Fine ME. Effects of Al on the microstructure and ductility of NiAl-strengthened ferritic steels at room temperature. Intermetallics. 2010;18(8):1437–1443.
[11] Teng ZK, Liu CT, Miller MK, Ghosh G, Kenik EA, Huang S, Liaw PK. Room temperature ductility of NiAl-strengthened ferritic steels: effects of precipitate microstructure. Mater Sci Eng A. 2012;541: 22–27.
[12] Teng ZK, Zhang F, Miller MK, Liu CT, Huang S, Chou YT, Tien RH, Chang YA, Liaw PK. Thermodynamic modeling and experimental validation of the Fe-Al-Ni-Cr-Mo alloy system. Mater Lett. 2012;71:36–40.
[13] Teng ZK, Zhang F, Miller MK, Liu CT, Huang S, Chou YT, Tien RH, Chang YA, Liaw PK. New NiAl-strengthened ferritic steels with balanced creep resistance and ductility designed by coupling thermodynamic calculations with focused experiments. Intermetallics. 2012;29:110–115.
[14] Teng ZK, Miller MK, Ghosh G, Liu CT, Huang S, Russell KF, Fine ME, Liaw PK. Characterization of nanoscale NiAl-type precipitates in a ferritic steel by electron microscopy and atom probe tomography. Scripta Mater. 2010;63(1):61–64.
[15] Guinier A, Fournet G. Small-angle scattering of X-ray. New York: Wiley; 1955.
[16] Ilavsky J, Jemian PR, Allen AJ, Zhang F, Levine LE, Long GG. Ultra-small-angle X-ray scattering at the Advanced Photon Source. J App Crystallogr. 2009;42(3):469–479.
[17] Huang EW, Liaw PK, Porcar L, Liu Y, Liu Y-L, Kai J-J, Chen W-R. Study of nanoprecipitates in a nickel-based superalloy using small-angle neutron scattering and transmission electron microscopy. App Phys Lett. 2008;93(16), Article ID 161904. http://scitation.aip.org/content/aip/journal/apl/93/16/10.1063/1.3002300.
[18] Hao SM, Takayama T, Ishida K, Nishizawa T. Misciility gap in Fe-Ni-Al and Fe-Ni-Al-Co systems. Metall Trans A. 1984;15A:1819–1828.

[19] Nembach E. Particle strengthening of metals and alloys. New York: John Wiley & Sons, Inc.; 1997.
[20] Martin JW. Micromechanisms in particle-hardened alloys. Cambridge: Cambridge University Press; 1980.
[21] Kocks UF. On the spacing of dispersed obstacles. Acta Metall. 1966;14:1629–1631.
[22] Ardell AJ. Precipitation hardening. Metall Trans A. 1985;16:2131–2165.
[23] Noebe RD, Bowman RR, Nathal MV. Physical and mechanical properties of the B2 compound NiA. Inter Mater Rev. 1993;38:193–232.
[24] Reppich B, Huhlein W, Meyer G, Puppel D, Schulz M, Schumann G. Duplex γ' particle hardening of the superalloy Nimonic PE 16. Mater Sci Eng. 1986;83: 45–63.
[25] Darolia R. NiAl alloys for high-temperature structural applications. JOM. 1991;43:44–49.
[26] Miracle DB. The physical and mechanical properties of NiAl. Acta Metall Mater. 1993;41:649–684.
[27] Anderson TL. Fracture mechanics: fundamentals and applications. Boca Raton: CRC Press; 2005.

33

The Suppression of Instabilities via Biphase Interfaces During Bulk Fabrication of Nanograined Zr

J.S. Carpenter[a]*, T.J. Nizolek[b], R.J. McCabe[a], S.J. Zheng[c], J.E. Scott[a], S.C. Vogel[d], N.A. Mara[a,c], T.M. Pollock[b] and I.J. Beyerlein[e]

[a]*Materials Science and Technology Division, Los Alamos National Laboratory, Los Alamos, NM 87545, USA;* [b]*Materials Department, University of California at Santa Barbara, Santa Barbara, CA 93106, USA;* [c]*Materials Physics and Applications Division, Los Alamos National Laboratory, Los Alamos, NM 87545, USA;* [d]*Los Alamos Neutron Science Center, Los Alamos National Laboratory, Los Alamos, NM 87545, USA;* [e]*Theoretical Division, Los Alamos National Laboratory, Los Alamos, NM 87545, USA*

Severe plastic deformation (SPD) is a common method to fabricate nano-grained metals. However for Zr, a structural metal for nuclear applications, obtaining a nanoscale grain structure via SPD has been problematic due to deformation twinning and phase transformations. Here, nanostructured hcp Zr is fabricated through a refinement process via the introduction of a biphase interface. Despite mechanical and thermal conditions known to chemically mix Zr and Nb, no intermixing is observed and the heterophase interfaces appear resistant to phase transformations and twinning. Increasing the density of chemically sharp Zr–Nb interfaces is a very different refinement mechanism than substructure development, stacking fault formation, or alloying.

Keywords: Severe Plastic Deformation, Zirconium, Nanomaterials, Interfaces

Nanomaterials have been found to exhibit remarkable strength and, in some cases, simultaneous ductility.[1–6] Nanolayered metallic composites often achieve these exceptional properties while retaining good thermal stability.[7–9] Realizing these outstanding material properties in structural applications requires methods of fabricating nanomaterials in the bulk form. Over the past few decades, severe plastic deformation (SPD) techniques, such as wire drawing,[10,11] high-pressure torsion (HPT),[12–15] equal channel angular pressing (ECAP),[16,17] and accumulative roll bonding (ARB),[18–23] have been employed to successfully make nanomaterials and nanolayered materials. The last two processes, ECAP and ARB, can potentially produce nanomaterials in sheets and rods in both sizes and quantities suitable for structural applications.[24–26]

Nanostructuring ($<$ 100 nm characteristic length scales) has been achieved primarily for cubic metals, such as face-centered cubic (fcc) and body-centered cubic (bcc) as well as composite combinations of the two, i.e. fcc–fcc and fcc–bcc.[17,22,27–30] Many technically relevant metals have a hexagonal close packed (hcp) structure and, to date, making bulk nano-hcp metals has been a challenge with SPD techniques.[19,20,23,31,32] In this work, we develop a method using ARB to produce nanolayered Zr–Nb composites with layer thicknesses h of $\sim$90 nm.

It has been notoriously difficult to make nanoscale hcp metals (not only Zr but also Mg and Ti) in engineering quantities. Prior attempts to make nanostructured Zr metals created materials in small quantities. For example, nanolayered Zr–Nb composites have been fabricated

*Corresponding author. Email: carpenter@lanl.gov

This material is published by permission of the Los Alamos National Laboratory, operated by Los Alamos National Security, LLC for the US Department of Energy under Contract No. DE-AC52-06NA25396. The US Government retains for itself, and others acting on its behalf, a paid-up, non-exclusive, and irrevocable worldwide license in said article to reproduce, prepare derivative works, distribute copies to the public, and perform publicly and display publicly, by or on behalf of the Government.

The work of John S Carpenter, Rod McCabe, Shijian Zheng, Jeffrey Scott, Sven Vogel, Nathan Mara and Irene Beyerlein was authored as part of their official duties as Contractors for the United States Government and is therefore a work of the United States Government. In accordance with 17 USC. 105, no copyright protection is available for such works under US Law.

Tom Nizolek and Tresa Pollock hereby waive their right to assert copyright, but not their right to be named as co-authors in the article.

using bottom-up approaches such as magnetron sputtering or other deposition methods.[33] Nanoscale Zr grains can be fabricated utilizing HPT.[13–15] However, HPT does not lend itself to the production of engineering scale quantities.[34] In addition, HPT can cause an allotropic phase transformation from hcp alpha-Zr to hexagonal omega-Zr and bcc beta-Zr.[13,15,35] Similarly, while ball-milling can create nanograin Zr (~5 nm), a fraction of Zr transforms into an fcc phase.[36] Earlier attempts to use ARB to make Zr–Nb layered composites ended with a layer thickness of $h = 4\,\mu m$, since further rolling led to shear banding.[37,38] It was found in the ARB of monolithic Zr that an ultra-fine-grained microstructure developed with equiaxed grains of diameter ~300 nm.[39,40] Cryo-rolling followed by annealing treatments has been utilized to fabricate bimodal distributions of ultra-fine-grained and nanocrystalline Zr in the past.[41,42] However, this nanocrystalline Zr was captured during the annealing treatment as opposed to during the refinement process and showed little thermal stability as evidenced by the sharp decrease (70%) in the volume fraction of nanograins with an increase of 50°C in the annealing temperature.[41]

To fabricate nanolayered Zr/Nb, we started with $h_0 = 2\,mm$ sheets of high-purity Zr [43] and reactor-grade, high-purity Nb (99.97% pure, ATI-Wah Chang) (see supplement for composition and initial microstructure). Both starting sheets showed measureable preferred texture and equiaxed grains with average sizes of 40 μm for Zr and 300 μm for Nb (see Figure S1 in supplement). The plastic deformation behavior of similar single-phase Zr has been well studied, and it has been found that it deforms primarily by slip with only a modest amount of twinning at room temperature [44–46] and by an interplay of slip and twinning at 77 K.[37,46,47] Nb similar to that used here has been found to deform by slip, activating both $\{110\}\langle 111\rangle$ and $\{112\}\langle 111\rangle$ slip at room temperature in rolling and uniaxial testing.[38,48] The ARB process (see supplement for more detail concerning this process) consists of a repetitive sequence of cutting, cleaning, stacking, and roll bonding sheets with the initial step in our current process consisting of sandwiching a 2 mm Nb sheet between two 1 mm Zr sheets. Using this cladding technique, new Nb–Zr interfaces are only formed during the first rolling pass. As the number of bonding steps is repeated, the number of layers increases and the individual layer thickness h decreases. The total strain as a function of h is given by

$$\varepsilon = \ln \frac{h_0}{h} \qquad (1)$$

Using the ARB process, we are able to obtain a final average layer thickness of 92 nm corresponding to an extreme strain level of 10.7. Figure 1(a) shows a representative transmission electron microscopic (TEM) image of the final nanostructure (see supplement for TEM methods). The interfaces are planar with modest morphological roughening. This is the first main achievement of this study—*fabrication of nanolayered Zr/Nb in quantities suitable for structural applications using deformation conditions amenable to mass production*.

While the ARB process is well known, a few key factors enabled success in this nanomaterial synthesis. Instabilities such as pinch-off ('island formation'), debonding, and shear banding can prevent layer refinement. An earlier mathematical mechanics model [49]

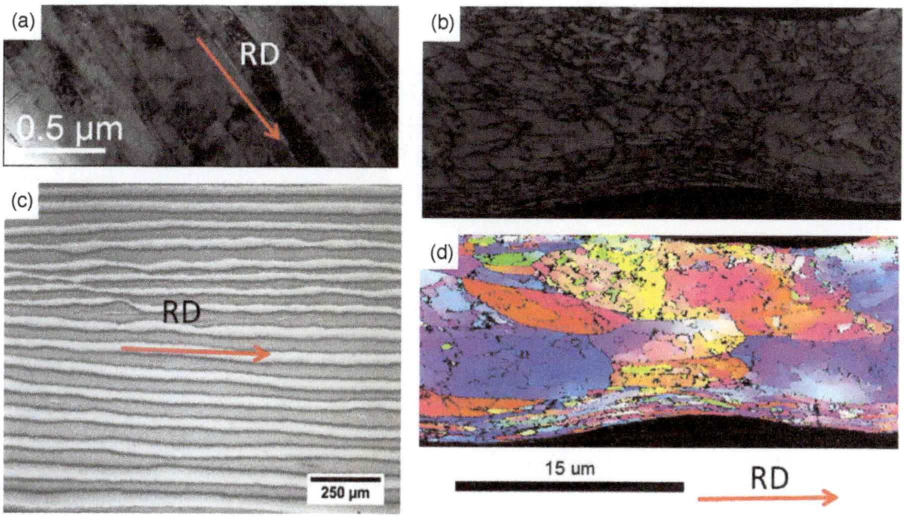

Figure 1. (a) Bright field TEM micrograph of Zr–Nb lamellar composite with $h = 92\,nm$. (b) Optical micrograph of Zr–Nb lamellar composite with $h = 59\,\mu m$, illustrative of the bulk nature of the material fabricated. (c) EBSD image quality map showing sub-grain boundaries for $h = 12\,\mu m$ layer of Nb. (d) Orientation map from EBSD scan on material from (c). Note that dislocation cell structure is present at sub-grain scales for $h = 12\,\mu m$.

suggested that these can be avoided with low ratios (< 5) in flow stress between the two constituents. On this basis, static annealing steps at 575°C for 60 min were performed in this work after every second roll bonding pass corresponding to incremental strain intervals of 1.4. The time and temperature combination was chosen in order to minimize intermixing of the Zr and Nb, which can occur above temperatures of around 500°C.[50]

To investigate the chemical sharpness of the Zr–Nb interfaces, energy-dispersive spectroscopy (EDS) was utilized within the TEM. EDS line scans on the $h = 92$ nm sample showed no signs of intermixing and showed chemically distinct, pure phases in the layers to within 4.5 nm from the interfaces. Figure 2 provides a representative high-angle annular dark field scanning TEM image with accompanying EDS line scan showing the sharp transitions between Zr and Nb. Parameters utilized in obtaining the EDS line scans are specimen thickness ≈ 50 nm, electron probe size of ≈ 1 nm, and electron beam energy of 300 keV.

According to its equilibrium phase diagram, Zr–Nb shows miscibility at relatively low (500°C) temperatures [50] and thus the intermediate annealing temperatures applied here could have led to interdiffusion between Zr and Nb. Furthermore, mechanical deformation from ball-milling and wire drawing has been known to induce chemical mixing even between immiscible metals.[51–53] Preservation of two phases allows the layer thickness to limit the grain size, thus facilitating

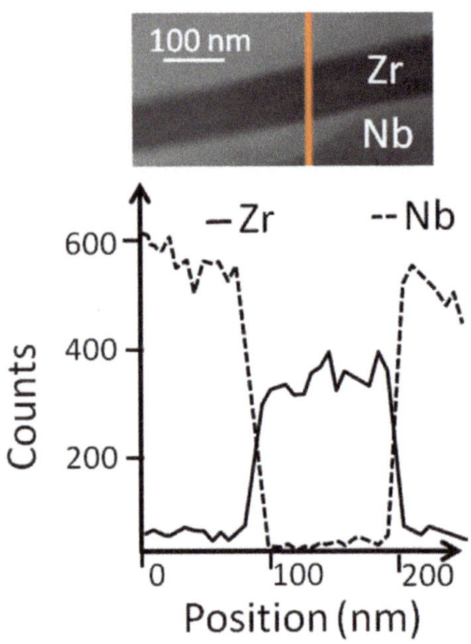

Figure 2. Zr–Nb composite micrograph where $h = 92$ nm with accompanying EDS-based line scan. The sharp transition between Zr and Nb exhibited indicates chemically distinct phases to within 4.5 nm from the interfaces.

grain refinement to the nanoscales. In addition, the thermal stability of the nanostructure is greatly enhanced by the presence of the Zr–Nb bimetal interfaces in comparison with the single-phase Zr nanograins in cryo-rolling studies.[41,42]

To detect possible changes in the primary deformation mechanisms and crystal structures of the phases, we measured the bulk textures for each phase at all length scales using neutron diffraction (NeD, see supplement). NeD was carried out at the High Pressure Preferred Orientation neutron time-of-flight diffractometer at the Los Alamos Neutron Science Center at Los Alamos National Laboratory.[54,55] Compared with other texture measurement techniques, NeD enables statistically relevant data without potential modifications due to surface preparation. Texture measurements also provide a robust indicator of the primary deformation mechanisms, such as when interface-driven plasticity,[28,56] grain boundary-mediated processes,[57] deformation twinning,[58–60] and partial-mediated slip [61] prevail. Figure 3 shows examples of textures at the micron scale and nanoscale in the form of pole figures, a compact way to present texture, as well as the standard texture presentation for Zr (albeit not for Nb). The analysis undeniably shows that the Zr retains an hcp crystal structure throughout the extreme deformation. Thus, unlike nanostructured Zr produced by HPT,[13,15,35] nanostructuring via ARB resulted in no measurable allotropic phase transformation.

We observe that a preferred texture develops at all length scales which is distinct from the initial texture (Figure S1), and indicates that the prevailing mechanism is a crystallographic slip. Zr's strong texture indicates that interface or grain boundary-mediated processes, which tend to randomize the texture, are not active.[57] More importantly, the textures are remarkably similar (apart from the usual modest variations) over the wide range of h for each phase. Abrupt transitions in texture would have significant changes in the predominant deformation mechanism when h crosses from the micron to nanoscale, such as twinning, partial dislocation-mediated slip, or interfacial sliding. In spite of the increased density of interfaces and nanoscale dimensions, such transitions are not present in the Zr–Nb composite when refining from the micron scale to the nanoscale.

Another significant finding is that the textures in Figure 3 are very similar to textures in coarse-grained Zr and Nb when rolled alone. In past studies of rolled single-phase Nb, a texture was observed consisting of α (001) and γ (111) fibers with particularly strong intensity associated with the $\{001\}\langle 110\rangle$ and $\{111\}\langle 110\rangle$ orientations.[62,63] Monolithically rolled Zr develops texture with strong intensities associated with $\{0001\}$ that are tilted 20–40° away from the normal direction (ND) about the rolling direction (RD).[40,64,65] The textures of our Zr–Nb nanolayered composites (Figure 2) clearly

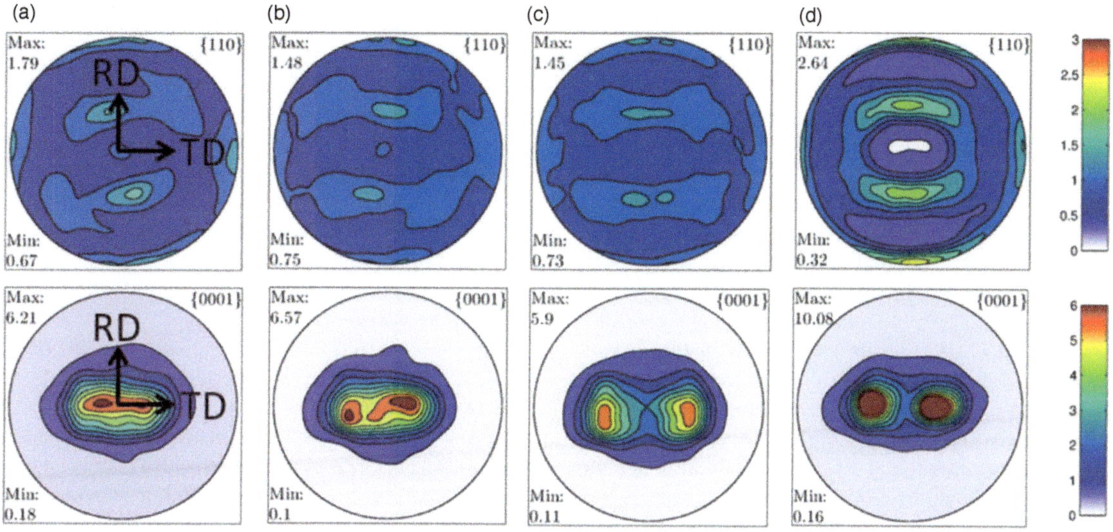

Figure 3. NeD-based pole figures for Nb $\{110\}$ (top) and Zr $\{0001\}$ (bottom) for length scales of $h =$ (a) 106 μm as rolled, (b) 26 μm as rolled, (c) 26 μm annealed and (d) 92 nm as rolled. Zr pole figures are plotted on a scale of 0–6 MRD while Nb pole figures are plotted on a scale of 0–3 MRD.

exhibit the same predominant features of monolithic rolling textures within normal variation. Moreover, prior modeling studies that calculated the relative contributions of the different deformation modes clearly showed that these Zr and Nb rolling textures are a consequence of dislocation slip.[37,38] The similarity of the nanoscale phase textures measured here to conventional rolling textures is another indication that interface-mediated processes are not the dominant mechanisms of plasticity, even at submicron layer sizes. It was not previously known that dislocation-mediated plasticity could persist in hcp Zr down to the nanoscale. *Persistent dislocation plasticity in Zr in the nanoscale regime is the second key finding of this work.*

As a highly influential factor facilitating nanostructuring, we investigated the effect of annealing on microstructure and texture development during ARB processing. Figure 3(b) and 3(c) presents the 26 μm textures before and after annealing. For these micron layers, Zr and Nb both respond to annealing in a similar fashion as severely rolled hcp and bcc materials in the monolithic form.[63] Specifically, Nb maintains its weak rolling texture prior to and after annealing and Zr shows a reorientation of the basal poles from 10–20° to 30–40° away from the ND about the RD after annealing. In order to assess accompanying changes in grain and layer morphology, multiple electron backscatter diffraction (EBSD) scans were conducted on samples before and after annealing. In Figure 4, we present the grain structure, where it is clearly seen that both Zr and Nb recrystallize during the annealing step. The low internal grain misorientations in the recrystallized grains in Figure 4 also provide evidence that the dislocation content was

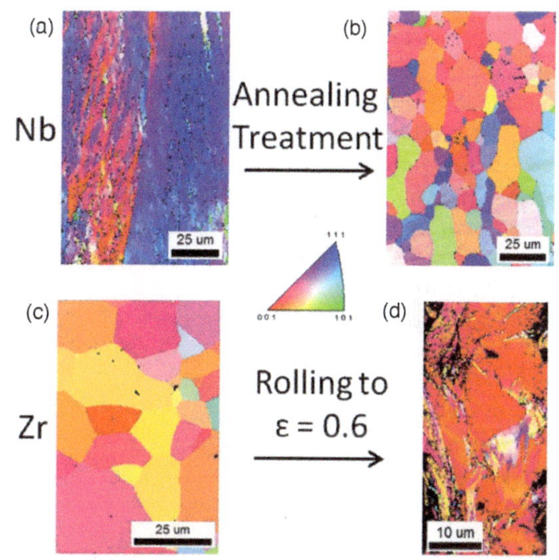

Figure 4. EBSD-based inverse pole figure maps for (a) Nb, $h = 466$ μm as-rolled, (b) Nb, $h = 466$ μm as-annealed, (c) Zr, $h = 106$ μm as-annealed, and (d) Zr, $h = 59$ μm as-rolled. Recrystallized microstructures are noted after annealing treatments.

sufficiently reduced during annealing. This reduction was likely instrumental in keeping the flow-stress ratio of Zr to Nb sufficiently low to suppress instabilities and enable nanoscale refinement.

The foregoing analysis warrants repeating for layer thicknesses approaching the submicron regime. Very recently in a study of ARB Cu–Nb composites, it was found that at finer layers $h < 1.9$ μm, a typical recrystallized microstructure did not develop after annealing.[66] A similar finding is revealed for the ARB Zr–Nb composites through an EBSD-based study. More than $20,000\,\mu m^2$ of Zr–Nb material ($h = 1.5\,\mu m$) was scanned in the as-rolled and annealed states with no difference in grain size noted. It should be noted that texture was consistent between NeD and the EBSD scans, implying that a representative area was scanned. This indicated that the annealing treatment had little microstructural and textural effects at this reduced length scale in comparison with Figure 4. Previously, it was proposed in [66] that the bimetal interfaces can act as sinks for dislocations, leading to a reduction in the dislocation content in the phases with strain and consequently a reduction in the driving force for recrystallization after deformation. Our analysis suggests that a similar phenomenon is operating in Zr–Nb. A reduction in dislocation content in the layers as they become single-grain thick is expected to be instrumental in continued layer refinement from the submicron to the nanoscales.

Another intriguing question concerns the apparent decoupled texture evolution of Zr and Nb in spite of the fact that the phases were clearly coupled microstructurally and mechanically. Microstructural coupling is clear from our EBSD and TEM analyses, which indicate that one grain spans the layer thickness in the submicron and nanolayered composites ($h < 1.5\,\mu m$). Therefore, all Zr and Nb grains are joined by bimetal interfaces. Mechanical coupling can be proven via a simple calculation, which indicates that the applied rolling reduction should have reduced both layers to a nominal $h = 92$ nm. Achievement of this average value of h by both phases in the actual material indicates that the Zr and Nb phases co-deformed. In spite of these strong links, each phase in the composite develops a classical rolling texture as if rolled alone. This texture remains stable into the nanoscale regime with no noticeable texture transitions. In contrast, microstructural and mechanical coupling in ARB Cu–Nb composites caused two abrupt transitions in texture evolution. The first occurred when the layers became fine enough that one grain spanned the layer thickness ($h < 2\,\mu m$). The second one occurred when the layers reduced below 100 nm, invoking deformation twinning in Cu. Based on this comparison and current texture analysis, it appears that in ARB Zr–Nb, the bimetal interfaces and nanoscale dimensions amazingly do not promote coupling in texture or deformation mechanisms different from those in the single-phase coarse-grained material.

In order to highlight this difference in more detail, Figure 5 compares the textures of the Nb phase in the Zr–Nb and Cu–Nb ARB composites in the submicron regime ($h = 676$–770 nm) where significant contrast in texture is expected to occur first. The textures are strikingly different, with an unusually sharp, atypical rolling Nb texture for the Cu–Nb composite and a typical but weak rolling Nb texture for the Zr–Nb composite. To elucidate this distinctive behavior at the nanoscale, we repeat this comparison for $h = 60$–90 nm in Figure 6. Again, we observe that the Nb phase within the present nanolayered Zr–Nb composite corresponds to a weak rolling texture, whereas the Nb phase within the nanolayered Cu–Nb composite is very sharp and strongly deviates from a typical rolling texture. Evidently, texture evolution was phase-driven in the Zr–Nb and interface-driven in the Cu–Nb.

This last point naturally brings us to the important issue of interface evolution. Interface character and properties in cubic systems have been found to play critical roles in dislocation and twin nucleation.[67–69] To date, the available data and numerical studies on hcp/bcc interfaces are insufficient for understanding their role in plasticity. Certainly, the sharpness and planarity of the Zr–Nb interfaces within the present layered Zr–Nb composites inspire further exploration of the interface character. The weak rolling texture in Nb (Figure 2) signifies that the characteristics of many grain orientations are dispersed about the ideal rolling components. Very strong, single-crystal like textures, however, are associated with development of a preferred interface character (as defined by five parameters describing the orientation relationship and interface plane), as seen in the Cu–Nb system.[70,71] It is, therefore, unlikely that a prevailing interface forms in Zr–Nb. We therefore conclude that two contrasting SPD behaviors are possible for nanoscale composites: one where texture is driven by grain orientation stabilization as for Zr–Nb composites and one driven by interface character stabilization as in the Cu–Nb composites.

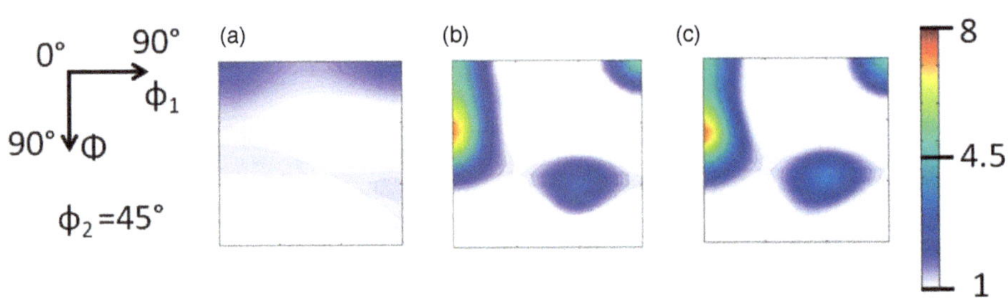

Figure 5. NeD-based orientation distribution function slices of $\varphi_2 = 45°$ for Nb in (a) Zr–Nb $h = 676$ nm as rolled, last annealed at $1.5\,\mu m$, (b) Cu-Nb $h = 770$ nm as annealed,[21] and (c) Cu–Nb $h = 714$ nm as rolled, with no prior annealing steps.[22]

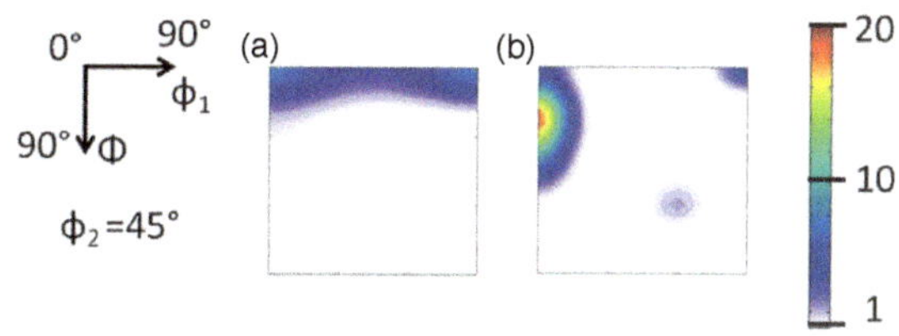

Figure 6. NeD-based orientation distribution function slices of $\varphi_2 = 45°$ for Nb in (a) Zr–Nb $h = 92$ nm as rolled, (b) Cu–Nb $h = 60$ nm as rolled, with no prior annealing steps.[22]

A key finding of this work is that nano-refinement arises as a result of the reduction in spacing between the bimetal interfaces. TEM results in Figure 1(a) provide a clear evidence that Zr and Nb are co-deforming as indicated by the similarity in layer thicknesses. Furthermore, our accompanying chemical analysis shows a lack of intermixing at the interface. Figure 1(c) and 1(d) both shows that thick Nb layers ($h = 12\,\mu$m) contain grain boundaries and a sub-grain dislocation cell structure, while Figure 1(a) shows that nanoscale layers exhibit little dislocation accumulation. Taken together, the results suggest that new grain boundary formation and intragranular dislocation wall formation are not driving the refinement process at the nanoscale. Instead the bimetal interface is driving the refinement process through deformation carried by dislocation slip within the grains. An additional outcome of the texture analysis is that twinning and phase transformations are suppressed through the use of a bimetal interface as a refinement mechanism.

Despite thermal and mechanical conditions that could lead to chemical mixing of Zr and Nb, these two elements are found to maintain separation at the nanoscale under ARB with intermittent annealing treatments to form sharp Zr–Nb interfaces. The results imply that the Zr–Nb interface is in a lower energy state compared with a chemically varying interface and hence limits instabilities such as twinning and grain boundary sliding which have been observed to prevent nanostructuring of hcp Zr via SPD in the past. The lack of deformation-induced instabilities may indicate that the associated changes such as twinning and grain boundary sliding to the interface would induce an energy penalty. Phase transformations observed under HPT [13–15,35] arose as a result of a combination of stress state and pressure making it unclear if the interface, itself, is responsible for limiting the phase transformations.

In summary, we demonstrate a way to exploit the ARB SPD process to produce a 50/50 volume fraction hcp/bcc nanostructured composite material.The material fabricated is a two-phase hcp Zr-bcc Nb composite possessing a lamellar architecture with single nanocrystals spanning layer thicknesses.NeD measurements of texture indicate that dislocation-mediated plasticity persists even at the nanoscale with no evidence of interface-driven processes. This is in remarkable contrast to cubic-based nanolayered systems produced by the same technique.Thus, critical length scales below which size-dependent or interface-driven mechanisms emerge in the hcp–bcc system, should they exist, must lie below 90 nm.Furthermore, based on the exceptional properties discovered for ARB Cu-Nb systems,[9,56,71–74] we envision that the Zr–Nb nanolayered material will exhibit many unusual and interesting properties, potentially far exceeding those of current coarse-grained Zr.This study opens up an innovative route to manufacturing hcp and bcc nanostructured metals as well as a novel material for pioneering hcp/bcc interface and nanoscale hcp research.

Acknowledgements JSC and RJM were supported by the Los Alamos National Laboratory Directed Research and Development (LDRD) project 20140348ER. TMP, IJB and NAM wish to acknowledge support by the UC Lab Fees Research Program # UCD-12-0045.15. TN was supported by the Department of Defense (DoD) through the National Defense Science & Engineering Graduate Fellowship (NDSEG) Program. Los Alamos National Laboratory is operated by Los Alamos National Security LLC under DOE Contract DEAC52-06NA25396. NeD results were collected on the High Pressure Preferred Orientation (HIPPO) beam line at the Los Alamos Neutron Science Center. Electron microscopy was performed at the Los Alamos Electron Microscopy Laboratory.

References

[1] Meyers MA, Mishra A, Benson DJ. Mechanical properties of nanocrystalline materials. Prog Mater Sci. 2006;51:427–556.

[2] Gleiter H. Nanocrystalline materials. Prog Mater Sci. 1989;33:223–315.

[3] Lu L, Chen X, Huang X, Lu K. Revealing the maximum strength in nanotwinned copper. Science. 2009;323:607–610.

[4] Misra A. Nanostructure control of materials. Cambridge: Woodhead Publishing Company; 2006.

[5] Wang YM, Chen MW, Zhou FH, Ma E. High tensile ductility in a nanostructured metal. Nature. 2002;419:912–915.

[6] Zhu YT, Liao XZ. Nanostructured metals - retaining ductility. Nature Mater. 2004;3:351–352.

[7] Misra A, Hoagland RG. Effects of elevated temperature annealing on the structure and hardness of copper/niobium nanolayered films. J Mater Res. 2005;20:2046–2054.

[8] Bellou A, Scudiero L, Bahr DF. Thermal stability and strength of Mo/Pt multilayered films. J Mater Sci. 2010;45:354–362.

[9] Carpenter JS, Zheng SJ, Zhang RF, Vogel SC, Beyerlein IJ, Mara NA. Thermal stability of Cu-Nb nanolamellar composites fabricated via accumulative roll bonding. Philos Mag. 2013;93:718–735.

[10] Vidal V, Thilly L, Lecouturier F, Renault PO. Effects of size and geometry on the plasticity of high-strength copper/tantalum nanofilamentary conductors obtained by severe plastic deformation. Acta Mater. 2006;54:1063–1075.

[11] Han K, Embury JD, Petrovic JJ, Weatherly GC. Microstructural aspects of Cu-Ag produced by the Taylor wire method. Acta Mater. 1998;46:4691–4699.

[12] Valiev R. Nanostructuring of metals by severe plastic deformation for advanced properties. Nature Mater. 2004;3:511–516.

[13] Perez-Prado MT, Gimazov AA, Ruano OA, Kassner ME, Zhilyaev AP. Bulk nanocrystalline omega-Zr by high-pressure torsion. Scr Mater. 2008;58:219–222.

[14] Nikulin S, Dobatkin S, Rogachev S. Iop. Nanocrystalline zirconium alloys obtained by severe plastic deformation. Names10: New Achievements Mater Environ Sci. 2013;416.

[15] Edalati K, Horita Z, Yagi S, Matsubara E. Allotropic phase transformation of pure zirconium by high-pressure torsion. Mater Sci Eng Struct Mater Properties Microstruct Process. 2009;523:277–281.

[16] Segal VM, Hartwig KT, Goforth RE. In situ composites processed by simple shear. Mater Sci EngStruct Mater Prop Microstruct Process. 1997;224:107–115.

[17] Valiev RZ, Langdon TG. Principles of equal-channel angular pressing as a processing tool for grain refinement. Prog Mater Sci. 2006;51:881–981.

[18] Saito Y, Utsunomiya H, Tsuji N, Sakai T. Novel ultra-high straining process for bulk materials - development of the accumulative roll-bonding (ARB) process. Acta Mater. 1999;47:579–583.

[19] Dehsorkhi RN, Qods F, Tajally M. Investigation on microstructure and mechanical properties of Al-Zn composite during accumulative roll bonding (ARB) process. Mater Sci Eng Struct Mater Prop Microstruct Process. 2011;530:63–72.

[20] Ohsaki S, Kato S, Tsuji N, Ohkubo T, Hono K. Bulk mechanical alloying of Cu–Ag and Cu/Zr two-phase microstructures by accumulative roll-bonding process. Acta Mater. 2007;55:2885–2895.

[21] Carpenter JS, Vogel SC, LeDonne JE, Hammon DL, Beyerlein IJ, Mara NA. Bulk texture evolution of Cu-Nb nanolamellar composites during accumulative roll bonding. Acta Mater. 2012;60:1576–1586.

[22] Carpenter JS, McCabe RJ, Zheng SJ, Wynn TA, Mara NA, Beyerlein IJ. Processing parameter influence on texture and microstructural evolution in Cu–Nb multilayer composites fabricated via accumulative roll bonding. Metall Mater Trans A. 2014;45:2192–2208.

[23] Perez-Prado MT, del Valle JA, Ruano OA. Grain refinement of Mg-Al-Zn alloys via accumulative roll bonding. Scr Mater. 2004;51:1093–1097.

[24] Ferrasse S, Segal VM, Alford F, Kardokus J, Strothers S. Scale up and application of equal-channel angular extrusion for the electronics and aerospace industries. Mater Sci Eng Struct Mater Prop Microstruct Process. 2008;493:130–140.

[25] Valiev RZ, Langdon TG. Achieving exceptional grain refinement through severe plastic deformation: new approaches for improving the processing technology. Metall Mater TransPhys Metall Mater Sci. 2011;42A:2942–2951.

[26] Beyerlein IJ, Caro A, Demkowicz MJ, Mara NA, Misra A, Uberuaga BP. Radiation damage tolerant nanomaterials. Mater Today. 2013;16:443–449.

[27] Zhu YT, Liao XZ, Wu XL. Deformation twinning in nanocrystalline materials. Prog Mater Sci. 2012;57:1–62.

[28] Beyerlein IJ, Mara NA, Bhattacharyya D, Alexander DJ, Necker CT. Texture evolution via combined slip and deformation twinning in rolled silver-copper cast eutectic nanocomposite. Int J Plast. 2011;27:121–146.

[29] Yasuna K, Terauchi M, Otsuki A, Ishihara KN, Shingu PH. Formation of nanoscale Fe/Ag multilayer by repeated press-rolling and its layer thickness dependence of magnetoresistance. Mater Sci Eng Struct Mater Prop Microstruct and Process. 2000;285:412–417.

[30] Han K, Embury JD, Sims JR, Campbell LJ, Schneider-Muntau HJ, Pantsyrnyi VI, Shikov A, Nikulin A, Vorobieva A. The fabrication, properties and microstructure of Cu–Ag and Cu–Nb composite conductors. Mater Sci Eng Struct Mater Prop Microstruct Process. 1999;267: 99–114.

[31] Al-Maharbi M, Karaman I, Beyerlein IJ, Foley D, Hartwig KT, Kecskes LJ, Mathaudhu SN. Microstructure, crystallographic texture, and plastic anisotropy evolution in an Mg alloy during equal channel angular extrusion processing. Mater Sci Eng Struct Mater Prop Microstruct Process. 2011;528:7616-7627.

[32] Jian WW, Cheng GM, Xu WZ, Yuan H, Tsai MH, Wang QD, Koch CC, Zhu YT, Mathaudhu SN. Ultrastrong Mg alloy via nano-spaced stacking faults. Mater Res Lett. 2013;1:61–66.

[33] Broussard PR, Mael D, Geballe TH. Specific-heat of niobium-zirconium multilayers. Phys Rev B. 1984;30: 4055–4056.

[34] Kawasaki M, Langdon TG. The significance of grain boundary sliding in the superplastic Zn-22 % Al alloy processed by ECAP. J Mater Sci. 2013;48:4730–4741.

[35] Perez-Prado MT, Zhilyaev AP. First experimental observation of shear Induced hcp to bcc transformation in pure Zr. Phys Rev Lett. 2009;102.

[36] Manna I, Chattopadhyay PP, Banhart F, Fecht HJ. Formation of face-centered-cubic zirconium by mechanical attrition. Appl Phys Lett. 2002;81:4136–4138.

[37] Knezevic M, Beyerlein IJ, Nizolek T, Mara N, Pollock TM. Anomalous basal slip activity in zirconium under high-strain deformation. Mater Res Lett. 2014;1:133–140.

[38] Knezevic M, Nizolek T, Ardeljan M, Beyerlein IJ, Mara N, Pollock TM. Texture evolution in two-phase Zr-Nb lamellar composites during accumulative roll bonding. Int J Plast. 2014;57:16–28.

[39] Sabirov I, Molina-Aldareguia JM, Jiang L, Kassner ME, Perez-Prado MT. Effect of accumulative roll bonding on plastic flow properties of commercially pure zirconium. In: Menary G, editor. 14th International Conference on Material Forming Esaform, 2011 Proceedings, Vol. 1353. Berlin: Springer-Verlag; 2011. p.487–492.

[40] Jiang L, Perez-Prado MT, Gruber PA, Arzt E, Ruano OA, Kassner ME. Texture, microstructure and mechanical properties of equiaxed ultrafine-grained Zr fabricated by accumulative roll bonding. Acta Mater. 2008;56:1228–1242.

[41] Guo D, Li M, Shi Y, Zhang Z, Zhang H, Liu X, Wei B, Zhang X. High strength and ductility in multimodal-structured Zr. Mater Design. 2012;34:275–278.

[42] Guo D, Li M, Shi Y, Zhang Z, Zhang H, Liu X, Zhang X. Effect of strain rate on microstructure evolutions and mechanical properties of cryorolled Zr upon annealing. Mater Lett. 2012;66:305–307.

[43] Cerreta E, Gray GT, Hixson RS, Rigg PA, Brown DW. The influence of interstitial oxygen and peak pressure on the shock loading behavior of zirconium. Acta Mater. 2005;53:1751–1758.

[44] Beyerlein IJ, McCabe RJ, Tome CN. Stochastic processes of {10(1)over-bar2} deformation twinning in hexagonal close-packed polycrystalline zirconium and magnesium. Int J Multiscale Comput Eng. 2011;9:459–480.

[45] Proust G, Kaschner GC, Beyerlein IJ, Clausen B, Brown DW, McCabe RJ, Tome CN. Detwinning of high-purity zirconium: in-situ neutron diffraction experiments. Exp Mech. 2010;50:125–133.

[46] Niezgoda SR, Kanjarla AK, Beyerlein IJ, Tome CN. Stochastic modeling of twin nucleation in polycrystals: an application in hexagonal close-packed metals. Int J Plast. 2014.

[47] McCabe RJ, Proust G, Cerreta EK, Misra A. Quantitative analysis of deformation twinning in zirconium. Int J Plast. 2009;25:454–472.

[48] Hansen BL, Carpenter JS, Sintay SD, Bronkhorst CA, McCabe RJ, Mayeur JR, Mourad HM, Beyerlein IJ, Mara NA, Chen SR, Gray GT, III. Modeling the texture evolution of Cu/Nb layered composites during rolling. Int J Plast. 2013;49:71–84.

[49] Steif PS. On deformation instabilities in clad metals subjected to rolling. J Appl Metalworking. 1987;4: 317–326.

[50] Massalski TB, Okamoto H. Binary alloy phase diagrams. Materials Park, OH: ASM international; 1990.

[51] Sauvage X, Renaud L, Deconihout B, Blavette D, Ping DH, Hono K. Solid state amorphization in cold drawn Cu/Nb wires. Acta Mater. 2001;49:389–394.

[52] Wu F, Bellon P, Melmed AJ, Lusby TA. Forced mixing and nanoscale decomposition in ball-milled Cu-Ag characterized by APFIM. Acta Mater. 2001;49:453–461.

[53] Yavari AR, Desre PJ, Benameur T. Mechanically driven alloying of immiscible elements. Phys Rev Lett. 1992;68:2235–2238.

[54] Wenk HR, Lutterotti L, Vogel S. Texture analysis with the new HIPPO TOF diffractometer. Nuclear Instrum Methods Phys Res SectA. 2003;515:575–588.

[55] Vogel SC, Hartig C, Lutterotti L, Von Dreele RB, Wenk HR, Williams DJ. Texture measurements using the new neutron diffractometer HIPPO and their analysis using the Rietveld method. Powder Diffr. 2004;19:65–68.

[56] Beyerlein IJ, Mara NA, Carpenter JS, Nizolek T, Mook WM, Wynn TA, McCabe RJ, Mayeur JR, Kang K, Zheng S, Wang J, Pollock TM. Interface-driven microstructure development and ultra high strength of bulk nanostructured Cu-Nb multilayers fabricated by severe plastic deformation. J Mater Res. 2013;28:1799–1812.

[57] Weissmuller J, Markmann J. Deforming nanocrystalline metals: new insights, new puzzles. Adv Eng Mater. 2005;7:202–207.

[58] Brown DW, Beyerlein IJ, Sisneros TA, Clausen B, Tome CN. Role of twinning and slip during compressive deformation of beryllium as a function of strain rate. Int J Plast. 2012;29:120–135.

[59] Yapici GG, Tome CN, Beyerlein IJ, Karaman I, Vogel SC, Liu C. Plastic flow anisotropy of pure zirconium after severe plastic deformation at room temperature. Acta Mater. 2009;57:4855–4865.

[60] Knezevic M, Beyerlein IJ, Brown DW, Sisneros TA, Tome CN. A polycrystal plasticity model for predicting mechanical response and texture evolution during strain-path changes: application to beryllium. Int J Plast. 2013;49:185–198.

[61] Skrotzki W, Eschke A, Joni B, Ungar T, Toth LS, Ivanisenko Y, Kurmanaeva L. New experimental insight into the mechanisms of nanoplasticity. Acta Mater. 2013;61:7271–7284.

[62] Raabe D, Lucke K. Rolling textures of niobium and molybdenum. Z Fur Metallkunde. 1994;85:302–306.

[63] Engler O, Randle V. Introduction to texture analysis. Boca Raton, FL: CRC Press; 2010.

[64] Jiang L, Ruano OA, Kassner ME, Perez-Prado MT. The fabrication of bulk ultrafine-grained zirconium by accumulative roll bonding. JOM. 2007;59:42–45.

[65] Philippe MJ, Serghat M, Vanhoutte P, Esling C. Modeling of texture evolution for materials of hexagonal symmetry .2. Application to zirconium and titanium alpha-alloys or near-alpha-alloys. Acta Metallurgica Et Mater. 1995;43:1619–1630.

[66] Lim SCV, Rollett AD. Length scale effects on recrystallization and texture evolution in Cu layers of a roll-bonded Cu–Nb composite. Mater Sci EngStruct Mater Prop Microstruct Process. 2009;520:189–196.

[67] Beyerlein IJ, Wang J, Kang K, Zheng SJ, Mara N. Twinnability of bimetal interfaces in nanostructured composites. Mater Res Lett. 2014;1:89–95.

[68] Beyerlein IJ, Wang J, Zhang R. Mapping dislocation nucleation behavior from bimetal interfaces. Acta Mater. 2013;61:7488–7499.

[69] Beyerlein IJ, Jian W, Ruifeng Z. Interface-dependent nucleation in nanostructured layered composites. APL Mater. 2013;1:032112. Available from: http://scitation.aip.org/content/aip/journal/aplmater/1/3/10.1063/1.4820424

[70] Carpenter JS, McCabe RJ, Beyerlein IJ, Wynn TA, Mara NA. A wedge-mounting technique for nanoscale electron backscatter diffraction. J Appl Phys. 2013;113:094304. Available from: http://scitation.aip.org/content/aip/journal/jap/113/9/10.1063/1.4794388

[71] Zheng S, Beyerlein IJ, Carpenter JS, Kang K, Wang J, Han W, Mara NA. High-strength and thermally stable bulk nanolayered composites due to twin-induced interfaces. Nature Commun. 2013;4:1696. Available from: http://www.nature.com/ncomms/journal/v4/n4/abs/ncomms2651.html

[72] Monclus MA, Zheng SJ, Mayeur JR, Beyerlein IJ, Mara NA, Polcar T, Llorca J, Molina-Aldareguia JM. Optimum high temperature strength of two-dimensional nanocomposites. APL Mater. 2013;1:052103. Available from: http://scitation.aip.org/content/aip/journal/aplmater/1/5/10.1063/1.4828757

[73] Han WZ, Cerreta E, Mara N, Beyerlein IJ, Carpenter J, Zheng SJ, Trujillo CP, Dickerson P, Misra A. Deformation and failure of shocked bulk Cu-Nb nanolaminates. Acta Mater. 2014;63:150–161.

[74] Han W, Demkowicz MJ, Mara NA, Fu E, Sinha S, Rollett AD, Wang Y, Carpenter JS, Beyerlein IJ, Misra A. Design of radiation tolerant materials via interface engineering. Adv Mater. 2013;25:6975–6979.

Permissions

All chapters in this book were first published in MRL, by Taylor & Francis Online; hereby published with permission under the Creative Commons Attribution License or equivalent. Every chapter published in this book has been scrutinized by our experts. Their significance has been extensively debated. The topics covered herein carry significant findings which will fuel the growth of the discipline. They may even be implemented as practical applications or may be referred to as a beginning point for another development.

The contributors of this book come from diverse backgrounds, making this book a truly international effort. This book will bring forth new frontiers with its revolutionizing research information and detailed analysis of the nascent developments around the world.

We would like to thank all the contributing authors for lending their expertise to make the book truly unique. They have played a crucial role in the development of this book. Without their invaluable contributions this book wouldn't have been possible. They have made vital efforts to compile up to date information on the varied aspects of this subject to make this book a valuable addition to the collection of many professionals and students.

This book was conceptualized with the vision of imparting up-to-date information and advanced data in this field. To ensure the same, a matchless editorial board was set up. Every individual on the board went through rigorous rounds of assessment to prove their worth. After which they invested a large part of their time researching and compiling the most relevant data for our readers.

The editorial board has been involved in producing this book since its inception. They have spent rigorous hours researching and exploring the diverse topics which have resulted in the successful publishing of this book. They have passed on their knowledge of decades through this book. To expedite this challenging task, the publisher supported the team at every step. A small team of assistant editors was also appointed to further simplify the editing procedure and attain best results for the readers.

Apart from the editorial board, the designing team has also invested a significant amount of their time in understanding the subject and creating the most relevant covers. They scrutinized every image to scout for the most suitable representation of the subject and create an appropriate cover for the book.

The publishing team has been an ardent support to the editorial, designing and production team. Their endless efforts to recruit the best for this project, has resulted in the accomplishment of this book. They are a veteran in the field of academics and their pool of knowledge is as vast as their experience in printing. Their expertise and guidance has proved useful at every step. Their uncompromising quality standards have made this book an exceptional effort. Their encouragement from time to time has been an inspiration for everyone.

The publisher and the editorial board hope that this book will prove to be a valuable piece of knowledge for researchers, students, practitioners and scholars across the globe.

List of Contributors

S. Pfeifer and P.R. Bandaru
Department of Mechanical and Aerospace Engineering, Materials Science Program, University of California, Room 258, Engineering 2, San Diego, La Jolla, CA 92093-0411, USA

Yudi Rosandi
Department of Physics, Universitas Padjadjaran, Jatinangor, Sumedang 45363, Indonesia
Fachbereich Physik und Forschungszentrum OPTIMAS, Universität Kaiserslautern, Erwin-Schrödinger-Straße, D-67663 Kaiserslautern, Germany

Fairoja Cheenicode Kabeer
Theoretische Physik, Universität Kassel and Center for Interdisciplinary Nanostructure Science and Technology (CINSaT), Heinrich-Plett-Straße 40, 34132 Kassel, Germany
Fritz-Haber-Institut der Max-Planck-Gesellschaft, Faradayweg 4–6, D-14195 Berlin, Germany

Yaroslav Cherednikov and Herbert M. Urbassek
Fachbereich Physik und Forschungszentrum OPTIMAS, Universität Kaiserslautern, Erwin-Schrödinger-Straße, D-67663 Kaiserslautern, Germany

Eeuwe S. Zijlstra and Martin E. Garcia
Theoretische Physik, Universität Kassel and Center for Interdisciplinary Nanostructure Science and Technology (CINSaT), Heinrich-Plett-Straße 40, 34132 Kassel, Germany

Nail A. Inogamov
Landau Institute for Theoretical Physics, Russian Academy of Science, Prospekt Akademika Semenova 1-A, 142432 Chernogolovka, Moscow, Russia

Marc Heggen and Michael Feuerbacher
Peter Grünberg Institut, Forschungszentrum Jülich GmbH, D-52425 Jülich, Germany

Rui Huang, Zhang-Jie Wang, Ling Huang and Zhi-Wei Shan
Center for Advancing Materials Performance from the Nanoscale (CAMP-Nano) & Hysitron Applied Research Center in China (HARCC), State Key Laboratory for Mechanical Behavior of Materials, Xi'an Jiaotong University, Xi'an 710049, People's Republic of China

Evan Ma
Center for Advancing Materials Performance from the Nanoscale (CAMP-Nano) & Hysitron Applied Research Center in China (HARCC), State Key Laboratory for Mechanical Behavior of Materials, Xi'an Jiaotong University, Xi'an 710049, People's Republic of China
Department of Materials Science and Engineering, Johns Hopkins University, Baltimore, MD 21218, USA

Qing-Jie Li
Department of Materials Science and Engineering, Johns Hopkins University, Baltimore, MD 21218, USA

Ju Lia
Department of Nuclear Science and Engineering, Massachusetts Institute of Technology, 77 Massachusetts Avenue, Cambridge, MA 02139, USA
Department of Materials Science and Engineering, Massachusetts Institute of Technology, 77 Massachusetts Avenue, Cambridge, MA 02139, USA

Ming-Hung Tsai
Department of Materials Science and Engineering, National Chung Hsing University, Taichung 40227, Taiwan

Jien-Wei Yeh
Department of Materials Science and Engineering, National Tsing Hua University, Hsinchu 30013, Taiwan

Sagi Chen, Daniel Rittel and Dan Mordehai
Department of Mechanical Engineering, Technion – Israel Institute of Technology, 32000, Haifa, Israel

A. Léger, L. Weber and A. Mortensen
Laboratory of Mechanical Metallurgy, Institute of Materials, École Polytechnique Fédérale de Lausanne (EPFL), Station 12, 1015 Lausanne, Switzerland

J. M. Molina-Jordá
Instituto Universitario de Materiales de Alicante, Universidad de Alicante, Apto 99, E-03080 Alicante, Spain

Michael Feuerbacher, Markus Heidelmann and Carsten Thomas
Peter Grünberg Institut PGI-5, Forschungszentrum Jülich GmbH, D-52425 Jülich, Germany

Richard E. Baumer
Research and Development, The Dow Chemical Company, Building 1776, Midland, MI 48674, USA
Department of Materials Science and Engineering, Massachusetts Institute of Technology, Cambridge, MA 02139, USA

Michael J. Demkowicz
Department of Materials Science and Engineering, Massachusetts Institute of Technology, Cambridge, MA 02139, USA

Xiaoyang Fang, Yuewu Zeng and Liang Meng
Department of Materials Science and Engineering, Zhejiang University, Hangzhou 310027, People's Republic of China

Jiabin Liu
Department of Materials Science and Engineering, Zhejiang University, Hangzhou 310027, People's Republic of China
College of Electrical Engineering, Zhejiang University, Hangzhou 310027, People's Republic of China

Lihui Zhang and Youtong Fang
College of Electrical Engineering, Zhejiang University, Hangzhou 310027, People's Republic of China

Wei Liu and Kangying Shu
Department of Materials Science and Engineering, China Jiliang University, Hangzhou 310018, People's Republic of China

Khaled M. Youssef
Department of Materials Science and Technology, Qatar University, PO Box 2713, Doha, Qatar

Alexander J. Zaddach, Changning Niu, Douglas L. Irving and Carl C. Koch
Department of Materials Science and Engineering, North Carolina State University, Raleigh, NC 27695-7907, USA

Y.W. Chang, M. Pozuelo and J.M. Yang
Department of Materials Science and Engineering, University of California Los Angeles, Los Angeles, CA 90095, USA

Liangfa Hu and Ankush Kothalkar
Department of Materials Science and Engineering, Texas A&M University, College Station, TX 77843, USA

Morgan O'Neil
Department of Mechanical Engineering, Texas A&M University, College Station, TX 77843, USA

Ibrahim Karaman and Miladin Radovic
Department of Materials Science and Engineering, Texas A&M University, College Station, TX 77843, USA
Department of Mechanical Engineering, Texas A&M University, College Station, TX 77843, USA

C. X. Huang
School of Aeronautics and Astronautics, Sichuan University, Chengdu 610065, People's Republic of China

W. P. Hu
Institute of Physical Metallurgy and Metal Physics, RWTH Aachen University, Aachen 52056, Germany

Q. Y. Wang
Institute for Mechanics of Advanced Materials and Structures, Chengdu University, Chengdu 610106, People's Republic of China

C. Wang and G. Yang
Central Iron and Steel Research Institute, Beijing 100081, People's Republic of China

Y. T. Zhu
Department of Materials Science and Engineering, North Carolina State University, Raleigh, NC 27695, USA
School of Materials Science and Engineering, Nanjing University of Science and Technology, Nanjing 210094, People's Republic of China

Xianghai An, Qingyun Lin, Shiding Wu and Zhefeng Zhang
Shenyang National Laboratory for Materials Science, Institute of Metal Research, Chinese Academy of Sciences, 72 Wenhua Road, Shenyang 110016, People's Republic of China

Kwai S. Chan, Michael A. Miller, and Carol Ellis-Terrell
Southwest Research Institute, 6220 Culebra Road, San Antonio, TX 78238, USA

Wuwei Liang
Southwest Research Institute, 6220 Culebra Road, San Antonio, TX 78238, USA
Math Works, 3 Apple Hill Drive, Natick, MA 01760, USA

C. Sun, E. Martínez, A. Caro, J.A. Valdez, K. Baldwin, Y. Xu, B.P. Uberuaga, O. Anderoglu and S.A. Maloy
Los Alamos National Laboratory, Los Alamos, NM 87545, USA

J.A. Aguiar
National Renewable Energy Laboratory, Golden, CO 80401, USA

L. Leclercq and L. Capolungo
George Woodruff School of Mechanical Engineering, Georgia Institute of Technology Lorraine, UMI 2958 Georgia Tech-CNRS, 57070 Metz, France

D. Rodney
Laboratoire de Science et Ingénierie des Matériaux et des Procédés, Grenoble Institute of Technology, CNRS, UJF, 38402 Saint Martin d'Hères, France
Institut Lumière Matière, Université Lyon 1, CNRS, UMR 5306, 69622 Villeurbanne, France

M. Komarasamy, N. Kumar and R.S. Mishra
Department of Materials Science and Engineering, Center for Friction Stir Processing, University of North Texas, Denton, TX 76203, USA

Z. Tang and P.K. Liaw
Department of Materials Science and Engineering, The University of Tennessee, Knoxville, TN 37996, USA

S. S. Singh, J. J. Williams, M. F. Lin and N. Chawla
Materials Science and Engineering, Arizona State University, Tempe, AZ 85287 6106, USA

X. Xiao and F. De Carlo
Advanced Photon Source, Argonne National Laboratory, Argonne, IL, USA

Josef Cobb and Judy Schneider
Mechanical Engineering Department, Mississippi State University, Starkville, MI 39762, USA

Shraddha Vachhani
Mechanical Engineering Department, Georgia Institute of Technology, Atlanta, GA 30332, USA

Robert M. Dickerson, Patricia O. Dickerson and John S. Carpenter
Materials Science and Technology Division, Los Alamos National Laboratory, Los Alamos, NM 87545, USA

Weizhong Han
Center for Advancing Materials Performance from the Nanoscale, State Key Laboratory for Mechanical Behavior of Materials, Xi'an Jiaotong University, Xi'an 710049, People's Republic of China
Center for Integrated Nanotechnologies, Los Alamos National Laboratory, Los Alamos, NM 87545, USA

Nathan A. Mara
Materials Science and Technology Division, Los Alamos National Laboratory, Los Alamos, NM 87545, USA
Center for Integrated Nanotechnologies, Los Alamos National Laboratory, Los Alamos, NM 87545, USA

R.L. Schoeppner
School of Mechanical and Materials Engineering, Washington State University, Pullman, WA 99164, USA
School of Materials Engineering, Purdue University, West Lafayette, IN 47907, USA

J.M. Wheeler, J. Zechner and J. Michler
Empa, Swiss Federal Laboratories for Materials Science and Technology, Laboratory for Mechanics of Materials and Nanostructures, Thun CH-3602, Switzerland

H.M. Zbib
School of Mechanical and Materials Engineering, Washington State University, Pullman, WA 99164, USA

D.F. Bahr
School of Materials Engineering, Purdue University, West Lafayette, IN 47907, USA

T. Miyauchi and E. Abe
Department of Materials Science & Engineering, University of Tokyo, Tokyo 113-8656, Japan

H. Kato
Institute for Materials Research, Tohoku University, Sendai 980-8577, Japan

K.Y. Yu
Department of Materials Science and Engineering, China University of Petroleum Beijing, Beijing 102249, People's Republic of China
Department of Materials Science and Engineering, Texas A&M University, College Station, TX 77843-3123, USA

Z. Fanc
Department of Mechanical Engineering, Texas A&M University, College Station, TX 77843-3123, USA

Y. Chen, M. Song and Y. Liu
Department of Materials Science and Engineering, Texas A&M University, College Station, TX 77843-3123, USA

H. Wang
Department of Electrical and Computer Engineering, Texas A&M University, College Station, TX 77843-3128, USA

M.A. Kirk
Materials Science Division, Argonne National Laboratory, Argonne, IL 60439, USA

M. Li
Nuclear Engineering Division, Argonne National Laboratory, Argonne, IL 60439, USA

X. Zhang
Department of Materials Science and Engineering, Texas A&M University, College Station, TX 77843-3123, USA
Department of Mechanical Engineering, Texas A&M University, College Station, TX 77843-3123, USA

J.Y. Li and X.J. Jin
State Key Lab of Metal Matrix Composite, School of Materials Science and Engineering, Shanghai Jiao Tong University, Shanghai 200240, People's Republic of China
Institute of Phase Transformation and Complex Microstructure, School of Materials Science and Engineering, Shanghai Jiao Tong University, Shanghai 200240, People's Republic of China

W. Li and M.J. Jin
Institute of Phase Transformation and Complex Microstructure, School of Materials Science and Engineering, Shanghai Jiao Tong University, Shanghai 200240, People's Republic of China

William Yi Wang
Department of Materials Science and Engineering, The Pennsylvania State University, University Park, PA 16802, USA
State Key Laboratory for Advanced Metals and Materials, University of Science and Technology Beijing, Beijing 100083, People's Republic of China

Shun Li Shang, Yi Wang, Zhi-Gang Mei and Zi-Kui Liu
Department of Materials Science and Engineering, The Pennsylvania State University, University Park, PA 16802, USA

Suveen N. Mathaudhu
Materials Science Division, U.S. Army Research Office, Research Triangle Park, NC 27709, USA

Kristopher A. Darling and Laszlo J. Kecskes
U.S. Army Research Laboratory, Weapons and Materials Research Directorate, RDRL-WMM-B, Aberdeen Proving Ground, MD 21005, USA

Xi Dong Hui
State Key Laboratory for Advanced Metals and Materials, University of Science and Technology Beijing, Beijing 100083, People's Republic of China

Q. Z. Tao, C. F. Hu, F. Z. Li, D. Qu and M. L. Wu
Ningbo Institute of Material Technology and Engineering, Chinese Academy of Sciences, Ningbo 315201, People's Republic of China

S. Lin and Y. P. Sun
Key Laboratory of Materials Physics, Institute of Solid State Physics, Hefei 230031, People's Republic of China

H. B. Zhang
Institute of Nuclear Physics and Chemistry, China Academy of Engineering Physics, Mianyang 621900, People's Republic of China

Y. Sakka
Advanced Materials Processing Unit, National Institute for Materials Science, Tsukuba, Ibaraki 305-0047, Japan

M. W. Barsoum
Department of Materials Science and Engineering, Drexel University, Philadelphia, PA 19104, USA

A. A. Leino, O. H. Pakarinen, F. Djurabekova and K. Nordlund
Helsinki Institute of Physics and Department of Physics, University of Helsinki, PO Box 43, Helsinki, FI-00014, Finland

P. Kluth and M. C. Ridgway
Department of Electronic Materials Engineering, Research School of Physics and Engineering, Australian National University, Canberra, ACT 0200, Australia

H. Mohseni, B. A. Mensah, N. Gupta, S. G. Srinivasan and T. W. Scharf
Department of Materials Science and Engineering, University of North Texas, Denton, TX 76207, USA

J.A. Sharon, K. Hattar and B.L. Boyce
Sandia National Laboratories, PO Box 5800, Albuquerque, NM 87185, USA

L.N. Brewer
Sandia National Laboratories, PO Box 5800, Albuquerque, NM 87185, USA
Naval Postgraduate School, 700 Dyer Rd., Monterey, CA 93943, USA

Ruslan Salikhov, Ulf Wiedwald and Michael Farle
Fakultät für Physik and Center for Nanointegration (CENIDE), Universität Duisburg-Essen, Duisburg 47057, Germany

Anna S. Semisalova
Fakultät für Physik and Center for Nanointegration (CENIDE), Universität Duisburg-Essen, Duisburg 47057, Germany
Faculty of Physics, Lomonosov Moscow State University, Moscow 119991, Russia Institute of Ion Beam Physics and Materials Research, Helmholtz-Zentrum Dresden-Rossendorf, Dresden D-01328, Germany

Andrejs Petruhins, Arni Sigurdur Ingason and Johanna Rosen
Thin Film Physics Division, Department of Physics, Chemistry and Biology (IFM), Linköping University, Linköping SE-581 83, Sweden

Zhiqian Sun, Gian Song and Peter K. Liaw
Department of Materials Science and Engineering, The University of Tennessee, 1508 Middle Drive, Knoxville, TN 37996, USA

Jan Ilavsky
X-ray Science Division, Advanced Photon Source, Argonne National Laboratory, Lemont, IL 60439, USA

J.S. Carpenter, R.J. McCabe and J.E. Scott
Materials Science and Technology Division, Los Alamos National Laboratory, Los Alamos, NM 87545, USA

T.J. Nizolek and T.M. Pollock
Materials Department, University of California at Santa Barbara, Santa Barbara, CA 93106, USA

S.J. Zheng
Materials Physics and Applications Division, Los Alamos National Laboratory, Los Alamos, NM 87545, USA

S.C. Vogel
Los Alamos Neutron Science Center, Los Alamos National Laboratory, Los Alamos, NM 87545, USA

N.A. Mara
Materials Science and Technology Division, Los Alamos National Laboratory, Los Alamos, NM 87545, USA
Materials Physics and Applications Division, Los Alamos National Laboratory, Los Alamos, NM 87545, USA

I.J. Beyerlein
Theoretical Division, Los Alamos National Laboratory, Los Alamos, NM 87545, USA

Index